Evolution of the Earth

Geodynamics Series

The Final Reports of the International Geodynamics Program sponsored by the Inter-Union Commission on Geodynamics.

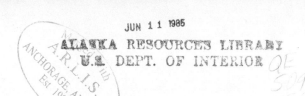
Evolution of the Earth

Edited by R.J. O'Connell
W.S. Fyfe

Geodynamics Series
Volume 5

American Geophysical Union
Washington, D.C.

Geological Society of America
Boulder, Colorado

1981

Final Report (Part B) of Working Group 5, Properties
and Processes of the Earth's Interior coordinated by
O.L. Anderson on behalf of the Bureau of the Inter-
Union Commission on Geodynamics.

American Geophysical Union, 2000 Florida Avenue, N.W.
  Washington, D.C. 20009

Geological Society of America, 3300 Penrose Place, P.O. Box 9140
  Boulder, Colorado 80301

Library of Congress Cataloging in Publication Data
Main entry under title:

Evolution of the earth.

  (Geodynamics series; v. 5)
  Includes bibliographies.
  1. Earth--Internal structure--Congresses. 2. Geo-
dynamics--Congresses. I. Anderson, O. L. (Orson L.)
II. Series.
QE509.E96   551   81-14864
ISBN 0-87590-506-4   AACR2

Printed in the United States of America

CONTENTS

After a decade of intense and productive scientific cooperation between geologists, geophysicists and geochemists the International Geodynamics Program formally ended on July 31, 1980. The scientific accomplishments of the program are represented in more than seventy scientific reports and in this series of Final Report volumes.

The concept of the Geodynamics Program, as a natural successor to the Upper Mantle Project, developed during 1970 and 1971. The International Union of Geological Sciences (IUGS) and the International Union of Geodesy and Geophysics (IUGG) then sought support for the new program from the International Council of Scientific Unions (ICSU). As a result the Inter-Union Commission on Geodynamics was established by ICSU to manage the International Geodynamics Program.

The governing body of the Inter-Union Commission on Geodynamics was a Bureau of seven members, three appointed by IUGG, three by IUGS and one jointly by the two Unions. The President was appointed by ICSU and a Secretary-General by the Bureau from among its members. The scientific work of the Program was coordinated by the Commission, composed of the Chairmen of the Working Groups and the representatives of the national committees for the International Geodynamics Program. Both the Bureau and the Commission met annually, often in association with the Assembly of one of the Unions, or one of the constituent Associations of the Unions.

Initially the Secretariat of the Commission was in Paris with support from France through BRGM, and later in Vancouver with support from Canada through DEMR and NRC.

The scientific work of the Program was coordinated by ten Working Groups.

WG 1 Geodynamics of the Western Pacific-Indonesian Region
WG 2 Geodynamics of the Eastern Pacific Region, Caribbean and Scotia Arcs
WG 3 Geodynamics of the Alpine-Himalayan Region, West
WG 4 Geodynamics of Continental and Oceanic Rifts
WG 5 Properties and Processes of the Earth's Interior

WG 6 Geodynamics of the Alpine-Himalayan Region, East
WG 7 Geodynamics of Plate Interiors
WG 8 Geodynamics of Seismically Inactive Margins
WG 9 History and Interaction of Tectonic, Metamorphic and Magmatic Processes
WG 10 Global Syntheses and Paleoreconstruction

These Working Groups held discussion meetings and sponsored symposia. The papers given at the symposia were published in a series of Scientific Reports. The scientific studies were all organized and financed at the national level by national committees even when multinational programs were involved. It is to the national committees, and to those who participated in the studies organized by those committees, that the success of the Program must be attributed.

Financial support for the symposia and the meetings of the Commission was provided by subventions from IUGG, IUGS, UNESCO and ICSU.

Information on the activities of the Commission and its Working Groups is available in a series of 17 publications: Geodynamics Reports, 1-8, edited by F. Delany, published by BRGM; Geodynamics Highlights, 1-4, edited by F. Delany, published by BRGM; and Geodynamics International, 13-17, edited by R. D. Russell. Geodynamics International was published by World Data Center A for Solid Earth Geophysics, Boulder, Colorado 80308, USA. Copies of these publications, which contain lists of the Scientific Reports, may be obtained from WDC A. In some cases only microfiche copies are now available.

This volume is one of a series of Final Reports summarizing the work of the Commission. The Final Report volumes, organized by the Working Groups, represent in part a statement of what has been accomplished during the Program and in part an analysis of problems still to be solved. The papers in this volume, the second from W.G. 5 (Chairman, O.L. Anderson), were presented at a symposium convened by W.S. Fyfe, E.A. Lubimova and R.J. O'Connell at London, Ontario in 1979. The volume was edited by W.S. Fyfe and O'Connell.

At the end of the Geodynamics Program it is clear that the kinematics of the major plate movements during the past 200 million years is well understood, but there is much less understanding of the dynamics of the processes which cause these movements.

Perhaps the best measure of the success of the Program is the enthusiasm with which the Unions and national committees have joined in the establishment of a successor program to be known as:

Dynamics and evolution of the lithosphere: The framework for earth resources and the reduction of the hazards.

To all of those who have contributed their time so generously to the Geodynamics Program we tender our thanks.

C. L. Drake, President ICG, 1971-1975

A. L. Hales, President ICG, 1975-1980

Members of Working Group 5:

O.L. Anderson
G. Barta
W.S. Fyfe
S. Akimoto
V. Babuska
E. Boschi
A.H. Cook
C. Froidevaux
K. Fuchs
I.G. Gass
P. Grew
D.G. Kautzleben

J.G. Negi
A. Nicolas
L.O. Nicolaysen
R.J. O'Connell
M.G. Rochester
C.H. Scholz
N.V. Sobolev
F.D. Stacey
H. Stiller
L.P. Vinnik
K. Yagi

PREFACE

During September 2-5, 1979 a group of about 80 earth scientists from many parts of the globe assembled in London, Ontario to attend a seminar devoted to discussion of the current (1979) state of geophysical and geochemical thought concerning the evolution of planet Earth. The papers included in this volume represent most of the views expressed at that conference. Unfortunately, participants in a few critical areas did not wish to submit manuscripts. The symposium was one of many organized to pay particular attention to advances in the earth sciences which had occurred during the period of the International Geodynamics program.

There is little doubt that the progress in understanding our planet of the decade 1970-80 is remarkable. The general picture of earth structure and internal convective motions that control the surface and our environment and our resources has been roughed out. But while progress is remarkable the unanswered details are vast. For example, we still know little of the core-mantle boundary and interactions across that boundary; there is still debate about whole mantle convection or convection restricted to upper and lower mantle. Many aspects of important tectonic processes such as subduction are virtually unknown. While we have reasonable models of the structure of the ocean floor lithosphere, the structure of continental crust and lithosphere is inadequate. And while we begin to understand modern earth processes the state of the young planet and ancient tectonic regimes are still largely a mystery.

Since the time of our conference there have already been spectacular advances. It will be interesting to compare views in this volume with those we will see by 1990.

We would like to acknowledge the support of all who attended and the University of Western Ontario for providing logistic support. Miss J. Ainge was responsible for most of the arrangements and her assistance is gratefully acknowledged. We also gratefully acknowledge financial assistance from the Inter-Union Commission on Geodynamics and a substantial grant from the Natural Science and Engineering Research Council of Canada which made participation by many possible.

O. Anderson
University of California, Los Angeles.

R. J. O'Connell
Harvard University, Massachusetts.

W. S. Fyfe
University of Western Ontario, Canada.

E. A. Lubimova
Acad. Sci., U.S.S.R.

GEODYNAMICS: THE UNANSWERED QUESTIONS

Anton L. Hales

Geosciences Program, The University of Texas at Dallas, Richardson, Texas, 75080

Many of the papers at this symposium of Working Group 5 will be concerned with what has been learned during the Geodynamics Project. There will be other opportunities for discussion of the achievements of the Geodynamics Project and so I shall talk here about the questions which the Geodynamics Project will leave unanswered. I do not intend to list all the questions still to be answered for that is a task for the new organization whatever that may be.

During the past three decades new concepts of the evolution of the crustal features of the Earth involving large horizontal motions of the continents relative to one another have developed and are now generally accepted. In brief, paleomagnetic studies during the fifties led to renewed interest in continental drift. Then the development of the magnetic polarity time scale and the studies of magnetic lineations at sea led to the recognition that virtually all the ocean floor was less than 200 million years old. Finally, the deep sea drilling programs confirmed the inferences from the magnetic lineations with regard to the age of the various parts of the sea floor. We now know the ages of the various parts of the ocean floor remarkably well and we have therefore a good picture of the kinematics of plate motions during the past 200 million years. But do we understand the dynamics?

I believe that understanding of the dynamics of plate motions is rooted in understanding of the interplay between zones of lower than average strength, or lower than average viscosity, in the upper mantle and the gravitational or inertial forces acting on the lithosphere. In fact, Hager and O'Connell have shown that many of the major features of plate motion can be accounted for in terms of a relatively simple model of upper mantle viscosity; the model may be simpler than the real Earth. There are some difficulties in understanding the motion of plates having continental as well as oceanic segments. As I see it, there are three prime requirements for more complete understanding of the processes. These are (1) more complete knowledge of the variation of anelasticity with depth and laterally in the upper

mantle; (2) more complete knowledge of the physical properties, and especially the rheology, of rocks at upper mantle temperatures; (3) more complete data on the history of plate motion over the more than three billion years of earth history before the Mesozoic.

In so far as (1) is concerned the recent finding that Q varies quite rapidly between 0.1 Hz and 3 Hz is a complicating factor. I think that broad band seismic systems covering the 0.1 to 3 Hz band will be needed to provide the data. You will not be surprised when I suggest that portable arrays of this type of instrument will be necessary.

The variation of Q with frequency also complicates the determination of the rheological and physical properties for now these properties must be measured at seismic frequencies. Stacey, a member of this Working Group, and his colleagues have shown how this can be done at room temperatures and pressure. The challenge is now to carry out similar experiments at high temperature and at pressures sufficient to keep the cracks closed.

The third requirement is important for before the dynamics can be understood completely it is necessary to know whether the processes during pre-Mesozoic time were similar in character to those which have occurred in the last 200 million years. This requirement can only be satisfied through careful paleomagnetic studies. A start has been made on the Precambrian, principally in Canada and Australia, but much more data is needed.

Of recent years the emphasis of earth science research has been on the horizontal motions of the plates. However, as Beloussov, and others, remind us from time to time, the geological record shows many examples of vertical movements of as much as 10 km. Some of these, in the plate collision regions for example, are obviously related to the horizontal motions and caused by them. In the case of the inland seas, such as the Caspian, or sedimentary basins such as those on the Witwatersrand, in the Karoo or in the Gulf Coast of the southern United States it is by no means obvious that the vertical movements are related to, or are caused by, horizontal movements. Pressure-temperature dependent phase transformations offer an intriguing possible mechanism in the case of the deep sedimentary basins for the increase in pressure at depth due to the deposition of sediment will be more or less in step with the increase in load. The increase in temperature due to the thermal blanketing of the sediments will lag behind the increase in load by many millions of years, the lag increasing with depth. Thus given the right pressure and temperature conditions one can envisage the phase transformation moving initially in the direction of increasing density and deepening of the basin due to pressure, and later to decreasing density and uplift due to the increase in temperature. Artushkov and Jantsin have suggested phase transformation and foundering of the lower crust as an explanation for the Caspian Sea. This amounts to oceanization of a continental crust. How widespread a phenomenon is oceanization?

The role of phase transformations in vertical movement was suggested many years ago by Kennedy and Lovering particularly with reference to the gabbro-eclogite transformation and the Mohorovicie discontinuity. Geophysical arguments led to the rejection of the hypothesis. Kennedy pointed out later that reactions such as the transformation from gabbro to eclogite were rate dependent processes, i.e., that the rate of the reaction depended exponentially on the difference between the temperature and some critical temperature. This dependence was such that the reaction might have a characteristic time of billions of years at one temperature and less than a million years at temperatures close to the critical temperature. Thus the geophysical argument which led to the rejection of Kennedy and Lovering's hypothesis might not be valid. The Artushkov-Jantsin hypothesis for the Caspian is a variant of the gabbro-eclogite hypothesis of Kennedy and Lovering.

Phase transformations may play a significant role in vertical movements of the crust and therefore further study of the pressure temperature dependence of these transformations is necessary. Since the relevant transformations may be sluggish under laboratory conditions difficult and time consuming experiments will be necessary.

Lin-Gun Liu has pointed out recently that there is another class of transformation (which he has called the chemical interaction transformations) that may have similar pressure-temperature dependent properties to the phase transformations and may account for seismic discontinuities which cannot be associated with known phase transformations, for example that at a depth of about 200 km.

Much of the basement-rock which one sees at the surface is metamorphic and it may be that a significant proportion of the layer which seismologists refer to as the "granitic layer" is metamorphic. I see as one of the challenges of the next decade understanding of the processes by which great sedimentary basins are formed, the diagenesis of the sediments in these basins and their evolution into metamorphic rocks, and, on occasion, into granites.

I think that there is much to be said for John Elliston's idea that the diagenesis of sediments, their evolution into metamorphic rocks, and the concentration of metals into ore bodies, all occurred while the sediments were wet. In some cases the heat generated during these processes was sufficient to cause melting of most of the material in the rock so that the end result was a granite rather than a metamorphic rock. Certainly some granites are derived from sediments for Williams has shown that the zircons in some types of granite are much older than the granites and has hypothesized that these older ages represent the provenance ages of the sediments from which the granites were formed.

Irrespective of whether Elliston's views are correct, or not, clearly there is a need for a program of study of the diagenesis of sediments on the scale of the program of experimental work on the petrology of igneous rocks during the past half century.

Science tends to make rapid progress at times when new tools become available for the study of old problems. This may well turn out to be the case of geodynamics during the next decade with the development of techniques of satellite geodesy. It is expected that it will be possible within five years to detect relative movements of 1 cm over distances of up to 200 km and relative movements of a few centimeters over continental distances. The potential of these new techniques for understanding of the creaking and groaning of the crust of the Earth is very great.

With a measurement accuracy of one cm over 200 km it should be possible to measure the secular variation of strain in a continental block such as Australia and attempt to answer difficult questions such as whether the motion of the plates is continuous or episodic.

My thoughts on the unsolved problems of geodynamics owe a great deal to stimulating discussions with colleagues in Australia. However, this acknowledgement should not be taken to mean that they subscribe to any of the views which I have expressed in this paper.

A GLOBAL GEOCHEMICAL MODEL FOR THE EVOLUTION OF THE MANTLE

Don L. Anderson

Seismological Laboratory, California Institute of Technology

Pasadena, California 91125

Abstract. Basalt, eclogite, and harzburgite, differentiation products of the Earth, appear to be trapped in the upper mantle above the 670 km seismic discontinuity. It is proposed that the upper mantle transition region, 220 to 670 km, is composed of eclogite, or olivine eclogite, which has been derived from primitive mantle by about 20% partial melting and that this is the source and sink of oceanic lithosphere. The remainder of the upper mantle is garnet peridotite, or pyrolite, the source of continental basalts and hotspot magmas. This region is enriched in incompatible elements by partial melts or hydrous and CO_2 rich metasomatic fluids which have depleted the underlying layers in the L.I.L. elements and L.R.E.E. The eclogite layer is internally heated. It may control the convection pattern in the upper mantle. Material can only escape from this layer by melting. The insulating effect of thick continental lithosphere may lead to partial melting in both the peridotite and eclogite layers. Hotspots and ridges would then mark the former locations of continents. Most of the basaltic or pyroxenitic fraction of the oceanic lithosphere returns to the eclogite layer.

In this model plate tectonics is intermittent. At a depth of 150-220 km the continental thermal anomaly triggers kimberlite and carbonatite activity, alkali and flood basalt volcanism, vertical tectonics and continental breakup. Hotspots remain active after the continents leave, building the oceanic islands. Mantle plumes rise from a depth of about 220 km. Midocean ridge basalts originate in the depleted layer below this depth. Material from this layer may also be displaced by subducted oceanic lithosphere to form back-arc basin magmas.

Introduction

Although convection plays an important role in plate tectonics and heat transport in the Earth, it has not succeeded in homogenizing the mantle. Magmas are still being produced from mantle reservoirs which have remained separate for the order of 1 to 2 x 10^9 years [e.g., De Paolo, 1979; Sun and Hansen, 1975]. Oceanic lithosphere is continuously returned to the mantle but the difference in element ratios in the reservoirs, e.g., Rb/Sr, U/Pb, Th/Pb and Sm/Nd, persists. If the depth of earthquakes in subduction zones can be used as a guide, oceanic lithosphere is presently being delivered to the region of the mantle between about 220 km and 670 km. The geochemical data can be satisfied if this is also the source region for midocean ridge basalts (MORB). This leaves the upper mantle or the lower mantle as the source region for continental flood basalts (CFB), hotspot magmas and ocean island basalts (OIB). The upper mantle low-velocity zone (LVZ), or asthenosphere, is a likely source region since temperatures there are closest to the melting point.

Ocean floor basalts have comparatively uniform and low $^{87}Sr/^{86}Sr$, $^{206}Pb/^{204}Pb$, and $^{144}Nd/^{143}Nd$ ratios whereas continental magmas and basalts from ocean islands not associated with island arcs have less uniform and higher ratios (De Paolo and Wasserburg, 1979). The latter magmas are also enriched in volatiles and the incompatible large-ion lithophile (LIL) elements (White and Schilling, 1978; Frey et al., 1978). The study of isotopes has introduced the time constraint that reservoirs with different element ratios -- Rb/Sr, U/Pb, Th/Pb, and Sm/Nd -- have existed for the order of 1 to 2 b.y. The source region for MORB has been providing uniform composition lavas for long periods of time. It must therefore be immense in size and global in nature (Schilling, 1975). The reservoir for continental and ocean island magmas appears to be less uniform but also global in extent. Its products may be mixed with varying amounts of MORB before being sampled.

There are two competing petrological viewpoints regarding the nature of the source regions. The common view is that all basalts represent various degrees of partial melting of a garnet peridotite. The alternative is that some basalts represent extensive melting of a deep garnet pyroxenite or eclogite source. Both eclogite and garnet peridotite inclusions are common in kimberlite pipes. The eclogite inclusions, although not rare, represent only about 20% of the total. This suggests that eclogite is either a less abundant component of the mantle or it occurs deeper than the garnet peridotite, as befits its higher density. Neither of the two types of fragments can represent primitive mantle (Allsop et al., 1969). They must therefore be a result of a previous differentiation event. It is therefore desirable to test the hypothesis that eclogite, peridotite and continental crust are the principle products of mantle differentiation and that xenoliths in kimberlites may be samples from the mantle source regions. If true, this would have considerable impact on our ability to model the composition and evolution of the mantle.

Chemical Stratification of the Mantle

The mantle is also heterogeneous in its seismic properties. It has not been clear, however, if or how the seismological and geochemical heterogeneities are related. The largest lateral variations in seismic velocities occur in the outer 200-250 km of the Earth and are related to such surface tectonic features as shields, trenches, rises, and volcanic belts. The mantle is also inhomogeneous radially with the lithosphere, asthenosphere, and transition zone (220-670 km) being the main subdivisions of the upper mantle.

A chemically layered upper mantle could provide distinct and isolated reservoirs and is more suitable in many ways than models involving isolated blobs or regional inhomogeneities (Hofmann et al., 1978). It has been proposed that the low-velocity zone is the depleted reservoir and the source of mid-ocean ridge basalts (Schilling, 1973). Plume basalts, i.e., magmas from the L.I.L. enriched reservoir, have been attributed to deeper sources. If the LVZ is enriched in volatiles, as proposed by Anderson and Sammis (1970) on geophysical grounds, then this explanation is untenable. Frey et al. (1978) have discussed other objections to this model. They argue that volatiles should have enriched the upper layers of the mantle.

On the basis of seismic velocities and seismicity patterns, Anderson (1979c) proposed that there were chemical discontinuities in the mantle at 220 and 670 km. The former is the base of the LVZ and near the maximum depth of earthquakes in continental collision zones and regions of subduction of young, <50 Ma, oceanic lithosphere. The latter is a sharp seismic discontinuity and is near the maximum depth of earthquakes. Only old oceanic lithosphere penetrates this deep. The seismic velocities between 220 and 670 km are consistent with eclogite.

The continental lithosphere extends no deeper than about 180 km (Anderson, 1979c). It may terminate at the boundary between granular and

sheared lherzolite nodules, ~150 km (Boyd and Nixon, 1975). We will assume that the sheared nodules are representative of the mantle below the lithosphere and above the Lehmann discontinuity at 220 km. The shallower granular nodules have apparently been subjected to basalt extraction since they contain less CaO and Al_2O_3 than the sheared variety. They may be an important, perhaps major, component of the continental lithosphere. Both varieties of nodules are enriched in the L.I.L. elements compared to oceanic crust, the inferred MORB source region and the minerals of eclogite inclusions. The fertile peridotite presumably rises to shallower depths under the oceans, of the order of 80 km. Thus, the average thickness of the fertile peridotite layer is about 120 km. Volumetrically, this is an adequate source region for continental and hotspot magmas but not for the more voluminous MORB.

We suggest that differentiation of the Earth has concentrated Al_2O_3, CaO and the LIL into the upper mantle, and that the lower mantle is residual peridotite. As the Earth cools the base of the original crust transforms to eclogite which sinks through the upper mantle. In this model, the present upper mantle is peridotite overlying a thick (450 km) garnet-rich layer. Partial melting in the garnet-rich, or eclogite section, allows material to escape and to melt extensively upon ascent. A deep eclogite layer may form by garnet and clinopyroxene fractionation in early Earth history (O'Hara et al., 1975). This is proposed as the source of oceanic crust.

The basaltic portions of the oceanic lithosphere return to the eclogite layer by subduction. The fertile peridotite layer of the upper mantle can partially melt and provide basalts when the upward convection of mantle heat is prevented by the insulation of continental lithosphere. Thus, both reservoirs are global and underlie both oceans and continents. It appears that the MORB source region can also provide magma to back-arc basins, perhaps when material is displaced by the descending slab.

The Eclogite Source Region

Pipe eclogites have a strong resemblance to oceanic tholeiites in both the major and trace elements. The Rb/K and other ratios in bimineralic eclogite closely resemble the corresponding ratios in abyssal tholeiite basalts. The similarities are even more pronounced if the eclogites are compared with estimates of the average composition of the oceanic crust.

The first column of Table 1 gives a composition which is representative of oceanic tholeiites. More likely estimates of the composition of the primary magma are the total composition of the oceanic crust (column 2) and basaltic komatiites (column 3). These compositions are remarkably similar and have appreciably more MgO and less Al_2O_3 and Na_2O than tholeiites which are

TABLE 1.

Possible Compositions of the Transition Zone Eclogite Layer

	(1)	(2)	(3)	(4)	(5)	(6)	(7)	(8)
SiO_2	50.3	47.8	46.2	45.7	49.5	46.6	45.5	47.2 ± 2.4
TiO_2	1.2	0.6	0.7	0.4	0.5	0.8	1.9	0.6 ± 0.3
Al_2O_3	16.5	12.1	12.6	17.9	8.5	13.7	12.4	13.9 ± 4.5
FeO	8.5	9.0	11.4	11.2	8.8	9.1	9.5	11.0 ± 3.6
MnO	0.1	0.1	0.2	0.2	0.2	0.2	0.2	0.2 ± 0.1
MgO	8.3	17.8	16.6	11.9	16.2	16.1	18.8	14.3 ± 3.0
CaO	12.3	11.2	10.5	7.4	10.6	11.8	9.7	10.1 ± 2.2
Na_2O	2.6	1.3	1.2	2.0	1.7	1.3	1.6	1.6 ± 1.1
K_2O	0.2	0.03	0.02	0.4	1.1	0.02	0.1	0.5 ± 0.4

(1) Oceanic tholeiite (Kay et al., 1970; Engel and Engel, 1964).
(2) Oceanic crust, calculated from ophiolite section (Elthon, 1979).
(3) Basaltic "komatiite", Gorgona Island (Gansser et al., 1979).
(4)
(5) } Mantle eclogites (Nixon, 1973).
(6) Possible eclogite extract in fractionation of primary magma in upper
     mantle (O'Hara et al., 1975).
(7) Picrite (Ringwood, 1975).
(8) Average bimineralic eclogite in kimberlite (Ito and Kennedy, 1974).

considered to be the last crystallizing liquid from a more primary magma. The bimineralic eclogites in kimberlite (columns 4, 5 and 8) are virtually identical to these estimates of the average composition of the oceanic crust. Trace element comparisons between kimberlite eclogites and abyssal tholeiites are given in Table 2. Again, the correspondence is remarkable.

It appears that material similar to eclogite inclusions in kimberlites is a suitable parent for the oceanic crust. The inclusions themselves may represent cumulates from mantle diapirs that were trapped in the continental lithosphere. Diapirs rising from such great depth would melt extensively if their ascent were unimpeded by the continent.

TABLE 2.

Estimates of Lithophile Element Concentrations (ppm) in Bulk Earth,
the Eclogite and Peridotite Source Regions and Various Products
of these Source Regions

	K	Rb	Sr	Rb/Sr	U	Ref.
Abyssal tholeiite	732	0.75	92	0.008	0.16	(1)
Kimberlite eclogite	820	0.7	95	0.007	0.17	(2)
Kimberlite peridotite	617	3.4	55	0.061		(2)
Kimberlite peridotite	483	2.0	59	0.035		(2)
"Plume" source	>468	2.5	60	0.042		(1)
Continental flood basalts	6400	17	320	0.053	0.3	(3,4)
Ocean island basalts	3160	5.3	231	0.023		(1)
Continental crust	15 000	33	370	0.089	0.7	(3)
Intergranular material in eclogites	16 000	48	550	0.087		(2)

(1) White and Schilling (1978)
(2) Allsop et al. (1969)
(3) Jacobsen and Wasserburg (1979)
(4) Carmichael et al. (1974)

The Garnet-Peridotite Layer

The K, Rb, and Sr contents of some kimberlite garnet peridotite inclusions are given in Table 2. Also given are estimates of CFB and of the "plume" source. Note the agreement between tholeiites and eclogites and between peridotites and the inferred plume source region. Another way to estimate the trace element content of a partial melt from peridotite is to assume that the difference in composition between fertile and barren peridotite is due to basalt removal. The trace element content of the resulting liquid is given in Table 3 and compared with continental and oceanic basalts. The peridotite compositions are from Rhodes and Dawson (1975) and it is assumed that the basalts represent 20% melting. These are extremely fresh peridotite xenoliths from the Lashame tuff-cone in northern Tanzania that have apparently come from a depth of ∼150 km. They are chemically and mineralogically similar to peridotite inclusions from kimberlites except that they appear to be relatively less contaminated. The $^{87}Sr/^{86}Sr$ ratios of these samples are about 0.705. The inferred melt is much higher in K, Rb, and Sr than oceanic tholeiites, a characteristic of continental basalts. The K/Rb and Rb/Sr ratios are also much different than abyssal basalts. Fertile garnet peridotite therefore seems a suitable source material for continental flood basalts but not for MORB. It also has the characteristics inferred for the "plume" source region (White and Schilling, 1978). This part of the mantle has probably been subjected to metasomatic enrichment of the incompatible trace elements (Lloyd and Bailey, 1975). Such enrichment has also been proposed for the source region of continental

(Boettcher and O'Neil, 1979) and plume basalts (White et al., 1979). The fact that enriched xenoliths are extensively sampled by kimberlites argues for the shallowness of the plume reservoir.

Midocean ridge basalts generally have $^{87}Sr/^{86}Sr$ ratios between about 0.702 and 0.704 while continental basalts are usually greater than 0.704 and range up to 0.710 (Carter et al., 1978; De Paolo, 1979). Basalts from oceanic islands are intermediate in value and may represent mixtures. The data on kimberlite xenoliths is sparse and equivocal (Allsop et al., 1969; Barrett, 1975; Simazu, 1975). Pipe peridotites have $^{87}Sr/^{86}Sr$ values of 0.7060-0.7075 and other characteristics appropriate for the source region of continental basalts. Eclogite xenoliths may have been brought into the continental lithosphere by deeper diapirs and evolved for some time in an environment different from PEL prior to pipe eruption. Whole rock measurements on eclogite xenoliths from S. Africa generally have high $^{87}Sr/^{86}Sr$ ratios (0.704-0.711). Allsop et al. (1969) estimated the ratio in "ideal" bimineralic eclogite as 0.702. A sample from Tanzania has a value of 0.7004; discrete diopside nodules give 0.7029-0.704 (Barrett, 1975).

Location of the Two Source Regions

As discussed earlier, at least part of the oceanic lithosphere seems to be returned to depths between 220 and 670 km. The mantle discontinuities at these depths are sharp and they are associated with changes in seismicity, as if they were acting as barriers to slab penetration. This could be due to density jumps caused by changes in mantle chemistry. The isotopic data,

TABLE 3.

Trace Elements in Midocean Ridge Basalts, Eclogite Xenoliths,
Partial Melt of "Fertile" Garnet Peridotite and Continental Flood Basalts

	Oceanic			Continental		
	(1)	(2)	(3)	(4)	(5)	(6)
K	820	732	700	2000-3600	4000	6400
Rb	0.7	0.75	1.1	11-20	20	17
Sr	95	92	134	30-500	400	320
K/Rb	1170	976	640	100-300	400	376
Rb/Sr	0.01	0.008	0.01	0.04-0.3	0.005	0.053

(1) "Ideal" eclogite xenolith (Allsop et al., 1969). Potential source region for oceanic tholeiites.
(2) Oceanic tholeiite (White and Schilling, 1978)
(3) Oceanic tholeiite (Table 1)
(4) Inferred partial melt (20%) product from fertile garnet peridotite xenoliths with sterile peridotite xenoliths as residual (Rhodes and Dawson, 1975)
(5) Karoo basalts (Carmichael et al., 1974)
(6) Continental flood basalts (Jabobsen and Wasserburg, 1979)

although useful in finger printing the source regions and giving age control does not provide information about major element chemistry or intrinsic density. This is where the kimberlite inclusions become useful.

Eclogite is appreciably denser than garnet peridotite and should therefore occur deeper in a gravitationally stable mantle. I have suggested that the Lehmann discontinuity at 220 km is the boundary between garnet peridotite and eclogite, and the discontinuity at 670 km is the boundary between eclogite in the garnetite assemblage and peridotite in the ilmenite plus spinel assemblage (Anderson, 1979b). The eclogite layer is perched (PEL) in the upper mantle and forms the transition region. In a convecting system composed of two superposed layers there is a thermal boundary layer, i.e., region of high thermal gradient, on each side of the interface. This is where temperatures are most likely to approach or exceed the solidus and where diapirs would originate. There is also a thermal boundary layer associated with the lithosphere-asthenosphere boundary. Temperatures at 670 km and below are likely to be well removed from the melting point.

Differentiation of a silicate planet results in two distinct products, a basaltic or picritic melt and residual peridotite. The melt, resulting from low pressure, high temperature partial melting of primordial mantle possibly resembling peridotitic komatiite, would be concentrated in a thick layer at the surface. Subsequent cooling results in crystallization, cumulate formation and conversion of basaltic crust to eclogite. This leads to a massive over-turning of the outer layers of the planet, subduction of the eclogite protoplate and destruction of the early geological record. This may explain the rarity of crustal rocks older than 3.8 Ga.

Given that planetary differentiation concentrates basalt in the outer layers and that the Earth has generated and subducted massive amounts of oceanic lithosphere what is the likely distribution of basaltic material in the interior? To answer this I have estimated the density as a function of depth for basalt and peridotite (Anderson, 1979a). Basalt below about 50 km converts to eclogite which is denser than normal mantle even after olivine has converted to spinel. Pyroxene and garnet react at higher pressures to form a garnet solid solution. Normal mantle also undergoes a series of phase changes but remains less dense than garnetite until ilmenite and perovskite structures become stable below 670 km. The eclogite cannot sink below this level. The addition of Al_2O_3 to $CaSiO_3$ and $(Mg, Fe)SiO_3$ expands the stability field of garnet and increases the pressure required for transformation to such dense lower mantle phases as perovskite and ilmenite. This means that eclogite cannot sink into the lower mantle. Whole mantle convection, which may have been possible prior to the establishment of the eclogite layer and the chemical discontinuities in the upper mantle, would be replaced by separate convection systems in the lower mantle, the eclogite layer and the upper mantle above 220 km.

The oceanic part of the plate tectonic cycle in this scheme is very simple (Figure 1). Heating of the eclogite layer causes partial melting and the rise of eclogitic diapirs. The latent heat for further melting is provided by adiabatic decompression. Oceanic crust forms from this eclogite liquid which may be picritic in composition. MORB forms the surface veneer and represents the latest freezing fraction. Subduction causes the crust to reinvert to eclogite and it sinks back to the PEL. The harzburgite part of the lithosphere remains in the upper mantle.

Primitive Mantle

The isotopic data indicates that the two source regions are the results of an early differentiation event. If we accept the 220 and 670 km discontinuities as its boundaries, the eclogite layer represents about 20% of the mass of the mantle. By assuming that the whole mantle was involved in this early differentiation we can obtain a spectrum of estimates of primitive mantle composition. Several of these are given in Table 4.

There are other approaches that have been used for estimating primitive bulk Earth chemistry. Ganapathy and Anders (1974) have provided a cosmochemical mixing estimate which is also given in Table 4. There is surprisingly good agreement between these estimates and the resulting compositions are distinct from any modern rock type. Peridotite komatiites are widespread in

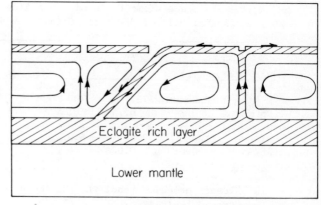

Fig. 1. Schematic of the primary plate tectonic cycle with the transition zone eclogite layer serving as the source and sink of the oceanic lithosphere. The harzburgite portion of the lithosphere remains in the upper mantle.

TABLE 4.

Inclusions in Kimberlites and Estimates of Primitive Mantle Composition

	Xenoliths			Mantle Compositions			
	(1)	(2)	(3)	(4)	(5)	(6)	(7)
SiO_2	47.2	47.3	47.3	47.3	48.0	46.6	44.8
TiO_2	0.6	0.05	0.2	0.2	0.3	0.3	0.2
Al_2O_3	13.9	1.6	4.1	5.3	5.2	3.0	5.3
FeO	11.0	5.8	6.8	7.4	7.9	10.4	10.3
MgO	14.3	43.8	37.9	35.0	34.3	34.2	34.3
CaO	10.1	1.0	2.8	3.7	4.2	4.8	4.4
Na_2O	1.6	0.2	0.5	0.6	0.3	0.2	0.4
K_2O	0.5	0.2	0.2	0.3	0.2	0.03	0.03

(1)   Average eclogite nodules (Ito and Kennedy, 1974)
(2)   Average garnet lhzerolite in kimberlite (O'Hara et al., 1975)
(3)   20% eclogite, 80% garnet lhzerolite.
(4)   30% eclogite, 70% garnet lhzerolite.
(5)   Primitive mantle (Ganapathy and Anders, 1974)
(6)   Average peridotitic komatiite, S. Aftica (Viljoen and Viljoen, 1969)
(7)   Peridotite with quench texture, W. Australia (Nesbitt, 1972)

early (<3.5 Ga) Precambrian terrains. Viljoen and Viljoen (1969) propose that these approximate the composition of primitive mantle. Indeed, these have Mg/Si ratios in the range of whole Earth estimates. They may represent primitive mantle that has left some garnet in its source region. I propose that a material similar to those in Table 4 was the parent from which the current mantle reservoirs were derived. These reservoirs are a shallow peridotitic layer and a deeper eclogite layer. In this scheme pyrolite would not represent primitive mantle but mantle which has already been depleted in a basaltic component.

It has long been recognized that the source region of MORB is depleted in LIL elements compared to alkali basalts, continental flood basalts, and hotspot magmas. One would expect, however, that the original primary differentiation would enrich the basalt/eclogite fraction relative to the residual peridotite. Whole rock analyses of pipe eclogite indeed show such enrichment. The major phases, omphacite and garnet are, however, depleted and the enrichment occurs in the intergranular material (Allsop et al., 1969). The intergranular material LIL content is similar to that of the continental crust (Table 2). The eclogite layer may have become depleted and the peridotite layer enriched by the upward transport of fluids as discussed by Frey et al. (1978), Boettcher et al. (1979) and Mysen (1979). These fluids may be kimberlitic in composition.

Tatsumoto (1978) and Hedge (1978) proposed a

model, based on lead and strontium isotopes that is similar to the present result, i.e., the LVL ("asthenosphere") is undepleted or enriched and supplies "hotspot" magmas; the underlying mantle ("mesosphere") is depleted and provides abyssal tholeiites. This is the opposite of Schilling's (1973) model. Continental and hotspot related magmas represent a wide range of partial melting, from about 4% to 25% (Frey et al., 1978; White et al., 1979). This suggests that they come from a wide range of (shallow) depths. MORB's are invariably tholeiitic, indicating extensive (>25%) melting and a consistently deep origin.

The incompatible trace elements in both source regions are enriched relative to recent estimates of bulk Earth composition (Ganapathy and Anders, 1974). For example, if the lower mantle is identical in composition to the peridotitic upper mantle and if the silicate portion of the planet is 21% eclogite and 0.5% continental crust then the major elements are in agreement with bulk Earth estimates such as Ganapathy and Anders (1974) but such elements as K, Rb, and Sr are about a factor of 2-1/2 higher. This can be accounted for if the mantle below 670 km has transferred its incompatible trace elements to the crust and upper mantle. This presumably occurred during the early differentiation of the Earth and accompanied melt extraction from primitive mantle. The calculated Rb/Sr ratio of the continental crust plus upper mantle (peridotite plus eclogite) is 0.028. This is also the value inferred for the bulk Earth

(Ganapathy and Anders, 1974; DePaolo and Wasserburg, 1976). This plus the complementary nature of the two source regions suggests that the material above 670 km may exhibit bulk Earth patterns of the LIL elements, and have a 2-3 times enrichment. The lower mantle would be extremely depleted in the LIL elements.

Evolution of the Mantle

The evolution of the Earth's mantle according to the present scheme is shown in Figure 2. The primitive mantle has roughly the composition given by a 1:4 mix of eclogite and garnet peridotite. Early differentiation processes lead to the development of a basaltic crust, cumulate layers and a residual peridotite mantle. As the basaltic crust cools it converts to eclogite which settles through the upper mantle and becomes perched near 670 km by a phase boundary in the peridotite mantle. Eclogite cumulates would suffer the same fate. Incompatible trace elements are removed into the continental crust and the upper mantle garnet peridotite regions by metasomatic, possibly kimberlitic, fluids. The eclogite layer becomes depleted in those components which are concentrated in the overlying layers including the continental crust. An eclogite cumulate layer may also be the result of crystallization in a deep magma ocean.

The highest temperatures in the mantle, relative to melting temperatures, are in the thermal boundary layer near the top of the PEL, ∿220 km depth. This is where the density contrast between eclogite and peridotite can be overcome by partial melting and where eclogite diapirs would originate. Peridotite diapirs originate near the top of the thermal boundary layer. Their shallower depth and the broad melting interval of peridotite leads to relatively small amounts of partial melting, a requirement of alkali basalt petrology (Frey

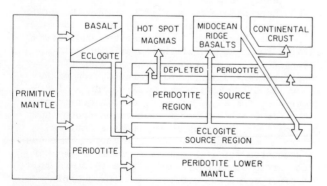

Fig. 2. Flow chart of mantle differentiation. The primitive differentiation results in picritic magmas and residual peridotite. LIL elements are transferred from the eclogite source region to the overlying layers.

et al., 1978). This is a persuasive but over-looked argument for a shallow location of the plume source region relative to the source region for oceanic tholeiites.

Continental and alkali basalts are usually emplaced at greater elevations than the oceanic tholeiitic basalts. This is sometimes taken as evidence that the MORB source region is shallower than the plume source region. Alternatively, the alkali basalts are emplaced at higher elevations because they are intrinsically less dense or more viscous than tholeiites. Eruption rates may also control the elevation.

The parameters of the thermal boundary layer depend on the thermal properties, viscosity and heat flow at the interface. The thickness is estimated to be about 20 km and the temperature rise, 300-600°C. This is comparable to the near surface gradient and brings the average temperature close to the melting point of mantle silicates at 220 km. The thermal perturbation by a stationery continent, or a large continent moving slowly, causes a further temperature rise and may be the trigger that initiates partial melting.

Depleted peridotite, the refractory product of partial melting of the garnet peridotite layer, is lighter than any other component of the mantle and becomes part of the continental lithosphere and the suboceanic harzburgite layer. The lower mantle need not be involved at all in the current magmatic cycle.

Mantle Metasomatism and the
Redistribution of Trace Elements

Evidence that the mantle has experienced metasomatism prior to the transport of samples to the Earth's surface by kimberlites has been provided by Ridley and Dawson (1975), Erlank and Shimazu (1977) and summarized by Mysen (1979). Parallel evidence from alkali basalts is given by Boettcher and O'Neil (1979). Carbonatites, kimberlites, "depleted" granular peridotites, alkali basalts and the interstitial phase in eclogite xenoliths all show extreme LREE enrichment. The source region of MORB on the other hand is LREE depleted. These observations all suggest that a vapor or fluid phase removes the REE from the eclogite layer and deposits them in the peridotite layer and the continental lithosphere. The other LIL elements also show a complimentary pattern between MORB and plume basalts.

Mass balance calculations for the incompatible elements suggest that the continental crust is not the only repository of LIL enriched material. The remainder we propose is in the continental lithosphere and the LVZ. This is supported by the evidence from material which has been sampled from these regions. The depleted nature of the MORB source region is an argument that it lies below the REE enriched regions, i.e., below the depth of generation of kimberlites and alkali basalts and the xenoliths they contain.

Plate Tectonics

A major source of mantle heat flow is the Perched Eclogite Layer (PEL). With tholeiitic concentrations it would provide 0.7 μcal/cm^{2}sec, about 1/3 of the global average surface heat flow. If tholeiites form by olivine extraction from a picritic parent magma then these estimates will be reduced. For 50% olivine fractionation the heat flow from the PEL is 0.35 μcal/cm^{2}sec. The scale length of convection in the PEL will be of the order of the layer thickness. Convection in the overlying peridotite layer may be driven by this non-uniform heating from below and would therefore be characterized by narrow ascending plumes. The high thermal gradient in the boundary layer leads to a large decrease in viscosity at the interface and the two convecting systems may be thermally coupled rather than coupled by viscous drag forces. That is, cold descending regions would occur in close proximity in the two layers. A descending slab, for example, may trigger detachment of the cold boundary layer in the PEL. This is shown schematically in Figure 3. Seismic waves will

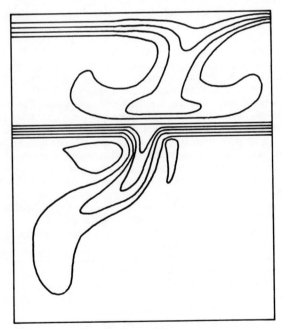

Fig. 3. Schematic illustration of isotherms for convection in a stratified system showing boundary layer detachment. The locations of descending plumes in the lower layer may be controlled by the locations of the cold isotherms in the upper layer. Material penetrating into the lower layer may initiate diapiric uprise from this region.

see a continuous cold region with high velocities. There may also be other regions of cold descending plumes in the PEL which are not directly related to slabs. Likewise, there may be regions of rising currents in both the PEL and the overlying layer, which do not express themselves in surface features such as mid-ocean ridges. The convective pattern, however, should be evident in detailed analyses of topography, gravity and seismicity.

Earthquakes do not extend below about 250 km at most convergent plate boundaries. In other regions there is a gap in the seismic zone between about 250 and 500 km. Even where the zone is continuous it is usually contorted near 250-350 km. In many cases the deeper zone is more-or-less spatially continuous with the upper zone but in Chile, Peru and New Zealand the two zones appear to be displaced. This is suggestive of the type of two-layered convection considered here.

There are several ways to estimate the lateral extent of the convection cells. We assume that convection in the eclogite layer controls convection in the thinner and shallower peridotite layer. Thiessen et al. (1979) suggested that the distribution of highspots in Africa reflects the underlying convection pattern. By comparison with laboratory data they inferred a vertical extent of convection of about 500 km. They proposed that this pattern could only be observed through a stationery continent.

Jordan (1978) showed that terrain, crustal thickness, and Bouguer gravity anomalies have correlation distances of the order of 550 km, remarkably similar to African highspot distances. This again suggests a scale length of convection comparable to the thickness of the transition layer.

Menard (1973) attributed depth anomalies in the eastern Pacific and the bobbing motion of drifting islands to convection cells in the upper mantle of half-wavelength 250-500 km. The depth anomalies, having amplitudes of ±300 meters can be explained by temperature differences in a 200 km thick layer of about 200°C. The depth anomalies seem to be fixed relative to hotspots. Menard believes that motion of plates over these bumps explain many aspects of vertical tectonics. The "bumpy" asthenosphere envisaged by Menard is a natural consequence of the convection pattern proposed here.

The pattern of convection in the PEL is probably fairly complicated. The migration of trenches and continents may change the locations of the descending plumes in an individual cell or groups of cells in the PEL but it seems likely that the cells themselves cannot move far relative to one another. This provides a rationale for a fixed hotspot reference frame and a mechanism for allowing the surface expression of a hotspot to wander on the order of 5°. The proposed upper mantle convection pattern is shown schematically in Figure 4.

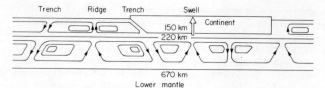

Fig. 4.   Illustration of flow in superposed
convecting layers. The lower layer is
internally heated and is characterized by
broad upwellings. The upper layer is mainly
heated from below and is characterized by
narrow ascending plumes.

Hotspots and Plumes

Morgan (1972) suggested that island and seamount chains are produced by plate motion over
convective plumes extending from near the core-
mantle interface to the base of the lithosphere.
Anderson (1975) proposed that plumes came from
a distinctly different source region than mid-
ocean ridge basalts and that kimberlites,
carbonatites and continental flood basalts were
all related to hotspot or plume activity.
I suggested that plumes were a result of a
thermal perturbation due either to continental
insulation or a diapir rising from the deep
mantle. In either case the plume source region
differed in chemistry from mantle providing
abyssal tholeiites and was the source region
for distinctive ocean island magmas and under-
saturated continental magmas. It was proposed
that this source region was rich in Ti, Ba, Sr,
Y, La, Zr, Nb, Rb, CO_2, P_2O_5, H_2O, etc., compared
to "normal" mantle (the source region of MORB).
These characteristics have since been found to be
typical of plume chemistry (Unni and Schilling,
1978; Schilling et al., 1976; Bonatti et al.,
1977; White and Schilling, 1978).

Although I initially favored a deep mantle
origin for plumes it now appears that they
originate above the 220 km mantle discontinuity
in a region of the mantle that has been enriched
in incompatible components by metasomatic
processes that have depleted the source region
of MORB.

Since hotspot activity is restricted both in
space and time we need a thermal anomaly to
initiate partial melting in the peridotite layer.
One possibility is thermal blanketing by the
thick conductive continental lithosphere
(Anderson, 1975). A large stationary continent
may thereby cause its own break-up and a
subsequent period of rapid plate motion. The
temperature anomaly and partial melt zone remains
after the continent leaves and it becomes an
oceanic hotspot. Hotspots will have a finite
lifetime which appears to be at least 200 Ma
after the continent starts to move off.

With a mantle heat flow of 0.7 μcal/cm^{2}sec
there are 2×10^9 cal/cm^2 delivered to the base
of subcontinental lithosphere in 10^8 years.

Even if only 20% of this is trapped this is
enough to heat a 50 km thick section of mantle
by 200°C and to melt it to the extent of 20%.
Continental blanketing therefore seems to be an
adequate mechanism for turning on melting spots
below the continental lithosphere.

The hotspot tracks in the Atlantic and Indian
Oceans can be traced back to continental
interiors. The timing of Mesozoic and Cenozoic
continental ·flood basalts in North and South
American, Africa, Europe and Siberia is
appropriate for their location over hotspots
when they formed (Morgan, 1979). If our
hypothesis concerning the origin of hotspots
under stationary continents is correct then we
would expect that a continent would have been
over Hawaii some 200 m.y. ago. The Hawaiian-
Emperor seamount chain disappears into the
Aleutian Trench so we cannot trace it beyond
about 70 m.y. By backing up the Pacific plate
we can infer that some of these continental
fragments may have been incorporated into
northwestern North America. The 210 m.y. old
flood basalts from central Alaska to Northern
Oregon, so-called Wrangallia (Jones et al., 1977),
originated near the equator (∿15°) and
subsequently moved north to become attached to
North America. Greenstones in central Japan
were formed near the equator in the late
Paleozoic (Hattori and Hirooka, 1979). Other
fragments are elsewhere around the Pacific Margin
(Nur and Ben-Avraham, 1979).

In the present scheme it is the thermal
perturbation caused by the deep (∿150 km)
continental lithosphere that is responsible for
the onset of hotspot activity. The hotspots
generate kimberlites, carbonatites, alkali
basalts, and lead to continental breakup. There
follows a period of rapid continental drift and
seafloor spreading. Continental igneous activity
wanes as the continents drift off their hotspots
but ridge and ocean island volcanism increase.
In this scenario hotspots play an important,
perhaps dominant, role in breaking up and driving
the plates. At the end of an interval of rapid
spreading they would tend to be centrally located
in the oceans, much as they are today. An
important force in plate tectonics may be the
"hotspot fleeing force". The thermal
perturbation by continents may also control the
location of midocean ridges (Nur and
Ben-Avraham, 1979).

Hotspot Propulsion

Elder (1976) has considered the propulsion of
continents by a horizontal temperature gradient.
In the absence of resisting forces at plate
boundaries the velocity is

$$u = \alpha g \Delta T h^3 / 3 \upsilon \ell$$

where α is the coefficient of thermal expansion
(3×10^{-5}/K), g is acceleration due to gravity

(10^3 cm/sec^2), ΔT is the temperature anomaly (200K), h is the layer thickness (100 km), υ is the viscosity (10^{20}P) and ℓ is the horizontal scale of continent (1000 km). The assigned values are just for the purpose of obtaining an order of magnitude estimate for u which turns out to be about 6 cm/yr.

Periods of extensive continental magmatism are correlated on a global basis and seem to last of the order of 0.3 to 0.4 Ga (Windley, 1977). They are separated by periods on the order of 0.7 to 1.0 Ga. Periods of rapid plate motion, or at least of rapid apparent polar wander, last for about 30-60 Ma (Gordon et al., 1979). At a velocity of 10 cm/yr. this would lead to total displacements of 3000 to 6000 km which are of continental and inter-continental distances. The continents, on the average, then would come to rest far from their own or other continents' hotspots. The gestation period for forming a hotspot appears to be 200 to 400 Ma and the lifetime estimated from the duration of continental magmatism and the duration of the subsequent hotspot track may be as much as 500 Ma. Since convection is more efficient through the oceanic mantle we expect that hotspots will start to dissipate as soon as their continents move off and that regions of high heat flow in the oceans would mark the previous locations of stationary or slowly moving continents.

Reconstruction of the continents indicate that most of the Atlantic and Indian ocean hotspots were beneath continents from about 100 to >350 m.y. ago. The present African hotspots were apparently beneath Europe at the earlier time.

Summary and Discussion

Isotopic evidence indicates that there are at least two source regions of basaltic magma in the mantle which have remained separate for the order of 1 to 2 Ga. The sub-continental and hotspot source region is enriched in incompatible components compared to the source region for midocean ridge basalts. Eclogite and garnet peridotite xenoliths in kimberlite pipes seem to have the appropriate characteristics to provide midocean ridge basalts, and continental basalts, respectively. The eclogite layer, the source of midocean ridge basalts, is denser, and therefore deeper than the enriched layer. Melting in both layers may result from the thermal insulation provided by the thick continental lithosphere.

The latent heat for extensive melting of eclogite diapirs is available if rapid ascent of 160 km or greater is possible (Yoder, 1976). This is less likely under thick continental lithosphere than in the oceanic asthenosphere. Therefore, oceanic tholeiites occur in the oceans and eclogite xenoliths occur under continents. The temperature rise required for the initiation of garnet peridotite diapirs may require a long period of continental insulation. Therefore,

hotspots start under continents and lead directly to uplift and breakup and provide the initial driving force for continental drift.

It is possible that continental insulation is also required to initiate melting in the eclogite layer. The difference in geometric style between ridges and hotspots would then reflect the difference in convection in the eclogite and peridotite layers; rising sheets in the internally heated eclogite layer and rising plumes in the overlying peridotite layer which is primarily heated from below.

One would expect the magma formed in the primary differentiation of the Earth to be LREE enriched and enriched in the incompatible elements. Present MORB is coming from a source region which has been depleted in these elements. The MORB source is, therefore, either a cumulate or has experienced a prior episode of melt removal. Continental and hotspot magmas are enriched in H_2O, CO_2, P, Cl, F, Rb, Sr, Ba, Ti, etc. Ultramafic mantle xenoliths are also generally enriched in these components, even those which have been depleted in a basaltic component. The minerals in eclogite xenoliths are low in the incompatible elements but the intergranular material is enriched in a material having abundances similar to the continental crust. We suggest that LIL elements were removed from the MORB reservoir not only by extraction of the continental crust but also by removal of a fluid or vapor phase which has enriched the continental lithosphere and the upper mantle peridotite layer. These fluids are possibly interstitial melts of a crystallizine eclogite cumulate.

A schematic continental geotherm with a thermal boundary layer is shown in Figure 5. The ascent path of an eclogite diapir is also shown. Yoder (1976) has estimated that eclogite must rise about 160 km from its source region in order to completely melt if the heat of melting is obtained by adiabatic rise. For a 220 km deep source region complete melting would be achieved at 60 km. Note that completely molten eclogite, i.e., basaltic magma, can be delivered to the surface without melting dry garnet peridotite. With the geotherm shown, wet peridotite diapirs can rise from the thermal boundary layer but they will be only partially molten even after ascent to the surface because of the broad melting interval. Extensively molten peridotitic diapirs are not possible today because of their high liquidus temperature and their relatively shallow origin. This suggests that komatiites could only form when peridotite diapirs could rise from depths greater than some 300 km, i.e., prior to the establishment of the thick eclogite layer. Only such deep diapirs could extensively melt (>60%) before they reach the surface. This assumes that the geotherm approaches the adiabat only at greater depths.

The gradual heating associated with continental insulation will mobilize mantle fluids before extensive melting occurs. Therefore, a

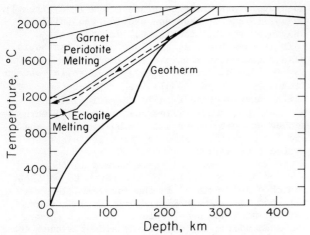

Fig. 5. Geotherm in a chemically stratified mantle showing the thermal boundary layer and the path of ascent of an eclogite diapir that starts to melt at about 250 km. Temperatures at the top of the eclogite layer must appreciably exceed the solidus before the diapir has sufficient buoyancy to rise through the peridotite layer. It will melt further if its upward escape is not prevented by an overlying continental lithosphere. Extrusion temperatures and depth of complete melting may be greater than shown here. Dry melting curves are from Wyllie (1971) and Yoder (1976).

metasomatic precursor and a redistribution of LIL can be expected prior to continental magmatism. Alkali basalt activity can also be expected to precede and accompany tholeiitic eruption from the deeper levels.

We have suggested that a thermal perturbation may be required to initiate the rise of diapirs from the MORB source region. Tholeiitic volcanism is also associated with back-arc spreading. This suggests that subduction can also trigger the rise of diapirs from the MORB source. They will not have been preheated to the extent of normal MORB magmas and may be more volatile rich. The spatial relationships of marginal basin and island arc volcanics, relative to the Benioff zone, suggest that the latter come from shallower depths than the back-arc tholeiites. Since island arcs are generally less than about 150 km above the Benioff zone (Ringwood, 1975) this suggests that marginal basin basalts come from somewhat deeper, consistent with our interpretation of the location of the MORB source region.

Acknowledgements. This work was supported by National Aeronautics and Space Administration Grant NSG-7610. I would like to thank G. J. Wasserburg, L. T. Silver, E. Stolper and F. Richter for helpful discussions. P. Wyllie, J.-G. Schilling, G. Ernst, R. Oxburgh and E. Stolper reviewed a preliminary draft and made helpful suggestions. Contribution No. 3296, Division of Geological and Planetary Sciences, California Institute of Technology, Pasadena, California 91125.

References

Akaogi, M., and S. Akimoto, Pyroxene-garnet solid solution, Phys. Earth Planet. Int., 15, 90-106, 1977.

Allsop, H. L., L. O. Nicolaysen, and D. Hahn-Weinheimer, Rb/K ratios and Sr-isotopic compositions of minerals in eclogite and peridotitic rocks, Earth Planet. Sci. Lett., 5, 231, 1969.

Anderson, D. L., Chemical plumes in the mantle, Geol. Soc. Amer. Bull., 86, 1593-1600, 1975.

Anderson, D. L., The upper mantle transition region; eclogite?, Geophys. Res. Lett., 6, 433, 1979a.

Anderson, D. L., Chemical stratification of the mantle, J. Geophys. Res., 84, 6297-6298, 1979b.

Anderson, D. L., Deep structure of continents, J. Geophys. Res., 84, 7555-7560, 1979c.

Anderson, D. L., and C. Sammis, Partial melting in the upper mantle, Phys. Earth Planet. Int., 3, 41-50, 1970.

Barrett, D. R., The genesis of kimberlites and associated rocks: strontium isotopic evidence, Physics and Chemistry of the Earth, 9, 637-654, 1975.

Boettcher, A. L., and J. R. O'Neil, Stable isotope, chemical, and petrographic studies of high-pressure amphiboles and micas; evidence for metasomatism in the mantle source regions of alkali basalts and kimberlites, Am. J. Sci., 1979.

Bonatti, E., C. G. A. Harrison, D. Fisher, J. Hannorez, and J-G. Schilling, Easter volcanic chain (southeast Pacific): A mantle hot line, J. Geophys. Res., 82, 2457-2478, 1977.

Boyd, R. R., and P. H. Nixon, Origins of the ultramafic nodules from some kimberlites of Northern Lesotho and the Monastery Mine, South Africa, Phys. and Chem. of the Earth, 9, 431-454, Pergamon Press, 1975.

Carmichael, I. S. E., F. J. Turner, and J. Verhoogen, Igneous Petrology, 739 pp, McGraw-Hill, New York, 1974.

Carter, S. R., N. Evensen, P. Hamilton, and R. K. O'Nions, Continental volcanics derived from enriched and depleted source regions: Nd and Sr-isotope evidence, Earth Planet. Sci. Lett., 37, 401-408, 1978.

DePaolo, D. J., Implications of correlated Nd and Sr isotopic variations for the chemical evolution of the crust and mantle, Earth Planet. Sci. Lett., 43, 201-211, 1979.

DePaolo, D. J., and G. J. Wasserburg, Inferences about magma sources and mantle structure from variations of $^{143}Nb/^{144}Nb$, Geophys. Res. Lett., 3, 743-746, 1976.

DePaolo, D. J., and G. J. Wasserburg, Neodymium

isotopes in flood basalts from the Siberian platform and inferences about their mantle sources, Proc. Natl. Acad. Sci., 76, 3056-3060, 1979.

Elder, J., The Bowels of the Earth, 222 pp., Oxford Press, 1976.

Elthon, D., High magnesia liquids as the parental magma for ocean floor basalts, Nature, 278, 514-518, 1979

Engel, A. E., and C. G. Engel, Composition of basalts from the Mid-Atlantic ridge, Science, 144, 1330-1333, 1964.

Erlank, A. J., and N. Shimazu, Strontium and strontium isotope distributions in some kimberlite nodules and minerals, Ext. Abs. Second Int. Kimberlite Conf., Santa Fe, New Mexico, 1977.

Frey, F., D. Green, and S. Roy, Integrated models of basalt petrogenesis: A study of quartz tholeiites to olivine melilitites from south-eastern Australia utilizing geochemical and experimental petrological data, J. Petrol., 19, 463-513, 1978.

Ganapathy, R., and E. Anders, Bulk compositions of the Moon and Earth estimated from meteorites, Proc. Lunar Sci. Conf., 5th, 1181-1206, 1974.

Gansser, A., V. J. Dietrich, and W. E. Cameron, Paleogene komatiites from Gorgona Island, Nature, 278, 545-546, 1979.

Gordon, R., M. McWilliams and A. Cox, Pre-tertiary velocities of the continents: A lower bound from paleomagnetic data, J. Geophys. Res., 84, 5480-5486, 1979.

Green, D. H. and A. E. Ringwood, The genesis of basaltic magmas, Contr. Mineral. Petrol., 15, 103-190, 1967.

Hattori, I., and K. Hirooka, Paleomagnetic results from Permian greenstones in central Japan and their geological significance, Tectonophysics, 57, 211-236, 1979.

Hedge, C. E., Strontium isotopes in basalts from the Pacific Ocean basin, Earth Planet. Sci. Lett., 38, 88-94, 1978.

Hofmann, A. W., W. M. White, and D. J. Whitford, Geochemical constraints on mantle models: The base for a layered mantle, Carnegie Inst. Wash. Yearbook, 77, 548-562, 1978.

Ito, K., and G. C. Kennedy, The composition of liquids formed by partial melting of eclogites at high temperatures and pressures, J. Geol., 82, 383-392, 1974.

Jacobsen, S. G., and G. J. Wasserburg, Nd and Sr isotopic study of the Bay of Islands ophiolite complex and the evolution of the source of mid-ocean ridge basalts, J. Geophys. Res., 84, 7429-7445, 1979.

Jones, D. L., N. J. Silberling, and J. Hillhouse, Wrangellia -- A displaced terrane in north-western North America, Can J. Earth Sci., 14, 2565-2592, 1977.

Jordan, S., Statistical model for gravity - topography and density contrasts in the Earth, J. Geophys. Res., 83, 1816-1824, 1978.

Kay, R., N. J. Hubbard, and P. W. Gast, Chemical characteristics and origins of oceanic ridge volcanic rocks, J. Geophys. Res., 75, 1585-1613, 1970.

Lloyd, F. E. and D. Bailey, Light element metasomatism of the continental mantle: The evidence and the consequences, Physics and Chemistry of the Earth, 9, 389-416, 1975.

Menard, H. W., Depth anomalies and the bobbing motion of drifting islands, J. Geophys. Res., 78, 5128, 1973.

Morgan, W. J., Plate motions and deep mantle convection, Geol. Soc. Amerc. Memoir, 132, 7-22, 1972.

Morgan, W. J., Hotspot tracks and the opening of the Atlantic and Indian oceans, in The Sea, in press, 1979.

Mysen, B., Trace element partitioning between garnet peridotite minerals and water-rich vapor: Experimental data from 5 to 30 kbar, Am. Mineral., 64, 274-287, 1979.

Nesbitt, R. W., Skeletal crystal forms in the ultramafic rocks of the Yilgarn Block, Western Australia; evidence for an Archaean ultramafic liquid, Geol. Soc. Australia Spec. Pub., 3, 331-350, 1972.

Nixon, P. H. (ed.), Lesotho kimberlites, Lesotho National Development Corp., Maseru, Lesotho, 1973.

Nur, A. and Z. Ben-Avraham, Speculations on mountain building and the lost Pacifica continent, J. Phys. Earth, 26, Suppl., 521-537.

O'Hara, M., M. J. Saunders, and E. L. P. Mercy, Garnet-peridotite, primary ultrabasic magma and eclogite: Interpretation of upper mantle processes in kimberlite, Phys. Chem. of the Earth, 9, Pergamon Press, 571-604, 1975.

Rhodes, J. M. and J. B. Dawson, Major and trace element chemistry of peridotite inclusions from the Lashine Volcano, Tanzania, Phys. and Chem. of the Earth, 9, Pergamon Press, 545-558.

Ridley, W. I. and J. B. Dawson, Lithophile trace elements data bearing on the origin of peridotite xenoliths, Ankaramite and Carbonatite from Lashaine Volcano, N. Tanzania, Phys. and Chem. of the Earth, 9, 559-570, 1975.

Ringwood, A. E., Composition and Petrology of the Earth's Mantle, McGraw-Hill, New York, 1975.

Schilling, J. G., Iceland mantle plume; geochemical evidence along Reykjanes Ridge, Nature, 242, 565-579, 1973.

Schilling, J. G., Azores mantle blob: rare-earth evidence, Earth Planet. Sci. Lett., 25, 103-115, 1975.

Schilling, J. G., Rare earth, Fe, and Ti variations along the Galapagos spreading center, and their relationship to the Galapagos mantle plume, Nature, 261, 108-113, 1976.

Shimizu, N., Geochemistry of ultramafic inclusions from Salt Lake Crater, Hawaii and from southern Africa kimberlites, Phys. and Chem. of the Earth, 9, 655-670, 1975.

Sun, S. and G. N. Hanson, Evolution of the mantle; geochemical evidence from alkali basalt, Geology, 3, 297-302, 1975.

Sun, S. S. and G. N. Hanson, Origin of Ross Island basanitoids and limitations upon the heterogeneity of mantle sources for alkali basalts and nephelonites, Contrib. Mineral. Petrol., 52, 77-106, 1975.

Tatsumoto, M., Isotopic composition of lead in oceanic basalt and its implication to mantle evolution, Earth Planet. Sci. Lett., 38, 63-87, 1978.

Thiessen, R., K. Burke, and W. S. F. Kidd, African hotspots and their relation to the underlying mantle, Geology, 7, 263-266, 1979.

Unni, C. K. and J. G. Schilling, Cl and Br degassing by volcanism along the Reykjanes Ridge and Iceland, Nature, 272, 19-23, 1978.

Viljoen, R. P. and M. J. Viljoen, Evidence for the composition of the primitive mantle and its products of partial melting, Spec. Publ. Geol. Soc. S. Africa, 2, 275-295, 1969.

White, W. M., M. Tapia and J.-G. Schilling, The petrology and geochemistry of the Azores Islands, Contrib. Mineral. Petrol., 69, 201-213, 1979.

White, W. M., and J. G. Schilling, The nature and origin of geochemical variations in Mid-Atlantic ridge basalts from the central North Atlantic, Geochim. Cosmochim. Acta, 42, 5101-1516, 1978.

Windley, B., The Evolving Continents, Wiley, London, 385 pp., 1977.

Wyllie, P. J., The Dynamic Earth, Wiley and Sons, Inc., New York, 416 pp., 1971.

Yoder, H. S., Jr., Generation of basalt magma, Natl. Acad. Sci., Washington, D.C., 265 pp., 1976.

TEMPERATURE PROFILES IN THE EARTH

Orson L. Anderson

Institute of Geophysics and Planetary Physics and Department of Earth and Space Sciences

University of California, Los Angeles, California 90024

Abstract. A geotherm from 80 to 5153 km depth is proposed. The temperature gradient in the upper mantle is found by using laboratory data on minerals in conjunction with seismic travel-time curves. The temperature at the beginning of the transition zone, 380 km, is taken to be 1400°C, following the results of the phase transitions in olivine, according to Akaogi and Akimoto (1979). Their data are also used to estimate the temperature gradient throughout the transition zone. The lower mantle temperature gradient is found by assuming adiabatic compression and taking γ to be γ_a (gamma acoustic). A small correction is made for the effect of apparent superadiabaticity (100°) throughout the lower mantle. The temperature gradient of the outer core is found by assuming adiabatic compression and by taking T to be constant ($\gamma = 1.4$). An estimate for the ΔT jump at thermal barriers at 670 km and 2885 km depth is needed. The result gives 4080°K at the inner-outer core boundary. The results are compared with several current thermal models of the Earth.

Introduction

One of the major goals of the geodynamics program of the Inter-Union Commission on Geodynamics has been the elucidation of the temperature profile of the Earth. Progress towards this goal has been slow compared to the definition of the seismological parameters of the Earth, with the associated understanding of mechanical properties. Nevertheless, some outstanding progress has been made in the past decade.

One major development has been an improvement in the understanding of melting. On the experimental side considerable data on this subject have been

Publication 1962, Institute of Geophysics and Planetary Physics, University of California, Los Angeles, California 90024 (USA)

Presented at the Conference on Physics and Chemistry of the Origin of the Earth, London, Ontario, Canada, September 2-5, 1979.

provided by the many papers of G. C. Kennedy and his coworkers, for example. On the theoretical side, the Lindemann Law has been given a sound thermodynamic basis by Stacey and Irvine (1977) for solids and by Stevenson (1979) for liquids. This law has now regained its former importance in theories of melting.

Another general development has been an improved understanding of the Grüneisen parameter, γ, which must be specified in order to relate the temperature profile to the density profile along an adiabat. While approaches differ, there is a convergence towards sets of solutions of γ versus depth that are yielding similar temperature gradients in the regions of the Earth where adiabaticity is a fair approximation.

With these major developments, the stage has been set for the construction of thermal models of the Earth controlled by rigorous thermodynamic principles and constrained by the seismological data. The thermal model proposed by Stacey (1977) is an outstanding example. Nevertheless, for some time in the future thermal models of the Earth will not have the certainty or detail that are found in seismological models. Regions of lateral and vertical inhomogeneity--now so clearly defined in seismological models--are more difficult to specify in thermal models because the thermal data are limited. One may expect thermal models to deal with the average Earth properties for the deep interior properties.

Many thermodynamic properties used in constraining Earth thermal models are essentially partial derivatives, so integrating constants are needed to constrain the temperature profile itself. There are three regions where constraints have been identified. The first arises from the assumption that the inner-outer core boundary is at the normal liquidus temperature of core material. By assuming the composition of the core, and by estimating the liquidus of that composition at 320 GPa, the temperature profile in the core has a fixed point. This assumption is the basis of Stacey's (1977) thermal model. The second constraint arises in the assumption

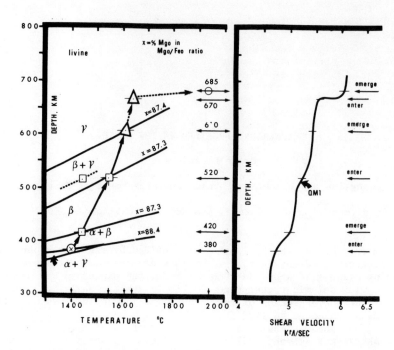

Fig. 1. Correspondence between the temperature and mantle depth of phase boundaries in the olivine system. On the left is the phase boundary in depth versus T for the three phases (α, β, γ) and their mixtures (Akaogi and Akimoto, 1979). On the right is the profile of depth versus v_s according to the QM1 seismic model (Hart, et al., 1976). Transition pressures are found from the velocity profile, but selecting the boundaries where the pure phases enter and emerge. These intersections on the phase diagram give the temperature.

that the transition zones of the upper mantle shown in the seismic shear velocity can be correlated to the phase transitions in olivine and possible other mantle mineral constituents. This amounts to a linking of seismology and experimental petrology. This assumption is the basis of the recent discussion of thermal profiles by Jeanloz and Richter (1979).

The third constraint comes from heat flow. Using surface heat-flow measurements, the temperature profile can be estimated downward. This method most recently has been used in the extrapolations down to depths of about 80 km by Lachenbruch and Sass (1978). Other constraints are possible.

In the present paper, the approach is to fix the temperature profile at the olivine-olivine modified spinel transition that, following Akaogi and Akimoto (1979), is 1400°C at the beginning of the transition (380 km). Thus the choice of the fixed temperature point used here follows the recommendation of Jeanloz and Richter (1979), but does not agree exactly with their number.

The temperatures are then computed in both directions: through the upper mantle to check with the lithospheric results of Lachenbruch and Sass (1978) for temperatures in the lithosphere; and to the lower mantle and outer core to compare with Stacey's and other estimates of T at the inner-outer core boundary.

General Results for the Transition Zone and Lower Mantle

The temperatures in the transition zone of the upper mantle (region C) have been estimated by correlating the shear velocity structure of the QM1 model by Hart et al. (1976) with the pressure-temperature stability diagram of the olivine (α) modified spinel and spinel (γ) transformation region in the $Mg_2SiO_4-Fe_2SiO_4$ system by Akaogi Akimoto (1979) (see Figure 1). In addition to 1400°C at 380 km, there results 1550°C at 520 km and 1610°C at 610 km. These results are equivalent to an adiabatic compression using a Grüneisen parameter of 2: γ(C) = 2. (See also Brown and Shankland, 1979.) The seismic model QM1 arises from Hart et al. (1976).

From these fixed points the temperature at 670 km, T(670), is estimated as 1640°C by adiabatic compression (Figure 1). The jump at ΔT at the 670-km mantle, ΔT(670), discontinuity must then be estimated. The sharpness of this discontinuity is estimated to be 12 km by D. L. Anderson (1976), which implies a very small dT/dP, or a small ΔS of the transition. I have used an emergent depth of 685 km for this transition, with a parameterization of ΔT(670) at three values: 0°C, 150°C, and 300°C. The latter two figures correspond roughly to heats of transition of 30 and 60 kcal/mole.

Extrapolating through a homogeneous lower mantle (region D'), the transition zone (region D''), and the outer core (region E) by methods described later, the temperature at the core-mantle boundary is 2544°K, 2731°K, and 2930°K, using the three parameters for ΔT at 670 km. This compares with 3157°K proposed by Stacey (1977) and 2850°K to 3300°K proposed by Usselman (1975) (Figure 2).

The temperature increase in region D'' (100 km above the core-mantle boundary) is taken to be ΔT(D'') = 300°, corresponding to boundary layer calculations made by Jeanloz and Richter (1979). This calculation is sensitive to the heat flux passing from the core into the mantle. Jeanloz and Richter assume one-fourth of the total terrestrial heat flow arises originally in the core. A large value of ΔT(D'') apparently was used by Stacey (1977). In another paper, ΔT (D'') was taken to be 500°, corresponding to an assumption of a larger heat flux from the core (Baumgardner and Anderson, 1980).

A correction is made for apparent superadiabaticity probably arising from several small phase boundaries in the lower mantle amounting to 200°. This is examined in detail in a later section.

Assuming adiabatic compression in the lower mantle and outer core and ΔT''(D'') = 300°, I find 4086°K at the solid-liquid boundary of the core for the middle cases of ΔT(670) (150°K). This value is compared with 4168°K which was used by Stacey (1977) as the fixed temperature. He obtained his figure by extrapolating Usselman's (1975) measurements of the iron-sulfur eutectic from 60 kbar to 3.2 mbar, using Lindemann's Law. Usselman's extrapolations of his own data led to 3750°K to 4050°K in the inner-outer core boundary. Thus, the general model of this paper agrees fairly well with models of Stacey or Usselman at the inner-outer core boundary depending upon the value chosen for ΔT(670) and also upon the inclusion of other possible effects in the lower mantle. If it is assumed that ΔT(D'') is larger, then T at the inner-outer core boundary is correspondingly found to be larger.

In the present paper, for the lower mantle $\gamma(D') = 1.5 \ (V/V_0)^{1.35}$ according to O. L. Anderson (1979), which yields a monatonic decrease of γ from 1.31 to 0.98. This formula is based on the acoustic Grüneisen parameter using basic Debye theory.

Stacey (1977) used γ(D') monatonically decreasing from 1.03 to 0.914, following the law $\gamma = 68.08 \ \rho^{-\frac{1}{2}}$ (Irvine and Stacey, 1979). Jeanloz and Richter (1979) used γ(D'') varying monatonically from 1.1 to 0.7 following their estimates of the γ-ρ dependence of MgO and stishovite. Brown and Shankland (1979) found γ(D') varying monatonically from 1.292 to 0.954, which agrees quite well with the gamma acoustic results found by O. L. Anderson (1979) by a different method. O. L. Anderson (1979) reported that the change in T due to adiabatic compression, $\Delta T_S(D')$, in the lower mantle was 662°. Stacey found $\Delta T_S(D')$ near 600°. Jeanloz and Richter reported $\Delta T_S(D')$

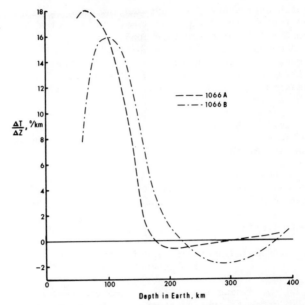

Fig. 2. Variation of dT/dZ for an olivine mantle, using the two seismic models 1066A and 1066B. Note the sharp change in the slope at 200 km.

= 460°. Brown and Shankland reported $\Delta T_S(D') = 576°$.

General Results for the Outer Core

γ was taken to be constant throughout the outer core at a value of 1.4. An argument can be made that γ should be a monatonic decreasing function with depth, but there is some uncertainty about the value of γ itself. Stacey used his theory for lattice γ (Irvine and Stacey, 1979) and then added a term for electronic contribution to the thermal pressure, γ_e. Now γ_e contributed about 0.1 to the total γ which is small compared to uncertainty in γ discussed below. Stacey's (1977) values for γ(E) for the outer core decreased monatonically from 1.419 to 1.265. The preliminary estimate by Irvine and Stacey (1979) indicated γ(E) ~ 1.4. Brown and Shankland (1979) found a monatonic decrease showing γ(E) from 1.683 to 1.023. A calculation by Jamieson et al. (1978) indicated γ(E) may range between 1.2 to 2.0. Kennedy and Higgins (1973) used γ(E) = 1.75 for their calculations. The value of gamma acoustic, γ_a, reported by O. L. Anderson (1979) is on the average 1.48 in the outer core.

All the previous estimates of γ(E) and the resulting $\Delta T_S(E)$ can be criticized since they rely on equations grounded in the theory of solids. Stevenson (1980) derived an equation for melting closely analogous to Lindemann's law by using modern theory of liquids

$$\partial \ln T_m / \partial \ln \rho = (\gamma C_V - k)/(C_V - 3/2 \ k)$$

which approaches the Lindemann law when $C_V \to 3k$. He further found the approximate values of γ to be 1.3–1.5 for outer core conditions.

McQueen et al. (1970) reported data on shocked iron, which was interpreted by Jamieson et al. (1970) to give $\gamma(E)$ between 1.1 and 3.1. Jeanloz (1979) found values of $\gamma \sim 1.3$ under shockwave conditions. Until the theory for γ for liquids at high pressures becomes more refined, I believe that an approximate constant value of $\gamma(E)$ near 1.4 for the outer core can be justified, and this value is used for T calculations.

In the present paper $\Delta T_S(E) = 953°$; Stacey reported $\Delta T_S(E) = 948°$; and Brown and Shankland reported $\Delta T_S(E) = 854°$; Jamieson et al. reported $\Delta T_S(E)$ varies from $900°$ to $1700°$. Higgins and Kennedy (1973) reported $\Delta T_S(E) = 0.33T(5000)$.

Superadiabaticity and Inhomogeneity in the Lower Mantle

In the discussion so far, no consideration has been given to superadiabatic gradients in the lower mantle.

It is recognized that for a convecting homogeneous phase, the temperature has to be slightly superadiabatic, but the ΔT from this effect need not be large, so its effect is not included in the calculation.

There is evidence for several small phase changes in the lower mantle (see, for example, Hales et al., 1980), and these may be associated with small ΔT jumps in the lower mantle.

Brown and Shankland (1980) estimated this effect from the PEM data by evaluating the residual

$$\Delta = (\Delta\rho/\rho - \Delta P/K_s)/\Delta P/K_s$$

throughout the mantle. This measures the departure from the Adams-Williamson requirement for a homogeneous mantle. By setting $(\Delta\rho/\rho - \Delta P/K_s)$ to $\alpha\Delta T_{sa}$, where α is the thermal expansion coefficient at depth, the superadiabatic component ΔT_{sa} is found. Over the whole mantle, T_{sa} was found by Brown and Shankland (1980) to be $203°C$. Jeanloz and Richter (1979) estimated this effect to be $100°C$ to $150°C$. In this report we assume the apparent superadiabatic component to be $100°C$ following Jeanloz and Richter's suggestion; a $25°T$ jump each is placed at 1246, 1546, 1770, and 2521 km which bound regions of homogeneity according to the Butler and D. L. Anderson (1978) model. In another paper, the ΔT_{sa} component of the lower mantle was taken to be $200°$, following Brown and Shankland's calculations (Baumgardner and Anderson, 1980).

The density profile in the lower mantle and core used in the calculations of T in the present paper were taken from the PEM solution of Dziewonski et al (1975). The uncertainties in the values of $\Delta T(670)$ and $\Delta T(D'')$ suggest that the use of a smoothed seismic model of the Earth, such as is the PEM, is more appropriate for

regions D' and E than is a detailed seismic solution. In region B of the upper mantle, discussed below, a seismic model which has not been smoothed by a polynomial is needed to examine the details of the temperature profile near the low-velocity zone (LVZ), so the PEM is not used.

Region B of the Upper Mantle

In region B the variation of shear velocity with density at various depths can be defined from the seismic data. If $\partial v/\partial\rho$ is constrained by the variation of velocity with temperature, $\partial v/\partial T$, then a solution is found for the thermal gradient dT/dZ at various depths. The equation results from a straightforward application of calculus (O. L. Anderson, 1980). The result for a homogeneous upper mantle in which melting is ignored is

$$dT/dZ = \rho g v_s'(v_s/K_e)(\partial v_s/\partial T)_P^{-1} + \rho g (\partial T/\partial P)_{v_s} \quad (1)$$

where $v_s' = \partial\ln v_s/\partial\ln\rho$, K, v_s, ρ, and g are determined at various depths from the seismic data. Now $(\partial v_s/\partial T)_P$ as well as

$$(\partial T/\partial P)_{v_s} = -(\partial v_s/\partial P)_T/(\partial v_s/\partial T)_P$$

are determined from experiments in mineral physics on candidate minerals. v_s' can be recognized as $\gamma_s - 1/3$, where γ_s is the acoustic shear mode Grüneisen constant. It should have a value of near unity in homogeneous compression.

In (1) the first term on the right is ordinarily negative, since v_s' is ordinarily positive, and the second term on the right is positive. Thus, dT/dZ depends on the difference of two terms, but the controlling parameter is v_s', which depends in turn upon seismic models. In the LVZ dv_s/dZ changes sign, but $d\rho/dZ$ changes sign at a different depth. Thus, for a region around 100 km, v_s decreases with Z while ρ increases with Z. This has the effect of causing v_s' to be anomalously negative and dT/dZ to diminish steadily with depth, finally becoming zero or even negative. Somewhere around 200 km, the rate of change of v_s with ρ is sufficiently large and positive that the first term on the right of (1) dominates the second term, and dT/dZ becomes positive and normal again.

The details of the v_s' profile are sensitive to the exact location and depth of the density minimum. In the model QM1 by Hart et al. (1976), the minimum in v_s is at 121 km, while in models 1066A and 1066B of Gilbert and Dziewonski (1975) the minimum in v_s is 160 km and 100 km respectively. Jordan and D. L. Anderson (1974) found the minimum in v_s to be near 190 km in their B1 model. Thus we see little agreement on the exact position of the minima, although most current models agree that there are minima.

Models 1066A and 1066B are used as the basic data for computing the temperature derivative distribution. These two models may unduly emphasize the magnitude of the minimum in v_s'. Many authors have pointed out that the resolving power of the seismic data is not good in the LVZ. Thus the calculations of the v_s' and dT/dZ can be considered only as tentative results.

The values of the material properties $(\partial v_s/\partial T)_P$ and $(\partial v_s/\partial P)_T$ calibrate the dT/dZ curve. Here we have assumed a garnet lherzolite upper mantle in which the dominant minerals are olivine, orthopyroxene, garnet, and clino pyroxene. We therefore require the values of velocity derivatives of these minerals at high T, say 1400°C, since that is the fixed temperature point at 380 km) which is the basis of the integration of the resulting dT/dZ distribution. $(\partial v_s/\partial T)_P$ changes with temperature from room temperature down to about 2θ, then becomes a constant value. The method of extrapolation is given by Soga (1967). We take $(\partial v_s/\partial P)_T$ to be independent of temperature, since isotherms of μ versus P are generally parallel. Neither parameter should vary significantly with pressure in the range of P in the upper mantle (25 to 130 kb). The most uncertainty is on the value of $(\partial v_s/\partial T)_P$ at 1400°, but fortunately this parameter has been measured at that temperature for pyrope and almandine-pyrope (Soga, 1967; Sumino and Nishizawa, 1978). Data for olivine (Kumazawa and Anderson, 1969) and for orthopyroxene (bronzite)(Frisillo and Barsch, 1972) exist near room temperature and were extrapolated to 1400°C by Soga's (1967) method. Suitable data for clinopyroxene could not be found. Thus I assumed for the upper mantle model that the data for clinopyroxene could be replaced by those of orthopyroxene.

The values of the material derivatives used in (1) are listed in Table 1.

The thermal gradient for an olivine upper mantle is shown in Figure 2 for the seismic models 1066A and 1066B. No account of possible melting or change in iron content is considered in this curve. The oscillation of the temperature gradient arises because of the oscillation of v_s' with depth. This in turn is due to the failure of v_s to be exactly proportional to ρ throughout the LVZ.

The thermal gradient is integrated to obtain the temperature distribution and the integrating constant is taken to be, as before in this paper, 1400°C at 380 km (Akaogi and Akimoto, 1979). Results for the olivine minerals orthopyroxene, and garnet for the 1066A model are shown in Figure 6 of Anderson (1980). It is found that the data for a garnet mantle require a large hump in T at

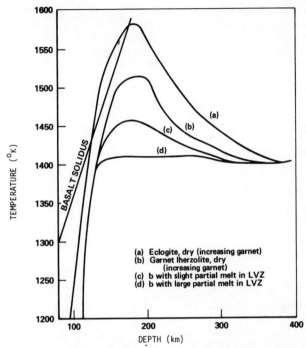

Fig. 3. The geotherms for four assumed models of the mantle using the 1066A model: (a) dry eclogite, (b) dry lherzolite, (c) lherzolite with a small amount of partial melting, (d) lherzolite with a larger amount of partial melting.

110-250 km, and further that T crosses the basalt solidus. On the other hand, T for an olivine mantle has a slight hump and does not cross the basalt solidus.

Two mantle petrological models, composed of mixtures of olivine (Ol), orthopyroxene (Op), and garnet (Ga) are considered: (1) (Ol) 10% at all depths, (Op) varying from 85% at 80 km to 70% at 380 km, (Ga) varying from 5% at 80 km to 20% at 380 km; (2) (Ol), 50% at 80 km to 47% at 380 km; (Op) 45% at 80 km to 93% at 380 km; (Ga) 5% at 60 km to 10% at 380 km. Model 1 is similar to eclogite; model 2 is similar to lherzolite. Assuming a linear mixing law which relates T of the mineral to T of the rock, profiles of T can be found in Figure 3.

It is seen that the eclogite-like model crosses the basalt solidus and has a 170° hump, whereas the lherzolite-like model does not cross the basalt solidus and has a 100° hump. Partial

TABLE 1: Values of $(\partial v_s/\partial T)_P$ and $(\partial v_s/\partial P)_T$ at 1400°C

	$(\partial v_s/\partial T)_P$ (km/sec/°)	$(\partial v_s/\partial P)_T$ (km/sec/kb)	$(\partial T/\partial P)_{v_s}$ (°/kbar)
Olivine	-4.4×10^{-4}	3.7×10^{-3}	8.5
Orthopyroxene	-6.29×10^{-4}	5.5×10^{-3}	8.7
Garnet	-3.4×10^{-4}	2.2×10^{-3}	6.4

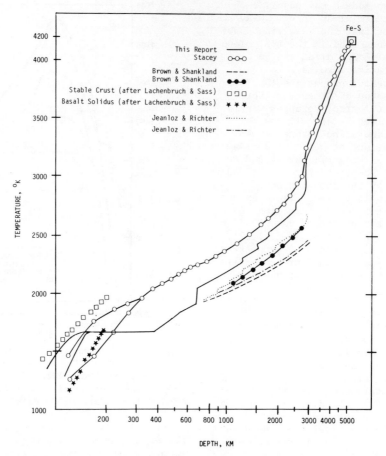

Fig. 4. The recommended T profile in the earth's mantle and outer core (solid line) compared with Stacey's model (1977) (circules), the solution of Jeanloz and Richter (1979) (dots), and the solution of Brown and Shankland (1980) (shaded). The square at the right top is Stacey's estimate of the Fe-S eutectic T. The bar below it is Usselman's estimate. On the lower left are solutions of the lithospheric T profile by Lachenbruch and Sass (1978).

melting lowers the hump (Anderson, 1980). This indicates that the thermal profile of the LVZ is quite sensitive to the composition model chosen.

There is some evidence that the average mass $\bar{m}/p$ decreases with depth in the upper mantle (Watt et al., 1975). If so, this would diminish the effect of hump. In the lithosphere the thermal profiles seen in Figure 3 agree well with the values found by Lachenbruch and Sass (1978), thus connecting the calculations of heat-flow-derived measurements from surface measurements to the calculations of the fixed T at pressures corresponding to 380 km, derived from solid-state transition measurements.

Suppose that the LVZ is compressed under an exactly adiabatic state and the Grüneisen parameter is a constant; then dlnT is proportional to dlnρ. But if ρ passes through a minimum in depth, then the thermal gradient would also exhibit a minimum as well as two regions where dT/dZ

vanishes. Thus even in the simplest thermodynamic case there is an irregularity in the temperature behavior in the LVZ unless ρ monatonically increases with Z throughout the LVZ.

The magnitude of the hump in T depends upon two conditions. The first is the degree to which the velocity minimum is shifted from the density minimum, and the second is the amount, if any, of partial melting in the LVZ.

Speaking to the first point, it must be recognized that the resolving power of seismology is not adequate to determine very accurately the derivative of dv_s/dZ and $d\rho/dZ$ through the LVZ. Thus the hump in T shown in Figure 4 may be an artifact arising from inadequate seismic models of the LVZ. Yet the general agreement of virtually all seismic models of the average earth that there is a minima in both v_s and ρ tends to show that there is an irregularity in the T profile somewhere in the LVZ.

On the second point there has recently been a

TABLE 2: Temperatures of the Mantle and Outer Core
According to This Model*

Upper Mantle

Radius (km)	Depth (km)	T (°K)
6271	100	1343
6251	120	1597
6231	140	1633
6211	160	1673
6171	200	1673
6131	240	1673
6091	280	1673
6051	320	1673
6011	360	1673
5991	380	1673
5951	420	1716
5911	460	1759
5891	480	1780
5851	520	1823
5811	560	1847
5771	600	1871
5731	640	1895
5701	670	1913
5686	685	2063

Lower Mantle

Radius (km)	Depth (km)	Density (gm/c³)	T (°K)
5671	700	4.395	2069
5600	771	4.437	2096
5400	971	4.453	2171
5200	1171	4.688	2255
5000	1371	4.779	2334
4800	1571	4.889	2425
4600	1771	4.996	2512
4400	1971	5.101	2570
4200	2171	5.203	2628
4000	2371	5.303	2680
3800	2571	5.401	2756
3600	2771	5.496	2804
3500	2871	5.543	2829
3485.7	2885.3	5.550	2831

Outer Core

Radius (km)	Depth (km)	Density (gm/c³)	T (°K)
3400	2971	10.043	3131
3200	3171	10.340	3261
3000	3371	10.611	3382
2800	3571	10.859	3493
2600	3771	11.084	3594
2400	3971	11.288	3688
2200	4171	11.473	3773
2000	4371	11.639	3850
1800	4571	11.789	3920
1600	4771	11.922	3983
1400	4971	12.042	4040
1300	5071	12.095	4065
1217.1	5153.9	12.139	4086

*Arrows in the low mantle indicate position of assumed transitions with corresponding jumps of 25°C. Jumps occur between regions of homogeneity according to Butler and D. L. Anderson (1978).

general agreement among workers in theory of the Earth to disregard the effect of partial melting on mechanical properties of the upper mantle, except beneath active provinces such as the ridges and volcanic belts. If this is true, partial melting can be ignored in a theory of the average Earth, but for applications to active provinces partial melting should be included (see especially comments of Goetze, 1977, p. 16).

It is well known that excess water vapor will depress the solidus of peridotite (Mysen and Boettcher, 1975) by several hundred degrees. But Goetze (1977) has questioned whether water can precipitate as a volatile in the mantle. He discusses a number of associated experiments and concludes "...These facts weaken the argument that water, if not accommodated within an identifiable hydrous phase, must form vapor or melt...Therefore the question of whether the solidus is significantly depressed by volatiles is still open."

It is apparently impossible to construct a temperature model which accounts for the laboratory data on velocities of the appropriate minerals at high T and the existing published data on average earth seismic models in the LVZ, and to obtain a monatonically increasing T profile throughout the LVZ.

A negative value of dT/dZ means that the temperature profile is in nonequilibrium. The time to reach a steady state for a mantle material may be greater than 10^{10} years. Thus, the upper mantle convection may not have reached steady state. A temperature hump in the LVZ cannot be definitely ruled out (see also Trubitzyn et al.,

this volume). Schubert (1979) shows a temperature hump is to be expected for a boundary layer in mantle flow, where a thermally insulated layer (which may be the lithosphere) bounds from below steady convection at high Rayleigh number. Using the results found by Schubert and Young (1976), who computed a two-cell model of mantle convection against a nonconducting lithosphere, it is seen that the negative gradient dT/dZ would be greatest in corners adjacent to an upwelling stream between cells (see also Trubitzyn et al., this volume, Figure 2). Thus near oceanic ridges and volcanic provinces we may expect a steep negative geotherm in the upper mantle, but under shields the effect would be minor or nonexisting.

It is difficult to say what the worldwide average T gradient really is but in any case it should be subadiabatic. For the recommended T profile of this paper, I shall assume the LVZ is isothermal.

The Proposed Geotherm, 80 to 5153 Km

The proposed geotherm for the average Earth is listed in Table II. It has the following characteristics.

Regions A and B: The geotherm is isothermal between 150 and 380 km. It has grazing incidence to the basalt solidus with a sharp knee and is midway between the Basin and Range and the Static Crust geotherms, below 100 km.

Region C: T(380) = 1400°C; T(520) = 1550°C; T(610) = 1610°C; T(670) = 1640°C; ΔT(670) = 150°; T(685) = 2063°K.

Region D': Adiabatic compression, $\Delta T_s(D') =$

662°. Four small transitions: $\Sigma^4 \Delta T_i = 100^{\circ}$; $T(2885) = 2825^{\circ}K$.

Region D'': $\Delta T(2885) = 300^{\circ}$; $T(2971) = 3125^{\circ}K$.

Region E: Adiabatic compression, $\Delta T_S(E) = 953^{\circ}$; $T(5153) = 4078^{\circ}K$.

The model has the virtue that it integrates the works of several disciplines. In particular,

o The findings of experimental petrology appropriate to region C in the upper mantle are incorporated

o The travel-time curves based on the seismology of the upper mantle are considered

o The data of mineral physics appropriate to region B of the upper mantle are incorporated

o The correlation with geotherms extrapolated from heat flow data to the bottom of the lithosphere is good

o The irregularity in the geotherm in the LVZ agrees with the concept of vigorous convection in the upper mantle

o The density profile of the PEM seismic model is incorporated into the T profile of the lower mantle and core

There are two major differences between the proposed thermal model and Stacey's thermal model. First, the temperature of Stacey's geotherm in Region C exceeds greatly the boundary conditions of Akaogi and Akimoto. Second, the value $T(685)$ in the proposed model is far below the lherzolite solidus, whereas Stacey assumed the solidus of Region C would be close to the value of the geotherm. In the proposed model T/T_s at 670 km is about 0.7, and the lower mantle never approaches close to the solidus.

Acknowledgments. I profited by advance copies of manuscripts sent to me by R. Jeanloz and T. Shankland. I also profited by discussions on various topics in this report with D. L. Turcotte, R. Jeanloz, F. D. Stacey, and T. Shankland. This research was supported by NSF grant EAR 77-04456 and EAR 80-08272.

Appendix: Equations Used

For the lower mantle, the empirical relation found for gamma acoustic is used (O. L. Anderson, 1979):

$$\gamma = \gamma_0 \, (\rho_0/\rho)^q \qquad (A-1)$$

where $\gamma_0 = 1.50$ and $q = 1.35$.

Using the basic relationship for an adiabatic process,

$$d \ln T = \gamma d \ln \rho, \qquad (A-2)$$

(A-1) and (-2) become

$$T_1 = T_2 \exp \left\{ \frac{\gamma_0}{q} \left[\left(\frac{\rho_0}{\rho_2} \right)^q - \left(\frac{\rho_0}{\rho_1} \right)^q \right] \right\} \qquad (A-3)$$

The value ρ_0 represents the uncompressed density of the lower mantle. Since the PEM model of Dziewonski et al. (1975) is used for the density distribution, their estimate of ρ_0, 3.991, found from their own data, is used. This value is appropriate to the temperatures of the upper mantle. When average mantle values of α (Brown and Shankland, 1979) and T are used, the correction indicates that $\rho_0 \simeq 4.15$ at standard conditions (see also Butler and D. L. Anderson, 1978).

For the outer core, where γ is taken as constant,

$$\frac{\Delta T}{T}_{av.} = \gamma \frac{\Delta \rho}{\rho}_{av.} \qquad (A-4)$$

or

$$\frac{T_1 - T_2}{T_1 + T_2} = \gamma \left[\frac{\rho_1 - \rho_2}{\rho_1 + \rho_2} \right] \qquad (A-5)$$

The derivation of the equation for dT/dZ used in the upper mantle comes from calculus. Ignoring effects of melting and changing iron content, the variation of v_s with ρ in the Earth is

$$\left(\frac{\partial v_s}{\partial \rho} \right)_e = \left(\frac{\partial v_s}{\partial P} \right)_T \left(\frac{\partial P}{\partial \rho} \right)_e + \left(\frac{\partial v_s}{\partial T} \right)_P \left(\frac{\partial T}{\partial \rho} \right)_e \qquad (A-6)$$

Since

$$\left(\frac{\partial T}{\partial \rho} \right)_e = \frac{dT}{dZ} \frac{dZ}{dP} \left(\frac{\partial P}{\partial \rho} \right)_e = \frac{dT}{dZ} \frac{dZ}{dP} \left(\frac{K_e}{\rho} \right) \qquad (A-7)$$

where K_e is the bulk modulus of the earth. Equation (A-6) can be solved for dT/dZ, and (1) is obtained.

References

Akaogi, M., and S. Akimoto, High-pressure equilibria in a garnet lherzolite with special reference to Mg^{2+}-Fe^{2+} portioning among constituent minerals, Physics of the Earth and Planetary Interiors, 19, 31-51 (1979).

Anderson, D.L., The 650-mantle discontinuity, Geophysical Research Letters, 3, 347-349.

Anderson, O.L., The high-temperature acoustic Grüneisen parameter in the earth's interior, Physics of the Earth and Planetary Interiors, 18, 221-231 (1979).

Anderson, O.L., The temperature profile in the upper mantle, Journal of Geophysical Research, 85, 7003-7010.

Baumgardner, J.R., and O.L. Anderson, Using the Thermal Pressure to Compute the Physical Properties of Terrestrial Planets, Advances in Space Exploration, Pergamon Press, in press (1981).

Brown, J.M., and T.J. Shankland, Thermodynamic parameters in the earth as determined from seismic profiles, Geophysical Journal of the RAS, in press (1981).

Butler, R., and D.L. Anderson, Equation of state fits to the lower mantle and outer core, Physics of the Earth and Planetary Interiors, 17, 147-162 (1978).

Dziewonski, A.M., A.L. Hales, and E.R. Lapwood, Parametrically simple earth models consistent with geophysical data, Physics of the Earth and Planetary Interiors, 10, 12-48 (1975).

Frisillo, A.L., and G.R. Barsch, Measurement of single-crystal elastic constants of bronzite as a function of P and T, Journal of Geophysical Research, 77, 6340-6344 (1972).

Gilbert, F., and A.M. Dziewonski, Application of normal mode theory in the retrieval of structural parameters and source mechanisms from seismic spectra, Philosophical Transactions of the Royal Society of London, Series A, 278, 187 (1975).

Goetze, C., A brief summary of present-day understanding of the effect of volatiles and partial melt on the mechanical properties of the mantle, in High-Pressure Research, edited by M. Manghnani and S. Akimoto, Academic Press, New York, p. 3-23 (1977).

Hales, A.L., K.J. Muirhead, and J.M.W. Rynn, A compressional velocity distribution for the upper mantle, Tectonophysics, 63, 309-348 (1980).

Hart, R.S., D.L. Anderson, and H. Kanamori, Shear velocity and density of an attenuating earth, Earth and Planetary Science Letters, 32, 25-34 (1976).

Irvine, R.D., and F.D. Stacey, Pressure dependence of the thermal Grüneisen parameter, with applications to the earth's lower mantle and outer core, Physics of the Earth and Planetary Interiors, 11, 157-169 (1979).

Jamieson, J.C., H.H. Demarest, Jr., and D. Schiferl, A re-evaluation of the Grüneisen parameter for the earth's core, Journal of Geophysical Research, 83, 5929-5935 (1978).

Jeanloz, R., Properties of iron at high pressures and the state of the core, Journal of Geophysical Research, in press (1979).

Jeanloz, R., and F.M. Richter, Convection, composition, and the thermal state of the lower mantle, submitted to Physics of the Earth and Planetary Interiors (1979).

Jordan, T.H., and D.L. Anderson, Earth structure from free oscillations and travel times, Geophysical Journal of the Royal Astronomy Society, 36, 411-419 (1974).

Kennedy, G.C., and G. Higgins, The core paradox, Journal of Geophysical Research, 78, 900-904 (1973).

Kumazawa, M. and O.L. Anderson, Elastic moduli, pressure derivatives, and temperature derivatives of single-crystal olivine and single-crystal forsterite, Journal of Geophysical Research, 74, 5961-5971 (1969).

Lachenbruch, A.H., and J.H. Sass, Models of an Extending Lithosphere and Heat Flow in the Basin Range Province, Geological Society of America Memoir 132, 209-250 (1978) (see Figure 9).

McQueen, R.G., S.P. Marsh, J.W. Taylor, J.N. Fritz, and W.J. Carter, The equation of state of solids from shockwave studies, in High Velocity Impacts, edited by R. Kinslow, pp. 293-417, Academic Press, New York (1970).

Mysen, B.O., and A.L. Boettcher, Melting of a hydrous mantle: I, Phase relations of natural peridotite at high temperatures with controlled activities of water, carbon dioxide, and hydrogen, Journal of Petrology, 16, 520-548 (1975).

Schubert, G., Subsolidus convection in the mantles of terrestrial planets, in Annual Review of Earth and Planetary Sciences, edited by F. Donath, F. Stehl, and G. Weatherill, 9, 289-342 (1979).

Schubert, G., and R.E. Young, Cooling of the earth by whole mantle subsolidus convection: A constraint on the viscosity of the lower mantle, Tectonophysics, 35, 201-214 (1976).

Soga, N., Elastic constants of garnet under pressure and temperature, Journal of Geophysical Research, 72, 4227-4234 (1967).

Stacey, F.D., A thermal model of the earth, Physics of the Earth and Planetary Interiors, 15, 351-358 (1977).

Stacey, F.D., and R.D. Irvine, Theory of melting: Thermodynamic basis of Lindemann's Law, Australian Journal of Physics, 30, 631-640 (1977).

Stevenson, D., Applications of liquid state physics to the earth's core, Physics of the Earth and Planetary Interiors, 22, 42-52 (1980).

Sumino, Y., and O. Nishizawa, Temperature variation of elastic constants of pyrope-almandine garnets, Journal of the Physics of the Earth, 20, 239 (1978).

Usselman, T.M., Experimental approach to the state of the core: Part II, Composition and thermal regime, American Journal of Science, 275, 291-303 (1975).

Watt, J.P., T.J. Shankland, and N. Mao, Uniformity of mantle composition, Geology, 3, 92 (1975).

THE STRUCTURE, DENSITY, AND HOMOGENEITY OF THE EARTH'S CORE

Bruce A. Bolt and R.A. Uhrhammer

Seismographic Station, University of California, Berkeley, California 94720

Abstract. In this paper, some recent independent results on the structure and physical state of the core are compared. Despite recent refinements, the present resolution of the P and S velocity gradients places considerable limitations on the sharpness of inferences on dynamical properties, density stratification and deviations from adiabatic conditions.

Recent core models have been based on limited sets of data on travel times and amplitudes of seismic core waves and the growing number of measurements of the eigenfrequencies of the Earth. A comparison shows differences between independently constructed models such as CAL6, PEM, and QM2 are almost everywhere less than 1 percent. The differences depend on weighting and selections of data and on assumptions on smoothing and optimality. Effects of damping on the eigenvibrations have not yet been fully taken into account in using observed eigenfrequencies to infer core models.

Recent seismological evidence has favored a simple two-layer core with a liquid outer core, region E, and a sharply-bounded solid inner core, region G. A thin transition layer just below the mantle-core boundary remains a question. The hypothesis of a sharply-bounded transition region F has fallen from favor because of the explanation of precursors to PKIKP in terms of scattering sources near the mantle-core boundary. Nevertheless, velocity jumps in F need be as small as only 0.01 km/sec to give rise to back-scattered waves with the observed amplitudes. No published work has yet precluded a contribution to the precursors from waves from second order scatterers around the inner core boundary. Strong constraints exist on the average velocities in E and G from observations of reflected core waves at nearly normal angles of incidence. The resulting P wave velocity distribution in the outer core shows that the P velocity between 1600 km and about 1200 km changes by only a few percent, i.e., there is near independence of sound speed over a pressure change of 400 kilobars. Thus, a region F remains.

The density distributions in E independently determined from eigenvibration and travel-time inversions, without using the assumptions of Adams and Williamson, confirm a density gradient close to the Adams and Williamson value (at least, in the central part of E). The conclusion is that zero deviations from homogeneity and adiabaticity through most of E are sufficient but not yet necessary to satisfy the data. Formal trade-off curves, however, indicate that the resolutions of density and the θ homogeneity index are not yet sharp. For an averaging kernel width of 400 km, the density estimates have standard errors of about 0.20 gm/cm^3 throughout the core.

Observations of the amplitude of PKiKP and PKIIKP entail that there must be a sharp discontinuity in density, or in compressibility, or both, at the inner core boundary. A jump in rigidity will not yield either of these reflections with the amplitudes observed. The times of the phase PKIIKP constrain the average P velocity in the inner core to a value of 11.14 $\pm$ 0.02 km/sec for a radius of 1216 km.

The probable large gradient in P velocity at the top of region G entails that the shear velocity increases rapidly with depth by about 0.3 km/sec from the inner core boundary to 950 km and then gently decreases by 0.15 km/sec toward the center. With values for the P and S velocities and density in the inner core given by recent models, the Poisson's ratio for the inner core has a peculiarly high value of about 0.40 compared with 0.5 for a liquid and 0.3 in the lower mantle. This condition, together with the low value of Q of about 450, suggests a highly viscous solid, perhaps near melting.

Introduction

This paper attempts to estimate the present resolution of structure and the degree of homogeneity in the Earth's core. The method is a critical comparison of three recent velocity models of the core. The stratification of K.E. Bullen will be adopted: the outer core is called region E. It surrounds a transition shell F, which in turn surrounds the inner core

G. We also mention E', a hypothetical boundary shell at the top of region E. We adopt for the inner core radius 1216 + 2 km; for the outer radius of F 1782 km and for the core radius 3486 + 2 km (Bolt and Uhrhammer, 1975).

The comparison indicates that recent studies have led to considerable convergence of physical properties of the core. Yet, the precision with which the seismic parameters are presently determined is still not sufficient for many geophysical purposes, such as those required for solid state and temperature gradient calculations (e.g., Jeanloz, 1979; Gubbins et al., 1979). Although refined geometries for convection within the core associated with the magnetic dynamo are now being considered theoretically, uncertainties on detailed stratification within the core from seismological inversions do not allow tests of uniqueness of dynamo models to be made.

There are two main difficulties in achieving the required resolution. First, Earth models of the present generation do not take into account in detail the effects of damping of surface waves and free oscillations (Jeffreys, 1965). This effect changes, for example, the eigenfrequencies of $_0T_2$ and $_0S_2$ by about 0.5 and 0.25 percent (13 and 8 sec), respectively (Randall, 1976; Hart et al., 1977). Therefore, estimation of physical properties based on perfectly elastic models constructed mainly from eigenfrequency data are subject to further adjustment when full results for visco-elastic models come to hand. (Models founded mainly on seismic body wave properties are not as defective in this regard.) Second, the fine structure of the core over narrow shells of order 100 km and including a possible boundary layer around its circumference and the detail of the transition shell F has not been closely explored.

We consider three core models. The first is the parameterized Earth model, PEM (Dziewonski, Hales and Lapwood, 1975), which has been widely used. The second model, parameterized in the form of cubic splines, is called CAL6 and was computed quite independently (Bolt and Uhrhammer, 1975; Uhrhammer, 1977). (CAL6 is the cubic spline equivalent of the model published in 1975.) These two models depend to a major but different degree on eigenperiod measurements, but neither takes into account the effect of damping on the oscillations. The third model, therefore, is the model QM2 published by Hart, Anderson, and Kanamori (1977) that allows approximately for the frequency shifts introduced by damping. They constructed QM2 by adjusting a set of observed eigenperiods for the predicted effect of the damping (up to 0.5 percent) and then inverting a model obtained from uncorrected data to determine a new spherically symmetric model. This inversion left the physical properties of the core of the original model essentially unchanged, so there remains a question about the procedure because damping in the inner core is large. All three models endeavor to fit the constraints of the total mass, moment of inertia, and time measurements from different sub-sets of eigenvibrations, body waves, and surface waves, within the observational uncertainties.

On the whole, more emphasis was given to eigenfrequencies in the inversion procedures to obtain PEM and more weight to the core body wave data to construct CAL6. For example, the important AB branch of PKP was not used for PEM although the corresponding rays bottom near the center of region E. It would be stressed that for inversion of velocities there is no basic superiority in the measurements of eigenfrequencies over travel times from body waves. Generally speaking, the core waves PKP of period about one second are well-recorded on short-period vertical seismometers and represent some of the most reliable seismological data available. The velocity distribution for CAL6 is close to the carefully constrained solution obtained from such readings by Qamar (1973).

In fact, calculated travel times of core waves from PEM and CAL 6 agree closely in most cases. Often, discrimination hinges on differences of a second or so, which is about the standard error of readings obtained from worldwide seismograms. Another important point is, in comparing the reliability of whole-Earth models, that the use of very large numbers of P readings to obtain a vast sample distribution does not necessarily provide seismological summary values for travel times which are superior to those obtained by carefully sampling from well-distributed stations with high reliability. As statistical sample theory shows, a small sample is able to provide a robust measure of central tendency which is closely equal to the population mean.

Core Velocity Models

Consider the differences in P wave velocity α between the models in Figure 1. The greatest differences in E amount to less than 0.5 percent over short distances and are mostly less than 0.2 percent. The PEM model has slightly lower velocities in region E (at the top of E, α = 8.09 km/sec in PEM). The velocities become equal towards the base of E as F is approached. From a radius of 1780 km, CAL 6 velocities become almost constant to the sharp boundary of the inner core at 1216 km, whereas the PEM α curve continues to rise, but more slowly than in E. There is a marked difference in the upper part of the inner core between the two models, with α for PEM being essentially constant throughout the inner core, whereas the CAL 6 model has a significant gradient in the outer part.

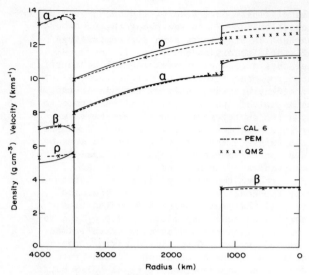

Fig. 1. A comparison of densities ρ and seismic velocities α and β estimated in three Earth models, called CAL6, PEM, and QM2.

The QM2 model has almost identical P velocities to those of PEM through E, but slightly higher velocities in F (0.15 km/sec at the base of F). Its P velocity is midway between PEM and CAL 6 at the top of G.

The reasons for the differences are the different sets of travel-time data on which these three models are based. The shear wave velocity in the inner core is close for all models, with a maximum difference of 0.08 km/sec. In order to remove any misconception, the α core distribution of Qamar is not significantly affected by the acceptance or rejection of hypotheses whether there are first order discontinuities associated with region F.

High-frequency seismic waves reflected steeply from the outer and inner core boundaries provide strong constraints. The differential travel times are, to the second order, independent of variations in structure of the crust and upper mantle and observing conditions, and allow a strong trade-off between the mean velocity in each shell and the shell radii. The onset times correspond to _impetus_ onsets, for which ray theory should hold to a close approximation. These differential times were especially included in the construction of CAL6 and given high weights to constrain the average velocities. For example, for the phases PKIIKP and PKIKP, the observed travel-time difference from FAULTLESS to LASA is 436.3 sec (Bolt, 1979). For a radius of 1216 km, the mean velocity in G is thus 11.14 km/sec. The mean α in G for PEM is 11.17 km/sec. An error analysis suggests an uncertainty of about 0.02 km/sec.

Now consider the seismic parameter in the core

$$\phi = \alpha^2 - \frac{4}{3} \beta^2 . \qquad (1)$$

Figure 2 shows this key observational function for the three models considered. There is close agreement in actual values of φ through most of the core, particularly through most of E. As with the P velocity distribution, the main differences are in E', F, and at the top of G. It is, however, the gradients of this parameters that are specially required for solid state, temperature, compositional, and convection arguments (see equation (8)). Figure 2 demonstrates that these gradients are not yet well-resolved. Near the boundaries of E and G, if equal weight is given to each model, the resolution of gradients is poor even when averages are made over spread of hundreds of kilometers. (A similar uncertainty in φ in region D'' appears at the base of the mantle from a comparison of these models in Figure 2.)

Region E

Let us now turn closer attention to the stratification in individual shells. Consider a hypothetical transition region E' between 10 and 100 km thick at the top of region E. Speculation about this region is allowable because of the lack of strict empirical control of the body wave inversion methods that have been used to obtain the velocity in this region. In the initial work of Jeffreys (1939), the travel times of the core waves were constructed by using SKS observations. SKS waves bottom in region E' and permit all but 35° of the time curve K for the core to be obtained. Jeffreys closed this gap by interpolation back to zero distance in the core. Thus, Jeffreys' velocity

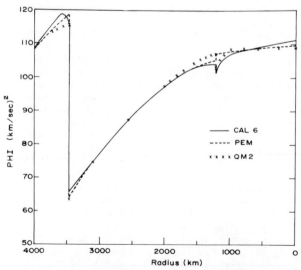

Fig. 2. A comparison of the seismic parameter φ calculated from three Earth models, called CAL6, PEM, and QM2.

distribution in E' is based on continuity and smoothness principles. The situation was improved by more recent work by Hales and Roberts (1970, 1971) in which both SKKS and SKS waves were used. This work left an observational gap of 14° through which smooth interpolation back to zero was required. The upshot of various published solutions has been a range of P velocities at the top of the outer core of $7.8 < \alpha < 8.3$ km/sec (Randall, 1973; Hales and Roberts, 1971). The resolution provided by these studies makes it hard to discount any value of α in this range.

We have performed ray-tracing to test the effect of a transition shell E' on SKS waves. Figures 3 and 4 show the SKS rays that are produced when a 50 km E' shell is introduced. In Figure 3, the starting velocity is 7.8 km/sec and the velocity curve connects smoothly onto a velocity of 8.3 km/sec at a depth of 50 km below the mantle-core boundary. In Figure 4, the surface velocity is 8.2 km/sec, which connects onto the same velocity at a depth of 50 km. The figures demonstrate that SKS rays are significantly affected by the introduction of such a shell. For the higher velocity E' shell, there is a triplication in SKS travel-time values and a shadow zone for $60° < \Delta < 75°$ along a profile of stations from a single earthquake, so that a boundary velocity as high as 8.2 km/sec appears

unlikely. The detection or otherwise of a transition zone E' in which α remained almost constant is important, both for temperature and stability estimates for E.

We must also mention a startling inference (Kind and Muller, 1977) on the possibility of a strong transition shell deeper within the liquid core. Analysis of phases SKS and SKKS suggested to these seismologists a marked change in velocity of about 0.2 km/sec over a range of 300 km centered about r = 2620 km. The reason that such a startling deviation from earlier seismological solutions could still be considered is that near 2620 km the corresponding travel times for PKP and SKS join with little overlap. The PKP times come from the extremity of the AB branch where there is less precision than in other parts of the PKP curve (Bolt, 1968). Thus, structural differences may have been overlooked by assuming continuity. On the other hand, it is unlikely that a structural change in the core occurs just at the radius which corresponds to the distance where the PKP and SKS travel-times curves happen to adjoin. Another explanation for the hypothesis of Kind and Muller is interference from other phases that arrive at the same time on the seismograms. For example, phases associated with PKIKS have about the same arrival times (P.G. Richards, personal communication).

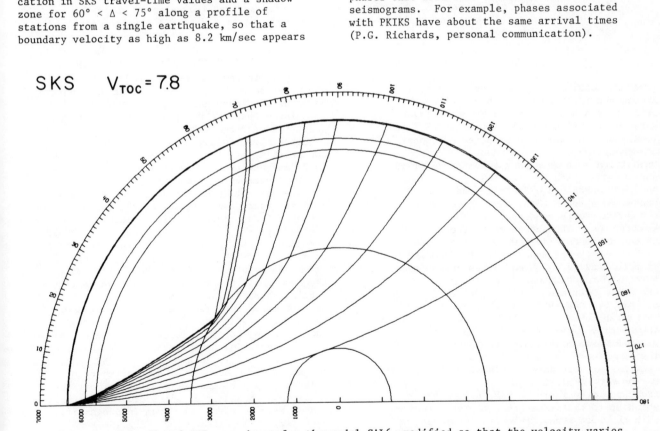

SKS $V_{TOC} = 7.8$

Fig. 3. A bundle of SKS rays drawn for the model CAL6, modified so that the velocity varies mostly through a 50 km thick shell at the top of the core from 7.8 to 8.3 km/sec. There is no shadow zone.

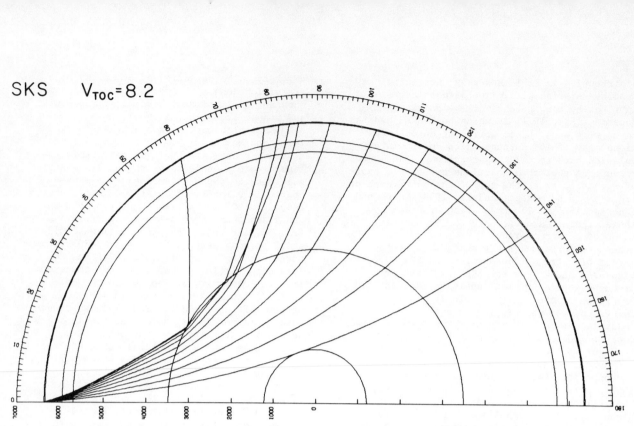

Fig. 4. A bundle of SKS rays for the model CAL6, modified so that the velocity varies smoothly through a 50 km thick shell at the top of the core from 8.2 to 8.3 km/sec. Note the broad shadow zone.

Region F

We now consider the evidence for a shell F around the inner core. Such a shell might arise from a non-convecting boundary layer whose state is intermediate to that of a true liquid or true solid (D.L. Loper, personal communication). The introduction of such a transition zone goes back to Jeffreys (1939) and Gutenberg and Richter (1939), who independently obtained α distributions that contained a transition; the former had a sharp inner core boundary, the latter a gradual one. The sharpness of the boundary of the inner core is now definitely established. High-frequency reflections PKiKP, at both grazing and high angles of incidence, have been recorded many times from this boundary (Bolt and O'Neill, 1964; Engdahl et al., 1970). In 1964, an existence theorem was established (Bolt, 1964) that the reduction in velocity given by Jeffreys in F could be removed while maintaining the sharp boundary of the inner core and not violating constraints from PKP travel times. This velocity distribution also explained some features of high-frequency precursors to PKP, called PKhKP, which up to that time had not been interpreted.

In the 1964 study, however, amplitudes were not considered. This was done for the first time by Qamar (1973). He showed that velocity

jumps in F of less than 0.01 cm/sec could give rise to reflected (back-scattered) waves of the observed amplitudes. Later, an alternative hypothesis (Cleary and Haddon, 1972) was formulated that the PKP precursors were waves scattered near the mantle-core boundary. This hypothesis was followed up in a number of papers (King et al., 1974; Doornbos and Husebye, 1972; and others). This work establishes that array measurements of the wave slownesses of the precursors, in many cases, do not agree with a first order discontinuity at the top of F. On the other hand, there is also evidence from arrays (Wright, 1975) that at least some PKhKP precursors are reflections from the transition zone.

The balance of the evidence now indicates that the hypothesis of scattering near the mantle-core boundary explains a wide range of properties of the precursors. A weakness of the hypothesis is that its proponents seem unsure as to the exact location and mechanism of the scattering, suggesting either scatterers at the base of the mantle or bumps on the mantle-core boundary. A more important point is that evidence of scattering near the mantle-core boundary does not prove that there cannot also be scattering from inhomogeneities of a similar magnitude around the inner core boundary. What is seen on seismograms may be superposition of these two effects.

Region G

No evidence has been found that contradicts the inference that the inner core of the Earth is solid, i.e., it has significant rigidity. This result is, however, difficult to confirm directly from body wave studies, such as definite identification of the phase PKJKP.

The graver modes and overtones of terrestrial eigenvibrations that have perceptible particle motions within the inner core provide the most direct seismological evidence of rigidity. The first estimate from eigenvibrations, given by Derr (1969), was $\bar{\beta} = 2.2$ km/sec. A more recent value using eigenfrequencies from additional modes (Dziewonski and Gilbert, 1972) is $\beta = 3.52$ km/sec. A quite independent estimate, based on Qamar's velocity distribution (Bolt, 1972) is β 3.1 km/sec at the center. As shown in Figure 1, inversions of full Earth models in the three cases, CAL6, PEM, and QM2 all converge to values of β in the inner core that are essentially constant at 3.5 km/sec. It would be premature, however, to assume that this value has a high precision; resolution and information matrices for α and β in the inner core have not yet been constructed and damping in the core is not included in any inversion.

The values of α and β given by CAL6 and PEM entail a Poisson's ratio as high as 0.4 or greater, which must be compared with 0.5 for a liquid and 0.3 in the lower mantle. This high value for Poisson's ratio suggests a state which allows large lateral contraction compared with longitudinal extension, suggesting a soft solid which is perhaps near melting. Under these conditions we might expect some variation of physical properties throughout the inner core. This result was in line with the low values for the damping parameter Q in G.

In the outer core, values between 6,000 and 10,000 for Q are well-established (Qamar and Eisenberg, 1974; Cormier and Richards, 1976). In marked contrast, Q is two orders of magnitude less in the inner core. Spectral analysis of the phases PKIIKP and PKiKP from the same record yield a mean Q of 450 ± 100 (Bolt, 1979). This estimate is in agreement with other recent measurements (Buchbinder, 1971; Sachs, 1971; Qamar and Eisenberg, 1974).

Density Distributions in the Core

The density distributions obtained from the model studies are shown in Figure 1. The largest differences in the estimates of density occur in the inner core between all three models. However, both CAL6 and PEM have about the same jump in density (0.6 g/cm³) at the inner core boundary. Models PEM and QM2 have essentially equal density distributions throughout the liquid outer core, which perhaps indicates that these density estimates are not

independent. However, CAL6 has a density about 0.2 g/cm³ greater, the difference tapering off from the inner core to the outer core boundaries. In all cases, the density distribution follows closely the Adams-Williamson law, but all the starting models assume this property. It is not clear what perturbations in the core were allowed in the developments of PEM and QM2, but in the calculations of CAL6, perturbations were allowed in three depths in E and two depths in G over ranges of several hundred kilometers. The inversions did not indicate any major change in the starting distribution of density.

For the first time, density trade-off curves have been calculated for the core. Thirty-nine selected eigenperiods, including the ten with the greatest particle motions in the core (Table 1; Bolt and Uhrhammer, 1975), and two constraints (mass and moment of inertia) were used to estimate density trade-off curves based on CAL6 structure. The mathematical development parallels Bolt and Uhrhammer (1975), equations 10-17. For computation purposes, the density model (represented parametrically

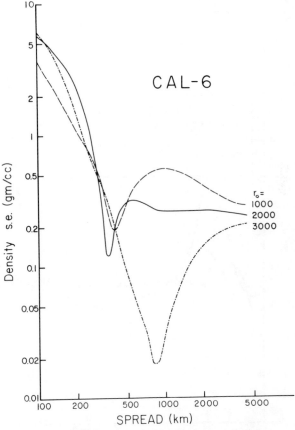

Fig. 5. Trade-off curves for mean density for the model CAL6, calculated for resolution kernels of various spreads centered at three radii, r_0 = 1000, 2000, and 3000 km.

by cubic splines) was initially divided into 41 discrete concentric shells. Because of numerical difficulties, this was later increased to 168 shells ranging from 5 km to 50 km in thickness.

Define the spread function to be

$$H(z,z_o,s) \propto e^{-\frac{(z-zo)^2}{2s^2}} , \qquad (2)$$

normalized so that

$$\int_0^1 Hdz \equiv 1 .$$

We then construct $H(z,z_o,s)$ from a linear combination of the date kernels G. We have

$$\begin{array}{ccc} G & \cdot & a & = & H(z,z_o,s) \\ (m \times n) & (n \times 1) & & (m \times n) \end{array}$$

where $m = 41$, $n = 168$. We proceed by decomposing G (Lanzos, 1961) so that

$$G^T = U\Lambda V^T .$$

Thus, the weights a are

$$a = U\Lambda^{-1}V^T H(z,z_o,s) \qquad (3)$$

and, by the addition theorem for the normal variate,

$$\overline{\delta\rho}(z_o,s) \cap N(0,\sigma^{*2} \Sigma a_i^2) . \qquad (4)$$

Therefore,

$$\sigma_\rho^2 = \Sigma a_i^2 \sigma^{*2} . \qquad (5)$$

Here, $\sigma^* = 2.45$ sec. A set of trade-off curves at various z_o's were computed by solving for the weights a_i for a range of spreads s. A set of 3 such curves ($r_o = 1000, 2000, 3000$ km) for CAL6 is given in Figure 5.

Note that the trade-off curves are not constrained to a classic retangular hyperbola in this work. Here the spread function H (equation 2) has been normalized to unity. Beyond $s = 1000$ km, the curves tend to a constant value because the normalized spread function tends towards a boxcar shape as the spread becomes larger. The curves show that for resolving kernels of spread 400 km the standard errors of the mean densities at radii of 1000, 2000, and 3000 km all are about 0.2 g/cm^3. In the inner core, the errors never

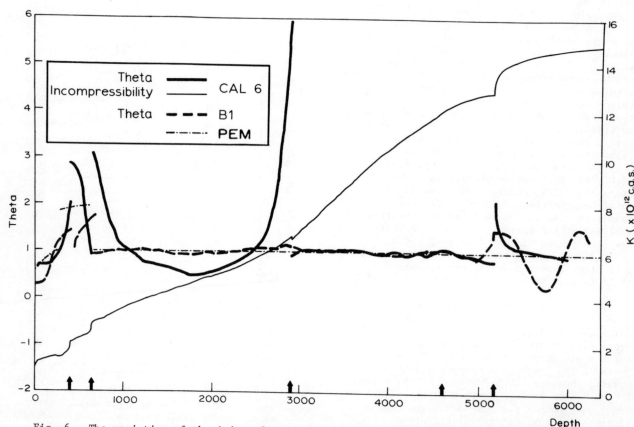

Fig. 6. The variation of the index of state θ from three Earth models, CAL6, PEM, and B1. The incompressibility k for CAL6 is also shown.

fall below this value while near the top of E standard errors fall to about 0.05 g/cm^3 where the resolution is over a shell of width about 600 km.

Density Jump at the Inner Core Boundary

High-frequency reflections PKiKP observed at almost vertical incidence do not depend on rigidity in the inner core. To get such sharp reflections, there must be a considerable mismatch in either ρ or k, or both, at a sharp inner core boundary which extends over only a few kilometers (Bolt and Qamar, 1970).

In this regard, the results of QM2 are unsatisfactory. If the central densities increased only to 12.7 g/cm^3, one can calculate that the jump in incompressibility would be too small to provide sufficient acoustic mismatch for observed amplitudes of reflected PKiKP waves. If we take the central density to be 13 g/cm^3, compatible with iron composition and amplitudes of PKiKP reflections (Bolt and Qamar, 1970), the corresponding jump in density at the inner core boundary is about 0.5 g/cm^3.

Both CAL6 and PEM have a significant jump of about 0.6 g/cm^3. Recently, Masters (1979) studied the question whether such a density jump is required by the set of observed eigenfrequencies taken along for inversion; and, if so, how uncertain is it? Masters calculated a resolving kernel for a discontinuity at this radius, and the results suggest that a density jump greater than 0.3 g/cm^3 is likely, even if the observation errors of the modes are doubled. This is a significantly larger jump than model QM2. Masters also argues that a density jump of 0.7 g/cm^3 seems to be indicated with an uncertainty of about 0.3 g/cm^3. This conclusion agrees with CAL6. The central densities of the three comparison models are 13.35, 13.01, and 12.57 g/cm^3 for CAL6, PEM, and QM2, respectively.

Index of Homogeneity of State

Consider the curve for the index of homogeneity θ, where

$$\theta(\eta, \tau) = -\frac{\phi}{\rho g} \frac{d\rho}{dr} \tag{6}$$

$$= -\frac{\phi}{g} \frac{d(\ln\rho)}{dr}$$

$$= \eta - \alpha_p \tau \phi/g \tag{7}$$

Here, α is the coefficient of expansion, τ the superdiabatic gradient, g the gravity, and η is Bullen's index

$$\eta = g^{-1} \frac{d\phi}{dr} + \frac{dk_s}{dp} \quad . \tag{8}$$

Note that

$$\text{var } \theta \simeq (\phi/g\rho)^2 \text{ var}(d\rho/dr). \tag{9}$$

The parameter ϕ was first introduced by Bolt, (1957, equation 12), but was later much developed by Bullen (1963) in the form η. As shown in (6), (7), and (9), θ and its variance are dependent on gradients in density and temperature, both poorly resolved over short distances. It is used here in the form θ originally proposed.

From Figure 6, $\theta \simeq 1$ through E, but varies widely in G for models CAL6 and B1. (B1 is an earlier model (Jordan and Anderson, 1974) that resembles QM2 in the core.) The model PEM assumes $\theta = 1$ exactly through the core so that it cannot be used to draw independent inferences on deviations from homogeneous conditions. The index θ has been used also by Masters (1979) to study chemical stratification. (Masters points out its relation to the stratification index of Pekeris and Accad (1972) and the Brunt-Vaisäla frequency in connection theory.)

While the three models in Figure 6 demonstrate that it is possible to fit satisfactorily observed eigenperiods for $\theta = 1$, the question remains as to the allowable range of perturbations about unity. Masters (1979) calculates trade-off curves in E of spread versus statistical error for various radii in the outer core. These show that with spreads of 500 km, the corresponding measurement errors in θ range from 0.06 to 0.10. For averaging kernels of less than 400 km spread, the statistical uncertainty grows rapidly (c.f., Figure 5). The errors are such that a partly stratified core cannot be excluded, but for broader averages $\theta \simeq 1$ in region E, thus excluding significant deviations from overall homogeneity. However, Master's calculations are based on models like PEM that have been smoothed rather arbitrarily. Also, his trade-off curves are based on a strong hyperbolic constraint, so that they fail to indicate the effects on the averaging procedures of the large discontinuities at the mantle core boundary.

Our analysis of uncertainties has not included the added uncertainties introduced by uncertainties in estimations of core temperatures. Verhoogen (1979) has lately summarized the present position and his analysis indicates how all approaches depend to some extent on knowledge of the seismic parameters.

What is needed to improve the estimates of core parameters is a more thorough explanation of the ways in which the inversions of seismological data are carried out. Some papers do not explain in full detail the actual numerical steps of the inversion. Those engaged in this work are well aware of the amount of trial and error involved and meaningful comparisons depend upon sufficient detail being available.

Acknowledgements. We would like to thank Professor J. Verhoogen for making available his text of the 1978 Arthur L. Day Lectureship of the National Academy of Sciences. Miss P. Murtha provided the plots of Figures 3 and 4 using a program kindly made available by Professor C.H. Chapman. The work was supported by NSF Grant EAR76-00118.

References

Bolt, B.A. Earth Models with Chemically Homogeneous Cores, M. Not. Roy. astr. Soc. Geophys. Suppl., 7, 372, 1957.

Bolt, B.A. The Velocity of Seismic Waves Near the Earth's Center, Bull. Seism. Soc. Am., 54, 191, 1964.

Bolt, B.A. Estimation of PKP Travel Times, Bull. Seism. Soc. Am., 58, 1305, 1968.

Bolt, B.A. The Density Distribution Near the Base of the Mantle and Near the Earth's Center, Phys. Earth Planet. Int., 5, 301, 1972.

Bolt, B.A. The Detection of PKIIKP and Damping in the Inner Core, Annali di Geofisica, XXX, N. 3-4, 1977 (published in 1980).

Bolt, B.A. and M.E. O'Neill. Times and Amplitudes of the Phases PKiKP and PKIIKP, Geophys. J. Roy. astr. Soc., 9, 223, 1964.

Bolt, B.A. and A. Qamar. An Upper Bound to the Density Jump at the Boundary of the Earth's Inner Core, Nature, 228, 148, 1970.

Bolt, B.A. and R. Uhrhammer. Resolution Techniques for Density and Heterogeneity in the Earth, Geophys. J. Roy. astr. Soc., 42, 419, 1975.

Buchbinder, G.G.R. A Velocity Structure of the Earth's Core, Bull. Seism. Soc. Am., 61, 429, 1971.

Bullen, K.E. An Index of Degree of Chemical Homogeneity in the Earth, Geophys. J. Roy. astr. Soc., 7, 584, 1963.

Cleary, J.R. and R.A.W. Haddon. Seismic Wave Scattering Near the Core-Mantle Boundary: A New Interpretation of Precursors to PKIKP, Nature, 240, 549, 1972.

Cormier, V.F. and P.G. Richards. Comments on "The Damping of Core Waves" by Anthony Qamar and Alfredo Eisenberg, J. Geophys. Res., 81, 3066, 1976.

Derr, J.S. Internal Structure of the Earth Inferred from Free Oscillations, J. Geophys. Res., 74, 5502, 1969.

Doornbos, D.J. and E.S. Husebye. Array Analysis of PKP Phases and Their Precursors, Phys. Earth Planet. Int., 5, 387, 1972.

Dziewonski, A.M. and F. Gilbert. Observations of Normal Modes from 84 Recordings of the Alaskan Earthquake of 1964 March 28, Geophys. J. Roy. astr. Soc., 27, 393, 1972.

Dziewonski, A.M., A.L. Hales, and E.R. Lapwood. Parametrically Simple Earth Models Consistent with Geophysical Data, Phys. Earth Planet. Int., 10, 12, 1975.

Engdahl, E.R., E.A. Flinn, and C.F. Romney. Seismic Waves Reflected from The Earth's Inner Core, Nature, 228, 852, 1970.

Gubbins, D., T.G. Masters, and J.A. Jacobs. Thermal Evolution of the Earth's Core, Geophys. J. Roy. astr. Soc., 59, 57, 1979.

Gutenberg, B. and C.F. Richter. On Seismic Waves (Fourth Paper), Gerlands Beiträge zur Geophysik, 54, 94, 1939.

Hales, A.L. and J.L. Roberts. The Travel Times of S and SKS, Bull. Seism. Soc. Am., 60, 461, 1970.

Hales, A.L. and J.L. Roberts. The Velocities in the Outer Core, Bull. Seism. Soc. Am., 61, 1051, 1971.

Hart, R.S., D.L. Anderson, and H. Kanamori. The Effect of Attentuation on Gross Earth Models, J. Geophys. Res., 82, 1647, 1977.

Jeanloz, R. Properties of Iron at High Pressures and the State of the Core, J. Geophys. Res., 84, 6059, 1979.

Jeffreys, H. The Damping of S Waves, Nature, 208, 675, 1965.

Jordan, J.H. and D.L. Anderson. Earth Structure from Oscillations and Travel Times, Geophys. J. Roy. astr. Soc., 36, 411, 1974.

Kind, R. and G. Müller. The Structure of the Outer Core from SKS Amplitudes and Travel-Times, Bull. Seism. Soc. Am., 67, 1541, 1977.

King, D.W., R.A.W. Haddon and J.R. Cleary. Array Analysis of Precursors to PKIKP in the Distance Range 128° to 142°, Geophys. J. Roy. astr. Soc., 37, 157, 1974.

Lanczos, C. Linear Differential Operators, De Van Norstrand and Co., 564 + xvi pages, 1961.

Masters, G. Observational Constraints on the Chemical and Thermal Structure of the Earth's Deep Interior, Geophys. J. Roy. astr. Soc., 57, 507, 1979.

Pekeris, C.L. and Y. Accad. Dynamics of the Liquid Core of the Earth, Phil. Trans. R. Soc. Lond. A, 273, 237, 1972.

Qamar, A. Revised Velocities in the Earth's Core, Bull. Seism. Soc. Am., 63, 1073, 1973.

Qamar, A. and A. Eisenberg. The Damping of Core Waves, J. Geophys. Res., 79, 758, 1974.

Randall, M.J. SKS and Seismic Velocities in the Outer Core, Geophys. J. Roy. astr. Soc., 21, 441, 1973.

Randall, M.J. Attenuative Dispersion and Frequency Shifts of the Earth's Free Oscillation, Phys. Earth Planet. Int. 12, 1, 1976.

Sacks, S.I. Anelasticity of the Inner Core, Annual Rept. Dir. Dept. Terr. Magn.,

1969-1970, <u>416</u>, Carnegie Inst., Washington, D.C., 1971.

Uhrhammer, R. Shear Wave Velocity Structure in the Earth from Differential Shear Wave Measurements, <u>Ph.D. Thesis</u>, University of California, Berkeley, 1977.

Verhoogen, J. <u>Energetics of the Earth</u>, Arthur L. Day Lectures, National Academy of Sciences, 1979.

Wright, C. The Origin of Short-Period Precursors to PKP, <u>Bull. Seism. Soc. Am.</u>, <u>65</u>, 765, 1975.

DEEP CRUSTAL STRUCTURE:
IMPLICATIONS FOR CONTINENTAL EVOLUTION

L.D. Brown, J.E. Oliver, S. Kaufman, J.A. Brewer, F.A. Cook, F.S. Schilt, D.S. Albaugh, G.H. Long*

Department of Geological Sciences, Cornell University, Ithaca, New York 14853

Abstract. Current concepts of the deep crust beneath the continents as compositionally heterogeneous and structurally complex represent a major refinement of earlier approximations of crustal structure as relatively simple two or three layer sequences of 'granitic' over 'gabbroic' rock. The increased recognition of the complexity of deep crustal structure has resulted in large part from the use of advanced seismic sounding methods. Simple refraction techniques, primarily useful for measuring gross crustal velocities and thicknesses, have been complemented by more elaborate data acquisition and analysis methods which use arrival time and amplitude observations of both reflected and refracted waves. The most significant recent development in mapping the crust has been the extensive application of modern multichannel reflection profiling to the study of deep structure. The largest program to collect deep seismic reflection data is directed by the Consortium for Continental Reflection Profiling. COCORP profiles have resolved large scale, deeply penetrating structures associated with the primary phases of crustal evolution, including: extensional tectonics in the Rio Grande rift of New Mexico, a Precambrian rift beneath the Michigan basin, and Mesozoic rift-related structures beneath the Atlantic coastal plain in South Carolina; transform faulting of the San Andreas system; and compressional tectonics in the Laramide and Appalachian orogenic belts. In establishing the role of compression in Rocky Mountain foreland deformation and demonstrating the existence of large scale, thin-skinned style overthrusting of crystalline crustal sheets in the southern Appalachians, COCORP results have made particularly important contributions to our understanding of crustal response to tectonic forces at convergent plate margins and the modes of continental accretion. Although our sampling of the variety of crustal tectonics is

still relatively limited, these and other deep seismic results are beginning to provide the critical observations linking surface geology, deep structure, and tectonic evolution.

Introduction

The nature of the deep crust is inferred primarily from four sources: direct observation of rocks from drill holes and of xenoliths in kimberlite and alkali basalts, direct observation of rocks in outcrop which were once at great depth, laboratory studies of possible mineral assemblages appropriate for the lower crust, and indirect observations of the lower crust by geophysical means. Unfortunately drill holes have rarely probed more than a few kilometers into the crystalline basement [Smithson and Ebens, 1971], providing little structural information at greater crustal depths. Although consideration is now being given to an extensive program of continental deep drilling which would greatly increase direct sampling of basement materials [U.S. Geodynamics Committee, 1979], frequent penetration to even mid-crustal depths is unlikely in the near future. Xenolith and laboratory studies, while providing considerable information on rock compositions in the lower crust, also tell us relatively little about deep structure [Kay and Kay, 1981; Fountain and Christiansen, 1975]. Perhaps the best geologic guides to structural style at depth are outcrops of certain metamorphic rock suites. The use of ophiolites as models of oceanic crust is well known; similarly, high grade gneiss terrains of the Archean may represent one possible structural style for the lower crust [Heier, 1973]. In general such terrains are quite complex, with isoclinal folding, shearing, and multiple intrusion evident at virtually all scales of observation [e.g. Bridgewater et al., 1976]. Younger, tectonically exposed high grade assemblages have also been considered as possible analogues for the lower crust. The best known of these is the Ivrea zone of N.E. Italy, interpreted as a

---

*Now at Cities Service Company, Tulsa, Oklahoma, 74102

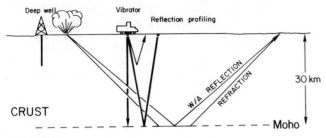

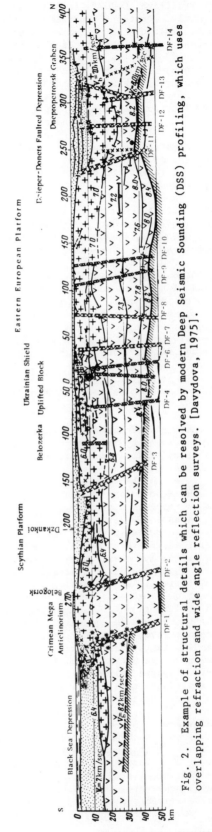

Fig. 1. Seismic methods for mapping crustal structure-refraction, wide-angle (W/A) reflection, and near-vertical reflection.

CRUST

MANTLE

cross-section of lower crust now exposed by Alpine thrusting [Mehnert, 1975].

Geophysical techniques provide indirect views of lower crustal structure at varying scales of resolution, ranging from low in the case of gravity and magnetics to relatively high with seismics. Our classic models of crustal structure have been based to a large degree on seismic refraction (Figure 1), which in its simplest implementation has been used to map variations in average crustal velocity and thickness. Refinement of refraction data acquisition and interpretation techniques, e.g. using overlapping profiles, wide-angle reflections, and synthetic seismograms [Mueller and Landisman, 1971; Giese et al., 1976], has led to a correspondingly more complex models for the crust (e.g. Mueller [1979] and Figure 2). However, the lateral averaging inherent in these large offset techniques, which often attempt to deduce the velocity structure of a crust only 30 km thick using seismic waves which have propagated (and averaged) over 3 times that distance, often restricts interpretations to varieties of layered velocity models. When compared with the complexity evident in relevant outcrops, it is understandable why these 'layer-cake' models tend to have limited usefulness in delineating the deep crustal roots of exposed structures.

Presently the geophysical technique with by far the highest resolution for structural detail is continuous seismic reflection profiling (Figure 1), a mainstay of oil exploration. Although modern reflection technology was developed by the oil industry for studying the sedimentary section, it has been successfully used to probe to lower crustal and upper mantle depths in many areas, particularly Germany, Canada, Australia, the Soviet Union, and the United States [see Dohr and Meissner, 1975, Oliver et al., 1976, and Brewer and Oliver, 1979, for relevant reviews]. The largest and most sustained effort to explore the continental crust with this method is directed by the Consortium for Continental

Fig. 2. Example of structural details which can be resolved by modern Deep Seismic Sounding (DSS) profiling, which uses overlapping refraction and wide angle reflection surveys. [Davydova, 1975].

Reflection Profiling (COCORP) in the U.S., which presently operates a continuous field effort to run multichannel seismic reflection surveys across crustal structures of particular geologic significance. The results of such surveys, in conjunction with relevant geologic and complementary geophysical evidence, have led to development of crustal models characterized by compositional heterogeneity and structural complexity [e.g. Smithson and Decker, 1974; Mueller, 1977]. Our understanding of the deep crust has progressed further than general statements on crustal heterogeneity, however - specific key structures are now being mapped almost routinely.

In this paper the main characteristics of deep crustal structure as inferred primarily from seismic reflection data are reviewed. Selected results from the COCORP project are used to illustrate how advances in mapping such features are leading to a better understanding of the evolution of continental crust. The descriptions of the data and the problems inherent in their interpretation will necessarily be brief; the reader is referred to the citations provided in the text and reviews by Brown et al. [1978], Schilt et al. [1979], and Brewer and Oliver [1980] for more complete treatments.

Seismic Characteristics and a General Model of
the Lower Crust

Until relatively recently, deep crustal structure was mapped by seismic refraction surveys in terms of layered variations of seismic velocity. The crust itself is usually defined in terms of the velocity jump at its base, i.e. the Mohorovicic discontinuity. With the increased resolution brought about by reflection technology, new criteria and approaches to characterizing the seismic response of the crust have become necessary. Sufficient reflection observations have now been made so that meaningful generalizations of these characteristics are possible.

As might be expected, seismic reflection sections of the continental basement look very different from those more commonly obtained for the sedimentary cover. In particular, reflections within the basement are often less uniform and more discontinuous than those from the overlying layered sedimentary sequences. Figure 3 illustrates this with a seismic reflection record obtained by COCORP in north-central Texas. Although strong, correlatable reflections are present (e.g. R1 and R2) most of the crust seems to be dominated by short discontinuous reflector segments whose density, dip, and amplitude vary laterally as well as vertically [Oliver et al., 1976]. Such discontinuous zones are perhaps the most ubiquitous seismic characteristic of the deep crust [e.g. Dohr and Meissner, 1975; Smithson,

1978]. Often the short reflection segments which make up these zones are dominantly sub-horizontal.

On many deep seismic sections are well-defined zones which contain few if any reflections. T in Figure 3 is an example. When reflections can be identified from below such a zone, it is plausible to infer that it is seismically transparent. Curved (hyperbolic) 'reflections' such as D in Figure 3 are common on some crustal seismic sections. Such events are usually identified as diffractions, energy backscattered from inhomogeneities of relatively limited extent. By proper analysis and/or processing (e.g. migration), these diffraction patterns can be 'collapsed' to locate the responsible structure [Schilt et al., 1980; Waters, 1978].

The interpretation of particular reflectors, discontinuous reflector zones, transparent zones, or diffraction patterns can be difficult, requiring careful consideration of the geology probed and the nature of the seismic techniques used. Deep seismic reflection sections are subject to the same interpretational ambiguities and pitfalls as conventional sections [Tucker and Yorston, 1976], if not more so. Artifacts such as multiple reflections, side reflections, reflected refractions, and distortion caused by shallow velocity variations can be misinterpreted as due to deep crustal structure. However, new approaches to data processing [e.g. Phinney and Jurdy, 1979] and increasing experience with deep seismic results are helping to reduce interpretational ambiguity.

Except in cases where events can be traced upward to outcrop or within reach of the drill, identification of specific reflectors is usually not possible. However reasonable, albeit speculative, geologic counterparts have been suggested to explain the reflector patterns seen on seismic records [Oliver et al., 1976; Smithson et al., 1977; Noponen et al., 1979; Smithson, 1979; Brewer and Oliver, 1980]. For example:

a) Relatively continuous reflectors or bands of reflectors may correspond to gouge or mylonite zones along major faults, metasedimentary layers, metavolcanic flows, interfingering sills or cumulate layering in mafic intrusives.

b) Zones of short discontinuous reflector segments may correspond to deformed and intruded metamorphic terrains, where folding, faulting, diapirism, boudinage etc. have disrupted reflector continuity. Variations in the amplitude of reflectors from such terrains could be due to focussing of energy reflected from concave geometries or constructive interference ('tuning') of reflections from properly spaced layering.

c) Seismically transparent zones may

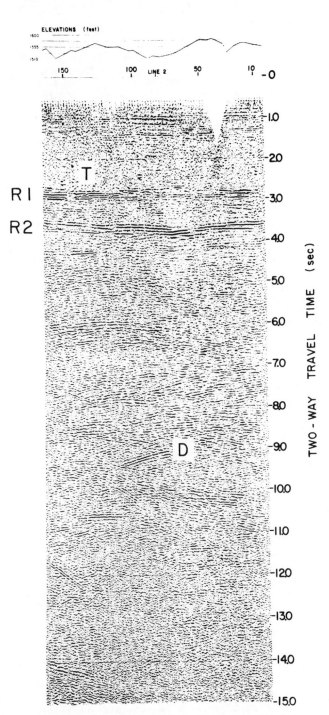

HARDEMAN CO.
LINE I

ELEVATIONS (feet)

150   100   LINE 2   50   10   -0

R1

R2

T

D

TWO - WAY TRAVEL TIME (sec)

Fig. 3. COCORP seismic reflection section
obtained in Hardeman County, Texas. Note
the scale of lateral and vertical variation
in reflection character. Letters refer to
features discussed in text. Horizontal and
vertical scales are approximately equal
over most of the section. Numbers on
horizontal scale refer to stations spaced
100 meters apart. To convert two-way travel
time to approximate depth, multiply by 3
km/sec. Letters refer to features discussed
in text. [After Schilt et al., 1977].

correspond to compositionally homogeneous (on
the scale of tens of meters) rock units, such
as granitic or unlayered mafic plutons, or
intense structural disruption which effectively
'homogenizes' a volume of rock. Cataclastic or
mylonitized regions around some major faults
might also exhibit this character.

d) Diffractions may represent
backscattered energy from intrusions which are
small relative to the seismic wavelengths used
or from abrupt truncations of structures by
faulting.

The subparallel, and often subhorizontal,
nature of basement reflections in many areas
raises the possiblity of a preferred structural
grain in the lower crust. If this alignment is
not due entirely to biases in the data
acquisition and subsequent processing, as seems
unlikely at this point, two possible
explanations come to mind [Smithson, 1978;
Phinney, 1978]: 1) horizontal movement during a
deformation event, or 2) viscous relaxation or
realignment under the influence of gravity.

Integrating these seismic characteristics
with evidence concerning the nature of the deep
crust gleaned from refraction measurements of
average seismic velocities, petrological
examination of lower crustal xenoliths and
outcrops of granulite facies rock suites, and
heat flow constraints on radioactive element
distribution, Smithson and colleagues have
presented a generalized model for the lower
crust (Figure 4) in a series of papers
[Smithson and Decker, 1974; Smithson and Brown,
1977; Smithson et al., 1977] The key feature of
this model is pronounced lateral and vertical
heterogeneity, with only a gross zonation of
the crust into an upper region invaded by
granites, a middle migmatitic zone, and a lower
zone of highly variable but more mafic
composition. Oliver [1978] points out that
extensive mid-crustal seismic velocity
discontinuities often inferred or extrapolated
from local seismic refraction studies tend to
lose significance when placed in the context of
the demonstrable complexity represented by
Figure 4. Although terms such as the Conrad or
Riel may have local applicability, a re-
evaluation of their general value in the light
of the growing body of seismic reflection
measurements is in order.

Lower crustal xenoliths of metasedimentary
nature [e.g. Padovani and Carter, 1975] and
interpretation of a sedimentary origin for
layered gneisses in the Ivrea complex [Mehnert,

M

Fig. 4. Generalized crustal model emphasizing lateral and vertical heterogeneity. M-Moho. [From Smithson and Decker, 1974].

1975] emphasize that the lower crust is not strictly derived from igneous additions but that supracrustal rocks of intermediate to silicic composition may be an important component at all levels of the crust. The important question of how such rocks get into the lower crust will be considered later, after presentation of some relevant reflection data.

Although the Moho in Figure 4 is represented by a sharp line, reflections from the crust-mantle transition zone indicate that it too is a complex entity, often appearing to be discontinuous and layered [Berry and Mair, 1977; Cook et al., 1979]. Meissner [1973] has suggested that the Moho in some areas is a zone of interfingered intrusions and/or partial melt. Berry and Mair [1977] opt for multiple, low-angle overthrusting as a means of laminating the Moho. However, in view of the compositional heterogeneity implied by Figure 4 for the overlying crust, it is likely that the base of the crust is defined by a variety of rock types and structures which may change radically over very short distances. If so, the nature of crust-mantle interaction (e.g. diapirism of mantle melts) could be strongly controlled by local rheology, density, and composition along this boundary. Most deep seismic reflection sections show a dramatic decrease in the number and/or strength of reflections from below the Moho. Whether this is evidence for greater homogeneity in the upper mantle or inadequate signal penetration is not yet clear. However, it is reasonable to expect that all crustal sounding techniques will be extended to probe the rest of the lithosphere [Hall, 1979].

Crustal Structure and Tectonic Evolution

Complex crustal structure implies complex tectonics. It is not surprising therefore that the increasing capability to recognize and map structure at depth has been accompanied by

attempts to infer the nature of the geologic processes responsible - such is the natural goal of structural geology. The study of the deep crust has now progressed beyond the point of establishing degrees of heterogeneity or the reasonableness of general models such a that in Figure 4 to such important questions as: How does the crust actually deform during continental breakup? How does the crust behave at active continental margins? What happens to the crust during continental collision? How do sediments and water get into the lower crust? How are magmas emplaced in the crust and how much do they contribute to continental growth during major tectonic events? What is the rheology of the lower crust on geologic time scales?

Within the plate tectonics framework, continental crust is modified at plate boundaries - divergent plate margins (rifts), where continents may split apart, convergent plate margins (subduction zones) where continental blocks may collide and be sutured together, and transform boundaries where continental blocks slide past each other. These boundaries are usually associated with extensional, compressional, and strike-slip tectonics respectively. It is logical, therefore, to consider crustal structure in these key tectonic environments as appropriate guides to crustal evolution. With this in mind, and avoiding for the moment the problem of differentiating plate boundary from truly intraplate tectonics, we will briefly review some of the seismic evidence bearing on deep crustal structure in areas associated with these tectonic regimes. Emphasis is placed on the COCORP results, which are presently the most extensive and uniform data set with which to attempt such an overview. To date COCORP has collected almost 3000 km of deep seismic reflection data in ten areas of the U.S. (Figure 5), many of which address geological problems related to tectonics in the above regimes (Table 1).

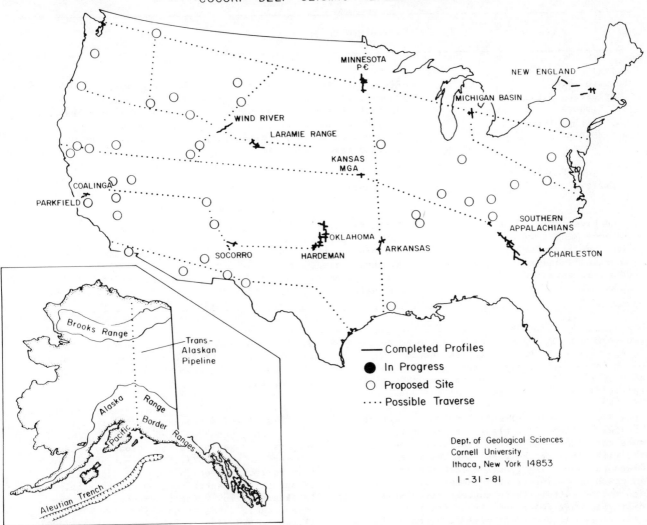

MINNESOTA
PЄ

NEW ENGLAND

MICHIGAN BASIN

WIND RIVER

LARAMIE RANGE

KANSAS
MGA

COALINGA

PARKFIELD

SOUTHERN
APPALACHIANS

OKLAHOMA

ARKANSAS

CHARLESTON

SOCORRO   HARDEMAN

Brooks Range

Trans-
Alaskan
Pipeline

Alaska   Range

Border

Pacific   Ranges

Aleutian Trench

—— Completed Profiles
● In Progress
○ Proposed Site
···· Possible Traverse

Dept. of Geological Sciences
Cornell University
Ithaca, New York 14853
1 - 31 - 81

Fig. 5. Status of the COCORP program in early 1981, showing profiles already surveyed (solid lines), suggested sites for future work (open circles), and possible routes for regional traverses (dotted lines).

Intracontinental Rifting

Continental breakup and fragmentation is usually associated with rifts. It is clear that during rifting the crust is modified by large scale normal faulting, volcanism, plutonism, and extensional thinning [Burke, 1979]. However, the details of how these processes lead to the formation of an ocean basin between two continental masses is still a matter of considerable speculation. Is crustal thinning accomplished by ductile flow or brittle faulting? Do the rift bounding faults flatten at depth? Where is magma generated and how is it emplaced in the crust? Is there an anomalous lower crust or upper mantle beneath rifts? What controls seismicity? What is the

role of prexisting fractures in guiding rift formation? Why do some rifts fail? Questions such as these have made rifts a favorite target for crustal geophysicists, and numerous studies have been made of deep structure in the Rhinegraben [Mueller et al, 1973], the Dead Sea rift [Fuchs et al, 1979], the Rio Grande rift [Riecker, 1979], and others [Braile et al., 1979]. Figure 6 shows the COCORP seismic sections obtained across the Rio Grande rift near Socorro, New Mexico [Brown et al., 1979; 1980]. Like the seismic section in Figure 3, the basement here seems to consist primarily of numerous short, discontinuous reflector segments. Large scale vertical faulting is suggested by reflector R which underlies a central basin divided by a major horst (H) into

TABLE 1. Some Tectonic Problems addressed by COCORP Profiling

Site	Structure or Event	Age
Rifts:		
Socorro, N.M.	Rio Grande Rift	Cenozoic
Charleston, S.C.	Opening of Atlantic	Mesozoic
S. Oklahoma	Anadarko Basin/Wichita Uplift	L. Precambr.-Paleozoic
Central Michigan	Buried Keweenawan rift	Precambrian
N.E. Kansas	Midcontinent Geophysical Anomaly (Buried Keweenawan rift)	Precambrian
Transform Faults:		
Parkfield, Cal.	San Andreas Fault	Cenozoic
Thrusts:		
Wind River Mtns., Wyo.	Laramide uplift	Tertiary
Laramie Range, Wyo.	Laramide uplift	Tertiary
Great Valley, Cal.	Paleo-subduction complex	Mesozoic
Georgia-Tennessee	S. Appalachians	Paleozoic
New York-Vermont	N.E. Appalachians	Paleozoic
Arkansas	Ouachitas	Paleozoic

two segments (BW and BE), each over 4 km deep. On the east is a shallow, faulted bench (BN) which is in fault contact with the uplifted eastern rift flank at VP 180 on Line 1. Beneath this surface fault, representing the eastern rift boundary, is a near-vertical, reflection free zone (F) which extends through the entire crustal section. This blank zone separates the more continuous reflection character beneath the rift flank to the east from the more disrupted appearing crust beneath the basin. It is tempting to interpret F as the rift boundary fault zone, within which intense structural disruption has destroyed any coherent reflectors. If so, it is clear that there is no flattening of this rift boundary fault. Alternatively F could be the expression of a pre-existing crustal shear zone along which Tertiary normal faulting was guided. However, the possibility that this zone is at least partly an artifact of seismic wave field distortion by drastic near-surface velocity variations can not be ruled out. There is no similarly penetrative zone at the western rift boundary, which intersects the surface at VP 420 on Line 1A. Instead the western boundary seems to be defined by a surprisingly low angle (about 40 degrees) reflector with 3 km of displacement. Whether this is a simple fault plane reflection [Cape et al., 1980] or the result of some more complex composite structure has not been determined. Within the Tertiary basin fill, the complexly faulted western edge of the bench (BN) is buried beneath a 4 km thick, west-dipping alluvial fan (A).

Perhaps the most interesting feature on these records is the strong, complex reflector at about 20 km depth (MB in Figure 6). This reflector can be traced from the center of Line 1A to the western edge of Line 1. Its depth corresponds with the top of a magma body inferred to exist beneath the rift in this area by Sanford et al. [1973] based primarily on the analysis of reflected shear waves from local earthquakes, but supported by other geophysical, geological and geodetic evidence [Sanford et al., 1977]. The apparent dip of this reflector seen on the east side of Line 1A is an artifact introduced by the increasing thickness of low velocity graben fill - the event itself seems to be virtually flat, with a slight northward tilt [Rinehart et al., 1979]. This reflector is also prominent on other COCORP sections in the rift. Concordance of the magma body reflector with overlying crustal structures (e.g. on Line 2A which crosses the profile in Figure 6; Brown et al., [1979]), its thin yet extensive geometry [Rinehart et al., 1979], and its emplacement at a depth corresponding to a gross change in crustal velocity [Toppozada and Sanford, 1976; Olsen et al., 1979] suggests that the magma was entrapped along the base of a prexisting midcrustal barrier of some sort. While magma accumulation along such barriers has been postulated on theoretical and experimental grounds [Ramberg, 1973; Elder, 1978], these seismic data represent a more direct constraint on how magma is emplaced and contaminated in the crust.

Other COCORP lines within the rift show prominent Moho reflections with a laterally complex, layered character. The absence of clear Moho reflections in Figure 6 is probably

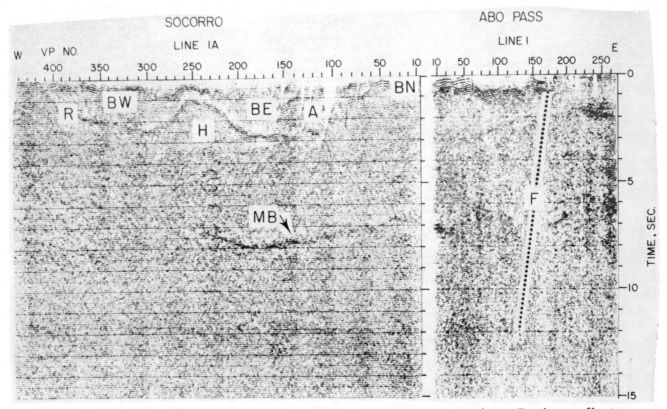

Fig. 6. COCORP reflection profile across the Rio Grande rift in New Mexico. The deep reflector at about 7 sec on the eastern half of Line 1A (MB) is believed to be a magma body. Station spacing is 134 m on Line 1A, 100 m on Line 1. Letters refer to features discussed in text. [After Brown et al., 1980].

due to the disruptive seismic effect of shallow structure on Line 1A. Reflection M on Line 1 is from a depth (about 36 km) appropriate for the Moho in this area [Olsen et al., 1979]. Seismic refraction studies of the southern part of the Rio Grande rift [Cook et al., 1979], together with similar results from the Rhinegraben [Mueller et al, 1973], indicate that the lowermost crust beneath these rifts has anomalously high seismic velocities, interpreted to be 'pillows' of modified lower crustal or upper mantle material.

COCORP profiles in other areas have also crossed rift-related structures. A COCORP study in the Atlantic coastal plain near Charleston, S.C., has mapped faulting possibly related to Mesozoic rifting of Africa away from North America [Schilt et al., 1979]. Holocene reactivation of some of these faults may account for the anomalous seismic activity in this region. There are some remarkable similarities in the seismic character of the crust beneath the Atlantic coastal plain and that beneath the Rio Grande rift. For example in both areas there are extensive transparent zones in the upper crust, lower crusts

dominated by discontinuous reflector zones, and complex, laminated Mohos [Schilt et al., 1979]. The significance of these resemblances is not clear at present, although both regions presumably were subject to similar magmatism. A COCORP survey in the Michigan basin has revealed a narrow rift-like trough, apparently filled by sediments and volcanics, in the Precambrian basement underlying the basin. This trough correlates with a major linear gravity and magnetic anomaly, and is interpreted as an extension of Keweenawan rifting [Jensen et al., 1979]. In this case reflections from beneath the rift fill are scarce and identifiable Moho reflections are absent altogether. Surveys across the Anadarko basin in southern Oklahoma, a possible Paleozoic aulacogen, and the Midcontinent Gravity High, presumably an aborted Precambrian rift, in Kansas have been initiated to learn more about rifts which fail to develop into ocean basins. Important information on rift structure is also provided by recent marine seismic surveys run by the U.S. Geological Survey, which have discovered rifted basins on the Atlantic continental margin [Hutchinson and Grow, 1978].

Fig. 7. COCORP reflection profile across the Wind River uplift. The Wind River thrust, indicated by arrows, can be traced to at least midcrustal depths as a moderate angle thrust. [After Brewer and Oliver, 1980].

Transform Faulting

Seismic studies of crustal structure at transform faults have been very limited. Reflection profiles have been carried out across the San Andreas fault in central California by COCORP [Long et al., 1978] and also by a group at the University of California at Berkeley [McEvilly and Clymer, 1978]. Uncertainties due to complex near surface structure and possible instrumental problems have hindered the interpretation of these data sets [Bronson et al., 1979]. However, both surveys indicate that below a depth of about 10 km the fault zone is characterized by a near vertical no-reflection zone about 4 km wide. Like the eastern boundary of the Rio Grande rift, this blank zone is interpreted as due to intense structural disruption. The deep San Andreas shear zone extends through to the base of the crust, with a substantial difference in the seismic character of the lower crust, including the Moho, evident on either side.

Compressional Orogeny

By far the most significant and least ambiguous COCORP contributions to our understanding of deep structure have resulted from seismic studies across major faults in the Rocky Mountain and Appalachian orogenic belts. Results from both areas have forced major

revisions in our concepts of the evolution of these classic mobile belts.

The mode of formation of the uplifted basement blocks in the Rocky Mountain foreland has been a subject of considerable interest and controversy. The central question is whether the mountain ranges were pushed up along steep faults by dominantly vertical forces or thrust up along less steep faults by compressive forces [Smithson et al., 1978]. The deep geometry of the major faults which flank these uplifts is therefore a key guide to deciding between these two possibilities. To help settle this question, COCORP ran a 160 km long deep seismic profile (Figure 7) across the Wind River uplift, the largest of these basement blocks. It indicates clearly that the Wind River thrust (marked by the arrows) extends as a moderately dipping fault to depths of at least 30 km, with no evidence of significantly increasing dip. Compressive forces must have been responsible for formation of this uplift, which resulted in at least 13 km of vertical and 26 km of horizontal displacement. If other foreland uplifts formed by a similar mechanism, substantial crustal shortening must have occurred during the Laramide orogeny, in addition to that inferred from the thin-skinned thrust belts to the west [Bally et al., 1966].

Fault plane reflections imply a significant acoustic impedance contrast, i.e. juxtaposition of materials with substantial differences in seismic velocity and/or density. For crystalline basement rocks this impedance contrast is perhaps provided by mylonitised rock within the fault zone [e.g. Brewer et al., 1980], although lab measurements have failed to confirm such an explanation [Jones et al., 1980]. The coherent appearance of the Wind River thrust to at least mid-crustal depths provides an important clue to lower crustal rheology , since it implies a degree of brittle behavior at those depths on the time scale of faulting involved. An extremely ductile lower crust would be expected to accomodate strain over a fairly wide volume instead of forming a well-defined reflection zone such as that indicated in Figure 7. On the other hand recent special processing of the Wind River data suggests that the fault plane may be a composite structure with many secondary splays, and that its dip may decrease in the lower crust [Lynn et al., 1979]. Furthermore, gravity data indicate that the Wind River thrust does not offset the Moho significantly [Smithson et al., 1978], indicating that perhaps ductile flow may be important in the lowermost portions of the crust.

Other, less pronounced, dipping reflections within the basement on this section may correspond to other thrust faults. Numerous deep reflections were recorded, including some at about 15 seconds beneath the Green River basin which may correspond to the Moho. Particularly distinctive reflector patterns beneath the Wind River Mountains have been interpreted as a nappe complex by Smithson et al. [1979]. The origin of some of the deep reflections beneath the Green River and Wind River basins have been the subject of some controversy, with Smithson et al. [1979] arguing that they may be multiple reflections generated within the thick sedimentary sections rather than true reflections from deep structure, while others have argued the reverse [e.g. Wallace, 1980].

Some of COCORP's most dramatic results were obtained during its survey of Appalachian structure in the southeastern U.S. The Appalachian system is believed by many to have resulted from a complex series of collisional interactions between a proto-North American continent and a protoAfrican/European continent, with or without complications from intervening microcontinents or island arcs [e.g. Bird and Dewey, 1970; Wielchowsky et al., 1978]. Thus the Appalachians are an appropriate subject for the study of continental accretion by suturing together continental fragments at a convergent plate boundary.

An initial target of COCORP work was the Brevard fault, a remarkably linear fault which extends over 500 km from Alabama to North Carolina. A large number of hypotheses concerning its origin have been put forward [e.g. Clark et al., 1978], including its interpretation as a transform fault and a suture zone between the colliding Paleozoic

continents. The first COCORP survey across this fault indicated that it is in fact a low angle thrust rooted in a much more profound decollement. Subsequent profiling identified this sole thrust as correlative with the Blue Ridge fault, which outcrops some 125 km northwest of the Brevard in the area of the survey. More importantly, this subhorizontal thrust was found to extend an impressive distance southeast beneath the Piedmont province (Figure 8). Accordingly, the entire Blue Ridge province - the Precambrian metamorphic core of the Appalachians - and much if not all of the Piedmont are allocthonous, part of a crystalline sheet which was thrust at least 250 km over a Paleozoic continental margin. Reflections immediately below this sheet are interpreted as continental margin sediments trapped beneath the thrust [Cook et al., 1979], perhaps having served to lower friction along the fault. The existence of these underthrust sediments raises the question of whether there is any heretofore unsuspected oil or gas potential in the Blue Ridge and Piedmont.

There is still some question as to where this thrust originates. The southeastern end of the COCORP profile is characterized by increasingly complex seismic patterns with numerous dipping reflectors and a very well defined Moho. It is possible that the thrust is rooted there - however, Cook et al. [1979] argue that the thrust can be traced all the way to the end of the seismic section and may be rooted much further to the southeast. Harris and Bayer [1979] speculate that it extends to beneath the continental margin. Recent extension of the COCORP survey confirms the existence of compressive basement structures beneath the Coastal Plain, although relationships are increasingly complex [Cook et al., 1981].In any case the scale of overthrusting involved is clearly impressive, regardless of how far away it will ultimately prove to be rooted or exactly how far along strike it can be legitimately extrapolated.

Implications and Speculations

The above examples are intended to illustrate how delineation of deep structure can lead to a much better understanding of crustal tectonics. By demonstrating that the Wind River fault is a moderate angle thrust penetrating to substantial depth, seismic profiling has established the importance of compressional forces in deformation of this unique orogenic foreland. Brewer et al. [1980] show how this compression may relate to subduction at a distant convergent plate margin then active along the west coast of North America. The geometry and coherence of the Wind River thrust implies relatively brittle behavior at depths often considered the domain of ductile flow.

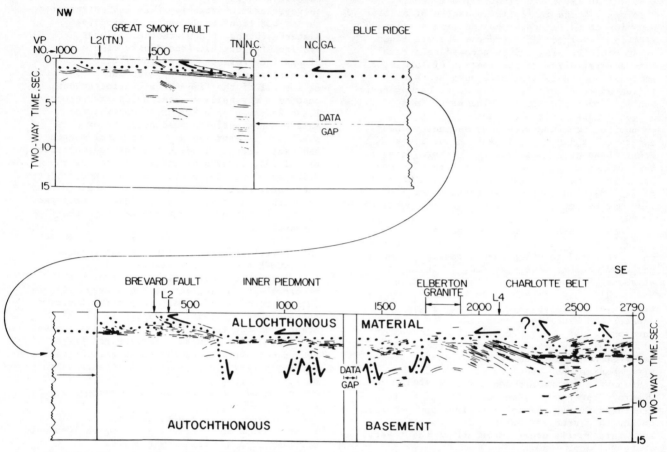

LINE I
COCORP SOUTHERN APPALACHIAN TRAVERSE
25 KM

Fig. 8. Line drawing of the COCORP seismic section across the southern Appalachians, demonstrating the allocthonous nature of the Blue Ridge and Piedmont. The base of Appalachian thrust sheet is indicated by the dotted line. [After Cook et al., 1979].

Meissner et al. [1980] discuss seismic reflection data in Germany which also indicate that major faults can penetrate the entire crust along well defined zones. Although more ambiguous, COCORP seismic results across faults in the Rio Grande rift and across the San Andreas fault support the notion of deeply penetrating shear zones. However, the reflection-free seismic appearance of these near-vertical fault zones differs considerably from the sharp reflections which have been obtained from moderate and low angle faults, due at least in part to the limitations of the seismic reflection method (as commonly practiced) in defining steeply dipping structures. The COCORP results within the Rio Grande rift which outline a mid-crustal magma

body and suggest that it was emplaced along a preexisting barrier to vertical migration are another example of how mapping deep structure can lead to important inferences regarding the mechanical behavior of the deep crust. New constraints on crustal rheology on various time scales is likely to become one of the more significant by-products from deep seismic profiling.

Crustal behavior during collisional orogeny has been much better defined by COCORP profiling in the southern Appalachians. The large-scale overthrusting discovered to dominate tectonics in this region, and which is similar in many respects to the 'flake' tectonics hypothesized by Oxburgh [1972] for the Alps, has obvious ramifications not only

48   BROWN ET AL.

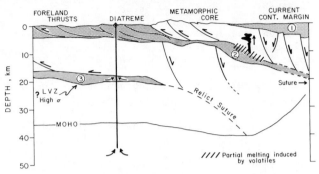

CONTINENTAL ACCRETION BY MULTIPLE OVERTHRUSTING

Fig. 9. Speculative model, based on results from the COCORP Appalachian survey, illustrating continental accretion by successive addition of thrust sheets during continental and/or island arc collisions. Stippled zones represent sediments and volatiles deposited at a contemporary passive continental margin (1), similar but older sedimentary rocks overthrust during a preceding orogeny (2), and even older metasedimentary shelf assemblages incorporated into the deep crust during a more ancient collisional event. Such zones of trapped (meta-) sedimentary rocks might explain some high conductivity ( ) and low velocity zones (LVZ) in the lower crust.

for unravelling the geologic history of mountain belts, but for understanding the mechanics of emplacing such large crystalline thrust sheets as well. On the more speculative side, the 'injection' of sediments, presumably wet, beneath such thrust sheets may be an important mechanism for incorporating metasediments into the lower crust, thus explaining their appearance among lower crustal xenoliths and exposed sections of the lower crust [Kay and Kay, 1981]. It is also plausible that such trapped layers of sediments and associated volatiles could explain at least some high conductivity zones and low velocity zones which have been inferred to exist in the lower crust from geomagnetic and seismic studies [e.g. Landisman et al., 1971; Porath, 1971]. Any water carried into the deep crust by such a mechanism could also be expected to have a significant influence on partial melting processes and resulting magma geochemistry. As is usually the case, these new results also raise new questions and re-emphasize old problems, such as what happened to the lithosphere which once underlay the large overthrust sheets.

If the overthrust geometry indicated by COCORP profiling in the Appalachians is representative of the mode of accretion of continental fragments onto cratonic nuclei

during collisional-type orogenies, clearly horizontal tectonics plays a major role in the development of crustal structure. A simple model for continental accretion by successive overthrusts of this kind is indicated in Figure 9. Allusion has already been made to the significance of the sediments trapped during these episodes. Horizontal trends in Archaean gneiss terrains [Bridgewater et al., 1974] suggests that such a process might have been operative in the earliest stages of crustal generation. Perhaps the sub-horizontal orientation of the numerous discontinuous reflectors seen on many deep crustal reflection sections represents a relict of this process.

Conclusions

The systematic collection of deep seismic reflection data is just beginning to provide some of the necessary, and long awaited, clues for understanding many aspects of the structural evolution of continental crust. As new seismic results are digested and integrated with complementary information from other fields of geology and geophysics, a more complete and coherent picture of this process will undoubtedly emerge.

The seismic sections presented here illustrate the complexity and heterogeneity of the continental crust. A proper understanding of many of the seismic characteristics of the lower crust is in many ways still in an early stage of development, although it is clear that meaningful variations in these characteristics can be defined. While the Moho is usually found to be a distinctive event, albeit often a complex, laterally varying transition zone, the concept of pervasive, coherent midcrustal discontinuities such as the Conrad loses significance in the face of this pervasive heterogeneity.

Some of the most significant and unambiguous results from deep seismic profiling have been obtained by tracing major thrust faults to depth. COCORP surveys across the Wind River uplift, establishing its compressional origin, and those across the southern Appalachians, which demonstrate the existence of an extensive allocthonous sheet emplaced by large-scale overthrusting, have made particularly salient contributions to our understanding of crustal evolution. Among the speculations inspired by the latter are models of continental accretion by successive overthrusting of crustal sheets, a mechanism of trapping sediments and volatiles at lower crustal depths, and a possible explanation for certain crustal low velocity and high conductivity zones.

Acknowledgments. COCORP's Executive Committee is composed of scientists from Cornell, Houston, Princeton, Texas, and Wisconsin

Universities and the USGS. The advice and
support of these individuals as well as the
members of COCORP's Site Selection and
Technical Advisory Committees, without which
the surveys upon which this paper is based
would not have been possible, is gratefully
acknowledged. The authors would also like to
express their appreciation to the Exploration
Services Division of Geosource, Inc and the
Compagnie General de Geophysique who carried
out the COCORP field work. The COCORP project
is supported by the National Science Foundation
under Grant No. 78-23672. This paper was also
supported by NSF Grant No. 78-23673. Cornell
contribution to geology No. 669.

References

Bally, A.W., P.L. Gordy, and G.A. Stewart,
Structure, seismic data and orogenic evolution
of the southern Canadian Rocky Mountains, Can.
Petrol. Geol. Bull., 14, 337-381, 1966.

Berry, M.J., and J.A. Mair, The nature of the
Earth's crust in Canada, in The Earth's Crust,
Geophys. Mongr. Ser., vol. 20, edited by J. G.
Heacock, pp. 319-348, AGU, Washington, D.C.,
1977.

Bird, J.M., and J.F. Dewey, Lithosphere plate-
continental margin tectonics and the evolution
of the Appalachian orogen, Geol. Soc. Amer.
Bull., 81, 1031-1060, 1970.

Braile, L.W., R.B. Smith, J. Ansorge, M.R. Baker,
C. Prodehl, J.H. Healy, S. Mueller, K.H.
Olsen, K. Priestley, J. Brune, The Yellowstone-
Snake River Plain profiling experiment: eastern
Snake River Plain, EOS Trans. AGU, 60, 941,
1979.

Brewer, J.A., and J.E. Oliver, Seismic reflection
studies of deep crustal structure, Ann. Rev.
Earth Planet. Sci., 8, 205-230, 1980.

Brewer, J.A., S.B. Smithson, J.E. Oliver, S.
Kaufman, and L.D. Brown, The Laramide Orogeny:
Evidence from COCORP deep crustal seismic
reflection profiles in the Wind River
Mountains, Wyoming, Tectonophysics, 62,
165-189, 1980.

Bridgwater, D., V.R. McGregor, and J.S. Myers, A
horizontal tectonic regime in the Archean of
Greenland and its implications for early
crustal thickening, Precambrian Res., 1,
179-197, 1974.

Bronson, M.A., R.L. Zawislak, R.A. Johnson, S.B.
Smithson, and E.K. Shumaker, A new study of
the COCORP Parkfield, California reflection
seismic line across the San Andreas fault, EOS
Trans. AGU, 60, 875, 1979.

Brown, L.D., P.A. Krumhansl, C.E. Chapin, A.R.
Sanford, S. Kaufman, and J. Oliver, COCORP
seismic reflection studies of the Rio Grande
rift, in Rio Grande Rift: Tectonics and
Magmatism, edited by R.E. Riecker, pp. 169-184,
AGU, Washington, D.C., 1979.

Brown, L., J. Brewer, F. Cook, S. Kaufman, P.
Krumhansl, G. Long, J. Oliver, and S. Schilt,
Mapping the continental lithosphere by seismic
reflection profiling, Arabian Jour. for Science
and Engineering, Special Issue, 1978.

Brown, L.D., C.E. Chapin, A.R. Sanford, S.
Kaufman, and J. Oliver, Deep structure of the
Rio Grande rift from seismic reflection
profiling, J. Geophys. Res., 85, 4773-4800,
1980.

Burke, K., Aulacogens and continental breakup,
Ann. Rev. Geophys. Space Physics, 5, 371-396,
1977.

Cape, C., S. McGeary, R.E. Bracken, L.D. Gagnon,
and G.A. Thompson, Cenozoic normal faulting and
the shallow structure of the Rio Grande rift
near Socorro, New Mexico, EOS Trans. AGU, 61,
1039, 1980.

Christiansen, N.I., and D.M. Fountain,
Constitution of the lower continental crust
based on experimental studies of seismic
velocities in granulite, Geol. Soc. Amer.
Bull., 86, 227-236, 1975.

Clark, H.B., J.K. Costain, and L. Glover,
Structural and seismic reflection studies of
the Brevard ductile deformation zone near
Rosman, North Carolina, Amer. J. Sci., 278,
419-441, 1978.

Cook, F.A., D.S. Albaugh, L.D. Brown, S. Kaufman,
J.E. Oliver, and R.D. Hatcher, Jr., Thin-
skinned tectonics in the crystalline southern
Appalachians; COCORP seismic-reflection
profiling of the Blue Ridge and Piedmont,
Geology, 7, 563-567, 1979.

Cook, F.A., L.D. Brown, J.E. Oliver, and S.
Kaufman, The nature of the Moho on COCORP
reflection data (abstract), EOS Trans. AGU, 59,
389, 1978.

Cook, F.A., D.B. McCullar, E.R.Decker, and S.B.
Smithson, Crustal structure and evolution of
the southern Rio Grande rift, in Rio Grande
Rift: Tectonics and Magmatism, edited by R.E.
Riecker, pp. 195-208, AGU, Washington, D.C.,
1979.

Cook, F.A., L.D. Brown, S. Kaufman, J.E. Oliver,
and T.A. Petersen, COCORP seismic profiling of
the Appalachian orogen beneath the Coastal
Plain of Georgia, Geol. Soc. Amer. Bull., in
press, 1981.

Davydova, N.I., DSS studies of deep faults, in
Seismic Properties of the Moho Discontinuity,
edited by N.I. Davydova, 138pp., Akademiya Nauk
SSSR, 1975.

Dohr, G., and R. Meissner, Deep crustal
reflections in Europe, Geophysics, 40, 25-39,
1975. of 2 Elder, J.W., Magma traps: Part 1,
Theory, Pageoph., 117, 3-14, 1978.

Giese, P., C. Prodehl, and A. Stein, Explosion
seismology in Central Europe, 429pp., Springer-
Verlag, New York, 1976.

Harris, L.D., and K.C. Bayer, Sequential
development of the Appalachian orogen above a
master decollement- A hypothesis, Geology, 7,
568-572, 1979.

Hall, D.H., Crustal studies: their implications
for the explorationist, Austral. Soc. Expl.

Geophysicists Bull., 53-59, 1076.

Heier, K., Geochemistry of granulite facies and problems of their origin, Phil. Trans. R. Soc. (Lond.), s273, 429-442, 1973.

Hutchinson, D.R., and J.A. Grow, Deep crustal reflectors and mantle returns from multichannel seismic profiles on the eastern United States Continental Shelf (abstract), Earthquake Notes, 49, 27, 1978.

Jensen, L.W., D. Steiner, L. Brown, S. Kaufman, and J.E. Oliver, COCORP deep crustal reflection studies in the Michigan basin (abstract), Michigan Geol. Soc. Ann. Meeting, 1979.

Jones, T, D. Murphy, A. Nur, and G. Thompson, Experimental investigations of the seismic nature of deep crustal fault zones (abstract), EOS Trans. AGU, 46, 1119, 1980.

Kay, R.W., and S.M. Kay, The nature of the lower continental crust: inferences from geophysics, surface geology, and crustal xenoliths, Rev. Geophys. Space Physics, in press, 1981.

Landisman, M., S. Mueller, and B.J. Mitchell, Review of evidence for velocity inversions in the continental crust, in The Structure and Physical Properties of the Earth's Crust, Geophys. Mon. Ser., vol. 20, edited by J.G. Heacock, pp. 11-34, AGU, Washington, D.C., 1971.

Long, G.H., L.D. Brown, and S. Kaufman, A deep seismic reflection survey across the San Andreas Fault near Parkfield, California (abstract), EOS Trans. AGU, 59, 385, 1978.

Lynn, H.B., L. Gagnon, E. Kjartansson, and D. Seeburger, Migrations and interpretations in laterally varying media, Wind River Thrust, Wyoming (abstract), Soc. Explor. Geophysicists Ann. Meeting Abst., 83, 1979.

Mathur, S.P., Gravity anomalies and crustal structure - a review, Austral. Soc. Explor. Geophys. Bull., 8, 111-117. 1977.

McEvilly, T.V., and R.W. Clymer, Seismic reflection profiling across the San Andreas Fault in San Benito County, California (abstract), Earthquake Notes, 49, 28, 1978.

Mehnert, K.R., The Ivrea zone; a model of the deep crust, Neues Jahrbuch fur Mineralogie, Abhandlungen, 125, 156-199, 1973.

Meissner, R., The "Moho" as a transition zone, Geophys. Surveys, 1, 195-965, 1973.

Meissner, R., H. Bartelsen, and H. Murawski, Seismic reflection and refraction studies for investigating fault zones along the geotraverse Rhenoherzynikum, Tectonophysics, 64, 59-84, 1980.

Mueller, S., A new model of the continental crust, in The Earth's Crust, Geophys. Mon. Ser., vol. 20, edited by J.G. Heacock, pp. 289-317, AGU, Washington, D.C., 1977.

Mueller, S., and M. Landisman, An example of the unified method of interpretation for crustal seismic data, Geophys. J. Royal Astron. Soc., 23, 365-371, 1971.

Mueller, S., E. Peterschmitt, K. Fuchs, D. Emter, and J. Ansorge, Crustal structure of the Rhinegraben area, Tectonophysics, 20, 381-392, 1973.

Noponen, I., P. Heikkinen, and S. Mehrotra, Applicability of seismic reflection sounding in regions of Precambrian geology, Geoexploration, 17, 1-9, 1979.

Oliver, J., Exploration of the continental basement by seismic reflection profiling, Nature, 275, 485-488, 1978.

Oliver, J., M. Dobrin, S. Kaufman, R. Meyer, and R. Phinney, Continuous seismic reflection profiling of the deep basement, Hardeman County, Texas, Geol. Soc. Amer. Bull., 87, 1537-1546, 1976.

Olsen, K.H., G.R. Keller, and J.N. Stewart, Crustal structure along the Grande Rift from seismic refraction profiles, in Rio Grande Rift: Tectonics and Magmatism, edited by R.E. Riecker, pp. 127-143, AGU, Washington, D.C., 1979.

Padovani, E.R., and J.L. Carter, Aspects of the deep crustal evolution beneath south central New Mexico, in The Earth's Crust, Geophys. Mon. Ser., vol. 20, edited by J.G. Heacock, pp. 19-56, AGU, Washignton, D.C., 1977.

Phinney, R.A., Interpretation of reflection seismic images of the lower crust (abstract), EOS Trans. AGU, 59, 389, 1978.

Phinney, R.A., and D.M. Jurdy, Seismic imaging of the deep crust, Geophysics, 4 44, 1637-1660, 1979.

Porath, H., A review of the evidence on low-resistivity layers in the earth's crust, in The Structure and Physical Properties of the Earth's Crust, Geophys. Mon. Ser., vol. 14, edited by J.G. Heacock, pp. 127-144, AGU, Washington, D.C., 1971.

Ramberg, H., Model studies in relation to intrusion of plutonic bodies, in Mechanism of Igneous Intrusion, pp. 261-272, Gallery Press, Liverpool, 1970.

Rinehart, E.J., A.R. Sanford, and R. M. Ward, Geographic extent and shape of an extensive magma body at midcrustal depths in the Rio Grande rift near Socorro, New Mexico, in Rio Grande Rift: Tectonics and Magmatism, edited by R.E. Reicker, pp. 237-251, AGU, Washington, D.C., 1979.

Sanford, A.R., O.S. Alptekin, and T.R. Toppozada, Use of reflection phases on microearthquake seismograms to map an unusual discontinuity beneath the Rio Grande rift, Bull. Seismol. Soc. Amer., 63, 2021-2034, 1973.

Sanford, A.R., R.P. Mott, Jr., P.J. Shuleski, E.J. Rinehart, F.J. Caravella, R.M. Ward, and T.C. Wallace, Geophysical evidence for a magma body in the crust in the vicinity of Socorro, New Mexico, in The Earth's Crust, Geophys. Mon. Ser., vol. 20, edited by J.G. Heacock, pp. 385-403, AGU, Washington, D.C., 1977.

Schilt, F.S., S. Kaufman, and G.H. Long, Analysis of 3-dimensional diffractions from COCORP data (abstract), EOS Trans. AGU, 58, 436, 1977.

Schilt, S., J. Oliver, L. Brown, S. Kaufman, D.

Albaugh, J. Brewer, F.A. Cook, L. Jensen, P. Krumhansl, G. Long, and D. Steiner, The heterogeneity of the continental crust: results from deep crustal seismic reflection profiling using the VIBROSEIS technique, Rev. Geophys. Space Phys., 17, 354-368, 1979.

Smithson, S.B., Modeling continental crust: structural and chemical constraints, Geophys. Res. Lttrs., 5, 749-752, 1978.

Smithson, S.B., Aspects of continental crustal structure and growth: targets for scientific deep drilling, Contrib. Geol. Univ. Wyo., 17, 65-75, 1979.

Smithson, S.B., and S.K. Brown, A model for the lower continental crust, Earth Planet. Sci. Lett., 35, 134-144, 1977.

Smithson, S.B., and E.R. Decker, A continental crustal model and its geothermal implications, Earth Planet. Sci. Lett., 22, 215-225, 1974.

Smithson, S.B., and R.J. Ebens, Interpretation of data from a 3.05 km borehole in Precambrian crystalline rocks, Wind River Mountains, Wyoming, J. Geophys. Res., 76, 7079-7087, 1971.

Smithson, S.B., P.N. Shive, and S.K. Brown, Seismic velocity, reflections, and structure of the crystalline crust, in The Earth's Crust, Geophys. Mon. Ser., vol. 20,

Smithson, S.B., J. Brewer, S. Kaufman, J. Oliver, and C. Hurich, Nature of the Wind River thrust, Wyoming, from COCORP deep reflection data and gravity data, Geology, 6, 648-652, 1978.

Smithson, S.B., J.A. Brewer, S. Kaufman, J.E. Oliver, and C. Hurich, Structure Laramide Wind River Uplift, Wyoming from COCORP deep reflection data and gravity data, J. Geophys. Res., 84, 5955-5972, 1979.

Toppozada, T.R., and A.R. Sanford, Crustal structure in central New Mexico interpreted from the Gasbuggy explosion, Bull. Seismol. Soc. Amer., 66, 877-866, 1976.

Tucker, P.M. and H.J. Yorston, Pitfalls in seismic interpretation, Soc. Expl. Geophysicists Mon. 2, 50pp., 1973.

U.S. Geodynamics Committee, Continental Scientific Drilling Program, 192 pp., U.S. National Academy of Sciences, Washington, D.C., 1979.

Wallace, M., Deep basement reflections in Wind River Line 1, Geophys. Res. Lttrs., 7, 729-732, 1980.

Waters, K.H., Reflection Seismology, 377 pp., John Wiley and Sons, New York, 1978.

Wielchowsky, C.C. et al., Construction, uses, and misuses of plate-tectonic models in the southern Appalachians (abstract), Geol. Soc. Amer. Abst. with Programs, 10, 202, 1978.

DYNAMICS OF THE EARTH'S CORE AND THE GEODYNAMO

F.H. Busse

Department of Earth and Space Sciences and Institute of Geophysics and Planetary Physics
University of California, Los Angeles, California 90024

Summary. Recent progress towards the understanding of the origin of the earth's magnetic field is reviewed. The dynamo mechanism of the generation of magnetic fields by fluid motions has become generally accepted as the cause of geomagmetism. But the properties of the actual mechanism operating in the earth's core are still not uniquely determined. Future progress in this field will depend on the development of numerical models based on the complete set of magnetohydrodynamic equations of the problem. In particular, those calculations will shed new light on the way in which the Lorentz force determine the magnetic field strength. The discovery and measurement of the magnetic fields of other planets has opened a new dimension for the research on the geodynamo. Since with minor modification any model of the geodynamo should apply to other planets as well, theories are now much better constrained than they have been in the past.

Introduction

No other geophysical phenomenon has given rise to as many scientific hypotheses about its origin as the phenomenon of geomagnetism. The book by Rikitake [1966] lists quite a few of them, but new hypotheses are still being proposed. Today the vast majority of geophysicists has accepted the dynamo hypothesis of the generation of the earth's magnetic field by motion in the liquid core of the earth. This view has been advocated primarily by Elsasser [1946] and Bullard [1949] and doubts about its theoretical feasibility were dispelled when Backus [1958] and Herzenberg [1958] showed that mathematically the dynamo process does indeed occur in a singly connected domain of fluid with uniform electrical conductivity. As geophysicists and geochemists have become more familiar with the dynamo hypothesis, the implication of this hypothesis for the physical properties and the dynamic state of the earth's core has received much attention. Since mechanical energy is converted into electromagnetic energy by the dynamo process and ultimately dissipated in the form of Ohmic heat, the question of the energy source is of particular importance. The energy derived from thermal

sources appears to be sufficient to drive the geodynamo, but not by a large margin. Thus many researchers in the field favor the buoyancy derived from gravitational separation of the chemical constitutents of the core as the driving force of motions in the core. The precession driven geodynamo advocated by Malkus [1968] does not seem very likely today, but in spite of the work of Loper [1975] and Rochester et al. [1975] the possibility of a dynamo caused by the turbulent flow originating from the differential precessional torques acting on mantle and core of the earth cannot be disregarded entirely. Since the energetics of the earth's core are discussed in detail in the article by Gubbins in this volumne, we shall not consider the subject in more detail.

The other requirement of the geodynamo is a sufficiently high level of motion in the core. Some indication of the flow in the earth's core can be obtained from the secular variation of the geomagnetic field. The westward drift of about 20° per century of the nondipole component of the field has been interpreted by Bullard et al. [1950] and others as a differential rotation. According to this view, the outer part of the liquid core exhibits a retrograde azimuthal velocity of about 3×10^{-2} cm sec^{-1}. This value exceeds by many orders of magnitude the velocity required for the dynamo process according to a simple condition derived by Backus [1958] and Childress [1969]. But for several reasons, this is not a very useful result. First, the condition of Backus and Childress is only a necessary criterion which is likely to underestimate the velocity or-- in a dimensionless term--the magnetic Reynolds number required for the realized geodynamo by a factor of the order 10 or more. Second, since alternative interpretations of the westward drift in terms of wave propagation have been proposed (Hide, 1966; Malkus, 1967; Acheson, 1972), the above mentioned value of 3×10^{-2} cm sec^{-1} is at best an upper bound for large-scale velocity fields in the core. Third, the geodynamo requires a vertical component of the velocity field in addition to the horizontal component estimated by the westward drift. A separate necessary condition for the vertical velocity field can be derived (Busse, 1975a), but unfortunately no reliable estimates for vertical velocities in the core are available.

Without motion in the core, the dipole component of the earth's magnetic field would decay exponentially with the characteristic time

$$\tau = r_o^2/\lambda\pi^2 \approx 2 \times 10^4 \text{ years} \qquad (1)$$

where r_o is the core radius and λ is the magnetic diffusivity of the core. Only the average of the velocity field over times of the order (1) is thus relevant for the generation of the magnetic field in the core. Oscillatory motions caused by internal waves in stably stratified regions of the core, as suggested by Bullard and Gubbins [1971], are not a likely cause of the dynamo process. On the contrary, the maintenance of a vertical motion over a considerable length of time requires that at least part of the core is well mixed and nearly adiabatic. It is this requirement which has given rise to the "core paradox" of Clark [1968], Higgins and Kennedy [1971], and Kennedy and Higgins [1973]. Subsequent studies have demonstrated that the paradox does not pose serious problems because of the presence of light elements such as sulphur or silicon in the outer core and because of the uncertainty of the thermodynamic parameters at high pressures. For a recent discussion of the subject, see Stacey [1977].

It is evident from the above discussion that the observations do not provide many constraints for the theory of the geodynamo nor has it been possible to derive much information from theoretical concepts about the physical properties of the core. This is typical for the early stages of a geophysical theory in which the contact between observational data and mathematical predictions is tenuous. As the theoretical models become more detailed, comparisons with a greater variety of data such as the geomagnetic secular variation became possible and provide further constraints for the models. This step-by-step approach will ultimately reveal the details of the geodynamo mechanism. The brief review of the theory given in the following will emphasize recent developments. From more detailed discussions, we refer to the book by Moffatt [1978] and recent reviews by Busse [1978], Gubbins [1974], Roberts and Soward [1972] and Soward and Roberts [1976].

The Magnetohydrodynamic Problem of the Geodynamo

The genertion of magnetic fields by fluid motions in an electrically conducting fluid can be regarded as an instability of the nonmagnetic flow. In the language of modern mathematics, the dynamo process is described as a bifurcation from the nonmagnetic solution of the problem. It is thus important to study first the problem without magnetic field. Because of its mathematical difficulties, this step has often been bypassed in the past and the attention has been focussed on the kinematic dynamo problem in which an arbitrary solenoidal velocity field is assumed.

But since widely differing velocity fields can give rise to the same magnetic field outside the conducting region, and because important parameters such as the amplitude of the magnetic field cannot be resolved by the kinematic dynamo problem, it has become obvious that the full magnetohydrodynamic dynamo problem must be attacked.

This problem is based on the equations of motion and the equation of magnetic induction. If, for simplicity, we assume thermal buoyancy as the driving force of motion in the core, we obtain as equations for the velocity vector $\underset{\sim}{v}$, for the deviation θ of the temperature field from its static solution T_s, and for the magnetic induction $\underset{\sim}{B}$

$$\frac{d}{dt}\underset{\sim}{v}+2\underset{\sim}{\Omega}\times v = -\nabla\pi-\theta\gamma g+\nu\nabla^2\underset{\sim}{v}+(\nabla\times B)\times B/\rho\mu \qquad (2a)$$

$$\nabla\cdot\underset{\sim}{v} = 0 \qquad (2b)$$

$$\frac{d}{dt}\theta + \underset{\sim}{v}\cdot\nabla T_s = \kappa\nabla^2\theta + q \qquad (2c)$$

$$\frac{d}{dt}\underset{\sim}{B} - \underset{\sim}{B}\cdot\nabla\underset{\sim}{v} = \lambda\nabla^2\underset{\sim}{B} \qquad (2d)$$

The Boussinesq approximation has been used in which the effects of compressibility have been neglected and in which all material properties are assumed to be constant except for the temperature dependence of the density ρ, which is taken into account solely in the gravity term. ν is the kinematic viscosity, γ is the coefficient of thermal expansion, κ is the thermal diffusivity, and λ is the magnetic diffusivity which is equal to the inverse of the product of the magnetic permeability μ and the electrical conductivity.

The Boussinesq approximation should be appropriate for the earth's core since the density scale height is considerably larger than the core radius. In applying equations (2), the temperature must be interpreted as the potential temperature, i.e., as the deviation of the actual temperature from an adiabatic reference state. The consideration of chemical bouyancy requires more complex equations, but the magnetohydrodynamic dynamo problem will be little affected since the energy providing forces are small in comparison with the Coriolis and Lorentz forces.

The analysis of the problem posed by equations (2) and appropriate boundary conditions proceeds in three steps. First, the problem without magnitic field is considered. Because of the nonlinear advection terms, this problem by itself is of considerable complexity, but its main features have been well understood and will be briefly described in the next section. In the second step, the linear dynamo equation (2d) is solved for the velocity field obtained in the first step. If the amplitude of the velocity field is sufficiently large such that dynamo action occurs, the strongest growing magnetic mode is selected. Otherwise, the parameters of the problem must be changed until growing magnetic fields are obtained. In the third step, the feedback provided by the Lorentz force is taken into account and the equilibrium is determined in which the velocity field is modified until the magnetic field obtained as solution of (2d) is stationary at least in the time averaged sense. Only perturbation results have been obtained for this third step (Busse, 1973, 1975); Soward, 1974), which are necessarily restricted to small amplitudes of the magnetic field. For planetary applications, the fully nonlinear problem must be considered. It is in this area that the most exciting discoveries must be expected in fiture research.

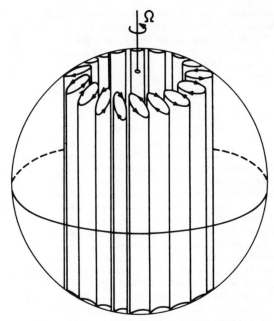

Fig. 1. Sketch of convection columns in a rotating, internally heated sphere for Rayleigh numbers close to the critical value.

Convection without Magnetic Field

The hydrodynamic problem of convection in the earth's core is described by equations (2a,b,c) without the Lorentz force term in the equation of motion. The dynamics of motions are governed in this case by the dominant Coriolis force, which can be balanced only by the pressure. A consequence of this geostrophic belance is the Proudman-Taylor theorem, which states that the motion becomes independent of the z-coordinate in the direction of the axis of rotation as the limit is approached in which viscous stresses and the time-dependence of the motion vanish. But the spherical boundary of the core prohibits z-independent motions unless the flow is purely azimuthal, in which case it could not be driven by buoyancy. Thus convective motion with a significant radial component must be slightly time dependent and the viscous stresses must reach a sufficient magnitude in order to support deviations from the Proudman-Taylor condition. The mathematical analysis of the problem (Busse, 1970) shows that the time-dependence is accomplished by a wavelike propagation of the convection columns in the prograde azimuthal direction, while viscous stresses become important owing to a small azimuthal length scale of the order $E^{1/3} r_o$, where

$$E \equiv \frac{\nu}{\Omega r_o^2} \qquad (3)$$

is the Ekman number which, in the earth's core, may be as small as 10^{-15}. A sketch of motions at the onset of convection in a uniformly heated sphere is shown in Figure 1. The flow in the form of columns conforms as much as possible to the Proudman-Taylor condition, while exhibiting a strong radial component, especially near the equatorial plane.

The small azimuthal length scale makes convection without magnetic field relatively inefficient in transporting heat. The increase in the characteristic scale of convection by the action of the Lorentz force is the basic physical cause of the generation of the substantial magnetic field. By balancing the Coriolis force, in particular the part that cannot be balanced by the pressure

gradient, the Lorentz force releases some of the constraint of rotation in the earth's core. But the Lorentz force is restricted by the dynamo process in several not yet well understood respects. One of them seems to be that the efficiency of the dynamo process drops rapidly when the scale of motion exceeds a certain fraction of the core radius according to the numerical kinematic models of Gubbins [1973] and Bullard and Gubbins [1977]. Thus, the actually realized length scale of convection in the earth's core is not known.

Since viscous forces are negligible in comparison with the Lorentz force in the earth's core, the actual value of the parameter E is of minor importance in numerical computations. In the numerical model of finite amplitude convection in rotating spheres developed by Cuong and Busse [1981] values as low as 10^{-4} have been used for E. The main purpose of these computations is to provide a physically realistic velocity field for the investigation of models of the geodynamo.

Magnetohydrodynamic Models of the Geodynamo

It is convenient to use the general representation for the magnetic field B in terms of poloidal and toroidal components,

$$\underset{\sim}{B} = \nabla \times (\nabla \times \underset{\sim}{r}h) + \nabla \times \underset{\sim}{r}g \qquad (4)$$

where r is the position vector with respect to the center of the core.

After introducing representation (4) into equation (2d), the r-component of the latter yields an equation for h, while the r-component of the curl of (2d) yields an equation for g. Here and in the following we use a spherical system of coordinates (r, θ, φ) with θ = 0 denoting the rotation of axis. It is useful to separate the poloidal potential h and the toroidal potential g into their axisymmetric and nonaxisymmetric parts

$$h = \bar{h} + \hat{h} , \qquad g = \bar{g} + \hat{g}$$

where the bar indicates the φ-average and $\bar{\hat{h}} \equiv 0$, $\hat{\bar{g}} \equiv 0$ by definition. Two basic dynamo mechanisms may be distringuished, depending on the role played by the axisymmetric part of the velocity field $\underline{v}$. A differential rotation can generate a mean toroidal field $\bar{g}$ from a mean poloidal field $\bar{h}$, but an axisymmetric velocity cannot generate a mean poloidal field by interacting with a toroidal field. The latter process must always be accomplished by the interaction of fluctuating magnetic and velocity fields which is also called the α-effect. If the mean poloidal and toroidal fields are both sustained against Ohmic dissipation by the α-effect, the dynamo is called an α^2-dynamo; if the toroidal field g is generated by a differential rotation, the dynamo is an αω-dynamo. There does not seem to exist the possibility of a purely toroidal or purely poloidal magnetic field generated by the dynamo process. But a mathematical proof of this hypothesis has not yet been given.

Another classification of dynamos in a sphere arises from the symmetry of the problem with respect to the equatorial plane. If the velocity field is symmetric with respect to this plane, as can be assumed in the case of convection, equation (2d) yields eigen solutions which are either symmetric or antisymmetric. In the latter case, the poloidal field h describes a field that is dipolar at large distances from the core. This class of fields is therefore called dipolar. In the symmetric case the quadrupolar component of the poloidal field h dominated at large distances and solutions of this kind are called quadrupolar. For reasons that have been illuminated in a recent paper by Proctor [1977], the critical magnetic Reynolds numbers required for both types of fields differ only slightly and meridional circulations seem to be necessary for providing a clear preference for one parity or the other. From the general point of view, it is actually surprising that all planetary fields discovered so far exhibit a dipolar character.

An important unknown quantity is the strength of the toroidal magnetic field in the earth's core. Since the mantle of the earth is essentially an insulator, the toroidal field cannot penetrate the core-mantle boundary and no direct information about its strength is available. Estimates about the vertical gradient of the toroidal field at the core-mantle boundary may eventually become available from methods such as that of Benton and Muth [1979]. But at the present time, the question cannot be decided whether a toroidal field of several hundred Gauss exists in the core as has

been assumed traditionally, or whether the toroidal field strength exceeds at best a few times the strength of about 10 G of the poloidal field, as suggested by the model of Busse [1975b]. This model relies on the α^2-dynamo effect of thermal convection, while a strong toroidal field

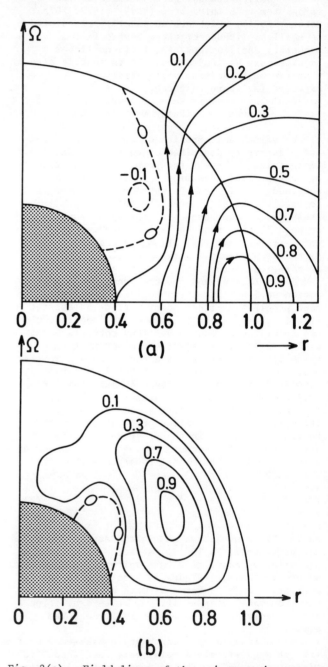

Fig. 2(a). Field lines of the axisymmetric component of the poloidal magnetic field generated by convection in a rotating spherical shell with insulating exterior (b) Lines of constant strength of the corresponding axisymmetric component of the toroidal field. (after Cuong and Busse [1891]).

requires an αω–dynamo. Considerable advances towards a complete magnetohydrohynamic model of the latter kind have recently been made by Braginsky [1975, 1978], and it should be possible in the near future to establish whether a αω–dynamo in the earth's core is feasible.

The numerical solutions of Cuong and Busse [1981] for convection in a rotating spherical shell have recently been used to obtain solutions of the magnetohydrodynamic dynamo problem in that geometry. For low amplitudes of convection and large values of the magnetic Prandt number ν/λ a quadrupolar solution represents the preferred magnetic field, while the dipolar solution becomes preferred as the amplitude of convection is increased with an associated decrease of the parameter ν/λ. A plot of the axisymmetric component of the magnetic field in the dipolar case is shown in Figure 2. As the magnetic field amplitude is increased, the Rayleigh number of convection increases, indicating a stable equilibrium balance. But these results are preliminary and extensions to higher magnetic field strengths are likely to exhibit new phenomena.

Concluding Remarks

Maps of the vertical intensity of the nondipole component of the geomagnetic field have often been compared with weather maps, and there are indeed dynamic similarities between the geomagnetic secular variation and the movement of weather systems. But it would be a mistake to deduce from this similarity that the earth's core is as complex a system as the atmosphere. The relevant parameter, the magnetic Reynolds number, is small compared to typical Reynolds numbers of atmospheric motions, and the kinetic energy of flows in the core is minute compared to the kinetic energy of the atmosphere. The fact that all planetary magnetic moments measured so far are aligned within 15° with the axis of rotation indicates that planetary dynamos possess a fairly regular structure. There are thus good reasons to expect that the mechanism of the geodynamo can be understood in detail despite that fact that direct observational evidence of many properties of the earth's core will never be available.

The discovery of the magnetic fields of other planets and the realization that the cores of the major planets are similar to the earth's core, but simpler in their physical properties, has given rise to the hope that all planetary dynamos can be described by a single model with only minor changes of the parameters. The apparent validity of a simple scaling law (Busse, 1976) tends to confirm this expectation. While the different planets provide indeed good constraints for the parameter dependence of the magnetic field strength, detailed information on important dynamic aspects such as reversals of the polarity is available only in the case of the earth. A magnetohydrodynamic model of the geodynamo capable of descri-

bing reversal events appears to be the most challenging problem of the next decade.

References

Acheson, D.J., On the hydromagnetic stability of a rotating fluid annulus, J. Fluid Mech., 52, 529–541, 1972.

Backus, G.E., A class of self-sustaining dissipative spherical dynamos, Ann. Phys., 4, 372–447, 1958.

Benton, E.R., and L.A. Muth, Estimates of meridional electric current density and vertical growth rate of zonal magnetic field at the top of Earth's outer core, Phys. Earth. Planet. Int., in press, 1979.

Braginsky, S.I., An almost axially symmetrical model of the hydromagnetic dynamo of the earth I, Geomag. Aeron., 15, 122–127, 1975.

Braginsky, S.I., Nearly axially symmetric model of the hydromagnetic dynamo of the earth, Geomagn. Aeron., 18, 225–231, 1978.

Bullard, E.C., The magnetic field within the earth, Proc. Roy. Soc. London, Ser. A, 197, 433–453, 1949.

Bullard, E.C., C. Freedman, H. Gellman, and J. Nixon, The westward drift of the earth's magnetic field, Phil. Trans. Roy. Soc. London, Ser. A, 243, 67–92, 1950.

Bullard, E.C., and D. Gubbins, Geomagnetic dynamos in the stable core, Nature, 232, 548–549, 1971.

Bullard, E.C., and D. Gubbins, Generation of magnetic fields by fluid motions of global scale, Geophys. Astrophys. Fluid Dyn., 8, 43–56, 1977.

Busse, F.H., Thermal instabilities in rapidly rotating systems, J. Fluid Mech., 44, 441–460, 1970.

Busse, F.H., Generation of magnetic fields by convection, J. Fluid Mech., 57, 529–544, 1973.

Busse, F.H., A necessary condition for the geodynamo, J. Geophys. Res., 80, 278–280, 1975a.

Busse, F.H., A model of the geodynamo, Geophys. J. Roy. Astron. Soc., 42, 437–459, 1975b.

Busse, F.H., Generation of planetary magnetism by convection, Phys. Earth Planet. Int., 12, 350–358, 1976.

Busse, F.H., Magnetohydrodynamics of the Earth's dynamo, Ann. Rev. Fluid Mech., 10, 435–462, 1978.

Childress, S., Théorie magnétohydrodynamique de l'effet dynamo, Report from Department Méchanique de la Faculté des Sciences, Paris, 1969.

Clark, S.P., Temperature gradients in the earth's core [abstract], Eos Trans. AGU, 49, 284, 1968.

Cuong, P.G., and F.H. Busse, Thermal convection and generation of magnetic fields in rotating spherical shells, Phys. Earth Planet. Int.,in pr.

Elsasser, W.M., Induction effects in terrestrial magnetism, Phys. Rev., 69, 106–116, 1946.

Gubbins, D., Numerical solutions of the kinematic dynamo problem, Phil. Trans. Roy. Soc. London, Ser. A, 274, 493–521, 1973.

Gubbins, D., Theories of the geomagnetic and solar dynamos, Rev. Geophys. Space Phys., 12, 137–154, 1974.

Herzenberg, A., Geomagnetic dynamos, Phil. Trans. Roy. Soc. London, Ser. A, 250, 543-585, 1958.

Higgins, G.H., and G.C. Kennedy, The adiabatic gradient and the melting point gradient in the core of the earth, J. Geophys. Res., 76, 1870-1878, 1971.

Kennedy, G.C., and G.H. Higgins, The core paradox, J. Geophys. Res., 78, 900-904, 1973.

Loper, D.E., Torque balance and energy budget for the precessionally driven dynamo, Phys. Earth Planet. Int., 11, 43-60, 1975.

Malkus, W.V.R., Precession of the earth as the cause of geomagnetism, Science, 160, 259-264, 1968.

Moffatt, H.K., Magnetic Field Generation in Electrically Conducting Fluids, Cambridge University Press, 1978.

Proctor, M.R.E., The role of mean circulation in parity selection by planetary magnetic fields, Geophys. Astrophys. Fluid Dyn., 8, 311-324, 1977.

Rikitake, T., Electromagnetism and the Earth's Interior. Elsevier Pub. Co., Amsterdam, 1966.

Roberts, P.H., and A.M. Soward, Magnetohydrodynamics of the earth's core, Ann. Rev. Fluid. Mech., 4, 117-154, 1972.

Rochester, M.G., J.A. Jacobs, D.E. Smylie, and K.F. Chong, Can precession power the geomagnetic dynamo?, Geophys. J. Roy. Astron. Soc., 43, 661-678, 1975.

Soward, A.M., A convection driven dynamo, I, The weak field case, Phil. Trans. Roy. Soc. London, Ser. A, 275, 611-645, 1974.

Soward, A.M., and P.H. Roberts, Recent developments in dynamo theory, Magnitnaya Gidrodinamika, 12, 3-51, 1976. English translation: Magnetohydrodynamics, 12, 1-36, 1977.

Stacey, F.D., A thermal model of the earth, Phys. Earth Planet. Int., 15, 341-348, 1977.

RADIOGENIC ISOTOPES AND CRUSTAL EVOLUTION

Donald J. DePaolo

Department of Earth and Space Sciences, University of California
Los Angeles, California 90024

Abstract. Recent studies of the geochemical
evolution of the continental crust using radiogenic isotopes of Sr, Nd and Pb have clarified
many aspects of the origin and history of the
crust. The times of formation of crustal rocks
from the mantle can be determined with relative
ease using Nd isotopes, which should allow a
more accurate description of the age and growth
history of the continents to be constructed.
Comparison with the moon suggests that the
absence of congenital crust on the earth is due
either to efficient recycling of the crust during
the first billion years or to a cataclysmic
destruction of the crust about 3.8-4.0 Æ ago.
The formation of new crust has involved a
considerable amount of reconstitution of preexisting crust in the Phanerozoic, but in contrast,
Archean crust appears to have been almost purely
mantle-derived. There is increasing evidence
of gross chemical layering in the crust which
forms within a few hundred million years after
the formation of new crust from the mantle.
Uncertainty about the composition and extent of
the lower crust leaves the overall composition
of the crust as still uncertain. Nd isotope
data, material balance considerations and transport models suggest that the crust has been
derived from only about 25% of the mass of mantle,
leaving this region severely depleted in many
elements, while the remainder of the mantle may
have retained a primitive composition. Future
isotopic studies will exploit the contrasting
geochemical behaviors of the Sm-Nd, Rb-Sr and
U-Th-Pb systems, which provide varied perspectives
on crustal evolution.

This paper focuses upon recent advances in the
application of long-lived natural radioactivities
to basic problems of the age, composition and
evolution of the earth's continental crust, and
will attempt to enumerate major unsolved problems
and the prospects for their solution in future
investigations. Many of the problems have been
recognized earlier (consult Wasserburg, 1966 for
an earlier discussion of geochronologic data
bearing on crustal evolution), but improvements
in analytical techniques (Papanastassiou and
Wasserburg, 1969), the use of previously
unexploited parent-daughter systems (Lugmair et
al., 1975; DePaolo and Wasserburg, 1976a), and
a tremendous increase in the amount of quality
major and trace element data to characterize
crustal rocks, have brought the solutions to
some problems within reach and show promise of
clarifying many others. In addition, the approach
to these problems has been catholicized over the
last decade as a result of the opportunity to
observe the crust of the moon, and in less detail,
those of Mars, Mercury and Venus, so that the
possible relationships between the size and
chemistry of planetary bodies and crustal history
has been more deeply appreciated.

Some Fundamental Questions

Whereas the age and remarkable regularity of
the oceanic crust are well documented, the age
of the continental crust is more subtle and
complex problem. It can be broken down into two
questions. First, has the mass of the crust
grown through time gradually or was it formed
mostly in a single event or in a short time
period (at or near the time of formation of the
earth)? Second, is the formation of continental
crust an irreversible process or does crustal
material have a finite mean life, with recycling
of crust back into the mantle? A second fundamental question concerns the composition of the
continental crust. Is the crust layered and, if
so, how does the composition vary with depth?
When did the layering form? Has the composition
of the crust changed through time? Related
questions apply to the composition, age, and
structure of the subcontinental mantle. If the
crust has been spawned from the mantle, then
crustal history must also be reflected in the
evolution of the mantle. A particular question
is whether the crust has been gleaned uniformly
from the entire interior of the earth, or instead,
has been derived from only, for instance, the
upper mantle, leaving the deeper mantle relatively
or virtually undifferentiated. Answers to these
questions could constrain many aspects of the
earth's thermal and degassing history, and may
also relate to the rheology of the mantle, and

the validity of the extrapolation of current tectonic patterns backward in time.

The application of radioactive isotopes to these problems is based upon a relatively simple principle. In any reservoir, the rate of accumulation of radiogenic daughter isotope is proportional to the parent/daughter ratio (P/D), so the abundance of the radiogenic isotope is an integrator of the P/D. In many cases P/D is closely related to other chemical parameters and therefore can be used to discriminate between major planetary-scale reservoirs (such as crust and mantle). The radiogenic isotope abundances in rocks are therefore diagnostic of their origin as well as their age, and are indicators of the petrogenetic processes involved in their formation. Of the parent-daughter systems currently utilized (U-Th-Pb, Rb-Sr, Sm-Nd), the behavior of Sm/Nd in petrogenetic processes is probably the best understood, Rb/Sr behavior is less well understood, and (U,Th)/Pb is least understood. For instance, it is quite clear that the continental crust is enriched in Sm and Nd relative to the original mantle abundances and that Nd has been preferentially enriched relative to Sm by a moderate amount, less than a factor of two. It is also known that magmas formed by melting of the mantle have roughly the same Sm-Nd characteristics as the continental crust. Hence, the zeroeth order interpretation of the crust as a low-melting fraction extracted from the mantle has definite and predictable consequences for the Sm/Nd ratio and the evolution of $^{143}Nd/^{144}Nd$ in the crust and mantle during earth history, which depend upon the details of the history of the formation of the crust. The situation is similar for Rb-Sr. The crust is known to be relatively more enriched in Rb than Sr, but whether by a factor of two or ten relative to the mantle is not at all clear. For (U,Th)/Pb it is not even certain whether the crust is relatively more enriched in U and Th or in Pb, although it is clearly highly enriched in all three elements. Background for the interpretation of Sm/Nd variations comes from numerous studies of rare-earth element variations, both theoretical and experimental (Haskin and Paster, 1979). Less extensive studies exist for Rb/Sr and very few exist for (U,Th)/Pb. However, experimental studies on the detailed behavior of these latter elements during magma genesis are now being undertaken (Tatsumoto et al., 1978; Benjamin et al., 1978).

The Age of the Crust

The first problem encountered when considering the age of the continental crust is the definition of crustal "age" itself. Here we will assume that the crust represents a differentiation product of the mantle and will define the age of a segment of crust as the time at which the material was extracted from the mantle. If the segment contains materials which were extracted

from the mantle at different times and then mixed together, we define the age as the mean age of the constituent parts, i.e., the sum of the ages of the different sub-segments weighted by their fractional masses.

Resolution of many questions regarding the age of the crust appears to be possible using the remarkably systematic variations of initial $^{143}Nd/^{144}Nd$ in crustal rocks (DePaolo and Wasserburg, 1979a). The basic model is illustrated in Figure 1. The growth of $^{143}Nd/^{144}Nd$ in the (mantle) source of crustal rocks, due to the decay of ^{147}Sm to ^{143}Nd with a half-life of 106 Æ, is described to a good approximation by the curve CHUR (CHondritic Uniform Reservoir). The rate of growth of $^{143}Nd/^{144}Nd$ in crustal rocks is much smaller than in the mantle (by a factor of ∿2), due to lower Sm/Nd in the crust. Thus the $^{143}Nd/^{144}Nd(\tau)$ vector of crustal rocks diverges from the CHUR curve at the time of their formation, and provides a marker which preserves information on the time of crust formation.

The Sm-Nd system is the only isotopic decay scheme where a mantle evolution curve is sufficiently well defined to allow this type of crustal age determination on a large range of crustal

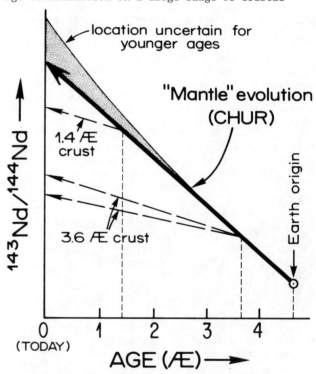

Fig. 1. Growth of $^{143}Nd/^{144}Nd$ with time in the mantle and in crustal segments derived from the mantle at 3.6 Æ and 1.4 Æ. The CHUR curve is well defined for T>1.5 Æ, but considerable uncertainty remains for more recent times. This model may allow crustal material to be "dated" in an unambiguous manner.

rock types. The "age" of a rock sample can be determined by measuring the present-day $^{143}Nd/^{144}Nd$ in the rock and its Sm/Nd. The latter gives the slope of the evolution curve of the rock (long dashed lines in Fig. 1) which can be extrapolated backward in time from the present $^{143}Nd/^{144}Nd$ until it intersects the CHUR curve. The time corresponding to the point of intersection is the model "crust formation age," referred to as T_{CHUR}. For mixtures of crust of different ages, the T_{CHUR} age obtained will not reflect the mean age weighted by mass, but rather the age weighted by the fraction of Nd atoms contributed by each sub-segment. Therefore, some ambiguity is present, but this may not be a major problem since rare-earth concentrations in crustal rocks, especially if averaged over sizeable volumes, are not highly variable. Another possibly important qualification regards the location of the mantle evolution curve at relatively recent times, as shown in Figure 1. Extraction of low-Sm/Nd crust must cause the Sm/Nd of the mantle to increase with time, hence its evolution curve should diverge from CHUR as shown. This problem is further complicated by the obvious evidence for isotopic heterogeneity in the present-day mantle. Nevertheless, the model appears to be fundamentally correct, and these problems may be only small perturbations. Furthermore, the model can be used to monitor the fate of crustal materials during erosion-sedimentation because clastic sedimentary rocks preserve information on the age of their source rocks, due to the virtual absence of Sm/Nd fractionation during sedimentation and diagenesis. In addition, only minor perturbations may occur during metamorphism. "Crust formation ages" can therefore be assigned to crustal segments regardless of whether they are dominantly igneous, sedimentary or metamorphic (DePaolo and Wasserburg, 1976b; McCulloch and Wasserburg, 1978).

The evidence to data clearly shows that continental crust formation has been a semi-continuous process from about 3.8 AE ago until the present. However, the distribution of ages within this time interval is still not known in detail. It has been suggested recently on the basis of Nd studies that a particularly large fraction of the crust was formed about 2.7 AE ago (McCulloch and Wasserburg, 1978). In contrast, Rb-Sr studies summarized a decade ago by Hurley and Rand (1969) suggested that the largest fraction of crust was formed since 2 AE ago and that the rate of crust production has increased through time. Existing geochronological data also indicate that a large amount of crust appears to be approximately 1.6-1.8 AE old. It appears that the mean age of the crust could be guessed at approximately 1.9 AE (or half of 3.8 AE), but this number could be off by as much as 0.5 AE. A more satisfactory solution to this problem may be possible using the Sm-Nd systematics, but insufficient data exist as yet.

The past ten years has resulted in the discovery

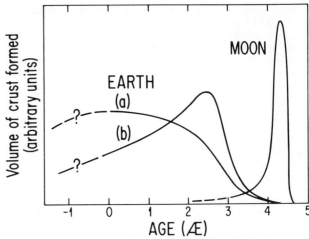

Fig. 2. Simplified schematic comparison of the age distribution of rocks of the lunar crust and the earth's continental crust. The curve for the moon is probably fairly accurate (cf. Wasserburg et al., 1977), but the earth curves shown are conjectural. The primary difference between the two bodies is the abundance of crust older than 4 Æ on the moon as compared to the apparent complete absence of such old crust on the earth. The lack of congenital crust on the earth is likely the result of efficient crust-destruction processes, a consequence of rapid convection during early earth history. The history of the moon can be thought of as grossly similar to the earth except that the crust-forming era on the moon was telescoped down into a narrow time window due to the reduced amount of gravitational energy available to fuel differentiation. The question posed by the two possible earth curves (a) and (b) is whether the earth has not yet reached--or is past--the peak of net crust production.

of rocks which are as old as 3.6-3.8 AE, and, interestingly, has also produced a strong conviction that crust older than 3.8 Æ does not exist. This conclusion has been based both upon Rb-Sr and Sm-Nd data (Moorbath, 1975; O'Nions et al., 1979), and raises the intriguing question of why there exists no record of the time interval from the earth's formation 4.55 AE ago until 3.8 AE ago. Possible answers are that no crust was produced during that time or that any crust which was being formed was also being continuously and efficiently destroyed and remixed with mantle. No resolution yet exists for this problem, but based on a comparison with the moon, the latter alternative appears to be more likely.

Comparison Between Earth and Moon

The crusts of the earth and moon can be compared in two ways--directly, by comparison of the apparent radiometric ages of the crustal rocks, and indirectly, by comparison of the isotopic evolution in the respective mantles. Lunar highland rocks have been dated by Rb-Sr to be as old as 4.5 Æ (Papanastassiou and Wasserburg, 1976), essentially identical to the age of the moon. Studies of the ubiquitous and uniform trace-element-rich KREEP rocks as well as other highland rock types suggest that the bulk of lunar crust formation may have extended over a period of a few hundred million years beginning at 4.5 Æ (Wasserburg et al., 1976; Carlson and Lugmair, 1979). If the mare basalt volcanism is included, crust formation continued at a reduced rate until 3.0 Æ ago or perhaps even to somewhat more recent times (Geiss et al., 1977; Head, 1976). Using this information, a schematic graph of volume versus age for the lunar crust can be constructed (cf. Wasserburg et al., 1976).

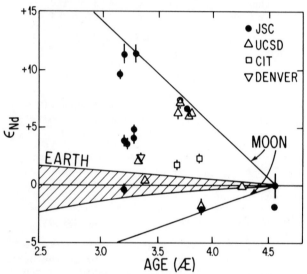

Fig. 3. A comparison of Nd evolution in the mantles of the earth and moon. The data points represent initial $^{143}Nd/^{144}Nd$ ratios of lunar mare basalts plotted against their crystallization age. ε_{Nd} is the deviation of these points from the chondritic evolution curve (CHUR) shown in Figure 1, in units of 10^{-4}. The ruled region gives the field of analogous data measured on terrestrial rocks. The data show that the lunar mantle became highly heterogeneous very early and that the layering persisted until at least 3.2 Æ ago and probably to the present. The mainly positive ε_{Nd} values are consistent with extraction of the low-Sm/Nd crust. The terrestrial data indicate an unfractionated, homogeneous early mantle, consistent with the lack of crust and efficient stirring of the mantle during this time. Lunar data are from a summary by Nyquist et al. (1979). Terrestrial data were summarized by DePaolo and Wasserburg (1979a).

Shown in Figure 2 is this curve and possible comparison curves for the earth. Curve (a) for the earth is adapted from the work of Hurley and Rand (1969), and curve (b) depicts the 2.7 Æ peak suggested by McCulloch and Wasserburg (1978). Regardless of the exact shape of the earth curve, or the possibility of a more irregular curve reflecting episodicity, it is clear that there is a drastic difference in crust-forming histories of the two bodies.

The contrast between the isotopic evolution in the mantles is shown in Figure 3. In the first half of earth history the initial Nd isotope ratios of mantle-derived rocks fall close to the chondritic evolution curve ($\varepsilon_{Nd}=0$). In contrast, the ε_{Nd} of mare basalts scatter, and some lie far away from $\varepsilon_{Nd}=0$, mostly in the positive direction. The lunar data show that high Sm/Nd reservoirs, probably complementary to the low-Sm/Nd lunar crust, were formed near the time of formation of the moon at about the same time as the lunar crust was being formed. The wide scatter in the lunar data further shows that the lunar mantle was highly heterogeneous. The terrestrial data show no evidence of fractionated reservoirs, consistent with the lack of any significant volume of crust. The narrow range of values suggests that either the earth's mantle was homogeneous at the time of its formation or that it was well mixed as a result of rapid convection.

Since much of the energy required to produce magmatic activity may be gravitational energy produced during planetary accretion, it is expected that the earth, due to its much larger mass, would have much more energy available for early differentiation than would the moon. This difference would be accentuated by early rapid core formation in the earth, since the moon has a much smaller dense core, if any (cf. Taylor, 1979). Since differentiated crust was produced on the moon very early, it seems likely that crust would also have formed on the earth soon after its formation. This argument, therefore, suggests that the lack of early crust is due to some process that either continuously destroyed crust as it was produced or destroyed all existing crust in a cataclysmic event about 3.8 Æ ago. This time, incidentally, coincides with the end (or sharp diminution) of the heavy meteorite bombardment recorded by the lunar surface, although there exists no other substantive evidence linking the two events. Since some early differentiation of the earth's mantle seems inescapable, the apparent mantle uniformity implied by Fig. 3 suggests the existence of some efficient stirring process which prevented fractionated reservoirs from surviving for any length of time. This same process may be that responsible for the destruction of the crust. Thus, a fluid, rapidly convecting early mantle may explain both the uniformity implied by the Nd isotope data and the lack of early crust, and would be consistent with the abundant energy available for melting in the earth as compared to the moon. However, the mechanisms for the remixing of low-density crust into the mantle by convection are far from obvious.

Arguments involving comparisons with the moon could be weakened if the moon had some potent early heat source--such as short-lived radio-nuclides (Lee, 1979; Runcorn, 1977)--which the earth did not have. This would not be expected unless the moon were accreted before the earth, although the time difference required would not be large for the case of ^{27}Al ($\tau_{1/2}=10^6$ yr.). On the other hand, if the lack of early terrestrial crust is interpreted to mean that there was no crust produced, it would imply that nearly 1 Æ passed before the earth became hot enough to begin melting. If this were true, it would require major revision of current models of early earth history (Hanks and Anderson, 1974; Ringwood, 1975).

Sources of Continental Crust

In determining the age of the continental crust it is necessary to determine the extent to which preexisting crust contributes to the ostensibly "new" continental crust formed in orogenic zones. Recent studies suggest that in the Proterozoic-Phanerozoic eras, preexisting crust may contribute a substantial fraction of the total volume of crust produced, although most of the material is newly added material derived from the mantle. Evidence for this is shown in Figures 4a and 4b. Figure 4a shows Nd and Sr isotopic data on granitic rocks from the Sierra Nevada and Peninsular Ranges batholiths (DePaolo, 1980b). These data suggest that the batholiths represent a mixture of mantle-derived materials, probably magmas, and preexisting crustal materials, probably sediments or meta-sediments comprising, or eroded from, the border-ing ∿1.8 Æ old craton. This interpretation is consistent with the extensive oxygen and strontium isotope data of Taylor and Silver (1978) shown in Fig. 4b, measured on rocks from the Peninsular Ranges of Baja California, Mexico. It has been estimated (DePaolo, 1981) that the crust within the geographic borders of the batholiths is composed of a 1:1 mix of 1.8 Æ old craton and sediments derived therefrom and Mesozoic magmas derived from the mantle. Consequently, the true "crust-formation age" of the batholiths would be about 1 Æ, as compared to the average radiometric or crystallization age of about 0.1 Æ. Such interpretations and conclusions depend upon a general understanding of the isotopic systematics of Sr, Nd and O, and could be subject to revision. Nevertheless, it is clear that these considerations are important in the construction of curves such as shown in Figure 2, and, therefore, in the understanding of crustal growth and its relationship to the thermal and tectonic history of the earth.

The Nd-Sr data shown in Figure 4a also indi-cate that the mantle source of continental crust, as represented by the batholiths, was previously depleted in elements which are enriched in the

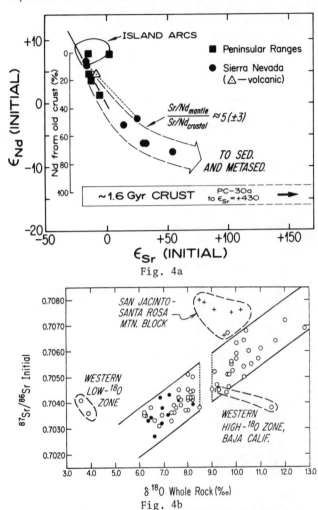

Fig. 4a

Fig. 4b

Fig. 4. (a) ε_{Nd}, ε_{Sr} data measured on gran-itic rocks from the Sierra Nevada and Penin-sular Ranges batholiths. The batholith rocks, which are taken to represent a new segment of continental crust formed in the Mesozoic, appear to be comprised of a mix-ture of an island-arc component derived from the mantle, and preexisting 1.6 Æ old crustal sediments or metasediments. Figure is taken from DePaolo (1980b).
(b) Oxygen and strontium isotopic data on Peninsular Ranges rocks, taken from Taylor and Silver (1978). The main correlation of $\delta^{18}O$ and $^{87}Sr/^{86}Sr$ is consistent with mixing of mantle and crustal components. The western part of the batholith shows evidence for exchange with hydrothermal fluids, also evident in the Sr-Nd data of Fig. 4a. The San Jacinto-Santa Rosa region gives evidence for a second type of crustal component, with lower $\delta^{18}O$.

continental crust, especially light rare-earth elements, Rb and K. This depletion may have occurred during previous episodes of crust formation, and suggests that crust may be repeatedly produced from the same limited volume of the mantle, which has become gradually depleted in these elements through earth history. The Nd and Sr isotopic composition of island arc volcanics, which may also represent continent building blocks, is also consistent with this interpretation. This secular change in the composition of the mantle may imply sympathetic shifts in the chemical composition of the crust.

In contrast to the Proterozoic-Phanerozoic, Archean crust appears to have been derived in most cases almost entirely from the mantle. Evidence for substantial contributions by older crust is rare (Moorbath, 1975; DePaolo and Wasserburg, 1976b). This may be a result of the presumably smaller amount of continental crust present at that time or possibly to a different process of crust formation.

The Composition of the Crust

The overall composition of the continental crust is still uncertain due to lack of sufficient knowledge of lower crust chemistry. Recently discovered Nd and Sr isotope systematics can be interpreted as suggesting that the crust has a substantially lower Rb/Sr (less than 0.10) than previously estimated (DePaolo, 1979). This would imply lower U, Th, K abundances in the crust, and would require that the "lower crust" be nearly devoid of these elements. However, this conclusion is based on a fairly simple model. In general, evidence for chemical layering in the crust with respect to minor elements is strong (Heier, 1965; Zartman and Wasserburg, 1969). The mechanisms and timing of formation of this layering are still poorly understood, as is the relationship of minor elements to major elements. Some evidence suggests that layering forms concurrently with crust formation (McCulloch and Wasserburg, 1978). Hamilton et al. (1979) interpret the Sm-Nd and Rb-Sr systematics of the 2.9 Æ old Lewisian gneisses to indicate that intra-crustal differentiation follows about 250 million years after formation of the crustal material from the mantle. This differentiation results in the upward concentration of alkalies, U, Th, and other large-ionic-radius elements leaving a granulite lower crust highly depleted in these elements. This process produces an upper crust with high Rb/Sr and a lower crust with low Rb/Sr. However, both upper and lower crust retain the same Sm/Nd. Similar data on younger rocks do not yet exist. The possibility of age "layering" in the crust which could be caused by magmatic underplating of existing crust has not really been addressed even though it would be critical for assessing the continental freeboard arguments as advanced by Wise (1974). The systematic Nd isotope variations in crustal rocks may allow

this to be explored through the use of T_{CHUR} ages as diagrammed in Figure 1, which could be determined for entire crustal sections in favorable localities.

A general model for the structure of the crust in terms of Sr and Nd isotopes (Figure 5) has been proposed by DePaolo and Wasserburg (1979b) and revised by DePaolo (1981). This model incorporates the observations that i) most crustal rocks were formed with initial $\varepsilon_{Nd} \geq 0$ and $\varepsilon_{Sr} \leq 0$, ii) most crustal rocks have Sm/Nd about 40% lower than estimated for the mantle ($f_{Sm/Nd} \approx -0.4$), regardless of whether they reside in the "upper" or "lower" crust, and iii) the crust is grossly layered into a high-Rb/Sr upper crust and a low-Rb/Sr lower crust. When these considerations are combined with the trend of ε_{Nd} and ε_{Sr} found in young oceanic basalts, shown by the shaded field in Fig. 5, a generalized blueprint emerges for the distribution of the <u>present</u> <u>values</u> of ε_{Nd} and ε_{Sr} in mantle and crustal reservoirs. This model may be an important aid in deciphering the age and compositional structure of the crust

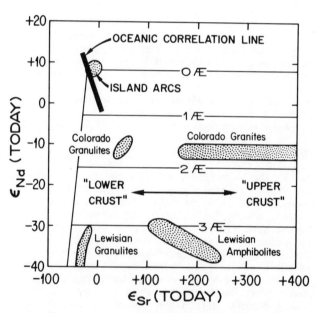

Fig. 5. Model for the distribution of present ε_{Nd} and ε_{Sr} values in mantle and crustal rocks (DePaolo, 1981a). It is assumed that crustal rocks have $\varepsilon_{Nd} \geq 0$ and $\varepsilon_{Sr} \leq 0$ when formed from the mantle and that Sm/Nd is roughly constant in crustal rocks. Rb/Sr is high in the upper crust and low in the lower crust. Rocks of a given age should lie in a horizontal zone with lower crustal rocks to the left and upper crustal rocks to the right. Examples are shown from the Precambrian of Colorado (DePaolo, 1981b) and the Lewisian of Scotland (Hamilton et al, 1978). This model may be useful in resolving questions related to the age and structure of the continental crust.

as well as the interactions between mantle-derived magmas and the crust.

Subcontinental Mantle

A problem which is intimately related to crustal evolution is the composition of the subcontinental mantle. Pb isotope data on young basalts suggest that age/composition provinces exist in the subcontinental mantle, and that the ages and boundaries of these provinces are similar to those of superjacent crustal provinces (Zartman, 1974). Sr and Nd data also indicate provinciality in the mantle (Leeman and Manton, 1971; DePaolo, 1978). Peculiarly, there is little evidence to indicate that the subcontinental mantle is depleted in those elements enriched in the crust (DePaolo and Wasserburg, 1976b; Leeman and Manton, 1971). Rather, studies of mantle-derived ultramafic xenoliths show that in many areas it appears to be enriched in comparison to oceanic mantle (Frey and Green, 1974; Basu and Tatsumoto, 1979).

Mantle-Crust Material Balance
and Crustal Recycling

Transport of material between the mantle and crust can be treated in terms of simple two-box models in order to evaluate the growth and evolution of the crust and its effect on the isotopic composition of the mantle. For the Sm-Nd system it can be assumed that the total crust + mantle has chondritic Sm/Nd. The $^{143}Nd/^{144}Nd$ evolution in the mantle can therefore be calculated if the age and Sm, Nd abundances of the crust can be estimated. The results for the most simple model are shown in Figure 6, which shows the evolution of ϵ_{Nd} in crust and mantle if the entire crust were extracted from the mantle in a single event either 1.5 or 4.5 AE ago. The solid arrows correspond to the case where the crust is uniformly extracted from the entire mass of the mantle. The dashed curve corresponds to the crust extracted from only the upper mantle, to about 400 km depth. For this calculation it was assumed that the Nd and Sm abundances in the original mantle were 1.5 time chondritic, and the abundances in the crust were 37.5 and 24 time chondritic, respectively, for Nd and Sm.

The present-day ϵ_{Nd} values of mid-ocean ridge basalts (MORB) average about +10. As shown in Fig. 5, if the entire mantle were uniformly depleted due to crust formation, a maximum value of $\epsilon_{Nd} \approx +5$ would result if the crust were 4.5 AE old. This is an absolute upper limit, since the crust clearly appears to be close to 1.5 AE old, as discussed above, which would give $\epsilon_{Nd} = +1.5$ in the mantle. The conclusion which results is that the upper mantle, as represented isotopically by MORB's, cannot be representative of the entire mantle and in fact must comprise only a small fraction of the mantle. The dashed arrows show that if the crust were extracted only from

the upper 400 km of the mantle, then extremely large ϵ_{Nd} effects would result, larger than those measured in oceanic basalts.

These conclusions can be generalized for arbitrary crustal growth histories and possible recycling of crust into mantle (Jacobsen and Wasserburg, 1979; DePaolo, 1980). It is found that in order to produce a mantle which is sufficiently fractionated with respect to Sm/Nd to account for the isotopic composition of the present oceanic mantle, it is necessary to limit the mass of mantle which interacts with the crust to about 25% of that of the whole mantle (DePaolo, 1978, 1980). The remaining 75% of the mantle would, therefore, have to be still essentially undifferentiated with respect to litho-

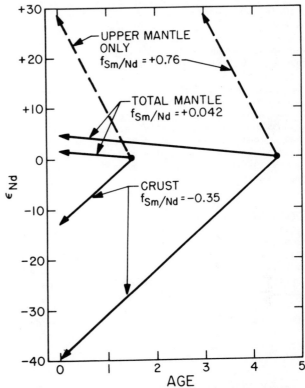

Fig. 6. The evolution of $\epsilon_{Nd}(T)$ in the crust and mantle if the crust were extracted from the mantle in a single event at 4.5 AE ago or 1.5 AE ago. The dashed lines give $\epsilon_{Nd}(T)$ in the upper mantle (to 400 km depth) if the crust were derived wholly from the upper mantle, leaving the lower mantle undifferentiated. Mid-ocean ridge basalts which have typically $\epsilon_{Nd}=+10$ and plot at T=0, clearly cannot be representative of the total mantle, regardless of the age of the crust. This material balance therefore suggests that the mantle must be layered. A depleted layer makes up about 25% of the mantle and is the source of MOR basalts, and the remainder of the mantle has a more primitive composition.

phile elements. This conclusion is consistent with the presence of primordial ^{3}He in some young basalts, which also indicates that primordial reservoirs may exist in the mantle (Craig and Lupton, 1976), but could be difficult to reconcile with whole-mantle-scale convection cells. These conclusions rest upon the assumption that the crust (to the Moho) is the sole long-term repository of fractionated (LREE-, K-, Rb-, U-, Th-enriched) material in the earth. If, as suggested above, the subcontinental mantle were also enriched to some extent in these elements and isolated from the rest of the mantle, a larger proportion of the mantle may be differentiated. Integration of these data with geophysical models for mantle convection will be of primary importance in furthering the understanding of the structure and evolution of the mantle.

This simple model can explain many aspects of the Sm-Nd and Rb-Sr data. For U-Pb and Th-Pb, however, the situation may be more complex. In particular, the low Th/U ratios of MORB's (Tatsumoto, 1978) are troublesome in that there is no known complementary reservoir with high Th/U. A possible, but non-unique, explanation is that U is preferentially returned to the mantle, perhaps in subduction zones, whereas Th may behave geochemically more like the rare-earths and therefore may be recycled much less efficiently. However, this would imply that present estimates of Th/U in either the crust or the total earth are substantially in error. A satisfactory solution to this problem may require more detailed knowledge of the relative behavior of U, Th and rare-earths in the weathering-sedimentation cycle (cf. Faure, 1977) and better estimates of the rate at which crustal material is returned to the mantle in subduction zones. This general problem of the extent to which crust is destroyed and returned to the mantle is poorly constrained by existing isotopic data.

Future Emphasis

Future isotopic studies will take advantage of the multiple available parent-daughter sytems U-Th-Pb, Rb-Sr, Sm-Nd and more recently Lu-Hf (Patchett and Tatsumoto, 1980). The usefulness of this approach lies in the radically different geochemical properties of these elements, which makes each system sensitive to different petrogenetic processes. For instance, the U-Pb budget of the earth may be dominated by near-surface processes, since the crust represents a large Pb reservoir and both U and Pb are mobile in the near-surface environment. In contrast, the mantle is the dominant repository of Sr and Nd in the earth, but while the crust is enriched in Rb relative to Sr, it is depleted in Sm relative to Nd. A correlation of ^{143}Nd/^{144}Nd and ^{87}Sr/^{86}Sr has been found in oceanic basalts, which suggests that these Sm-Nd

and Rb-Sr systems may have had a grossly similar evolutionary history in the mantle. No simple correlation involving Pb isotopes has been sufficiently documented, and the development of comprehensive models encompassing all three systems remains as a formidable task.

Several other isotopic systems could be useful in the study of the evolution of the crust, but have not yet been exploited to any great extent. Of particular interest is ^{176}Lu → ^{176}Hf. The refractory-lithophile nature of these elements suggests that this system may exhibit systematics similar to those found for Sm-Nd. Recent pioneering work by Patchett and Tatsumoto (1980) confirms this and demonstrates that Lu-Hf will yield yet another valuable perspective on crustal evolution. Use of ^{187}Re → ^{187}Os appears to be further in the future. ^{40}K → ^{40}Ca, ^{138}La → ^{138}Ba, ^{138}La → ^{138}Ce are presently tractable (Russell et al., 1978) and may be useful geochronologic tools in some cases. Implementation of these tools may require interdisciplinary approaches, perhaps involving the development of new instrumentation.

Acknowledgments. The present state of knowledge of crustal evolution as deduced from isotopic studies is a result of extensive diligent work carried out in a large number of laboratories throughout the world. This short review cannot do justice to all who have contributed, and the fundamentally personal nature of the emphasis and ideas should be recognized. The author acknowledges the support of the National Science Foundation (grant EAR78-12966).

References

Armstrong, R.L., Taubeneck, W.H. and Hales, P.O., Rb-Sr and K-Ar geochronometry of Mesozoic granitic rocks and their Sr isotopic composition, Oregon, Washington, and Idaho, Geol. Soc. Am. Bull. 88, 397-411, 1977.

Basu, A.R. and Tatsumoto, M., Samarium-Neodymium systematics in kimberlites and in the minerals of garnet lherzolite inclusions, Science 205, 398-401, 1979.

Benjamin, T., Heuser, W.R., and Burnett, D.S., Laboratory studies of actinide partitioning relevant to ^{244}Pu chronometry, Proc. Lunar Planetary Sci. Conf. 9th, 1393-1406, 1978.

Carlson, R.W. and Lugmair, G.W., Sm-Nd constraints on early lunar differentiation and the evolution of KREEP, Earth Planet Sci. Lett. 45, 123-132, 1979.

Craig, H. and Lupton, J.E., Primordial neon, helium and hydrogen in oceanic basalts, Earth Planet. Sci. Lett. 31, 369-385, 1976.

DePaolo, D.D., Study of magma sources, mantle structure and the differentiation of the earth using variations of ^{143}Nd/^{144}Nd in igneous rocks, Ph.D. Thesis, California Institute of Technology, 360 p., 1978.

DePaolo, D.J., Implications of correlated Nd and Sr isotopic variations for the chemical evolution of the crust and mantle, Earth Planet. Sci. Lett. 43, 201–211, 1979.

DePaolo, D.J., Crustal growth and mantle evolution: inferences from models of element transport and Nd and Sr isotopes, Geochim. Cosmochim. Acta 44, 1185–1196, 1980a.

DePaolo, D.J., The sources of continental crust: Nd isotope evidence from the Sierra Nevada and Peninsular Ranges, Science 209, 684–687, 1980b.

DePaolo, D.J. and Wasserburg, G.J., Nd isotopic variations and petrogenetic models, Geophys. Res. Lett. 3, 249–252, 1976a.

DePaolo, D.J. and Wasserburg, G.J., Inferences about magma sources and mantle structure from variations of $^{143}Nd/^{144}Nd$, Geophys. Res. Lett. 3, 743–746, 1976b.

DePaolo, D.J. and Wasserburg, G.J., Sm–Nd age of the Stillwater complex and the mantle evolution curve for Neodymium, Geochim. Cosmochim. Acta 43, 999–1008, 1979a.

DePaolo, D.J. and Wasserburg, G.J., Petrogenetic mixing models and Nd–Sr isotopic patterns, Geochim. Cosmochim. Acta 43, 615–627, 1979b.

Doe, B.R. and Zartman, R.E., Plumbotectonics, in Barnes, H.L., ed., Geochemistry of Hydrothermal Ore Deposits, Vol. II, Wiley and Sons, 1979.

Faure, G., Isotope Geology, John Wiley and Sons, Inc., 464 p., Chapter 16, 1977.

Frey, F.A. and Green, D.H., The mineralogy, geochemistry and origin of lherzolite inclusions in Victorian basanites, Geochim. Cosmochim. Acta 38, 1023–1059, 1974

Geiss, J., Eberhardt, P., Grogler, N., Guggisberg, S., Maurer, P. and Stettler, A., Absolute time scale of lunar mare formation and filling, Phil. Trans. Roy. Soc. Lond. 285A, 151–158, 1977.

Hamilton, P.J., Evensen, N.M., O'Nions, R.K. and Tarney, J., Sm–Nd systematics of Lewisian gneisses: implications for the origin of granulites, Nature 277, 25–28, 1979.

Hanks, T.C. and Anderson, D.L., The early thermal history of the earth, Phys. Earth Planet. Interiors 2, 19–29, 1969.

Haskin, L.A. and Paster, T.P., Geochemistry and mineralogy of the rare earths, in Gschneidner, K.A. and Eyring, L., eds., Handbook on the Physics and Chemistry of Rare Earths, North Holland Publishing Company, 1979.

Head, J.W., Lunar volcanism in space and time, Rev. Geophys. Space Phys. 14, 265–300, 1976.

Heier, K.S., Metamorphism and the chemical differentiation of the crust, Geol. Fören. Stockh. Förh. 87, 249–256, 1965.

Hurley, P.M. and Rand, Jr., R., Pre-drift continental nucleii, Science 164, 1229–1242, 1969.

Kistler, R.W. and Peterman, Z.E., Variations in Sr, Rb, K, Na, and initial $^{87}Sr/^{86}Sr$ in Mesozoic granitic rocks and intruded wall-rocks in central California, Geol. Soc. Am. Bull. 84, 3489–3512, 1973.

Lee, T., New isotopic clues to solar system formation, Rev. Geophys. Space Phys. 17, 1447–1473, 1979.

Leeman, W.P. and Manton, W.I., Strontium isotopic composition of basaltic lavas from the Snake River plain, Southern Idaho, Earth Planet. Sci. Lett. 11, 420–434, 1971.

Lugmair, G.W., Scheinin, N.B. and Marti, K., Sm–Nd age of Apollo 17 basalt 75075: Evidence for early differentiation of the lunar exterior, Proc. Lunar Sci. Conf. 6th, 1419–1429, 1975.

McCulloch, M.T. and Wasserburg, G.J., Sm–Nd and Rb–Sr chronology of continental crust formation, Science 200, 1003–1011, 1978.

Moorbath, S., The geological significance of early Precambrian rocks, Proc. Geol. Assoc. Lond. 86, 259–279, 1975.

Nyquist, L.E., Shih, C.-Y., Wooden, J.L., Bansal, B.M., and Wiesman, H., The Sr and Nd isotopic record of Apollo 12 basalts: Implications for lunar geochemical evolution, Proc. Lunar and Planetary Sci. Conf. 10th, 77 – 114, 1979.

Papanastassiou, D.A. and Wasserburg, G.J., Rb–Sr age of troctolite 76535, Proc. Lunar Sci. Conf. 7th, 2035–2054, 1976.

Patchett, P.J. and Tatsumoto, M., Hafnium isotope variations in oceanic basalts, Geophys. Res. Lett. 7, 1077–1080, 1980.

Ringwood, A.E., Composition and petrology of the earth's mantle, McGraw-Hill, Inc., 618 p., 1975.

Runcorn, S.K., Early melting of the moon, Proc. Lunar Sci. Conf. 8th, 463–469, 1977.

Russell, W.A. and Papanastassiou, D.A., and Tombrello, T.A., Ca isotope fractionation on the Earth and other solar system materials, Geochim. Cosmochim. Acta 42, 1075–1090, 1978.

Tatsumoto, M., Isotopic composition of lead in oceanic basalt and its implication to mantle evolution, Earth Planet. Sci. Lett. 38, 63–87, 1978.

Taylor, S.R., Structure and evolution of the moon, Nature 281, 105–110, 1979.

Taylor, H.P. and Silver, L.T., Oxygen isotope relationships in plutonic igneous rocks of the peninsular ranges batholith, Southern and Baja California, U.S. Geol. Survey Open-File Report 78-701, 423–426, 1978.

Wasserburg, G.J., Geochronology and isotopic data bearing on the development of the continental crust, in Advances in Earth Science, M.I.T. Press, 1966.

Wasserburg, G.J., Papanastassiou, D.A., Nenow, E.V. and Bauman, C.A., A programmable magnetic field mass spectrometer with on-line data processing, Rev. Sci. Instr. 40, 288–295, 1969.

Wasserburg, G.J., Papanastassiou, D.A., Tera, F., and Huneke, J.C., Outline of a lunar chronology, Phil. Trans. Roy. Soc. Lond. A285, 7–22, 1977.

Zartman, R.E., Lead isotopic provinces in the Cordillera of the western United States and their geologic significance, Econ. Geol. 69, 792–805, 1974.

Zartman, R.E. and Wasserburg, G.J., The isotopic composition of lead in potassium feldspars from some 1.0 b.y. old North American igneous rocks, Geochim. Cosmochim. Acta 33, 901–942, 1969.

THE CARBON-OXYGEN SYSTEM AT PRESENT AND IN THE PRECAMBRIAN

Erich Dimroth

Sciences de la Terre, Université du Québec à Chicoutimi, 930 est, rue Jacques-Cartier, Chicoutimi, Qué. Canada G7H 2B1.

Abstract. The geochemical cycle of carbon and oxygen can be considered to be a complex feedback system; rates of transfer from one reservoir to the other in general are functions of reservoir levels. Oxygen and organic matter are produced and consumed in the biologic subsystem. Small fractions of the oxygen and organic matter are withdrawn from the biologic subsystem and enter the exoxenic subsystem where they are used in various redox reactions and from where they return to the biologic subsystem in form of CO_2. A small fraction of oxygen and organic matter entering the exogenic subsystem is subduced and, thus, is withdrawn to the mantle. It reappears in form of mantle-derived CO_2 exhaled by volcanos.

Preserved reservoirs of oxidized species, of organic matter, and reduced species provide the evidence for the state of the carbon-oxygen system in the distant past. Large exogenic reservoirs of oxidized species (Fe^{3+}, SO_4^{2-}) existed since Archean time. The oldest well documented red beds are 2750 m.y. old. The oldest sulfate evaporites have an age of 3300 m.y. Oxidized, palagonitized volcanic glass, and ferric oxide crusts on basalt pillows exist since at least 2750 m.y. and must have constituted a huge reservoir of oxygen. Reservoirs of carbon, and of reduced species (Fe^{2+}, S^{2-}) also are similar to present reservoirs, neither excessively large, nor particularly small.

It is not likely that the feedback relations in the carbon-oxygen system changed with time. Consequently, the sedimentary evidence suggests that the state of the carbon-oxygen system has remained stable since the Archean.

Introduction

The geochemical cycle of carbon and oxygen consists of several subcycles, namely a biologic, an exogenic and an endogenic subcycle (Fig. 1). In the biologic subcycle, oxygen and organic matter are produced from CO_2 by plant photosynthesis and are consumed by organic decay, according to reaction (1):

$$CO_2 + H_2O \xrightleftharpoons[\text{decay}]{\text{production}} CH_2O + O_2 \qquad (1)$$

Fluxes in the biologic subcycle are very high and the residence time of organic matter in the biomass does not exceed a few years.

A very small part of the organic matter and oxygen produced, presently less than 1 per cent, is withdrawn from the biologic subcycle and enters the exogenic subcycle. Oxygen is consumed by oxidation of carbon, and of reduced species (Fe^{2+}, S^{2-}) during weathering and early diagenesis. Organic matter is sedimented and parts of it reacts with oxidized species (Fe^{3+}, SO_4^{2-}) during diagenesis. Carbon and oxygen eventually return to the biologic subcycle in form of CO_2 produced by diagenetic reduction of Fe^{3+} and SO_4^{2-} and by oxidation during weathering of carbon contained in older sediments.

A small part of the carbon and oxygen entering the exogenic system is withdrawn again, by deep burial and subduction into the mantle. These are the sinks of Fig. 1. Equivalent amounts of CO_2 and of reduced species (Fe^{2+}, S^{2-}) derived from the deeper crust and mantle are added to the exogenic and biologic subsystem by weathering of igneous rocks and in form of volcanic gases.

The geochemical cycle of carbon and oxygen forms a complex feedback system, the components of which are in dynamic interaction. Fluxes within the system and within its subsystems are related to reservoir levels by feedback. Unfortunately, little is presently known about the feedback relations, particularly within the exogenic and endogenic subsystems. The three know feedback relations are shown in Fig. 1, using the notation of Forrester (1968): (1) rates of oxidation during organic decay, weathering and diagenesis depend on ambient oxygen pressures; thus the rate of oxygen consumption decreases with decreasing levels in the oxygen reservoir. (2) Rates of biologic productivity increases with increasing level of the CO_2 reservoir. (3) Phosphorus is liberated during organic decay and weathering; it serves as fertilizer and organic productivity increases with increasing levels of the P_2O_5 - reservoir.

Sedimentary Cycles of Sulfur and Iron

The sedimentary cycles of sulfur and iron are closely linked to the carbon-oxygen cycle (Fig. 2).

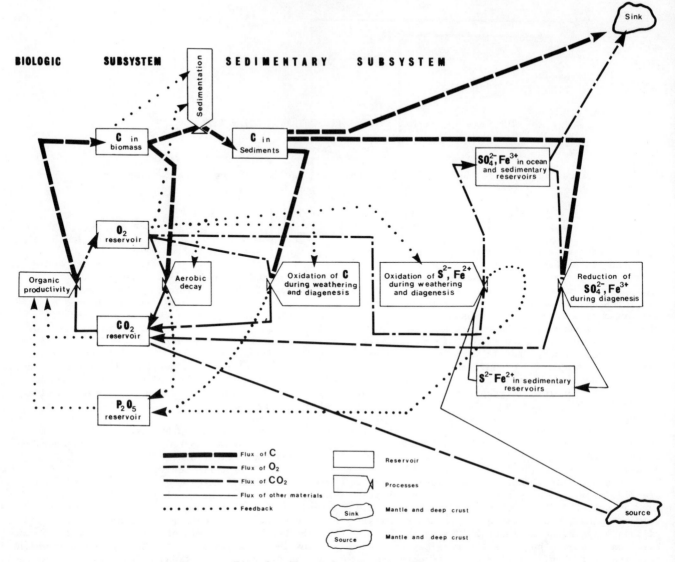

Fig. 1.   The carbon-oxygen system.

Feedback relations in this system are basically unknown except for (1) control of oxidation rates by ambient oxygen pressure and (2) control of rates of sulfate reduction by availability of organic matter and sulfate. Fig. 2 is largely self-explanatory, however some comments on the sedimentary recycling of sulfides are in place.

Sulfides are not instantly oxidized. Presently, detrital pyrite is known from fluvioglacial sands where it has survived about 10,000 years. It is known from the alluvium of Moselle river in Germany (Muller and Negendank, 1974). Sulfides placers are known from Chile (Clemmey, 1978). Thus, pyrite is not automatically destroyed during weathering and can survive transport for certain distances. Its preservation potential then depends on the rate of burial of the sediment. Detrital py-

rite can survive transport and early diagenesis, if erosion rates are high, the distance of transport fairly small, if burial rates are high, and if the pyrite is buried together with some organic matter.

Other Processes Producing Organic Matter

In addition to plant photosynthesis (reaction 1), there exist several other processes producing organic matter. The most important of these is bacterial photosynthesis according to reaction (2);

$$CO_2 + 2H_2S \longrightarrow CH_2O + H_2O + 2S \quad (2)$$

Furthermore, several types of bacteria can utilize chemical energy (for example oxidation of H_2S or Fe^{2+}) for the synthesis of organic matter.

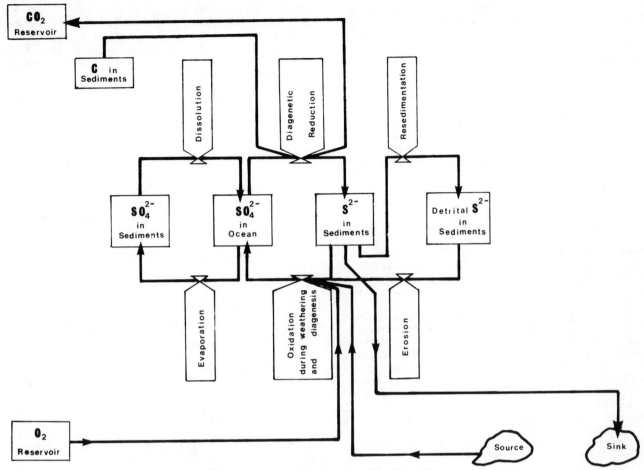

Fig. 2. The exogenic cycle of sulfur.

Such processes play a very minor role in the present carbon-oxygen cylce and, therefore have been deleted from Fig. 1. Bacterial photosynthesis can take place only in eutrophic waters, where the photic zone is polluted by H_2S. Chemosynthetic bacteria live close to their sources of energy, in particular close to hydrothermal springs, supplying H_2S and/or Fe^{2+}.

Geological Evidence

The preserved parts of the reservoirs of carbon, and of oxidized (Fe^{3+}, $SO_4{}^{2-}$) and reduced (Fe^{2+}, S^{2-}) species in Precambrian exogenic rocks provide the evidence for the antiquity of the carbon-oxygen system. The following pages are partly briefed from Dimroth and Kimberley (1976) to whom the reader is referred for more extensive discussion and for references to work published before 1976. The evidence indicates that the carbon-oxygen system has been essentially stable since at least 2700 m.y. and probably since at least 3300 m.y.

There is no evidence that bacterial photosynthesis or chemosynthesis produced a significant part of organic matter during the last 2700 (or probably 3300) m.y. Bacterial photosynthesis on a major scale would be possible only if the whole ocean, up to the photic zone, were polluted by H_2S. This would have major sedimentological consequences: all marine sediments, including those of shallows marine evironments, would then be euxinic; furthermore, all sea-floor weathering would then take place at reducing conditions, in the presence of H_2S. This has demonstrably not been the case.

The contribution of chemosynthesis also must have been small. Chemosynthetic organisms are present mainly in the discharge area of hydrothermal brines. Such discharge areas are well known from the 2750 m.y. old Archean sequence at Rouyn-Noranda, Québec, Canada. Organic remains generally are absent from these discharge areas and from their immediate vicinity.

Carbon Reservoirs

Carbon contents of present sediments are mainly controlled by four factors: (1) organic productivity in the larger environment, (2) grain size of the sediment, (3) sedimentation rate and (4) depth of deposition. Organic productivity in the larger environment determines the amount of organic matter available for sedimentation. Most organic detritus is fine-grained and has low density and, thus, accumulates with the clay fraction. Diffusion of oxygen into the sediment is inhibited by high rates of sedimentation and, consequently a larger part of deposited organic matter survives early diagenesis in fine-grained sediment. The ocean has an oxygen minimum at between 500 and 1500 meters depth and decay rates are low in this depth zone.

There are several exceptions to these rules: (1) Permeable organogenic rocks (for example stromatolitic limestones) generally contain little organic matter due to high rates of decay. (2) Oil reservoirs and their equivalents (tar sands, bituminous sandstones) are sediments into which organic matter hs been introduced by secondary processes. (3) Caustobioliths form under exceptional conditions particularly in coastal swamps and lacustrine swamps.

So far as presently known there appears to be no difference of carbon distribution in Precambrian and Recent rocks. Average carbon contents are comparable, and carbon is concentrated in clay-rich rocks. A correlation of sedimentary environment and carbon content also is apparent and suggests a correlation of organic sedimentation rate and productivity. It is not possible to verify the relation of carbon content to depth of deposition and sedimentation rate in ancient sediments. One of the three exceptions also can be verified: Precambrian stromatolites generally are low (<0.2 per cent) in carbon; this is true also for the Archean stromatolites known from several localities in Canada and Zimbabwe.

Reservoirs of Oxidized Species

Soils, the products of sea-floor weathering, red beds, and sulfate evaporites are the principal reservoirs of oxidized species. Oxidized soils (spodsols) are now known to underlie parts of the 2300 m.y. old Huronian sequence in Canada (Gay and Grandstaff, 1980). Archean soils have not yet been described. Dimroth and Lichtblau (1978, 1979) compared the products of Archean and Cenozoic sea-floor weathering: Archean (age 2750 m.y.) basalts and andesites were strongly palagonitized, and the palagonite is strongly oxidized; ferric oxide crusts formed during prolonged exposure of basalts to sea-floor weathering; iron, manganese, and potash are enriched in the parts of pillows exposed to oxidative alteration during sea-floor weathering (Dimroth and Rocheleau, 1979).

Detailed descriptions of Precambrian red beds have been presented by Dimroth (1978) and by Shegelski (1980). Detrital grains in red arkoses and sandstones are coated by hematite. Mafic minerals

have been partly dissolved, and have been replaced by hematite. Ratios of Fe_2O_3/FeO are much increased over the source rock (Shegelski, 1980). The red beds described by Dimroth are older than 1850 m.y. (Dressler, 1975), the red beds of Shegelski (1980) have a minimum age of 2700 m.y.

Silicified sulfate evaporites have now been described from the Archean Onverwacht Group in South Africa (age 3300 m.y.) and from the Archean of the Pilbara block (age 3500 m.y.). Archean cherts from the Onverwacht Group contain sabkha cycles (Lowe and Knauth, 1977, 1979). In the Archean sequence of the Pilbara block, sabkha evaporites have been replaced by silica and baryte (Barley et al., 1979; Dunlop, 1979). Typical relict textures of evaporites are preserved. Evidence for the presence of former evaporites exists in many Early Proterozoic sequences, for example in the iron formation of the Labrador trough (Dimroth, 1978) and in the Gunflint Formation (Markun and Randazzo, 1980).

Reservoirs of Reduced Species

I will only discuss the distribution of pyrite since Fe^{2+} is the predominant form of iron in all sedimentary rocks containing small amounts of organic matter. Just as to-day, most Precambrian pyrite is of biogenic origin, and occurs in carbonaceous shales. However, the volume of such pyritic carbonaceous shales is not very great and probably is not larger than in the Cenozoic or Paleozoic. Hydrothermal brines are the second source of sulfides; the Archean massive pyrite-Zn-Cu deposits are in all respects similar to the Miocene Kuroko deposits and to the Paleozoic pyrite-Cu deposits of southern Spain and Portugal.

There has been much speculation in the past, on the significance of the detrital pyrite in the Witwatersrand Supergroup and in the basal part of the Huronian Sequence (Roscoe, 1969, Pretorius, 1976). I have recently drawn attention to the fact that much of the pyrite in the Witwatersrand appears to be recycled diagenetic pyrite (Dimroth, 1979). This has been confirmed by further ore petrographic studies (Gochnauer, Dimroth and Guha, unpublished data) which confirmed that most of the detrital pyrite in the investigated samples is derived from biogenic pyrite and has been preserved in various stages of recrystallization. The investigated rocks of the Witwatersrand sequence also contain a substantial component of biogenic sulfide which formed in-situ (Dimroth, 1979; Gochnauer et al., unpublished data).

The presence of detrital pyrite should be interpreted in the light of data on transport and preservation of pyrite in the Recent. As noted above pyrite survives transport at certain conditions. Laid down at high sedimentation rates, together with organic matter, as in the Witwatersrand Supergroup, detrital pyrite does have a certain preservation potential.

Discussion

The geological evidence outlined above leaves no doubt that all present-day reservoirs in the carbon-oxygen cycle existed since at least 2700 m.y. and probably since at least 3500 m.y., that is since the oldest well preserved sedimentary and volcanic rocks formed. One must conclude, then, that the processes filling up those reservoirs took place; in other words, the geological evidence indicates that the present-day carbon-oxygen system has great antiquity. Presently available evidence leaves very little room for non-uniformitarian speculation. In particular, we can exclude that bacterial photosynthesis or chemosynthesis produced a significant portion of organic matter at any time in the last 2700 (3500 m.y.): the oxidation of volcanic rocks during sea-floor weathering and the essentially non-euxinic character of Archean sediments exclude that significant parts or the ocean were anoxic, which in an essential condition to large-scale bacterial photosynthesis.

Presently available evidence does not permit precise statements on the relative size of the reservoirs of carbon, and of oxidized (Fe^{2+}, S^{2-}) species. The sizes of various reservoirs depend in part on endogenic processes. For example, most red beds formed in epicratonic fault basins, and most evaporites in epicontinental inland seas. The absence of large continental areas, thus, make the Archean a period most unfavorable to the formation of red beds and of evaporites. On the other hand, interaction of volcanic rocks with sea water could have played a greater role in the Archean than to-day. Thus, the main exogenic reservoir of Fe^{3+} in the Archean may not have been red beds, but may have been volcanic rock, weathered at the sea-floor.

Absence of precise information on the relative sizes of reservoirs excludes precise statements on the relative rates of processes. However it is quite clear that the essential features of the feedback system were in place since the Archean. Oxidation may have taken place more at the bottom of the sea than at land; interaction of volcanic rocks and sea water may have played a greater role than at present. These, however, are relatively minor modifications of the state of the feed-back system described in Fig. 1, they are not basic modifications of the feed-back relations within the system.

Acknowledgements

I gratefully acknowledge many discussions on the theme of this paper with Dr. M.M. Kimberley and many other collegues, and with Dr. Gérard Woussen on the role of feedback systems in geology. Many of my students and assistants contributed to the factual basis presented in this paper. Marc Tremblay drafted the figures. This research has been supported by NSERC grant No. A 9145 and by successive G.S.C. research agreements.

References

Barley, M.E., Dunlop, J.S.R., Glover, J.E. and Groves, D.I., Sedimentary evidence for a distinctive Archean shallow-water volcanic-sedimentary facies, Eastern Pilbara Block, Western Australia. Earth Planetary Sci. Letters, 43, 74-84, 1979.

Clemmey, H., Implications of copper sulphide placers in Chile. J. Geol. Soc (London), 135, 5, 1978.

Dimroth, E., Labrador trough area. Quebec Dept. Natural Resources, Geol. Rept. 193, 398 p., 1978.

Dimroth, E., Significance of diagenesis for the origin of Witwatersrand-type uraniferous conglomerates. Phil. Trans. R. Soc. Lon., A291, 277-287, 1979.

Dimroth, E. and Lichtblau, A.P., Oxygen in the Archean ocean: comparison of ferric oxide crusts on Archean and Cenozoic pillow basalts. N. Jb. Min. Abh., 133, 1-22, 1979.

Dimroth, E. and Lichtblau, A.P., Metamorphic evolution of Archean hyaloclastites, Noranda, Quebec. Part I: comparison of Archean and Cenozoic sea-floor metamorphism. Can. J. Earth Sci., 16, 1315-1340, 1979.

Dimroth, E. and Rocheleau, M., Volcanology and sedimentology of Rouyn-Noranda area. Field trip A1, Geological Association of Canada. Quebec 1979, 204 p., 1979.

Dressler, B., Lamprophyres of the north-central Labrador trough, Quebec, Canada. N. Jb. Miner. Mh. 1975, 268-280, 1975.

Dunlop, J.S.R., The North Pole chert-barite deposit - a replaced Archean carbonate-evaporite sequence within the Pilbara Block of Western Australia. Geol. Assoc. Canada, Programs with Abstracts, 4, 48, 1979.

Forrester, J.W., Principles of Systems. Text and workbook. 2nd preliminary edition. Wright-Allen Press, Inc. Cambridge, Mass., 1968.

Gay, A.L. and Grandstaff, D.E., Precambrian paleosols at Elliot Lake. Precambrian Research, 12, 349-374, 1980.

Lowe, D.R. and Knauth, L.P., Sedimentology of the Onverwacht Group (3.4 billion years), Transvaal South Africa, and its bearing on the characteristics and evolution of the early earth. J. Geol., 85, 699-723, 1977.

Lowe, D.R. and Knauth, L.P., Petrology of Archean carbonate rocks in the Onverwacht and Fig Tree Groups, South Africa, Geol. Assoc. Can., Program with Abstracts, 4, 64, 1979.

Markun, C.D. and Randazzo, A.F., Sedimentary structures in the Gunflint Iron Formation, Schreiber Beach, Ontario, Precambrian Research, 12, 287-310, 1980.

Muller, M.J. and Negendank, J.F.W., Untersuchungen and Schwermineralien in Moseldedimenten. Geol. Rundschau, 63, 998-1034, 1974.

Pretorius, D.A., The nature of the Witwatersrand gold-uranium deposits. In K.H. Wolf (ed.) Handbook of strata-bound and stratiform ore deposits. 7, 29-88. Amsterdam, Elsevier, 1976.

Roscoe, S.M., Huronian rocks and uraniferous conglomerates in the Canadian Shield. Geol. Surv. Can., Pap. 68-40, 205, 1969.

Shegelski, R.J., Geologic evolution of the Shebandowan-Thunder Bay greenstone terrain. Precambrian Research, 12, 331-348, 1979.

STRONTIUM ISOTOPE COMPOSITION OF VOLCANIC ROCKS: EVIDENCE FOR CONTAMINATION OF THE KIRKPATRICK BASALT, ANTARCTICA

Gunter Faure

Department of Geology and Mineralogy and Institute of Polar Studies,
The Ohio State University,
Columbus, Ohio 43210

Abstract. Volcanic rocks have been used to derive information about the chemical composition and age of magma sources in the upper mantle. However, tholeiite basalt flows of Jurassic age in the Transantarctic Mountains (Kirkpatrick Basalt) have chemical and isotopic compositions that are not compatible with orthodox petrology and draw into question their reliability as a source of information about the upper mantle. The lava flows on Storm Peak and Mt. Falla in the Queen Alexandra Range are deficient in Al_2O_3, TiO_2, MgO and CaO compared to average tholeiite basalts and are enriched in SiO_2, FeO (total iron) and K_2O. In addition, they have high initial $^{87}Sr/^{86}Sr$ ratios (0.7115 $\pm$ 0.0002), high Rb/Sr ratios (0.46 $\pm$ 0.04), and low K/Rb ratios (231 $\pm$ 98). The concentrations of all major elements of the flows as well as their initial $^{87}Sr/^{86}Sr$ ratios vary systematically up section. The resulting correlations between chemical and isotopic parameters cannot be produced by fractional crystallization of magma, even if it was derived from anomalously old and Rb-rich sources in the mantle. However, the data are compatible with contamination of a normal mantle-derived magma with radiogenic ^{87}Sr and other chemical elements derived from the granitic gneisses of the continental crust that underlies the Transantarctic Mountains. If crustal contamination is a geologic fact here and elsewhere in the world, such volcanic rocks cannot be used to study the mantle.

Introduction

The study of volcanic rocks has contributed greatly to our knowledge regarding the mineralogical, chemical, and isotopic composition of the upper mantle where magma forms by partial melting. Such studies are generally based on the assumption that the chemical and isotopic composition of the resulting magma is directly related to the minerals that participated in the melting. For example, the isotopic composition of strontium of volcanic rocks is thought to be identical to that of the upper mantle whence the magma originated (Faure

and Powell, 1972). Another widely held view is that all specimens from a comagmatic suite of igneous rocks have the same $^{87}Sr/^{86}Sr$ ratio at the time of crystallization. These ideas, once held with great conviction, must now be revised in the light of greater understanding of the complexity of igneous petrogenesis.

The Kirkpatrick Basalt in the Transantarctic Mountains has anomalous chemical and isotopic properties that challenge orthodox concepts of petrogenesis. The relationships between anomalously high initial $^{87}Sr/^{86}Sr$ ratios and chemical composition, displayed by these flows at each of two localities in the Transantarctic Mountains, indicate that radiogenic ^{87}Sr and certain chemical elements, derived from the granitic rocks of the continental crust, have been added to the magma. Such interactions between mantle-derived magma and the rocks of the overlying crust compromise the reliability of the resulting volcanic rocks regarding information about the chemical composition of magma sources in the mantle and their chemical heterogeneity and age. In addition, such rocks cannot be dated by the Rb-Sr isochron method because they do not have the same initial $^{87}Sr/^{86}Sr$ ratio.

The Kirkpatrick Basalt

The continents of the southern hemisphere contain very large deposits of tholeiite basalt whose formation seems to have coincided with the breakup of Gondwana during the Mesozoic era. The largest of these basalt provinces are the Paraná Basin of Brazil, the Karroo Dolerites of South Africa, and the flows and sills of the Ferrar Supergroup of Antarctica which includes the Kirkpatrick Basalt. Measurements of initial $^{87}Sr/^{86}Sr$ ratios of volcanic rocks from the Paraná Basin have been reported by Compston et al. (1968) and by Halpern et al. (1974). Data for the Karroo basalts and dolerites have been published by Compston et al. (1968) and by Manton (1968). The dolerites of Tasmania (McDougall, 1962; Heier et al., 1965) probably belong to the Ferrar petrologic province of Antarctica.

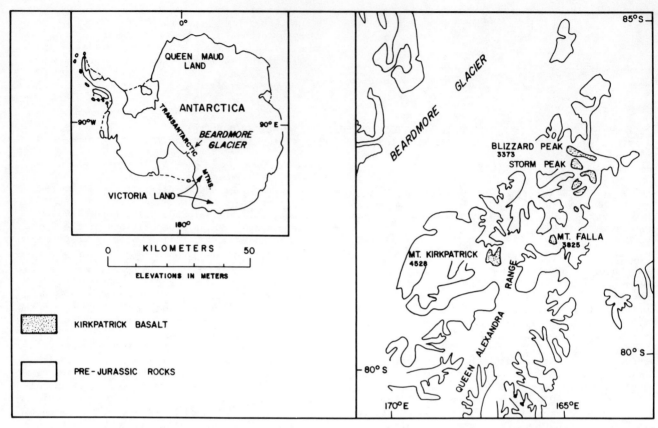

Fig. 1.   Location of Kirkpatrick Basalt in the Transantarctic Mountains.

The Ferrar Supergroup is recognized over a distance of 2500 kilometers along the strike of the Transantarctic Mountains (Figure 1) from Northern Victoria Land to the Dufek Massif of the Pensacola Mountains (Ford, 1976). It includes abundant dolerite sills and dikes that were intruded into the gently-dipping sedimentary rocks of the Beacon Supergroup and into the igneous and metamorphic rocks of the late Precambrian to early Paleozoic basement complex (Hamilton, 1965; Gunn, 1966). Lava flows of the Kirkpatrick Basalt occur primarily in the Mesa Range of Northern Victoria Land (Gair, 1967) and in the Queen Alexandra Range along the Beardmore Glacier (Elliot, 1972). The age of the rocks of the Ferrar Supergroup is lower Jurassic (about 170 million years) as indicated by isotopic dating (McDougall, 1963; Elliot, 1970; Fleck et al., 1977; Hall et al., 1979; Faure et al., 1981) and by paleontological evidence (Norris, 1965; Elliot and Tasch, 1967; Schaeffer, 1972). Basalt flows of similar age in Queen Maud Land (Kirwan Volcanics)differ from the Kirkpatrick Basalt in chemical composition and have low initial $^{87}Sr/^{86}Sr$ ratios similar to those of the Karroo dolerites (Faure and Elliot, 1971; Faure et al., 1972; Faure et al., 1979).

Two suites of specimens representing the flows of Kirkpatrick Basalt at two localities in the Queen Alexandra Range have been studied. The localities are Storm Peak (Faure et al., 1974) and Mt. Falla (Faure et al., 1981), separated by a distance of about 24 kilometers. On Storm Peak there are twelve flows with a total thickness of about 250 meters whereas Mt. Falla is capped by fourteen flows amounting to about 450 meters of section. The chemical compositions of the flows at both localities (Table 1) are similar but deviate from those of average tholeiites by having high concentrations of SiO_2, total iron as FeO, and K_2O. Their concentrations of TiO_2, Al_2O_3, MgO, and CaO are unusually low. The petrography of the flows on Storm Peak was described by Elliot (1970). The rocks are generally aphanitic with minor amounts of phenocrysts of iron-rich pyroxene and weakly zoned plagioclase having labradorite cores ($An_{70}-An_{65}$) and more sodic rims (An_{56}) (Kyle and Elliot, 1979). The matrix is composed of glass and devitrified glass that has partially recrystallized to a quartz-feldspar mesostasis.

In addition to having somewhat unusual chemical compositions, the flows on Storm Peak and Mt. Falla have anomalously high initial $^{87}Sr/^{86}Sr$, Rb/Sr and K/Rb ratios. The average values for all 26 flows at both localities are: Initial $^{87}Sr/^{86}Sr$ = 0.7115 $\pm$ 0.0002, Rb/Sr = 0.46 $\pm$ 0.04,

TABLE 1. Average Chemical Composition of Kirkpatrick Basalt on Storm Peak and Mt. Falla, Queen Alexandra Range in Weight Percent.

	Storm Peak[a]	Mt. Falla[b]	Average Tholeiite[c]
SiO_2	56.60	55.86 (14)	51.1
TiO_2	1.28	1.35 (2)	1.6
Al_2O_3	12.92	13.55 (10)	16.2
Fe_2O_3	4.50	5.00 (2)	3.1
FeO	7.52	6.69 (2)	7.6
FeO[d]	11.59	11.51 (14)	10.4
MnO	0.22	0.18 (2)	0.17
MgO	3.44	3.49 (14)	6.2
CaO	7.91	8.12 (2)	9.9
Na_2O	2.29	2.23 (14)	2.5
K_2O	1.29	1.42 (2)	0.7
P_2O_5	0.16	0.152(2)	0.22
H_2O+	1.62	1.22 (2)	0.7
H_2O-	–	0.46 (2)	–
CO_2	–	0.40 (2)	–
Total	99.75	100.12	99.99

[a] Average of twelve flows reported by Faure et. al. (1974).
[b] The numeral in brackets indicates the number of flows represented by each value as reported by Faure et al. (1981).
[c] Manson (1967).
[d] Total iron as FeO

K/Rb = 231 $\pm$ 98. All errors are one standard deviation of the mean. Similar values for these parameters were reported by Compston et al. (1968) for rocks of the Ferrar Supergroup. They pointed out that such values are typical of crustal rocks rather than of material derived from the upper mantle.

The high initial $^{87}Sr/^{86}Sr$ ratios, compared to basaltic rocks elsewhere (Faure and Powell, 1972; Faure, 1977), are explainable by means of three alternative petrogenetic models: 1) Magma originated from anomalous sources in the upper mantle that have maintained a high Rb/Sr ratio for a sufficient period of time to acquire high $^{87}Sr/^{86}Sr$ ratios; 2) Magma was generated by partial melting of rocks of Precambrian age at the base of the continental crust, and 3) Magma was generated in the upper mantle having "normal" chemical and isotopic properties and was subsequently contaminated by interaction with rocks of the continental crust. The anomalous-mantle hypothesis has been advocated by Brooks et al. (1976), Brooks and Hart (1978) and by Kyle (1980). The crustal contamination model was considered by Compston et al. (1968) and has been developed by Faure et al. (1972, 1974, 1981). The resolution of this question has important implications regarding

the apparent chemical and isotopic heterogeneity of the upper mantle and the ages of such heterogeneous regions as revealed by the so-called mantle isochrons of Brooks et al. (1976). This controversy also affects the concept of magmatic differentiation by fractional crystallization. If crustal contamination of mantle-derived magma is a geologic fact, then the chemical and isotopic composition of differentiated igneous rocks may be determined by both fractional crystallization and by interactions with the country rock. The evidence in favor of this view has been strengthened by the results of isotopic studies of strontium and neodymium in igneous rocks from many different geologic settings (Carter et al., 1978; Hawkesworth and Vollmer, 1979; DePaolo and Wasserburg, 1979).

The average chemical compositions of the flows on Storm Peak and Mt. Falla (Table 1) obscure the existence of important stratigraphic variations of the major oxides and the initial $^{87}Sr/^{86}Sr$ ratios. Figure 2 shows the systematic variations of SiO_2 and MgO in the flows at the two localities (Faure et al., 1974, 1981). In general, the initial $^{87}Sr/^{86}Sr$ ratios and the concentrations of SiO_2, total iron as FeO, Sr, and alkali metals

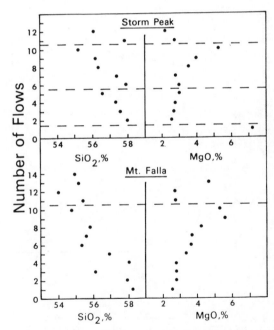

Fig. 2. Stratigraphic variation of the SiO_2 and MgO concentrations of successive flows of the Kirkpatrick Basalt on Storm Peak and Mt. Falla, Queen Alexandra Range (Faure et al., 1974, 1981). The flows at both localities can be subdivided into sets as suggested by the horizontal lines. Consecutive sets appear to repeat the pattern of variation displayed by all major elements.

decrease up section whereas TiO$_2$, Al$_2$O$_3$, MgO, CaO and MnO increase. The range of variation of the initial ^{87}Sr/^{86}Sr ratio is from 0.7094 to 0.7133 at Storm Peak and from 0.7100 to 0.7131 at Mt. Falla and amounts to a difference of greater than 0.003. The reproducibility of our measurements is ± 0.0003 based on replicate analyses. Therefore, the systematic variation of the initial ratio is real. Differences in the initial ^{87}Sr/^{86}Sr ratio cannot be the result of crystal fractionation in the magma chamber because the isotopes of strontium are not measurable fractionated by melting and crystallization of silicate minerals or by any other process in nature. Fractional crystallization does not explain the variation of the initial ^{87}Sr/^{86}Sr ratios even if the magma originated from an old and Rb-enriched region of the lithospheric mantle that was sufficiently old to acquire a high ^{87}Sr/^{86}Sr ratio.

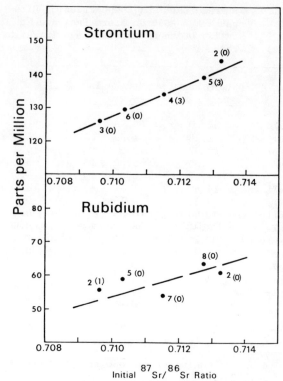

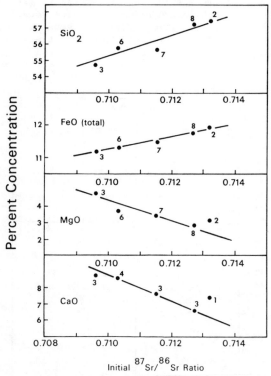

Fig. 3. Correlation of major oxide concentrations and initial ^{87}Sr/^{86}Sr of the Kirkpatrick Basalt flows on Storm Peak and Mt. Falla in the Queen Alexandra Range. The data for both localities were merged by averaging in increments of 0.0010 for the initial ^{87}Sr/^{86}Sr ratio. The numerals beside each data point indicate the number of flows represented by each. The data are from Faure et al. (1974) for Storm Peak and Faure et al. (1981) for Mt. Falla.

Fig. 4. Correlation between strontium and rubidium concentrations of the Kirkpatrick Basalt flows on Storm Peak and Mt. Falla, Queen Alexandra Range, and their initial ^{87}Sr/^{86}Sr ratios. The data were averaged in increments of 0.0010 for the initial ^{87}Sr/^{86}Sr ratio. The first numeral beside each point indicates the number of flows represented in the average, the second (in brackets) is the number omitted. The data are from Faure et al. (1974) and Faure et al. (1981).

The systematic stratigraphic variations of chemical and isotopic parameters discussed above are expressed as linear correlations between the initial ^{87}Sr/^{86}Sr ratios and concentrations of SiO$_2$, FeO, MgO and CaO in Figure 3 and with strontium and rubidium in Figure 4. These diagrams are based on the data from both Storm Peak and Mt. Falla which were averaged in increments of 0.001 for the initial ^{87}Sr/^{86}Sr ratio. The validity of these correlations is evident by inspection. These relationships are explainable by means of a mixing process between a basaltic magma that originated in the mantle and a "contaminant" derived from biotite-bearing granitic rocks of the continental crust (Faure et al., 1974; Faure et al. 1981). The contamination process raised the strontium content and ^{87}Sr/^{86}Sr ratio of the magma. In addition, the concentrations of SiO$_2$, total iron, alkali metals, and P$_2$O$_5$ increased whereas the concentrations of TiO_2, Al_2O_3, MgO, CaO and MnO <u>decreased</u>. The chemical composition of this hypothetical contaminant was estimated by Faure et al. (1974, 1981) by fitting a mixing equation to the Sr data in coordinates of the initial $^{87}Sr/^{86}Sr$ ratios and the reciprocals of the strontium concentrations. The Sr concentrations of the basalt magma and the contaminant were calculated from the mixing equation using reasonable estimates of their $^{87}Sr/^{86}Sr$ ratios. The concentrations of the other chemical elements in the end members were then obtained by extrapolating the linear data arrays between Sr and other elements. The resulting chemical compositions of the basalt magma and the contaminant are shown in Table 2.

The interpretation of the data in terms of crustal contamination indicates that the basalt magma was a tholeiite. The contaminant may have formed by partial melting of a biotite-bearing granitic rock. Partial fusion of such rocks is known to produce glass enriched in iron, alkali metals, and silica with an elevated $^{87}Sr/^{86}Sr$ ratio (Pushkar and Stoeser, 1975). The contaminant may be represented by the abundant glassy matrix that characterizes the flows on Storm Peak. Kyle and Elliot (1979) reported that the amount of glassy matrix correlates positively with the initial $^{87}Sr/^{86}Sr$ ratios. They also noted that a broad-beam microprobe analysis of the glass from one specimen of flow 7 on Storm Peak is similar to the composition of the contaminant.

Summary

The high $^{87}Sr/^{86}Sr$ ratios and the systematic variation of these ratios in sequentially extruded lava flows of the Kirkpatrick Basalt are

TABLE 2. Estimated chemical compositions of the basalt magma and the contaminant based on data from Storm Peak (Faure et al., 1974) and Mt. Falla (Faure et al., 1981), Queen Alexandra Range.

| | Basalt Magma | | Contaminant | |
	A	B	A	B
Sr, ppm	117	92	173	264
SiO_2, %	52.9	51.9	66.7	68.5
Al_2O_3, %	15.4	16.5	7.1	5.8
FeO, %	9.0	6.9	16.6	22.4
MgO, %	4.9	7.7	0	0
CaO, %	11.0	14.2?	1.2	?
Na_2O, %	1.7	1.7	4.5	3.5
K_2O, %	0.76	1.0	3.1	2.7

A = Storm Peak; B = Mt. Falla

explainable as a product of crustal contamination of basalt magma derived from the upper mantle. Direct derivation from old and Rb-rich mantle sources seems unlikely for two reasons: 1) The time-integrated Rb/Sr ratio of the mantle source would have to be greater than 0.066 which is unreasonably high; 2) Lava flows extruded sequentially through the same vent would have to be derived from different mantle sources having different Rb/Sr ratios or ages. In addition, the degree of melting or the composition of the sources would have to vary in order to generate the observed correlations between initial $^{87}Sr/^{86}Sr$ ratios and chemical composition. It seems unreasonable to demand this much order in the process of magma formation.

The contamination model leads to plausible conclusions regarding the origin and composition of the basalt and its contaminant which may be represented in part by the glassy matrix. However, several important questions regarding crustal contamination remain to be answered: 1) How was the contaminant formed and how did it mix with the basalt? 2) How could this process occur on such a vast scale? 3) How does the formation of tholeiite basalt relate to the break-up of Gondwana? Answers to these questions may eventually be found after crustal contamination of mantle-derived magmas is first accepted as a geologic fact. In that case, it will become apparent that we cannot study the mantle using rocks that have been contaminated.

<u>Note Added in Proof</u>. Following my presentation of this paper at the Geodynamics Conference at the University of Western Ontario, Dr. Joachim Hoefs (a participant of the Conference) and I decided to study the oxygen isotope composition of the lava flows on Mt. Falla. Dr. Hoefs made the measurements which we interpreted in a joint paper (Hoefs, Faure and Elliot, 1980). The results suggested that the magma was derived from a source or sources which had an $^{87}Sr/^{86}Sr$ ratio of 0.7093 during the Jurassic period. This magma was subsequently contaminated by interactions with crustal rocks which raised the $^{87}Sr/^{86}Sr$ ratios of some of the flows up to 0.7131. Therefore the thesis of this paper is confirmed by these results. The variation of the initial $^{87}Sr/^{86}Sr$ ratios of these volcanic rocks, and presumably those of others like them elsewhere in the world, cannot be attributed solely to the heterogeneity of magma sources in the mantle.

<u>Acknowledgements</u>. I would like to thank D. H. Elliot, J. R. Bowman, K. E. Pace, and R. L. Hill for participating in our long-standing effort to understand the origin of the Kirkpatrick Basalt. Philip R. Kyle made many helpful comments for which I am grateful. This work was supported by the National Science Foundation through Grant DPP77-21505. This paper is Contribution No. 51 of the Laboratory for Isotope Geology and Geochemistry (Isotopia).

References

Brooks, C. and S. R. Hart, Rb-Sr mantle isochrons and variations in the chemistry of Gondwanaland's lithosphere. Nature, 271, 220-222, 1978.

Brooks, C., D. E. James and S. R. Hart, Ancient lithosphere: Its role in young continental volcanism. Science, 193, 1086-1094, 1976.

Carter, S. R., N. M. Evensen, P. J. Hamilton and R. K. O'Nions, Neodymium and strontium isotope evidence for crustal contamination of continental volcanics. Science, 202, 743-747, 1979.

Compston, W., I. McDougall and K. S. Heier, Geochemical comparison of the Mesozoic basaltic rocks of Antarctica, South Africa, South America and Tasmania. Geochim. Cosmochim. Acta, 32, 129-149, 1968.

DePaolo, D. J. and G. J. Wasserburg, Petrogenetic mixing models and Nd-Sr isotopic patterns. Geochim. Cosmochim. Acta, 43, 615-628, 1979.

Elliot, D. H., Major oxide chemistry of the Kirkpatrick Basalt, central Transantarctic Mountains. In: R. J. Adie, ed., Antarctic Geology and Geophysics, 413-418. Universitetsforlaget, Oslo, 1972.

Elliot, D. H., Jurassic tholeiites of the central Transantarctic Mountains, Antarctica. In: E. H. Gilmour and D. Stradling, ed., Second Columbia River Basalt Symposium, 301-325, Eastern Washington State College Press, 1970.

Elliot, D. H. and P. Tasch, Lioestheriid conchostracans: A new Jurassic locality and regional Gondwana correlations. J. Paleontology, 41, 1561-1563, 1967.

Faure, G., K. E. Pace and D. H. Elliot, Systematic $^{87}Sr/^{86}Sr$ ratios and major element concentrations in the Kirkpatrick Basalt of Mt. Falla, Queen Alexandra Range, Transantarctic Mountains. In: C. Craddock, ed., Antarctic Geosciences, University of Wisconsin Press, 1981.

Faure, G., J. R. Bowman and D. H. Elliot, The initial $^{87}Sr/^{86}Sr$ ratios of the Kirwan Volcanics of Dronning Maud Land: Comparison with the Kirkpatrick Basalt, Transantarctic Mountains. Chem. Geol., 26, 77-90, 1979.

Faure, G., Principles of Isotope Geology. Wiley and Sons, New York, 464, 1977.

Faure, G. J. R. Bowman, D. H. Elliot and L. M. Jones, Strontium isotope composition and petrogenesis of the Kirkpatrick Basalt, Queen Alexandra Range, Antarctica. Contrib. Mineral. Petrol., 48, 153-169, 1974.

Faure, G. and J. L. Powell, Strontium isotope geology. Springer Verlag, Heidelberg and New York, 166, 1972.

Faure, G. R. L. Hill, L. M. Jones and D. H. Elliot, Isotope composition of strontium and silica content of Mesozoic basalt and dolerite from Antarctica. In: R. J. Adie, ed., Antarctic Geology and Geophysics, 617-624, Universitetsforlaget Oslo, 1972.

Faure, G. and D. H. Elliot, Isotope composition of strontium in Mesozoic basalt and dolerite from Dronning Maud Land. Bull. Brit. Antarct. Surv., 25, 23-27, 1971.

Fleck, R. J., J. F. Sutter and D. H. Elliot, Interpretation of discordant $^{40}Ar/^{39}Ar$ agespectra of Mesozoic tholeiites from Antarctica. Geochim. Cosmochim. Acta, 41, 15-32, 1977.

Ford, A. B., Stratigraphy of the Dufek Intrusion, Antarctica. U.S. Geol. Surv. Bull., 1405D, 1-36, 1976.

Gair, H. S., The geology of the upper Rennick Glacier to the coast, Northern Victoria Land, Antarctica. New Zealand J. Geol. Geophys., 10, No. 2, 309-344, 1967.

Gunn, B. M., Modal and element variation in Antarctic tholeiites. Geochim. Cosmochim. Acta, 30, 881-920, 1966.

Hall, B. A., J. F. Sutter and H. W. Borns, Jr., The inception and duration of Mesozoic volcanism in the Allan Hills - Carapace Nunatak area, Victoria Land, Antarctica. In: C. Craddock, ed., Antarctic Geosciences. University of Wisconsin Press, 1981.

Halpern, M., U. G. Cordani, and M. Berenholc, Variations in strontium isotopic composition of Paraná Basin volcanic rocks of Brazil. Revista Brasileira de Geosciencias, 4, 223-227, 1974.

Hamilton, W., Diabase sheets of the Taylor Glacier region, Victoria Land, Antarctica. U.S. Geol. Surv. Prof. Paper 456-B, 71, 1965.

Hawkesworth, J. C. and R. Vollmer, Crustal contamination versus enriched mantle: $^{143}Nd/^{144}Nd$ and $^{87}Sr/^{86}Sr$ evidence from the Italian volcanics. Contrib. Mineral. Petrol., 69, 151-166, 1979.

Heier, K. S., W. Compston, and I. McDougall, Thorium and uranium concentrations, and the isotopic composition of strontium in the differentiated Tasmanian dolerites. Geochim. Cosmochim. Acta, 28, 643-659, 1965.

Hoefs, J., G. Faure and D. H. Elliot, Positive correlation of $\delta^{18}O$ and initial $^{87}Sr/^{86}Sr$ ratios of Kirkpatrick Basalt, Mt. Falla, Transantarctic Mountains. Contrib. Mineral. Petrol., 75, 199-203, 1980.

Kyle, P. R., Development of heterogeneities in the subcontinental mantle: Evidence from the Ferrar Group, Antarctica. Contrib. Mineral. Petrol., 73, 89-104, 1980.

Kyle, P. R. and D. H. Elliot, Mineral chemistry of Kirkpatrick Basalt. In: C. Craddock, ed., Antarctic Geosciences. University of Wisconsin Press, 1981.

Manson, V., Geochemsitry of basaltic rocks: Major elements. In: H. H. Hess, ed., Basalts, vol. 1, 213-269, Interscience, New York, 1967.

Manton, W. I., The origin of associated basic and acid rocks in the Lebambo-Nuanetsi igneous province, southern Africa, as implied by strontium isotopes. J. Petrology, 9, 23-29, 1968.

McDougall, I., Differentiation of the Tasmanian dolerites: Red Hill Dolerite - granophyre association. Geol. Soc. Amer. Bull., 73, 279-316, 1962.

McDougall, I., Potassium-argon age measurements on dolerites from Antarctica and South Africa,

J. Geophys. Res., 68, No. 5, 1535–1545, 1963.

Norris, G., Triassic and Jurassic miospores and acritarchs from the Beacon and Ferrar Groups, Victoria Land, Antarctica. New Zealand J. Geol. Geophys., 8, 236–277, 1965.

Pushkar, P. and D. B. Stoeser, $^{87}Sr/^{86}Sr$ ratios in some volcanic rocks and some semifused inclusions of the San Francisco volcanic field. Geol., 3, 669–671, 1975.

Schaeffer, B., A Jurassic fish from Antarctica. American Museum Novitates, No. 2495, 17, 1972.

GEOSPHERE INTERACTIONS AND EARTH CHEMISTRY

W. S. Fyfe

Department of Geology, The University of Western Ontario

Abstract. The past decade of research into
plate tectonic processes has revealed the importance of mixing of surface regions of the
planet with the deep interior. While fluxes
out of the mantle at ocean ridges are rather
well quantified fluxes back into the mantle at
subduction zones are not quantified. Recent
work appears to indicate that there may be massive subduction of sediments and even continental
erosion or decretion. Until such processes are
quantified the exact chemistry of the mantle must
remain uncertain. Conclusions regarding the
evolution of continental crust also depend on
understanding the subduction process and recent
work has shown the need for more exact seismic
studies of the nature of the crust.

Introduction

It is interesting to look back one or two
decades at the views of geochemists on the planet
earth, how it was formed and how it changed with
time. Most of these earlier views were that the
planet accumulated rapidly and differentiated into rather well defined geospheres. Once having
attained a gravitationally stable configuration
most of the dynamic processes influencing earth
chemistry were associated with the crust-
hydrosphere-atmosphere system which moved rock
from place to place resulting in a cycle involving weathering → sedimentation → metamorphism →
melting. Thus we read in Mason (1958) that the
crust was a "separate system" and the major
geochemical cycle envisaged at that time involved the outer layers only with a minor injection of some primary mantle material. Igneous
activity was considered "a small-scale, local
phenomenon in relation to the earth as a whole,
and it is concentrated along zones of general
tectonic activity, specially the margin of the
Pacific Ocean". At that time most writers
underestimated the mass of igneous activity by
one or two orders of magnitude and the concept
of a feedback system to the mantle was not considered except by a view inspired workers like
Arthur Holmes. At this time there was almost a
total lack of knowledge about the crust under the
oceans.

For decades and since the classic work of V.M.
Goldschmidt, geochemistry was concerned with
description of the chemistry of earth materials
and geochemists were much concerned with the
development of the analytical techniques to describe where the nuclides were sited and the
general features of their concentrations. In
general, this branch of earth science has been
slow to respond to the "New View of the Earth"
(Uyeda, 1978), a view of a vigorously convecting
planet with 70% of its surface covered by young
crust, a discovery that must lead us to reject
the essentially "dead earth" models of the past.

While the subduction or return flow process is
not well quantified or understood at the present
time, we do know it is massive and models of one-
way differentiation of the interior must be re-
assessed and previous ideas on the composition
and evolution of the mantle must also be
reconsidered.

Convection is a mixing process. One of the
great geochemical challenges of the next decade
must be to quantify this mixing process and to
deduce from the crustal record how the process
has changed through time.

Given that the earth is cooling, given that the
earth is mixing now, it is clear that we must re-
consider models of one-way differentiation of the
planet. For example, given a hotter interior in
the past it is reasonable to suggest that light
volatile and easily fusible materials would
rapidly move to the surface. But if the earth
mixes as it cools it is not necessary that the
early configuration be maintained. To take a
simple example, we now know that water is subducted. We know that there are stable hydrated
phases that could form at considerable depths
in the mantle. If the outer layers are cooling,
then more volatiles may be fixed in the upper
mantle. Hence the question, is the hydrosphere
growing or shrinking? The answer to such questions must depend on the competition between
cooling rates and mixing rates. It is certain
that the present configuration of silica-rich
crust, hydrosphere and mafic relatively anhydrous
mantle is not thermodynamically stable for a cold
mixed body. But will the earth ever reach a
stable configuration?

According to many workers we are still moving away from the cold equilibrium state. In these brief notes I wish to comment on some of the present problems and our general state of ignorance on the problems of recycling or geosphere exchange. I think we can expect great changes in this knowledge over the next decades for geochemists are now beginning to consider the influence of the new global tectonics. The analytical tools to attack the problem are available but sampling is not adequate in many regions. In this essay I wish to concentrate on present processes and not to become involved in ancient processes which are a subject of great debate (see Fyfe, 1980). But a quantitative understanding of present mixing processes is essential to understanding the past when mixing could hardly have been less vigorous.

Geochemical Cycle Stage 1: The ocean ridge

New basaltic crust comes from the mantle at a rate of a little more than 10 km^3 a^{-1}. As the rate of mantle magmatism can hardly have been less in the past, the accumulated mass of the product of this process is of the order of 10% of the mass of the mantle. There is little evidence that the nature of this mantle extract has much changed with time except for the ancient high temperature peridotite lavas. Most of this new crust forms in a submarine environment just as it did in the past for most ancient rocks back to the beginning of the record at 3.8 Ga are submarine. Basalt liquid of density 2.7 g cm^{-3} forms rock of density 3.0 and contracts and cracks as it cools. As it is hot, and becomes permeable during cooling, water cooling is inevitable (Lister, 1977) and so begins the first major geochemical exchange.

Sea water is not in equilibrium with hot or cold basalt and a series of exchange processes occur which lead to most significant change in the mantle fusion product. It is now well documented that sea water convection strongly influences the heat flow patterns at ridges and as shown by Wolery and Sleep (1976) about 50% of the thermal energy focussed in ridge events may be transferred to sea water circulating deep in the crust. On average, the rock may interact with something like 30x its own volume of sea water (Fyfe and Lonsdale, 1980). As Anderson et al. (1979) have stated more than one-third of the entire surface of the world's ocean floor crust contains presently active geothermal systems. The depth to which flow occurs is a matter of debate but all agree it influences several km of thickness of the upper sea floor crust and it has been suggested (Lewis and Snydsman, 1977) that sea water may penetrate to near the Moho interface. Spectacular observation of the discharge systems have been made during the past year or so by submersibles (Francheteau et al., 1979; Corliss et al., 1979) and we now have documented observation that this phenomena

leads to important ore deposits (see Hutchinson et al., 1980).

This convective flow of oxygenated sea water through hot basalt leads to spectacular chemical change producing the altered rock commonly called spilite by petrologists. The general nature of the change has been summarized by Fyfe and Lonsdale (1980) and includes significant enrichment in the content of species such as H_2O, O_2, CO_2, S, Na, K, U while the isotopes of oxygen and strontium are changed. Strontium isotope ratios trend toward sea water values and O^{18} is enriched. Bloch and Bischoff (1979) have shown that about half the potassium added to oceans by erosion may be fixed in spilites while Aumento (1979) has shown that the amount of uranium may be doubled. Spooner (1976) has shown that in highly altered basaltic material on Cyprus, the strontium isotope ratios are changed to near sea water values. Thus hydrosphere-atmosphere-and continental crust species are added to basalt in significant quantities which are ultimately limited by sea water chemistry and the thermal energy available to drive the mass flow.

Geochemical Cycle Stage 2: Subduction

The age systematics of sea floor basalts clearly shows that the subduction process about balances the new crust formation at ocean ridges. But what is stressed here is that the material subducted (old ocean lithosphere with a spilite-serpentine, etc. capping) is not the same as material that formed the new lithosphere. Further, the old ocean floor crust carries sediments, some pelagic but with a variable component derived nearer the continents.

In two recent papers, Uyeda and Kanamori (1979), Uyeda (1979) have summarized much of the recent findings on the nature of the subduction process. In particular these workers stress that what is subducted may depend critically on the bending angle of the descending lithosphere. They propose two types of subduction processes, the low-angle Chilean type where much of the sedimentary cover is stripped off and added to the continent and the steep-angle Mariana type where sediments are subducted and even the continental margin may be eroded by so-called subduction erosion (see also Molnar and Gray, 1979).

Pelagic sediments (see Sibley and Vogel, 1976) may be subducted at a rate of about 1 $km^3 a^{-1}$. On average such sediments contain 3% K_2O, and 3 ppm uranium and are aluminous. Fyfe (1980) has commented on the lack of the mineral kyanite (Al_2SiO_5) in blueschist terraines. This mineral is rare in metamorphosed assemblages of fossil subduction zone sediments but should be common if all pelagic sediment was scraped off. This simple observation further confirms Uyeda's suggestions.

It is worth considering what these figures may mean in more detail. The water content of the upper layers of ocean floor crust, subducted at a rate of about 3×10^{16}g a^{-1} must be about 5%. Thus the water subduction rate is of the order of 1.5×10^{15}g a^{-1}. As the ocean mass is 1.4×10^{24} g, this implies recycling of the ocean mass is a billion years or so at the present rate of subduction. If pelagic sediments are subducted this recycling time is almost halved.

For potassium, subduction of 1 km^3 of pelagic sediment per year would recycle 6×10^{13}g a^{-1} and given the potassium added in spilites the quantity is probably close to 10^{14}g a^{-1}. The continental mass of K is about 2×10^{23}g so that all continental K would be recycled in about 2 billion years. Again for uranium, with pelagic sediments containing 3 ppm, the mantle would receive about 10^{10}g of uranium per year. The mantle contains about 5×10^{19}g of uranium (Gast, 1974; Fyfe, 1979). Again the recycling rate is significant and more U may be subducted than is moving back in ocean ridge basalts.

It is quite clear from the above figures that _if_ there is significant subduction of sediments and subduction erosion, and with the subduction of spilites, there must be an efficient return flow towards the surface to maintain continental crust and hydrosphere. But for a cooling earth it seems logical that incremental amounts of water and other volatiles will become progressively fixed in the upper mantle along with elements like K in mafic minerals like phlogopite. It is also clear that this process must confuse arguments on the outgassing of the earth, for the planet is also ingassing! Recent papers have discussed accretion and decretion processes at active margins (Leggett, 1980) while Wang (1980) has discussed the influence of sediments on frictional sliding in subduction zones.

The return flow of subducted materials is generally associated with the production of andesite magmas. Essentially andesites are of similar composition to basalts except for a higher content of continental crust elements such as U and K and SiO$_2$.

Recently Kay (1980) and Karig and Kay (1980) have carried out one of the first significant attempts to balance andesite chemistry against subduction chemistry. They clearly show that contributions to andesite chemistry must come from recycling of continental material. Armstrong (1980) has similarly examined Pb, Sr, and Nd isotopes and is again led to a conclusion of recycling of continental material. Karig and Kay suggest a rate of andesite magma production of 1.8×10^{15}g a^{-1}. This is close to the possible pelagic sediment subduction rate and points to a high degree of recycling and to a close balance between growth and removal of continental crust.

But we are only beginning to attack these problems. If we follow the logic of D.L. Anderson (this volume and 1980) some subducted material could be swept through a complete cycle and be carried back to an ocean ridge. The hot spot mantle magmas with their more K-Na rich nature (and perhaps even kimberlites,

carbonatites and the potassic mantle magma products like leucitites) could simply be chunks of subducted materials carried deep under continental regions and becoming unstable as thermal equilibrium is achieved at depth.

Finally, recent work of Eichelberger (1978) has shown how mantle magmas rising above subduction zones may lead to the heat flow necessary for continental crust fusion and that the two magma types, essentially basalt and granite, may actually mix to produce intermediate magma types. And as many oxygen isotope studies around batholiths have clearly shown (Taylor, 1978), in the region of plutons large geothermal systems are created which lead to extensive exchange with ground waters in the continental regime. These processes of mixing subduction products with mantle, and mantle magmas with crust must confuse the isotopic systematics commonly used to indicate crust or mantle sources. And if the subduction mixing process itself is part of the mantle magma generation process, it becomes even more difficult to define these systematics in "uncontaminated" mantle.

Geochemical Cycle Stage 3: Erosion

The next major process which is fundamental to the present state of the earth involves erosion. It is one of the most profound chemical transport processes on earth and as it leads to mass transport it must lead to compensating tectonic motions. It must also be stressed that many of the solution and mineralization processes in this part of the cycle are due to biological phenomena.

Again we may consider one or two simple processes. The average potassium content of continental runoff is 2.3 ppm and such water is delivered to the oceans at a rate of 3.6×10^{19} g a^{-1}. Thus the K flux to the oceans is about 10^{14}g a^{-1} almost exactly that required to balance the K in spilites and pelagic sediments. Garrels and Lerman (1977) give the rate of erosion in Phanerozoic time as $6-10 \times 10^{16}$g a^{-1}. At this rate the continental crust of mass 1.6×10^{25}g would be reworked in less than a billion years. These figures indicate that high elevations on earth in regions where tectonic forces no longer operate are transient. We should note that this overall erosion rate is at least double the rate of new crust formation at ridges. The ocean floor spreading process may sweep these sediments back to continental edges while some may be subducted. Garrels and Lerman suggest that subduction could involve 25% of the sediment mass.

The History of Continents

One of the present great debates concerns the formation of the continents and continental crust and the variation of the type and mass of this crust through time. But at the outset it must be stressed that our knowledge of the structure of continental crust and the processes that lead to its formation are still quite inadequate. Two examples may suffice to illustrate such processes.

The Proterozoic shield of Saudi Arabia has been the subject of intense examination over the past decade (see Al Shanti, 1979). The region is dominated by volcanic-plutonic material of the diorite-granite family produced over the period 1.1 - 0.6 Ga. The chemical and isotopic data strongly suggest that little or no older continental debris is present. The products are typical of those to be expected from magmatism above a subduction zone or series of zones in an ocean environment. This product has been swept onto the margin of the African continent. Thus a new marginal phase is developed by a mechanism essentially modern in style.

Recent work in the Appalachians (see Cook et al., 1980; and Brown et al., this volume) has revealed unexpected details of crust construction in this mobile belt. Here it is proposed that the present structure was formed by collision of an arc of ocean islands and is associated with major overthrusting of older crystalline rocks onto younger less metamorphosed sediments. The overthrust layers are thin. These writers have proposed that such thin-skinned thrusting may be a major mechanism of crust formation in many of the presently active major thrust belts of the earth. But what is important is that the rocks exposed at the surface may bear little obvious relation to rocks at depth. Oliver (1978) has stressed that where detailed studies have been carried out, continental crust structure is more complicated than many might have been supposed and even features such as the Moho boundary may be very complex. Such observations again lead us to conclude that continental crust can be thickened by thrusting and possibly by underplating by mantle magmas. These are often denser than continental crust. Thus present tectonic processes may add to continental crust by the sweeping together of volcanic arcs, and thicken continents by thrust tectonics. Erosion and subduction may remove continents. The balance of these major competing processes is not known and we are not sure if the mass of continental crust is growing or shrinking. Until we have a great increase in information on the structure of continental crust, discussion on its origin is likely to be frustrating.

Conclusions

Geochemical models of the earth must be developed in relation to the known convective fluxes. The rate of mantle magma addition at active spreading centres is well known. The alteration of new ocean crust by sea water is now a subject of intense study but is not yet well quantified.

The subduction process and in particular the exact nature of what is subducted is yet to be established. Study of the return flow of subducted materials is even less understood. All these processes are important to our understanding of mantle chemistry, particularly with regard to its volatile content and content of heat producing elements, uranium and potassium.

Detailed seismic studies of continental crust are revealing the inadequacies of extrapolation of surface observations to the mantle boundary. Until more is known about continental crust forming processes, models of the evolution of continents are likely to be highly speculative. Perhaps the greatest geochemical discovery of the past decade is that geosphere mixing involving all the parts, atmosphere-hydrosphere-biosphere-crust-mantle occur on a much larger scale than previously envisioned.

References

Al-Shanti, A.M.S. (Editor), Evolution of mineralization of the Arabian-Nubian Shield, I.A.G. Bulletin, vols. I-IV, Pergamon Press, 1979.

Anderson, D.L., Early evolution of the mantle, Episodes, 1980, 3-7, 1980.

Anderson, R.N., M.A. Hobart, and M.G. Langseth, Geothermal convection through oceanic crust and sediments in the Indian Ocean, Science, 204, 828-832, 1979.

Armstrong, R.L., Radiogenic isotopes: the case for crustal recycling on a near-steady-state, no-continental-growth earth, Phil. Trans. R. Soc. London, (in press), 1980.

Aumento, F., Distribution and evolution of uranium in the oceanic lithosphere, Phil. Trans. R. Soc. London, A., 291, 423-431, 1979.

Bloch, S. and J.L. Bischoff, The effect of low-temperature alteration of basalt on the ocean budget of potassium, Geology, 7, 193-196, 1979.

Cook, F.A., L.D. Brown, and J.E. Oliver, The southern Appalachians and the growth of continents, Sci. American, 243, 156-168, 1980.

Corliss, J.B., J. Dymond, L.I. Gordon, J.M. Edmond, R.P. Von Herzen, R.D. Ballard, K. Green, D. Williams, A. Bainbridge, K. Crane, and T.H. Van Andel, Submarine thermal springs on the Galapagos Rift, Science, 203, 1073-1083, 1979.

Eichelberger, J.C., Andesitic volcanism and crustal evolution, Nature, 275, 21-27, 1978.

Francheteau, J., H.D. Needham, P. Choukroune, T. Jutean, M. Seguret, R.D. Ballard, P.J. Fox, A. Carranza, D. Corduba, J. Guerrero, C. Rangin, H. Bongault, P. Camban, and R. Hekinian, Massive deep-sea sulphide ore deposits discovered on the East Pacific Rise, Nature, 277, 523-528, 1979.

Fyfe, W.S., The evolution of the Earth's crust: modern plate tectonics to ancient hot spot tectonics, Chem. Geol., 23, 89-114, 1978.

Fyfe, W.S., The geochemical cycle of uranium, Phil. Trans. R. Soc. London, A.291, 433-445.

Fyfe, W.S., Crust formation and destruction, in The Continental Crust and Its Mineral Deposits, edited by D.W. Strangway, Geol. Assoc. Canada, Special Paper 20 (in press), 1980.

Fyfe, W.S. and P. Lonsdale, Ocean floor hydrothermal activity, in The Sea, vol. 5, edited by C. Emiliani, John Wiley (in press), 1980.

Garrels, R.M. and A. Lerman, The exogenic cycles: reservoirs, fluxes and problems, in Global Chemical Cycles and Their Alterations by Man, edited by W. Stumm, Abakon Press, Berlin, 23-32, 1977.

Gast, P.W., The chemical composition of the earth, moon and chondritic meteorites, in The Nature of the Solid Earth, edited by E.C. Robertson, McGraw-Hill, New York, 19-40, 1974.

Hutchinson, R.W., W.S. Fyfe, and R. Kerrich, Deep fluid penetration and ore deposition, Min. Sci. Eng., (in press), 1980.

Karig, D.E. and R.W. Kay, Fate of sediments on the descending plate at convergent margins, Phil. Trans. R. Soc. London, (in press), 1980.

Kay, R.W., Volcanic arc magmas: implications of a melting-mixing model for element recycling in the crust-upper mantle system. J. Geol. (in press), 1980.

Leggett, J.K., Subduction (old and nes) examined. Geotimes, 25, 20-22, 1980.

Lewis, B.R.T. and W.E. Snydsman, Evidence for a low velocity layer at the base of the oceanic crust, Science, 266, 340-344, 1977.

Lister, C.R.B., Qualitative models of spreading-centre processes, including hydrothermal penetration, Tectonophysics, 37, 203-218, 1977.

Mason, B., Principles of Geochemistry, John Wiley and Sons Inc., New York, 1958.

Molnar, P. and D. Gray, Subduction of continental lithosphere: some constraints and uncertainties, Geology, 7, 58-63, 1979.

Oliver, J., Exploration of the continental basement by seismic reflection profiling, Nature, 275, 485-488, 1978.

Sibley, D.F. and T.A. Vogel, Chemical mass balance of the earth's crust: the calcium dilemma and the role of pelagic sediment, Science, 192, 551-553, 1976.

Spooner, E.T.C., The strontium isotopic composition of sea water and sea water-oceanic crust interaction, Earth Planet. Sci. Letters, 31, 167-174, 1976.

Taylor, H.P., Oxygen and hydrogen isotope studies of plutonic granitic rocks, Earth Planet. Sci. Lett., 38, 177-210, 1978.

Uyeda, S., The New View of the Earth, W.H. Freeman Co., San Francisco, 1978.

Uyeda, S., Subduction zones: facts, ideas and speculations, Oceanus, 22, 52-62, 1979.

Uyeda, S. and H. Kanamori, Back-arc opening and the mode of subduction, J. Geophys. Res., 84, 1049-1061, 1979.

Wang, C., Sediment subduction and frictional sliding in a subduction zone, Geology, 8, 530-533, 1980.

Wolery, T.J. and N.H. Sleep, Hydrothermal circulation and geochemical flux at mid-ocean ridges, Jour. Geol., 84, 249-275, 1976.

MAGNETOMETER ARRAYS AND GEODYNAMICS

D. Ian Gough

Institute of Earth and Planetary Physics, University of Alberta
Edmonton, Alberta, Canada T6G 2J1

Abstract. The use of large two-dimensional arrays of recording magnetometers, to observe time-varying magnetic fields in the period range 15-150 minutes, has led to the discovery and mapping of several large, conductive structures in the crust and upper mantle. In the western United States the observations are well fitted by induced currents flowing in the upper mantle, and the regions of high conductivity correlate well with high heat flow and with seismological evidence of partial melting. In contrast to these upper mantle structures, related to partial melt, a narrow conductor at crustal depths joins the northern end of the Southern Rockies in Wyoming to the Black Hills of South Dakota and to the Wollaston Domain of north central Saskatchewan. This conductor may be associated with an interplate boundary of Proterozoic age, and the high conductivity with graphite and possibly other metamorphic minerals. In southern Africa, two major crustal conductors have been well mapped by four array studies. One conductor revealed a hitherto unrecognized southwestward extension of the Luangwa-Zambezi Rift and saline water in fractures may provide the high conductivity. The other conductor underlies the tip of the continent south of 30°S. This body correlates with a major magnetic anomaly, with an isostatic gravity anomaly and with geochronologic and geologic information, in a manner consistent with a hypothetical underthrust of oceanic crust in Proterozoic time. Here, on this interpretation, hydrated ophiolitic rocks would constitute the good conductor.

These examples illustrate both the exploratory potential of magnetometer arrays and the various possible causes of high conductivity. Correlations with other geophysical and geologic data sometimes allow discrimination between partial melting in the mantle, conductive minerals in the basement, and saline water in the upper crust.

Introduction

Magnetic fields in the period range 10 minutes to 1 day, associated with transient electric currents in the magnetosphere and ionosphere, penetrate the Earth's crust and upper mantle, and there induce secondary currents which flow preferentially in the more conductive rocks. An array of magnetometers at the Earth's surface observes the sum of the fields of external and internal currents, and makes it possible to map concentrations of electric current and so, by inference, conductive structures in the Earth. The resolution of such maps is determined by the magnetometer spacing in the array: the arrays used in work considered in this paper involved 24 to 56 magnetometers at spacings of 50 to 150 km., so that large conductors in the crust or upper mantle are delineated with a precision of order 50 km. on the map. Depth resolution is less satisfactory, and long-line four-electrode resistivity sounding or magnetotelluric sounding are superior in this regard. In a few cases two-dimensional forward model calculations can be used in interpretation, but the crust and upper mantle are electrically extremely heterogeneous and an array may observe concentrations, by conductive structures under the array, of currents induced elsewhere by unknown fields in conductors of unknown shapes. In such cases quantitative modelling is impossible, but the current concentrations can still be accurately mapped.

Magnetometer array studies are subject to a depth limit of about 400 km., because there is, in the depth range 400-600 km, a general, planetary scale rise in conductivity to values in excess of 1 S/m (Banks 1972) which effectively limits further penetration of magnetic fields with periods of one day or less. High electrical conductivities occur in this near-surface shell in several ways. In the top few kilometres, where rock pores or cracks are connected, saline water may produce high conductivities. At all crustal depths, conductive minerals such as graphite or sulphides may occur along metamorphic belts. Hydrated minerals in ophiolitic rocks may similarly provide compositional conductors. In the upper mantle, partial melting of one to ten percent may raise the conductivity two orders of magnitude, if the melt is in interconnected spaces. This list of possi-

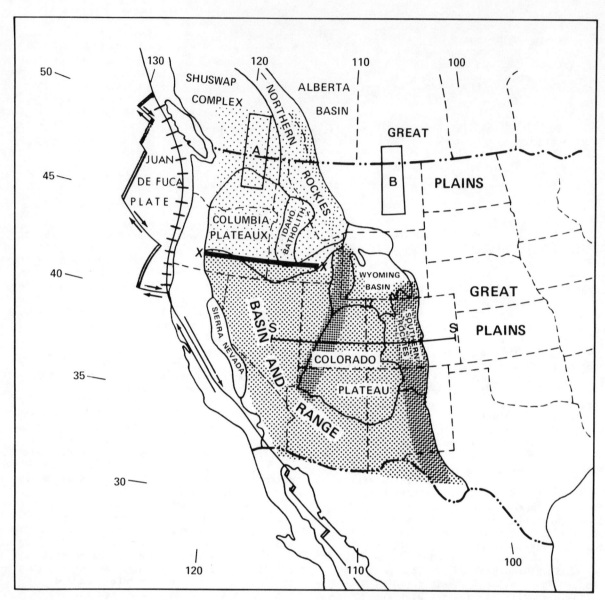

Fig. 1. Distribution of electrical conductivity in the upper mantle and tectonic provinces of western North America. The density of stippling gives a qualitative indication of variations of conductivity.

bilities is not complete, but shows that a high conductivity anomaly is an ambiguous object. Usually other geophysical or geological evidence is invoked to reduce the possibilities, sometimes to one. By contrast, a region of conductivity of order 10^{-4} S/m can only consist of dry crystalline rock at temperatures well below its solidus temperature. The range of conductivities found in the outermost 400 km of the Earth is from 10^{-4} to 4 S/m, the latter value for seawater.

Reviews of magnetometer array methods are given by Gough (1973) and Frazer (1974).

Western North America

Three large array studies, each with about 50 magnetometers deployed in four east-west lines, were made by the University of Texas at Dallas and the University of Alberta in the years 1967, 1968 and 1969. These arrays extended from the Great Plains, which lie on the craton, across much of the western superprovince with its Tertiary and later volcanics and diverse evidence of recent tectonic activity. Schmucker (1964,1970), working with linear magnetometer arrays, had discovered a conductivity anomaly at the Rio Grande boundary between the Great Plains and Basin and Range physiographic provinces, and our work extended his to two dimensions and to a larger area. Results from these arrays revealed much structure in the electrical conductivity west of the Great Plains, as represented schematically in Fig. 1. Highly conductive upper mantle underlies the Basin and Range tectonic province, with still greater thickness (or higher conductivity) of the conductive rock under the Wasatch Front and Southern Rocky Mountains, both of which have been uplifted in post-Cretaceous time. The anomalies across these tectonic provinces were the only ones the writer has encountered, in some ten array studies, in which the currents observed by the array were, in the main, induced directly in the underlying structures, so that two-dimensional modelling could be used to fit the observed magnetovariational anomalies. One conductivity model which fits the array data along latitude 38.5°N (SS' in Fig. 1) is shown in Fig. 2 (Porath 1971). In this model the depth of the conductor, which is indeterminate from the magnetovariational data, has been fitted to the low-velocity layer of seismology. A point made by this model is the thickness of the conductive layer, which is 100 km or more everywhere under the western region. This thickness assumes a high conductivity of 0.5 S/m: with a lower conductivity the thickness becomes still greater. The structure has to be in the mantle because the crust cannot contain it.

Since the conductive structures revealed by the magnetometer arrays correlate well with high heat flow and with the areas where there is seismological evidence of partial melting (Gough 1974), the stippling in Fig. 1 can be regarded as a first-order map of partial melting. It is worth noting that the Yellowstone plume is at the northern end of our Wasatch Front conductive ridge in the uppermost mantle. Geothermal exploration along the whole Wasatch fault belt, by means of closely-spaced magnetometer arrays of higher resolution, might be productive.

An important feature of the conductive structure is the abrupt reduction of the thickness-conductivity product at the northern boundary of the Basin and Range Province, near 42°N (XX' in Fig. 1). North of this boundary the conductive layer thickness falls to about 10 km (at constant conductivity). Presumably the partial-melt structure under western North America is related to the complex pattern of subduction and associated ascent of melt in Tertiary and Quaternary times, in ways difficult to unravel. It is interesting to find an order of magnitude less partial melt above the still-active subduction of the northwestern United States and southwestern Canada, than under the western U.S.A., where underthrusting ceased 10 to 20 My ago. The total geophysical-geological information leaves the origin of the partial-melt structures under western North America as a fascinating enigma.

The North American Central Plains Conductive Anomaly

In the 1967 array study we encountered a very large anomaly in the variation fields in the north-east corner of the array. It was clear that there was a concentration of current in the general area of the Black Hills of South Dakota. The 1969 array, which was intended mainly to investigate the Cordillera of northwestern U.S.A. and southwestern Canada, was stretched eastward to cover the Black Hills. As a result, one of the largest known anomalies in magnetic variation fields was discovered. In South Dakota the currents and the conductive structure must be less than 38 km deep, because the width at half-amplitude of the anomaly gives this depth for a line current. The conductive body is therefore crustal, and may well be near the top of the crystalline basement. Graphitic basement rocks are plentiful in the Black Hills, and a basement map of South Dakota compiled by Lidiak from other geophysical and borehole data shows a metamorphic belt which coincides with the conductive structure (Gough and Camfield, 1972). An array study was made in 1972 by Earth Physics Branch and the University of Alberta, with 40 magnetometers deployed to map the conductor northward and southward (Alabi, Camfield and Gough 1975). Fig. 3 shows a set of maps of amplitudes and phases of Fourier transforms, at a period of 68 minutes, of the magnetic field of a polar magnetic sub-

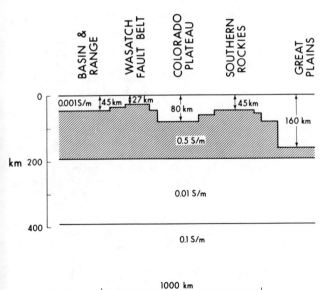

Fig. 2. A model of conductivity which fits magnetic variation anomalies recorded by the 1967 array. The section is at latitude 38.5°N., along SS in Fig. 1. Conductivities are in Siemens/metre. After Porath, 1971.

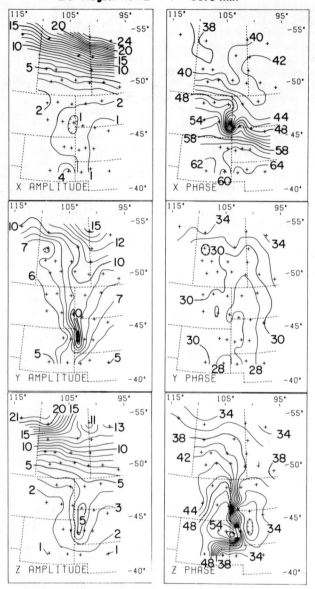

Fig. 3. Fourier transform amplitudes (nT) and phases (min) at period 68 min from a substorm on 1972 August 28, near the North American Central Plains conductive structure. After Alabi, Camfield and Gough, 1975.

storm. The external current system in this type of event includes an electrojet along the auroral zone, which is just north of the array, and the maps of amplitudes of X (northward horizontal) and Z (vertical) components, north of the Canadian-U.S. border, show mainly this auroral electrojet. The Y (eastward) amplitude and Z phase maps however show the currents channelled in the North American Central Plains (N.A.C.P.) conductor. From several such sets

of maps, from other maps of induction arrows representing transfer functions, and from inspection of the original magnetograms, the course of the conductive zone was traced to the northern limit of the array, 50 km south of the exposed shield of northern Saskatchewan, and southwestward from the Black Hills to the northern end of the Southern Rockies in Wyoming (Figs. 4,5).

Camfield and Gough (1977) have examined geological and geophysical information of several kinds to attempt an interpretation of the conductive structure. In the exposed shield (Fig. 5(a)) the Wollaston Domain, within the Churchill Province of the Shield, lies on strike and in line with the northern end of the mapped part of the N.A.C.P. conductor. The intensely, often isoclinally, folded schists and gneisses of the Wollaston Domain contain many conductivity anomalies found in airborne electromagnetic surveys. These conductivity anomalies strike northeast-southwest, along the fold belt and parallel to the N.A.C.P. conductor; and many have been identified with graphitic garnet-biotite gneiss units (Camfield and Gough, 1977). Well-developed sheets of graphite in nearly vertical limbs of folds are seen in outcrop in the Wollaston Domain. Graphite had already been suspected as a conductive mineral in the Black Hills section of the conductor, and may be involved along much of the 1800 km length of the N.A.C.P. conductor.

At the southern end the conductor skirts the east side of the Black Hills and turns southwest to run along the Hartville Arch and the Whalen Fault to the southern limit of the array in the Laramie Mountains at the northern end of the Southern Rockies (Fig. 5(b)). Exposures of Precambrian basement occur along the crest of the Hartville Arch, some showing graphite. Geochronologic work by Houston and others, referenced by Camfield and Gough (1977), together with geological data, indicate an age boundary along the line of the Whalen Fault (and the N.A.C.P. conductor). Continental granitic rocks of ages 2,500 My or more are found in central Wyoming to the northwest of the boundary, and intrusives of ages 1,800 My or less are the oldest rocks known southeast of it. The boundary is marked by a major fracture zone across the Medicine Bow and Sierra Madre fingers of the Southern Rockies province. Hills and others (1975) have proposed a Proterozoic plate boundary along the fracture zone and geochronologic boundary, with subduction under Nebraska and Colorado along the Whalen Fault stopping with an intercontinental collision about 1,700 My ago. Camfield and Gough (1977) extended this hypothesis with the suggestion that the whole N.A.C.P. conductor marks a large fracture zone which may be a proterozoic plate boundary, though without implying that subduction was involved along its whole length. Graphite is common in major fracture zones,

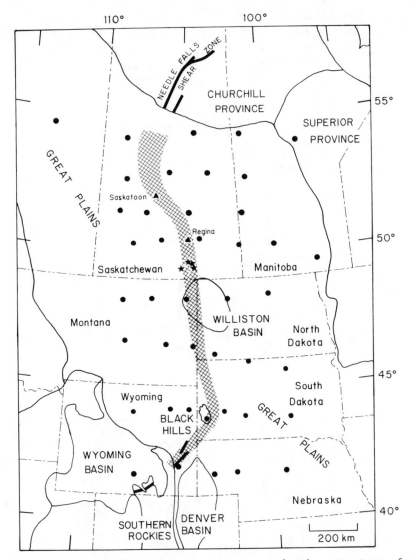

Fig. 4. Location of the North American Central Plains conductive structure, from the results of the 1972 array. Stars locate earthquake epicentres. After Alabi et al., 1975.

and has been observed at the northern end in the Wollaston Domain and at places already noted in South Dakota and Wyoming.

It is quite possible, though unproven, that the N.A.C.P. conductive structure continues along the Wollaston Domain to Hudson Bay, and the southern end down the subducted plate of Hills et al. (1975) to the conductive ridge due to partial melt in the upper mantle beneath the Southern Rockies. Currents concentrated in it may conceivably be induced partly by auroral-zone and polar-cap external fields in Hudson Bay, and partly by mid-latitude fields in the upper mantle under Colorado. The problems of modelling induction processes of such complexity are far beyond present computers, even if the input parameters could be specified. It is no

surprise that Porath, Gough and Camfield (1971) were unsuccessful in modelling the observed anomalous fields above the N.A.C.P. conductor, in a two-dimensional calculation for induction in the crustal conductor itself.

A Rift Anomaly in Southern Africa

Well-known magnetovariational and magneto-telluric work has been done on the Gregory Rift of Kenya, by Banks and his associates and by Hutton (1976) and hers. The present discussion considers results of two investigations with two-dimensional magnetometer arrays in southern Africa which have discovered a possible extension of the East African rift system across the continent to the Atlantic coast. The first

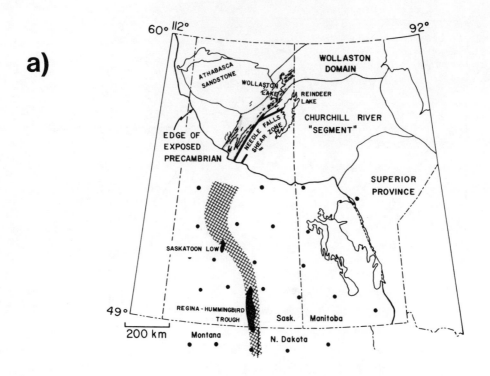

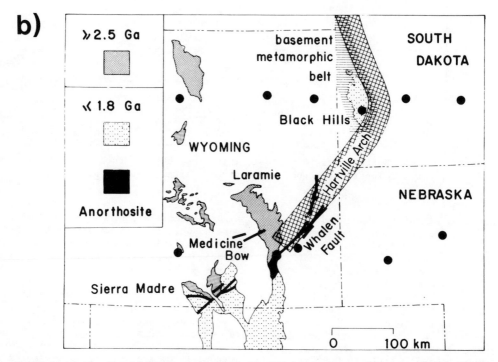

Fig. 5. Geology of the North American Central Plains conductor, after Camfield and Gough 1977.
a) Northern end. Trend lines in the Wollaston Domain join conductors mapped in airborne
electromagnetic surveys. b) Southern end. The fault zone marks a geochronologic boundary.
There are faults, not shown, with other strikes.

array study, in 1972 (de Beer et al., 1976) employed 25 magnetometers in South-West Africa, Botswana and Rhodesia to look for possible conductivity anomalies associated with the Damara Geosyncline in South-West Africa, and with the seismically active belt running south-westward from the Middle Zambezi rift (in which Lake Kariba lies) to the Okavango Delta of northern Botswana. In the event a single conductor was found, linking the Damara Geosyncline to the Okavango Delta and Middle Zambezi (de Beer, Gough and van Zijl 1975). This unforeseen result of the 1972 array was combined by de Beer et al. (1975) with mapped fractures and lineations, and with the seismicity of the region, in the hypothesis that the Luangwa/Middle Zambezi Rift continues south-westward to the Okavango Delta and thence westward across South-West Africa to the Damara. Earthquakes of magnitudes up to 6.7 have been recorded from the Okavango Delta since the inception of the South African seismological observatory system in 1949. In the Middle Zambezi, Lake Kariba triggered a large swarm of earthquakes (Gough and Gough 1970), the main shock having a normal-faulting mechanism (Sykes 1967). Scholz et al. (1976) have used micro-earthquake locations and mechanisms to show that rifting is in progress, with normal-faulting seismicity, along the Tanganyika/Luangwa/Middle Zambezi/Okavango Rift.

The high conductivity observed along the Zambezi-Okavango Rift is probably at crustal depth (de Beer et al. 1976) and may well be associated with saline water in rift fractures.

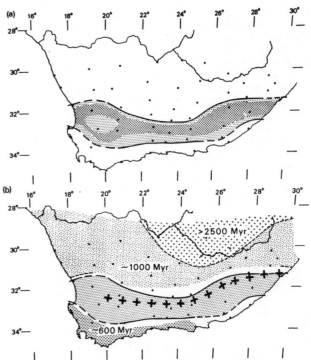

Fig. 7. (a) The Southern Cape Conductive Belt with the current concentrations indicated by density of stippling, from the magnetometer array indicated by dots, after de Beer and Gough (1980). (b) The conductive belt in relation to basement age provinces (Burger and Coertze 1973). The line of crosses marks the axis of the Beattie static magnetic anomaly.

De Beer et al. (1975) suggested that its westward extension might mark an extension of the rift structure along old weak zones in the lithosphere.

A second array was deployed in 1977 over the northern half of South-West Africa to investigate the western end of the anomaly at closer spacing and to the coast. A preliminary report of this investigation has been given by de Beer (1979). This work confirmed the results of the 1972 array east of 18°E., the western limit of the earlier array, but showed that the conductor bends to the southwest between 17° and 18°E. and then runs parallel to the fractures in the Damara Geosyncline. Fig. 6 shows the position of the conductive anomaly, as revealed by the two arrays, in relation to the African rifts and shields.

Schlumberger soundings over the conductive zone near its western end, made by the Geophysics Division of the South African National Physical Research Laboratory, show a drop in resistivity from 700 ohm.m to 20 ohm.m at a depth of only 3 km. The high conductivity is therefore unlikely to be related to high tem-

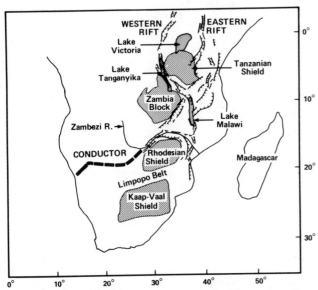

Fig. 6. African rifts, shields and zones of weakness in relation to the Botswana-South-West Africa conductive structure. From de Beer, Gough and van Zijl 1975 modified at the west end from de Beer 1979.

peratures. It could be associated with saline water in fractures or with conductive minerals. De Beer (1979) points out that the conductive belt joining the Damara and Zambezi belts could be a result of a suture containing oceanic lithosphere wedged between continental lithosphere slabs, on the model proposed by Burke, Dewey and Kidd (1977), but at a different site from that proposed by those authors. It would be reasonable to suppose that under a different stress field, the continent would later fracture along the old suture. Where the rifting is more active, part of the induced current could flow in the oceanic rocks and part in groundwater in the new fractures.

No less than 15 of 16 known calc-alkaline igneous complexes lie within the conductivity anomaly, 14 in the Damara Geosyncline and one in northwestern Rhodesia. This supports the argument for a zone of weakness, following the discussion by Sykes (1978). Sykes also suggests a relation between this zone of weakness and transform faults in the Atlantic floor.

The Southern Cape Anomaly

An array study in South Africa in 1971, intended to look for a possible change in crustal or upper mantle conductivity at the boundary between the 2,500 My-old Kaapvaal craton and the 1,000 My-old Namaqua-Natal mobile belt, discovered instead a very large conductivity anomaly elongated east-west and underlying the southern limit of the Karoo Basin and the Cape Fold Belt (Gough et al., 1973). Only one station, at the southern corner of the triangular array of 1971, lay above the conductor. In 1977 an array of 53 magnetometers was placed over the southern end of the continent, south of 30°S., in order to map and study this Southern Cape anomaly. As a result de Beer and Gough (1980) have been able to map a well-defined zone of high conductivity which they have named the Southern Cape Conductive Belt (Fig. 7). Half-widths of the magnetovariation anomalies indicate a maximum depth in the lower crust or upper mantle. The conductive structure correlates closely with the Beattie positive anomaly in the static magnetic field: if the same rock is magnetic as well as electrically conductive, it must lie above the Curie isotherm and so no deeper than 38 km. Marine crustal rocks containing either salt water or hydrated minerals such as serpentine, can be highly conductive and also highly magnetic. Further, the conductive belt lies outside the edge of Proterozoic Gondwanaland (Fig. 7(b)). Heat flow is not increased above it. Provisionally, de Beer and Gough (1980) associate the Southern Cape Conductive Belt with an accumulation of oceanic crustal rocks in either a Proterozoic zone of subduction, or a marginal sea of Proterozoic Gondwanaland.

Conclusions

These examples show both the capabilities and the limitations of magnetometer arrays. As a mapping technique the method is excellent and has revealed many structures of both present and past tectonic significance. Depth estimation and quantitative modelling are difficult and uniqueness far off, but an array study, involving two or three scientists for a year or so, has usually proved to reveal interesting structure in the top 400 km of the Earth, even though the structure discovered may be unexpected and unrelated to that the array was intended to investigate. Other methods can then be used to estimate depths. The ambiguity in the cause of high electrical conductivity in rocks is great, but can usually be resolved in favour of some one of the alternatives, as these examples have shown, by recourse to other geophysical, geochemical, geochronologic and geological information.

Acknowledgements. The inception of two-dimensional magnetometer arrays followed a discussion between Anton Hales and myself, of Schmucker's pioneer work, in 1964. In array studies over the last fourteen years I have benefitted from working with my colleagues J.S. Reitzel, H. Porath, C.W. Anderson III, D.W. Oldenburg, P.A. Camfield, A.O. Alabi, F.E.M. Lilley, M.W. McElhinny, J.H. de Beer, J.S.V. van Zijl, V.R.S. Hutton, J.M. Sik, G. Rostoker and J.R. Bannister, some of whom have been my students and all of whom have contributed to my understanding of the phenomena. I thank J.H. de Beer for allowing me to include a figure from a paper by him and myself.

References

Alabi, A.O., P.A. Camfield, and D.I. Gough, The North American Central Plains conductivity anomaly, _Geophys. J.R. astr. Soc._, _43_, 815-833, 1975.

Banks, R.J., The overall conductivity distribution of the Earth, _J. Geomag. Geoelectr._, _24_, 337-351, 1972.

Burger, A.J., and F.J. Coertze, Radiometric age measurements on rocks from southern Africa to the end of 1971, _Geol. Survey of S. Africa Bull._, _58_, 46 pp., 1973.

Burke, K., J.F. Dewey, and W.S.F. Kidd, World distribution of sutures - the sites of former oceans, _Tectonophysics_, _40_, 69-99, 1977.

Camfield, P.A., and D.I. Gough, A possible Proterozoic plate boundary in North America, _Can. J. Earth Sci._, _14_, 1229-1238, 1977.

de Beer, J.H., D.I. Gough, and J.S.V. van Zijl, An electrical conductivity anomaly and rifting in southern Africa, _Nature_, _225_, 678-680, 1975.

de Beer, J.H., J.S.V. van Zijl, R.M.J. Huyssen,

P.L.V. Hugo, S.J. Joubert, and R. Meyer, A magnetometer array study in South-West Africa, Botswana and Rhodesia, Geophys. J.R. astr. Soc., 45, 1-17, 1976.

de Beer, J.H., The tectonic significance of geomagnetic induction anomalies in Botswana and South-West Africa, Proc. Conference on The role of geophysics in the exploration of the Kalahari, Lobatse, Botswana, Feb. 1979.

de Beer, J.H., and D.I. Gough, Conductive structures in southernmost Africa: a magnetometer array study, Geophys. J.R. astr. Soc., 63, 479-495, 1980.

Frazer, M.C., Geomagnetic deep sounding with arrays of magnetometers, Rev. Geophys. and Space Phys., 12, 401-420, 1974.

Gough, D.I., Electrical conductivity under western North America in relation to heat flow, seismology and structure, J. Geomag. Geoelectr., 26, 105-123, 1974.

Gough, D.I., and P.A. Camfield, Convergent geophysical evidence of a metamorphic belt through the Black Hills of South Dakota, J. Geophys. Res., 77, 3168-3170, 1972.

Gough, D.I., J.H. de Beer, and J.S.V. van Zijl, A magnetometer array study in southern Africa, Geophys. J.R. astr. Soc., 34, 421-433, 1973.

Gough, D.I., and W.I. Gough, Load-induced earthquakes at Lake Kariba - II, Geophys. J.R. astr. Soc., 21, 79-101, 1970.

Hills, F.A., R.S. Houston, and G.V. Subbarayudu, Possible Proterozoic plate boundary in southern Wyoming, Geol. Soc. America Abstr. with Programs, 7, 614, 1975.

Hutton, R., Induction studies in rifts and other active regions, Acta Geodaet. Geophys. Montanist, 11, 347-376, 1976.

Porath, H., Magnetic variation anomalies and seismic low-velocity zone in the western United States, J. Geophys. Res., 76, 2643-2648, 1971.

Porath, H., D.I. Gough, and P.A. Camfield, Conductive structures in the northwestern United States and southwestern Canada, Geophys. J.R. astr. Soc., 23, 387-398, 1971.

Schmucker, U., Anomalies of geomagnetic variation in the south-western United States, J. Geomag. Geoelectr., 15, 193-221, 1964.

Schmucker, U., Anomalies of geomagnetic variations in the south-western United States, Bull. Scripps Inst. Oceanography, 13, 1-165, 1970.

Sykes, L.R., Mechanism of earthquakes and nature of faulting on the mid-oceanic ridges, J. Geophys. Res., 72, 2131-2153, 1967.

Sykes, L.R., Intraplate seismicity, reactivation of zones of weakness, alkaline magmatism and other tectonism postdating continental fragmentation, Rev. Geophys. Space Phys., 16, 621-688, 1978.

INTRAPLATE STRESS ORIENTATIONS FROM ALBERTA OIL-WELLS

J.S. Bell* and D.I. Gough**

*BP Canada Ltd., 333 5th Avenue S.W., Calgary, Alberta, Canada T2P 3B6

**Institute of Earth and Planetary Physics, University of Alberta
Edmonton, Alberta, Canada T6G 2J1

Abstract. Many oil-wells in Alberta are non-circular as a result of spalling of their walls. Such break-outs occur over a wide depth range covering the stratigraphic column from Devonian through Cretaceous, and in siltstones, sand-stones, limestones, dolomites and one shale. The azimuths of greatest elongations are given by modern four-arm dipmeter logs, and are tightly grouped in a northwest-southeast direction. They are unrelated to dip of the strata. We have elsewhere advanced the hypothesis that the spalling is caused by concentration of stress at the walls of the holes, in a general stress field having large, unequal horizontal principal stresses. The observed break-out orientations would require the larger horizontal principal stress to be directed northeast-southwest, normal to the Rocky Mountains fold axes. As a normal stress field (σ_1 vertical) would give horizontal stresses too small to produce break-outs at depths of a few hundred metres, it is believed that σ_1 is horizontal and NE-SW. In the present paper, new break-out orientation data are added, and hydraulic fracturing data are examined. In the Pembina area there is evidence that water pressure produces vertical fractures NE-SW. If these form normal to σ_3, the implied stress orientation gives σ_1 in agreement with that inferred from the oil-well break-outs. Steam injection near Cold Lake in east central Alberta gives similarly oriented vertical fractures. There is thus strong evidence that σ_1 is consistently northeast-southwest throughout the Alberta Plains. The data from oil-well break-outs do not discriminate between a thrust-type stress field (σ_3 vertical) and a strike-slip field (σ_2 vertical) but the hydraulic fractures suggest the latter. The stress field proposed could be related to present tractions on the plate, to postglacial rebound of Hudson Bay, to residual stresses from the Laramide folding of the Rockies, or to some combination of these. If our hypothesis is validated it may lead to determinations of stress orientations in many
sedimentary basins which have been drilled for hydrocarbons, where the four-arm dipmeter has been used in logging the wells.

Introduction

Substantial progress has been made in recent years in the study of the state of stress in the continental lithosphere (Sbar et al., 1972; Sbar and Sykes, 1973; McGarr and Gay, 1978). Orientations, and in some cases magnitudes, of the stress tensor have been estimated from over-coring measurements in mines (Obert, 1962), from hydrofracturing in wells (Kehle, 1964), and from fault-plane solutions of earthquakes (Raleigh et al., 1972; Sbar and Sykes, 1973). More qualitative indications of the present state of stress have come from man-made earth-quakes, oil industry hydraulic fracturing, recent movements on faults, post-glacial folding, pop-ups and various other types of rock failure. The resulting data have revealed the existence of high horizontal compressive stresses at shallow depths at many localities, and show that relatively uniform regional stress patterns exist in some areas of 10^5 km^2 or more.

Such evidence of uniformly oriented, large horizontal stresses, over considerable parts of a plate, encourage the hope that intraplate stress patterns may help in the elucidation of plate dynamics. But a simple stress pattern need not imply simple causation of the stress. A plate may be simultaneously under shear stress on its underside from mantle convection, shear and compressive stresses on its edges from adjoining plates, internal horizontal compression associated with post-glacial uplift, and its elastic response to its own weight; and yet exhibit a coherent and simple pattern of intraplate stress. Further, residual stress may be stored in crustal rocks and may reflect past tectonic episodes rather than present tractions on the plate. It would thus be unreasonable to hope that intraplate stresses, alone, will ever reveal plate drive mechanisms. Rather, intraplate stress data should be regarded as an important part of the total data base, from many geophysical, geochemical and geological disciplines, which will eventually elucidate the forces that move the plates.

We have recently suggested (Bell and Gough, 1979) that systematically oriented elongations in oil-wells in Alberta result from spalling of the walls of the wells in a stress field in which the largest principal stress is horizontal and oriented NE-SW. In this paper we present additional evidence in support of such a stress field, which appears to be essentially uniform over much of the Alberta sedimentary basin. If substantiated by direct measurements of the in situ stress, our hypothesis suggests that oil-well logs from other basins may contain evidence of large horizontal compressive stresses and of their orientations. While overcoring measurements and hydraulic fracture data remain superior in giving estimates of stress magnitudes, the oil-well break-outs may prove of special value in providing many estimates of the orientations of the horizontal principal stresses, as functions of both depth and position, in sedimentary basins which have been drilled for hydrocarbons.

Measurement of Borehole Ellipticity

The modern four-arm high resolution dipmeter tool was introduced in the late 1960's, and is used by the oil industry to obtain subsurface structural information and to record hole deviation (Allaud and Ringot, 1969; Cox, 1970). As the tool is raised up a well, it records its inclination and orientation while measuring rock resistivity by means of four electrodes which are pressed against the side of the borehole by hydraulically actuated caliper arms spaced ninety degrees apart. These four caliper arms also measure the width of the hole and, where it is not circular, record the lengths and azimuths of its greatest and smallest axes. Intervals of ellipticity can be identified from the caliper log and the recorded caliper bearing data permit one to establish the azimuths of the long axes of these intervals.

Asymmetric boreholes can arise from a variety of causes such as non-uniform mudcaking on the hole walls (Babcock, 1978), or because the well deviated significantly and the rotating drill-pipe partially eroded the "bend". The most frequently measured borehole asymmetry, however, is hole elongation caused by preferential spalling of the walls which occurs shortly after drilling the well. Figure 1 illustrates a typical case of such ellipticity. The curves on the far right of the log record the measurements of the two pairs of opposed calipers. The four noisy curves display the vertically variable resistivity of the rocks. Correlation of these curves, together with information on the relative vertical displacement which this involves, permits bedding attitudes to be computed. Tool orientation is documented on the left of the log, where the solid curve records the azimuth of caliper 1 with respect to magnetic North.

Consider the interval from 9527 ft. to 9494 ft. Calipers 1 and 3 record a hole width of approximately 9.5 inches, equivalent to the diameter of the drill bit. Calipers 2 and 4 record a varying diameter ranging between 10 and 11 inches. Inspection of the curve recording the compass azimuth of caliper 1 shows that the dipmeter tool had been rotating clockwise as it was drawn up the well. At 9527 ft. tool rotation ceased, calipers 2 and 4 became fixed within the elongated break-out, and the dipmeter was drawn up the well with caliper 1 oriented at an average azimuth of 200°. In the area where this well was drilled, the magnetic declination is +23°, so that the true geographic bearing of caliper 1 is 223°. However, it is calipers 2 and 4 which are recording hole elongation, thus this break-out zone has its long axis oriented at 133°, approximately northwest-southeast. As can be seen from the log (Figure 1), additional break-outs with a similar orientation also occur higher in the well.

Many wells exhibit numerous break-out zones and Cox (1970) noticed that in Western Canada, borehole elongation directions were amazingly uniform within and between wells. Babcock (1978) later extended and confirmed Cox's conclusions. Between them, they clearly demonstrated that, within a single well, the direction of maximum elongation is statistically consistent regardless of depth or lithology and that, throughout the Alberta Plains, the maximum elongations show a preferred northwest-southeast orientation. Moreover, borehole elongation orientations are independent of bedding attitudes and are not related to mild hole deviation (Cox, 1970; Babcock, 1978).

Cause of Borehole Elongations

The elongations discussed above are the result of caving or spalling of the walls of the boreholes shortly after drilling. In any given well, the break-outs are discrete, begin and end abruptly in depth and are separated by uncaved zones. In Alberta, Babcock (1978) noted that the orientation of the long axes of the break-outs corresponded to the orientation of one population of joints which he had measured in surface outcrops (Babcock, 1973; 1974; 1975; 1978), and proposed that the break-outs were caused by related subsurface fractures intersecting the well bores. Three other well-defined sets of joints, however, exhibit orientations that are markedly different from those of most of the break-outs (Babcock, op. cit). The present authors found Babcock's explanation

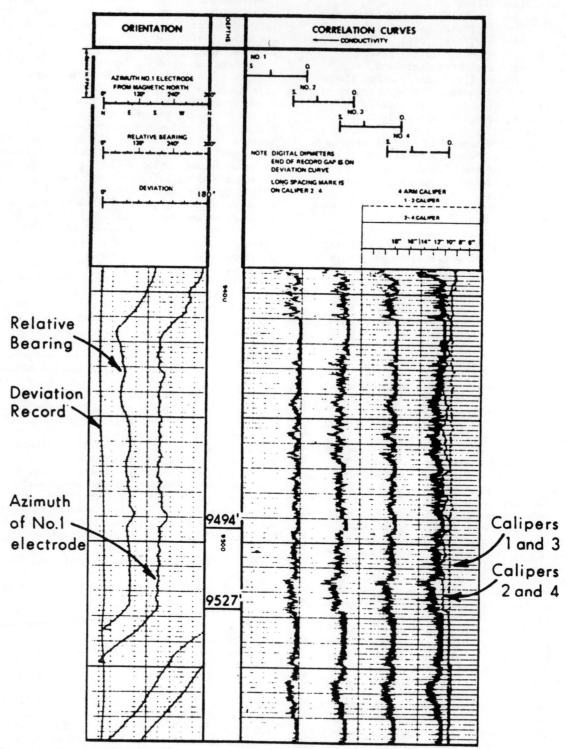

Fig. 1. Uncomputed four-arm dipmeter log showing zones of borehole elongation or break-outs.

difficult to accept and have proposed that the break-outs are a response to a regional stress field involving large, unequal horizontal compressive stresses which are concentrated by the boreholes themselves in an azimuth determined by the orientation of the stress tensor (Bell and Gough, 1979).

The stress concentrated around a vertical hole through a horizontally stressed body of rock can be approached by considering a large rectangular plate with a small hole in it which is subjected to uniaxial stress S (Figure 2). As Kirsch first demonstrated, the stress components at (r, θ) are given by the following equations:

$$\left.\begin{aligned}
\sigma_r &= \frac{S}{2}\left(1 - \frac{a^2}{r^2}\right) + \frac{S}{2}\left(1 + \frac{3a^4}{r^4} - \frac{4a^2}{r^2}\right)\cos 2\theta \\[2mm]
\sigma_\theta &= \frac{S}{2}\left(1 + \frac{a^2}{r^2}\right) - \frac{S}{2}\left(1 + \frac{3a^4}{r^4}\right)\cos 2\theta \\[2mm]
\tau_{r\theta} &= -\frac{S}{2}\left(1 - \frac{3a^4}{r^4} + \frac{2a^2}{r^2}\right)\sin 2\theta
\end{aligned}\right\} \quad (1)$$

for $r \geq a$, the radius of the hole, where σ_r, σ_θ and $\tau_{r\theta}$ represent, respectively, the radial, tangential and shear stresses.

At the hole boundary, where $r = a$,

$$\sigma_r = \tau_{r\theta} = 0, \quad \sigma_\theta = S - 2S\cos 2\theta \qquad (2)$$

and the tangential stress σ_θ has maximum value

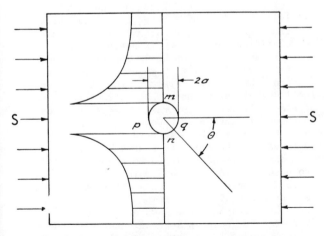

Fig. 2. Stress near a small circular hole in a large plate under uniaxial compression. The curves show the radial variation of the stress component σ_θ, along the diameter transverse to the applied compression S. Here σ_θ is parallel to S.

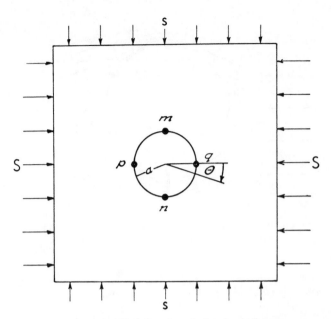

Fig. 3. Small circular hole in a large plate under biaxial compression.

$3S$ at $\theta = \frac{\pi}{2}$ or $\frac{3\pi}{2}$, that is at the ends m, n of the diameter perpendicular to S (Fig. 2). Along extensions of this diameter, where $\cos 2\theta = -1$, (1) gives

$$\tau_{r\theta} = 0, \quad \sigma_\theta = \frac{S}{2}\left(2 + \frac{a^2}{r^2} + \frac{3a^4}{r^4}\right). \qquad (3)$$

Clearly the stress concentration is very local for, at $r = 3a$, σ_θ has fallen to 1.074 S. The variation of σ_θ with r is shown in Figure 2. It may be noted that at p and q, the ends of the diameter parallel to S,

$$\sigma_r = \tau_{r\theta} = 0, \quad \sigma_\theta = -S$$

and σ_θ is a tensile stress equal in magnitude to the applied compressive stress S. The concentration of tangential stress near m and n is illustrated in terms of stress trajectories in Figure 4 (Jaeger, 1962).

In crustal rocks all principal stresses are compressive, except at very shallow depths. A closer approximation to the geological situation is shown in Figure 3, where unequal compressions S and s, $S > s$, are applied orthogonally to a plate with a small hole in it. From superposition it follows at once that, at the hole boundary, $r = a$

$$\sigma_r = \tau_{r\theta} = 0, \quad \sigma_\theta = S + s - 2(S-s)\cos 2\theta$$

so that σ_θ varies between maximum value $(3S-s)$

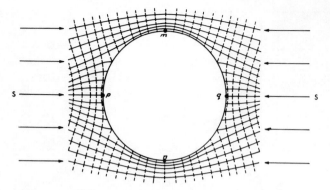

Fig. 4. Stress trajectories around a circular hole in an infinite solid subjected to uniaxial stress (after Jaeger, 1962).

when $\theta = \frac{\pi}{2}, \frac{3\pi}{2}$ (at m,n) and minimum value $(3s-S)$ when $\theta = 0, \pi$ (at p,q).

In the case of equal horizontal stresses $\sigma_\theta = 2S$ and is independent of θ. The hole doubles the horizontal stress and spalling may occur, but without preferred azimuth. However, if large and substantially unequal horizontal stresses are present, stress concentrations on the walls of a borehole may result in statistically well-aligned preferential caving, with the mean axis of borehole elongation at right angles to the larger principal horizontal stress.

Borehole break-outs give no indication of the magnitudes of the principal stress. They are unlikely to occur at depths of a few hundred metres in a normal stress field (σ_1 vertical), if the vertical principal stress approximates the overburden pressure, because the horizontal stresses would then be too small to produce spalling in competent rocks. We have therefore inferred that the borehole elongations in Alberta indicate either a thrust stress field (σ_3 vertical) or a strike-slip stress field (σ_2 vertical), with σ_1 horizontal and oriented northeast-southwest (Bell and Gough, 1979).

Failures of tunnel roofs in Ontario have been explained by Lo and Morton (1976) in terms of interaction of the tunnel with large horizontal compressive stress transverse to the tunnel. Here, the stress concentration by the hole is greatly increased if the tunnel roof is near the surface and if the rock is elastically anisotropic.

Crustal Stress in Alberta, Canada

No published determinations of *in situ* orientations and magnitudes are currently available for Alberta, but there are indications that unequal horizontal compressive stresses prevail in several areas of the province. Dusseault (1977) has noted that vertical fractures propagate in northeast-southwest vertical planes in western Alberta, which implies that the least principal stress is horizontal and oriented northwest-southeast.

In the Cold Lake heavy oil deposits of eastern Alberta, Imperial Oil engineers have generated northeast-southwest oriented vertical fractures in sandstones of the Clearwater Formation as a result of steam injection (Imperial Oil Ltd., 1978). Fracturing occurred at depths of approximately 450 m. at pressures above approximately 9 MPa (90 bars or 1300 psi) after which steam injection increased greatly. The preferential fracture trend was identified through early steam breakthrough into wells northeast and southwest of injection wells. Temperature observations indicated that the fractures were vertical. Information made public by Imperial Oil (1978) suggests that induced vertical fracture orientations vary from N 30°E to N 45°E. This implies that the minimum principal stress is horizontal and oriented between 120° and 135° (Hubbert and Willis, 1957). In consequence, the larger horizontal stress would be oriented between 30° and 45°.

Imperial Oil ran no four-arm dipmeter logs in wells on their Cold Lake property (Imperial Oil, personal communication), and the authors know of only one nearby well in which this tool was run and the caliper extensions recorded. This well contains one well-documented break-out between 477 and 479 m., in the Clearwater Formation sandstone, for which the mean azimuth of elongation is 131°, suggesting that the major horizontal stress direction is 41°.

An even better correspondence between stress orientation inferences from induced fractures and hole elongations exists in western Alberta. In a discussion about secondary recovery of oil from the Upper Cretaceous Cardium Formation sandstones within the J Lease of the Pembina oilfield, McLeod (1977) noted that permeability was greatest in a northeast-southwest direction. On water-flooding, oil-wells northeast or southwest of water injection wells experienced an initial surge in oil flow rates followed by premature water breakthrough. This behaviour implied that there was either a fracture system or else a permeability trend oriented northeast-southwest (Figure 5). However, Neilson (1957) shows Cardium sandbody isopach axes trending northwest-southeast in this area, so that any permeability trends related to depositional processes should run NW-SE. Therefore, the authors believe that McLeod's observations can be better explained in terms of vertical fractures oriented northeast-southwest. Such fractures are once again consistent with NW-SE orientation of the least principal stress σ_3, and therefore with stress control of the borehole break-outs (Figure 5).

Several wells near the J Lease in the Pembina oilfield exhibit borehole elongation. Figure 5 shows the locations of six wells in which four-arm dipmeter logs have been run within the

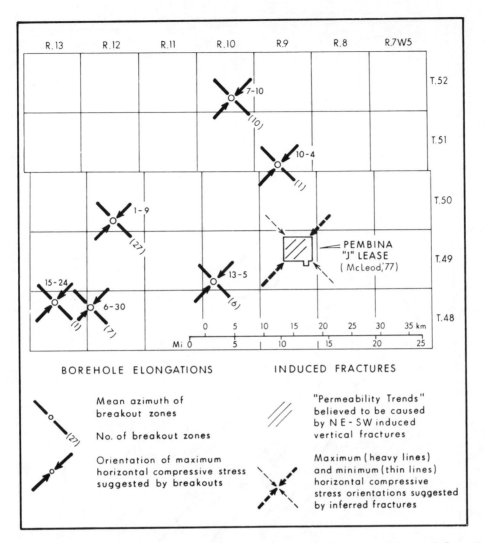

Fig. 5. Horizontal stress orientations inferred from borehole elongations and from induced fracturing, in the Pembina area, western Alberta.

Paleozoic section over depths ranging from 2500 to 4500 m. Azimuths of hole elongation are tightly clustered with mean values ranging between 132° and 141° (Figure 5). In individual wells the break-out azimuths vary over approximately 20°. Such dispersion of azimuths is consistent with stress control of the break-outs (Figure 4).

The regional picture of horizontal stress orientations inferred from borehole elongation data in the Alberta Plains is remarkably uniform. Few wells have yet been logged with four-arm dipmeter tools and only a small sample of those have been examined for break-outs (Cox, 1970; Babcock, 1978; this study). Figure 6 illustrates break-out azimuths from 36 wells distributed across the Alberta Plains. Break-outs are aligned largely in a northwest-southeast direction parallel to the strike of

thrust faults in the Rocky Mountains. The inference is clear that the horizontal principal stresses throughout the crust of the Alberta Plains are large and unequal, with the larger approximately northeast-southwest (Figure 7). For reasons already given this principal stress is believed to be σ_1. The vertical hydraulic fractures suggest that σ_3 is horizontal and that the stress field is of strike-slip or transcurrent faulting type.

The uniform stress field throughout the Alberta Plains, suggested by Figure 7, could have arisen in a number of ways. Contemporary tractions on the underside and edges of the North American plate, associated with whatever mantle motions drive the plate, may contribute. Hudson Bay lies northeast from the Alberta Plains, and is rising in post-glacial rebound. Presumably this contributes to σ_1 oriented

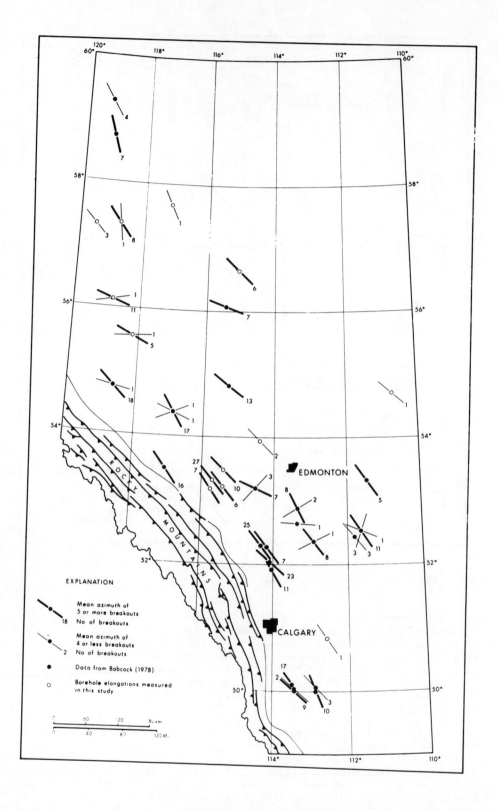

Fig. 6. Mean azimuths of borehole elongations for 36 wells in Alberta (partly after Babcock, 1978).

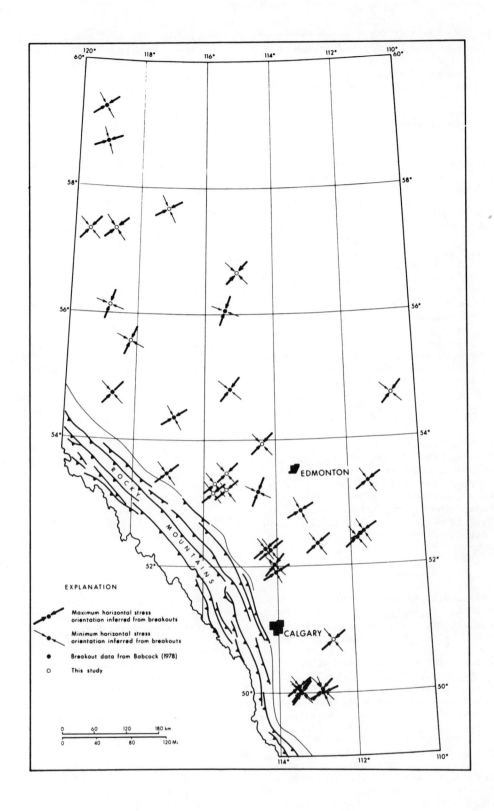

Fig. 7. Horizontal compressive stress orientations in Alberta inferred from four-arm dipmeter data in 35 wells.

NE-SW. Finally, residual stresses (Friedman, 1972; Engelder and Sbar, 1977) may remain in the crust from the mid-Tertiary folding and thrusting of the Rocky Mountains. All three processes could contribute to the stress field proposed.

Concluding Remarks

We hope to test our hypothesis by examining four-arm dipmeter logs run in wells drilled near sites of *in situ* stress measurements. It is clearly important that direct measurements of the stress tensor be made at several locations in the Alberta Plains. If stress measurements validate the hypothesis presented here and enunciated earlier (Bell and Gough, 1979), four-arm dipmeter logs may contribute much significant information about contemporary crustal stress in many parts of the world.

Acknowledgements. J.S.B. became aware of the systematic orientation of borehole elongations in Alberta, through discussions with Mr. J.W. Cox and Dr. E.A. Babcock. The idea that large, unequal horizontal stresses, concentrated near such holes, could cause unidirectional spalling occurred to D.I.G. after we had together discussed failure of tunnel roofs, in a large transverse stress field, with Dr. K.Y. Lo. Dr. M.B. Dusseault guided the authors to the cited references on induced subsurface fractures at Pembina and Cold Lake.

The authors have benefited from discussions with and encouragement from E.A. Babcock and J.W. Cox. Dr. Babcock also reviewed the paper and suggested improvements. D.I.G.'s research is supported by the Natural Sciences and Engineering Research Council of Canada. J.S.B. thanks BP Canada for permission to participate in this study and publish conclusions.

References

Allaud, L.A., and J. Ringot, The high resolution dipmeter tool, Log Analyst, 10, 3-11, 1969.

Babcock, E.A., Regional jointing in central Alberta, Can. J. Earth Sci., 10, 1769-1781, 1973.

Babcock, E.A., Jointing in southern Alberta, Can. J. Earth Sci., 11, 1181-1186, 1974.

Babcock, E.A., Fracture phenomena in the Waterways and McMurray Formations, Athabasca oil sands region, northeastern Alberta, Bull. Can. Petroleum Geol., 23, 810-826, 1975.

Babcock, E.A., Measurement of subsurface fractures from dipmeter logs, Amer. Assoc. Petroleum Geol. Bull., 62, 1111-1126, 1978.

Bell, J.S., and D.I. Gough, Northeast-southwest compressive stress in Alberta: evidence from oil wells, Earth Planet. Sci. Lett., 45, 475-482, 1979.

Cox, J.W., The high resolution dipmeter reveals dip-related borehole and formation characteristics, Paper presented at the Eleventh Annual Logging Symposium, Soc. Prof. Well Log Analysts, 1970.

Dusseault, M.B., Stress state and hydraulic fracturing in the Athabasca oil sands, J. Can. Petroleum Technology, 16, 19-27, 1977.

Engelder, T., and M.L. Sbar, The relationship between in situ strain relaxation and outcrop fractures in the Potsdam sandstone, Alexandria Bay, New York, Pure Appl. Geophys., 115, 41-55, 1977.

Friedman, M., Residual elastic strain in rocks, Tectonophysics, 15, 297-330, 1972.

Hubbert, M.K., and D.G. Willis, Mechanics of hydraulic fracturing, AIME Petroleum Trans., 210, 153-166, 1957.

Imperial Oil Ltd., The Cold Lake Project, Report to Energy Resources Conservation Board of Alberta, Canada, 1978.

Jaeger, J.C., Elasticity, Fracture and Flow, 2nd edition, Wiley, 212 pp., 1962.

Kehle, R.O., The determination of tectonic stress through analysis of hydraulic well fracturing, J. Geophys. Res., 69, 259-273, 1964.

Kirsch, G., Die Theorie der Elastizität und die Bedürfnisse der Festigkeitslehre, Zeitschrift des Vereines Deutscher Ingenieure, 42, 707, 1898.

Lo, K.Y., and J.D. Morton, Tunnels in bedded rock with high horizontal stresses, Can. Geotech. J., 13, 216-230, 1976.

McGarr, A., and N.C. Gay, State of stress in the Earth's crust, Ann. Rev. Earth Planet. Sci., 6, 405-436, 1978.

McLeod, J.G.F., Successful injection pattern alteration, Pembina J Lease, Alberta, Paper presented at 28th Ann. Mtg. of Petroleum Soc. of Can. Inst. Mining in Edmonton, 1977.

Nielsen, A.R., Cardium stratigraphy of the Pembina field, Alberta Soc. Petrol. Geol. Bull., 5, 4, 64-72, 1957.

Obert, L., In situ determination of stress in rock, Mining Engineer, 14, 51-58, 1962.

Raleigh, C.B., J.H. Healy, and J.D. Bredehoeft, Faulting and crustal stress at Rangely, Colorado, in Flow and Fracture of Rocks, Am. Geophys. Union Mon., 16, 1972.

Sbar, M.L., M. Barazangi, J. Dorman, C.H. Scholz, and R.B. Smith, Tectonics of the intermountain seismic belt: Micro-earthquake seismicity and composite fault plane solutions, Geol. Soc. America Bull., 83, 13-28, 1972.

Sbar, M.L., and L.R. Sykes, Contemporary compressive stress and seismicity in eastern North America: an example of intraplate tectonics, Bull. Geol. Soc. Amer., 84, 1861-1881, 1973.

PLANETARY MAGNETISM AND THE THERMAL EVOLUTION OF PLANETARY CORES

David Gubbins

Department of Earth Sciences, University of Cambridge, Madingley Rise, Cambridge CB3 0E7, England

Abstract. The power required to fuel planetary dynamos comes from radioactive heat and the gradual thermal evolution and differentiation of the planetary cores. Recent developments in the thermodynamics of cooling, self gravitating bodies have shown how the magnetic field strengths may be related to the cooling and differentiation processes. Several detailed models of the thermal history of the Earth's core have been proposed which lead to reasonable estimates of magnetic field strengths. In the future these calculations may be incorporated into a thermal history of the whole Earth, and applied to other planets which have, at least at some stage, possessed a magnetic field.

Introduction

The Earth's magnetic field has electric currents associated with it. These currents flow, for the most part, in the electrically conducting core, and the resistance of core material is sufficiently small to cause them to die away in about 10,000 years. Not only is a dynamo process necessary to sustain the magnetic field at its present strength over geological time, but work must be done by the fluid motions in order to replace the energy lost as heat by the electric currents. This Ohmic heating is quite large, estimates ranging from about 10^8W (Parker 1972) to 5.10^{12}W(Braginsky 1964), the larger figure being the more likely, and approaches in magnitude the Earth's surface heat flux of 4.10^{13}W (Sclater et al. 1980). Therefore some consideration should be given to the magnetic field when discussing the Earth's thermal history. The same statement cannot be made for the other planets. The magnetic energy of the Sun is a minute fraction of its total energy budget, and the same applies to Jupiter and the large planets. The magnetic fields of the terrestrial planets and satellites are all so small (see e.g. Busse 1976) that their heating effects are negligible. In fact these magnetic fields may all be due to magnetised rocks as Stephenson (1976) has suggested for the planet Mercury. While it is not possible to rule out an internally generated field for Mercury as yet (see Gubbins 1976), it

will eventually become possible to make observations to answer this question. As regards its magnetic field, as in so many other aspects, the Earth is unique.

The actual power requirements of the Earth's dynamo turn out to be much greater than just the Ohmic heating. In order to make any quantitative calculations for the overall energy requirements one must first specify how the dynamo is driven. If the liquid core is stirred by thermal convection then most of the heat will be convected away without any magnetic field being generated at all. Bullard & Gellman (1954) surmised that the fraction of heat energy converted to Ohmic dissipation is given by the Carnot efficiency factor $\Delta T/T$, where ΔT is the temperature difference across the convecting liquid and T is the temperature. This view has been promulgated by Metchnick et al (1974) who included the effect of heat conduction down the adiabatic gradient, which is rather steep in the core. A major step forward was taken by Backus (1975) and Hewitt et al (1975) who adopted a full thermodynamic approach to the problem. A crucial point is that the Ohmic losses remain inside the liquid core as an additional heat source. The global expression for conservation of energy does not involve the magnetic field in any way but is simply an equality (in steady state) between the sum of the heat sources and the heat flowing out into the mantle. Arguments involving the energy required to replace that lost to Ohmic heating are utterly false. The magnetic field does, however, enter into the entropy balance (see, for example, Backus 1975) and one can argue on the basis of entropy balance. Dissipative effects such as electrical or viscous heating always lead to positive entropy contributions. It is a consequence of the second law of thermodynamics that the diffusion constants be positive, which ensures that the corresponding entropy changes are positive definite. These gains must be offset by entropy losses arising from the various heat sources. The entropy depends not only on the quantity of heat supplied but also on the temperature at which it is supplied. Heating the top of the core, where it is coldest, will inhibit convection. Heating the bottom of the core where the fluid is hottest leads to the most efficient magnetic field generation. Backus (1975) and Hewitt et al (1975) arrive at a simple inequality:

$$\frac{\Phi}{Q} \lesssim \frac{\Delta T}{T_{min}} \qquad (1)$$

where Q is the total heat flowing out of the core and Φ is the Ohmic heating. The right hand side differs from the Carnot efficiency factor, which is $\Delta T/T_{max}$.

Gubbins (1976) has used (1) to obtain estimates of the heat required for the Earth's dynamo. Using a lower bound for Φ, based on Parker's (1972) approach, the heat flux, Q, is found to exceed the lower bound of 5.10^{11}W. In fact for reasonable estimates of core parameters this lower bound is greatly exceeded and Q is probably nearer 10^{13}W. This high value has led to a search for alternative energy sources.

The Thermodynamic Approach

The major sources of heat in the core are radioactivity, latent heat of freezing of the liquid as it cools, and drop in temperature of the whole core. Another source is the rotational energy of the Earth using precessional or tidal forcing. The interested reader is referred to the original papers of Malkus (1963, 1968) and the recent discussion by Rochester et al (1975), Loper (1975) and Gubbins & Masters (1979). Another source of energy is the gravitational energy released whenever the Earth contracts or there is rearrangement of material within it. The energy exchange of a cooling, contracting and self-gravitating body is very complex (Lapwood 1952). Consider the following model which incorporates most of the physical effects contained in recent discussions of this question. The outer core is composed of a mixture of liquid iron and some other, lighter, element such as sulphur (see e.g. Jacobs 1975). The inner core is supposed solid (Dziewonski & Gilbert 1971) and to contain a higher proportion of iron. Density data from seismology and shock wave experiments lend some credence to this view. The whole core cools slowly, the heat being carried out by convection. A heavy fraction of the liquid freezes and accretes on the inner core, releasing latent heat and leaving an excess of light liquid at the bottom of the core. This light material rises up and is redistributed throughout the core by the convection it helps to create. The gravitational energy released by this differentiation is converted into heat by the dissipative processes such as Ohmic heating.

This model can be analysed in terms of the thermodynamic approach set out by Backus (1975) and Hewitt et al (1975) and further developed by Gubbins (1977) and Gubbins et al (1979). It has been found crucial to retain all the terms relevant to a two-component self-gravitating

liquid. For example Gubbins (1977) found that the gravitational energy lost by compression of the core material, rather than by differentiation, is taken up as energy of formation of the body (i.e. work done by hydrostatic pressure forces during the compression) and is not available to drive convection. A similar phenomenon was noted in core formation (Flasar & Birch 1973). In a similar way Gubbins et al (1979) found that the molecular diffusion energy was also related to the gravitational energy release itself, and therefore cannot be self-consistently omitted, even though at first sight the molecular diffusion effect seems somewhat superfluous.

The two useful thermodynamic relations are conservation of energy and the equation relating exchange of entropy. Most authors have followed the notation of either Landau & Lifshitz (1959) or Malvern (1969). Energy conservation simply equates the heat flux out of the core to the energy sources within. Heat sources are radioactivity, latent heat, specific heat from drop of temperature, and heat of reaction between the components of the mixture. The gravitational energy released by differentiation is converted to heat and the changes in pressure lead to some adiabatic heating. Changes in the composition of the liquid also lead to a change in internal energy, depending on the chemical potential and its variation with depth in the core. There is also some gravitational energy lost by contraction of the liquid as it freezes, which appears as heat (Häge and Müller 1979). The entropy equation involves the major dissipative mechanisms: Ohmic, viscous, molecular diffusion of one component through another in the liquid, and thermal conduction. All dissipative terms give positive entropy gains to the system. Other entropy changes may be either positive or negative. For example if the reaction between iron and the light component in the core is exothermic, then heat will be absorbed during the dissociation at the inner core boundary and be released on re-mixing elsewhere in the liquid core. This amounts to heating the fluid from above which will inhibit convection, and the corresponding entropy change is positive. Other entropy changes are due to heating, gravitational energy release and change of internal energy. Roughly estimating each term in these equations gives an idea of whether the process will be able to generate a significant magnetic field.

This discussion is presented as a background for the review of recent models to be presented in the next section. More detailed mathematical derivations are given by Gubbins et al (1979) and Gubbins & Masters (1979).

Recent Models of a Cooling Core

Most of the recent work has assumed the core to be devoid of significant amounts of radioactive elements. Verhoogen (1961) was first to point out that heat released from freezing and cooling of the core was sufficient to balance the Ohmic heat losses of quite a large magnetic field. Recent calculations put the available heat at a lower value, and when the low thermodynamic efficiency of conversion of heat to magnetic energy and the loss of heat by conduction down the adiabatic gradient are taken into account the available heat does not seem adequate to generate any magnetic field (Gubbins et al 1979). The problem is that if the core cools sufficiently rapidly to drive the dynamo, the inner core grows to its present size in a time much shorter than the age of the Earth. The shortfall in energy is helped by the gravitational energy released by a small change in volume of the liquid on freezing. Häge & Müller (1979) have performed a careful study of this situation and conclude that this gravitational energy is released as heat at the inner core boundary, effectively augmenting the latent heat by another 60 - 70%. The statement by Gubbins and Masters (1979), that this gravitational energy is not available to assist convection, is incorrect.

Braginsky (1963) was first to propose that differentiation of the components of the outer core mixture was important in making the fluid convect. Gravitational energy released in this way contributes a disproportionately large factor to the entropy balance when compared to heat. In other words gravitational energy is much more effective than heat in generating magnetic fields. The reason for this is not hard to see. Light material rising up will drive fluid motions directly and all of this potential energy must be converted to heat via one of the dissipative processes before the heat escapes to the mantle. With heat-driven convection, however, most of the energy is simply carried away by hot fluid rising to the top of the core. It is the attractive proficiency of differentiation for driving the dynamo that has provoked so much recent interest in Braginsky's ideas. Gubbins (1977) extended Backus's (1975) thermodynamic treatment to include gravitational energy, latent and specific heat, and obtained an inequality analogous to (1)

$$\frac{\Phi - (G-P)}{Q} \lesssim \frac{\Delta T}{T_{min}} \qquad (2)$$

where G is the total release of gravitational energy and P is the work done by pressure forces during contraction. Gubbins concludes that the gravitational energy could give a modest field of 50 Gauss.

Loper (1978 a) has produced a more quantitative model for the gravitationally powered dynamo and concludes that a field of 700 Gauss or more could be maintained if the inner core had grown at a constant rate over the age of the Earth. The calculation equates gravitational energy of differentiation with Ohmic heating. Using another quantitative model and the full thermodynamic approach, Gubbins et al (1979) conclude

that the gravitational energy can produce only 150 Gauss. The main reason for the discrepancy - a factor of over 20 in the entropy integrals - is the dominance of the entropy gain due to thermal conduction over other dissipative mechanisms. In another paper Loper (1978 b) makes the important point that convection may be driven by the differentiation even though heat sources are inadequate to maintain the adiabatic gradient. Heat is actually convected downwards in this doubly diffusive system, and the total heat flowing out of the core will be less than that conducted down the adiabatic gradient. However, a price is paid in this because the entropy loss due to gravitational energy must balance entropy gain due to thermal conduction along the adiabat as well as Ohmic dissipation. Equating gravitational energy with Ohmic heating is only valid in the unlikely event that subsidiary heat sources are adequate to maintain an adiabatic gradient and no more. The model of Gubbins et al (1979) assumes a constant rate of drop of temperature at the core mantle boundary and a variety of plausible numerical values for the parameter. In most cases the heat flux is less than that conducted down the adiabatic gradient so that the doubly diffusive regime is pertinent. Typical field strengths are of the order of a hundred Gauss and heat fluxes, about 5.10^{12}W.

The dynamics of the core fluid will depend on the temperature gradient. In some regions this may be sub-adiabatic and vertical motion will be suppressed. This could easily be caused by the core cooling rapidly with the temperature at the base of the mantle remaining fairly constant. Consequences of such a mode of cooling have been examined by Gubbins et al (1981).

Another possibility is that the temperature in the outer core may touch the solidus at some point. The appropriate mode of convection is then two phase convection in which a slurry of solid particles and liquid rains down towards the inner core (Elsasser 1972, Busse 1972, Malkus 1973). Malkus (1973) stresses that the convection will depend critically on the size of the particles, which in turn depend on the reaction time of the freezing process. Details of this process must be very difficult to work out with any certainty. The existence of a slurry was postulated by Braginsky (1963). He required partial solidification to occur in a region near the inner core boundary and identified it with the seismological F layer (see for example Bullen 1963). The seismological evidence for the F layer has now largely disappeared (Haddon & Clearly 1974), and Gubbins (1978) has argued that the time constant for the melting process is too small to produce any attenuation of seismic waves. However, the idea of a compositional boundary layer near the inner core has regained some popularity with the recent interest in core energetics. Loper (1978 b) has given a thorough discussion of the various possibilities that can occur when the adiabatic gradient exceeds or falls below the melting curve, and also when the composition of light material exceeds that of a eutectic mixture. He concludes that Braginsky's (1963) scheme, in which a eutectic mixture freezes out leaving excess light material, is unworkable because a stably stratified layer of variable composition is set up around the inner core which inhibits heat transfer.

Clearly a whole new branch of geophysical fluid dynamics has been opened up in which phase changes and compositional variations are important. Loper & Roberts (1978) have made an ambitious start in setting out the equations governing such a system and plan further studies.

Conclusions and Future Developments

The thermodynamic approach to the energetics of the core has given a new viewpoint to the subject. Estimates of the power required by the dynamo, that actually available, and the heat flowing out of the Earth's surface, all give comparable numbers. This means that reasonably accurate calculations for the energy supply are desirable. These calculations must be done taking full account of the thermodynamics or gross numerical errors may result. The fluid mechanics of the core will be exceedingly complex because of the presence of many phases and components. We cannot safely ignore any effect that may arise just because it seems exotic, unless there is observational evidence for doing so. The boundary layers at the base of the mantle and at the bottom of the core will be important.

To press any of the thermal evolution calculations further it will be necessary to tie them to a thermal history of the whole Earth. This is quite easy to do once a satisfactory model for mantle convection and evolution has been found (e.g. Sharpe & Peltier 1979). The calculation can than be applied to the other terrestrial planets.

The most desperate need is for some hard observational evidence either supporting or eliminating some of the possible behaviour that is suggested by the energy calculations. Seismological studies are already revealing information about the D" region (Julian & Sengupta 1973, Dziewonski et al 1977), and there may be hope for a "re-discovery" of the F layer. Improved normal mode data may show up stably stratified regions in the core (Masters 1979). Observations of Q in the inner core tell us how close it is to its melting point (Doombos 1974) which would be useful in helping to establish the compositional difference between inner and outer cores.

The fluid dynamics of mixtures and the theory of the dynamo problem are both active research areas that will give a clearer picture of details of the convection in the core. There has so far been very little effort at examining iron meteorite samples to deduce the cooling behaviour of the parent body in relation to present day

cooling of the core. Such a study might well give clues as to the important fluid dynamical effects to be expected in the core.

Acknowledgements. This work was partially supported by Natural Environment Research Grant GR3/3475.

References

Backus, G.E., Gross thermodynamics of heat engines in deep interior of Earth, Proc. Nat. Aca. Sci. U.S.A.,72, 1555-1558, 1975.

Braginsky, S.I. Structure of the F layer and reasons for convection in the Earth's core, Dokl. Akad. Nauk. SSSR, 149, 1311-1314, 1963.

Braginsky, S.I., Magnetohydrodynamics of the Earth's core, Geomagn. Aeron., 4, 698-712, 1964.

Bullard, E.C. and Gellman, H., Homogeneous dynamos and terrestrial magnetism, Phil. Trans. Roy. Soc. Lond. A, 247, 213-278, 1954.

Bullen, K.E., An introduction to the theory of seismology, Cambridge Univ. Press, London, 381 pp., 1965.

Busse, F.H., Comments on "The adiabatic gradient and the melting point gradient in the core of the Earth" by G. Higgins and G.C. Kennedy, J. Geophys. Res., 77, 1589-1590, 1972.

Busse, F.H., Generation of planetary magnetism by convection, Phys. Earth Planet. Interiors, 12, 350-358, 1976.

Doornbos, D.J., The anelasticity of the inner core, Geophys. J.R. Astr. Soc., 38, 397-417, 1974.

Dziewonski, A.M. and Gilbert, F., Solidity of the inner core of the Earth inferred from normal mode observations, Nature, 234, 465, 1971.

Dziewonski, A.M., Hager, B.H. and O'Connell, R.J., Large scale heterogeneities in the lower mantle, J. Geophys. Res., 82, 239-255, 1977.

Elsasser, W.M., Thermal stratification and core convection (abstract), EOS Trans. AGU, 53, 605, 1972.

Flasar, R.M. and Birch, F., Energetics of core formation: a correction, J. Geophys. Res., 78, 6101-6103, 1973.

Gubbins, D., Observational constraints on the generation process of the Earth's magnetic field, Geophys. J.R. Astr. Soc., 47, 19-39, 1976.

Gubbins, D., Energetics of the Earth's core, J. Geophys., 43, 453-464, 1977.

Gubbins, D., Attenuation of seismic waves in an iron slurry, Geophys. Astrophys. Fluid Dynamics, 9, 323-326, 1978.

Gubbins, D., and Masters, T.G., Driving mechanisms for the Earth's dynamo, in Advances in Geophysics, 21, ed. B. Saltzman, 1979.

Gubbins, D., Masters, T.G. and Jacobs, J.A., Thermal evolution of the Earth's core, Geophys. J. R. Astr. Soc., 59, 57-100, 1979.

Gubbins, D., Thomson,C.J.,and Whaler, K.A., Stable regions in the Earth's liquid core, to appear, Geophys. J.R. Astr. Soc., 1981.

Haddon, R.A.W., and Cleary, J.R., Evidence for scattering of seismic PKP waves near the mantle core boundary, Phys. Earth Planet. Interiors, 8, 211, 1974.

Hage, H., and Müller, G., Changes in dimensions, stresses and gravitational energy of the Earth due to crystallisation at the inner core boundary under isochemical conditions, Geophys. J. R. Astr. Soc., 58, 495-508, 1979.

Hewitt, J.M., McKenzie, D.P. and Weiss, N.O., Dissipative heating in convective flows, J. Fluid. Mech., 68, 721-738, 1975.

Jacobs, J.A., The Earth's Core, Academic Press, London, 253 pp., 1975.

Julian, B.R., and Sengupta, M.K., Seismic travel time evidence for lateral inhomogeneity in the deep mantle, Nature, 242, 443-447, 1973.

Landau, L.D., and Lifshitz, E.M., Course of Theoretical Physics, Vol. 6, Fluid Mechanics, Pergamon, London, 536 pp., 1959.

Lapwood, E.R., The effect of contraction in the cooling by conduction of a gravitating sphere, with special reference to the Earth, Mon. Not. R. Astr. Soc. Geophys. Supp. 6, 402-407, 1952.

Loper, D., Torque balance and energy budget for precessionally driven dynamos, Phys. Earth Planet. Int., 11, 43-60, 1975.

Loper, D., The gravitationally powered dynamo, Geophys. J. R. Astr. Soc., 54, 389-404, 1978 a.

Loper, D., Some thermal consequences of a gravitationally powered dynamo, J. Geophys. Res., 83, 5961-5970, 1978 b.

Loper, D., and Roberts, P.H., On the motion of an iron-alloy core containing a slurry, I. General Theory, Geophys. Astrophys. Fluid Dynamics, 9, 289-321, 1978.

Malkus, W.V.R., Precessional Torques as the cause of Geomagnetism, J. Geophys. Res., 68, 2871-2886, 1963.

Malkus, W.V.R., Precession of the Earth as the cause of geomagnetism, Science, 160, 259-264, 1968.

Malkus, W.V.R., Convection at the melting point: A thermal history of the Earth's core, Geophys. Fluid Dyn., 4, 267-278, 1973.

Malvern, L.E., Introduction to the mechanics of a continuous medium, Prentice-Hall Inc., Englewood Cliffs, N.J., 713 pp., 1969.

Masters, T.G., Observational constraints on the chemical and thermal structure of the Earth's deep interior, Geophys. J. R. Astr. Soc., 57, 507-534, 1979.

Metchnik, V.I., Gladwin, M.T., and Stacey, F.D., Core convection as a power source for the geomagnetic dynamo - thermodynamic argument, J. Geomag. Geoelect., 26, 405-415, 1974.

Parker, R.L., Inverse theory with grossly inadequate data, Geophys. J. R. Astr. Soc., 29, 123-138, 1972.

Rochester, M.G., Jacobs, J.A., Smylie, D.E. and Chong, K.F., Can precession power the geomagnetic dynamo?, Geophys. J. R. Astr. Soc., 43, 661-678, 1975.

Sclater, J.G., Jaupart, C., and Galson, D., The heat flow through oceanic and continental crust and the heat loss of the Earth. Rev. Geophys. Space Phys., 18, 269-312, 1980.

Sharpe, H.N., and Peltier, W.R., A thermal history model for the Earth with parameterised convection, Geophys. J. R. Astr. Soc., 59, 171-204, 1979.

Stephenson, A.L., Crustal remanence and the magnetic moment of Mercury, Earth Planet. Sci. Letters, 28, 545-458, 1976.

Verhoogen, J., Heat balance of the Earth's core. Geophys. J., 4, 276-281, 1961.

ISOTOPIC COMPOSITION OF THE OCEAN - ATMOSPHERIC SYSTEM IN THE GEOLOGIC PAST

J. Hoefs

Geochemisches Institut, Goldschmidtstrasse 1, 3400 Güttingen, Germany

Abstract. One of the most promising tools in tracing back the history of the ocean - atmosphere system seems to be the analysis of stable isotope ratios of suitable samples, having preserved their primary isotope ratios since formation and deposition. Besides the oxygen isotopic composition of ocean water this review concentrates on the sulfur isotopic composition of ocean sulfate, the carbon isotope composition in the system ocean - atmosphere - biosphere and the strontium isotopic composition of ocean water. It is shown that there are definite fluctuations in the isotopic composition of these elements during the earth's history, especially during the Paleozoic. These changes cannot be seen as isolated phenomena and have far-reaching implications not only for the ocean - atmosphere system. They can only be explained if very important geologic parameters have fluctuated, two of which are a) the distribution of land and sea (due to changes in tectonic and volcanic activity), b) the atmospheric oxygen concentration, maybe due to varying rates of photosynthesis or carbon burial and sulfide oxidation.

Introduction

The question if the chemical composition of ocean water has been constant through the geological history or if there have been changes (gradual or "catastrophic") is very old and has been discussed quite frequently in the literature. Two of the more recent classical papers are those by Rubey (1951) and Holland (1972). The current view about the chemical composition of the ocean water in the geological past can be described by the principle that the sources are equal to the sinks or in other words that the input, mainly through rivers, is balanced by the output, mainly through formation of sediments. Since it is clear that the residence time of all dissolved substances in the ocean is much smaller than the age of the earth, it is a fundamental condition of ocean chemistry that the chemical outputs are more or less balanced by the chemical inputs. Furthermore from paleoecological studies we can deduce that ocean water should not have changed its chemical composition very drastically since marine organisms can only tolerate relatively little chemical changes in their marine environment. The similarity of the mineralogy of sedimentary rocks during the past 3 billion years and the restricted thermodynamic stability of some salt mineral paragenesis strengthens the conclusion that the chemical composition of ocean water should not have varied too much. However, in this crude view, changes in chemical composition by a factor of two or so might be possible and are undetected by conventional methods such as balance calculations (Rubey, 1951), or analysis of evaporite mineral assemblages (Holland, 1972). One of the most promising tools in looking for such possible changes seems to be the analyses of isotope ratios, which are one of the most sensitive tracers in geochemical research. Therefore, in the following, a review about the present knowledge of isotopic variations during the ocean's history together with new unpublished measurements are given. Furthermore, an attempt is made to combine these different pieces of evidence. In this connection, we have to ask which kind of geological samples or evidence gives us relevant information about the chemical composition of ocean water in the geological past or, in other words, what kind of geological record represents true information about the primary chemical composition. To illustrate this problem: today we know that the analysis of fluid inclusions in sedimentary rocks (Kramer, 1965) or the analysis of connate water (Chave, 1960) fails to give us the right information, because of sedimentary diagenetic change of the water after its entrapment. The consequences of these findings are that only those geological materials are suitable for an investigation like this which have not been altered since their time of formation. However, we must be fully aware that indications of secondary alteration are not at all clear and well known in all cases. This is the major restriction for the following discussion.

Sulfur

Maybe the best documented variation trend concerns the sulfur isotope distribution in marine sulfate. In 1964 Nielsen and Ricke and Thode and Monster published independently the first two "age curves" which showed that the isotopic composition of gypsum and anhydrite in marine evaporites was different at different times in the geological past. Since then, this curve has been updated with many more analyses (see Holser and Kaplan, 1966, Nielsen, 1972 and Holser, 1977)

demonstrating that especially during the Phanerozoic several pronounced maxima and minima in $\delta^{34}S$ exist, the extremes lying close to +10 (Permian) and +30 (Cambrian) (see Figure 1). Since the isotopic fractionation between the sulfate-containing evaporite and the sulfate in ocean water is almost negligible, the observed trend in evaporite sulfate should closely reflect fluctuations in the sulfur isotope composition of marine sulfate through geologic time. The present $\delta^{34}S$ value of marine sulfate is primarily the result of the activity of sulfate-reducing microorganisms. Below the sediment-ocean water interface anoxic conditions

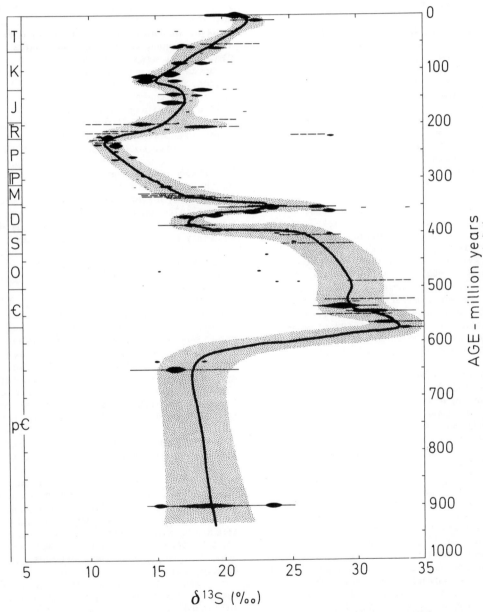

Fig. 1. $\delta^{34}S$ age curve of oceanic sulfate (from Holser, 1977).

are the rule rather than the exception. Bacterial sulfate reduction extracts isotopically "light" H_2S from the ocean reservoir which is fixed mainly as pyrite in newly formed sediments. Therefore, it is quite plausible that the exogenous sulfur is divided into a "light" fraction mainly incorporated in argillaceous sediments and into a "heavy" fraction, mainly consisting of the dissolved ocean water sulfate and of the evaporite sulfate. Changes in the $\delta^{34}S$ of marine sulfate during the geologic past may be caused by major changes in the budget between the individual reservoirs: on the one hand, periods of high biological sulfate-reduction, which should take place under favorable paleographic conditions - for example marginal basins - increases the $\delta^{34}S$ of ocean sulfate. On the other hand, periods of extended weathering introduce additional light continental sulfur into the ocean which decreases the $\delta^{34}S$ of ocean sulfate. Such periods of extended weathering are geologically plausible in periods of high tectonic, mountain-building activity.

Early models by Nielsen (1965) and Holser and Kaplan (1966) discussed qualitatively the consequences of such changes in the weathering rate and bacterial reduction rate. These earlier models did not, however, consider the possible influence of varying rates of evaporite formation on the $\delta^{34}S$-values of marine sulfate. REES (1970) was the first to point to the importance of sulfur extraction by evaporite formation and postulated that "δ (ocean) should have tended to high values in periods when evaporite formation was of minor importance and to low values in periods of major evaporite formation. This is qualitatively the case for the Cambrian and for the Permian..."

In the past few years a further mechanism of sulfate extraction became quite evident, that is the annual cycling of large quantities of seawater through midocean ridges which can have a remarkable effect on the chemistry of ocean water.

From the experiments by Mottl and Holland (1978) and Ohmoto et al. (1976) it is not only obvious that during basalt-seawater interaction oceanic sulfate is removed and reduced to pyrite and/or pyrrhotite but this sulfate removal is also accompanied by a sulfur isotope fractionation, favoring the light isotope in the sulfides and thus making the remaining sulfate heavier.

The sulfur isotope age curve is characterized by some steep increases in the ^{34}S-content of oceanic sulfate over a short geologic time. Holser (1977) recognized three such "events", in which the rise in $\delta^{34}S$ is so steep that an extremely and unreasonably high rate of sulfide precipitation would be required to change the whole ocean over a geologically short time. To overcome this difficulty Holser (1977) postulated the formation of local brines in deeps of a mediterranean basin. Underneath the brine

pyrite precipitation builds a store of brine heavy in $\delta^{34}S$ (sulfate). "Catastrophic" mixing of the brine and the surface ocean, is the source of the sharp rise in the sulfur isotope curve.

Whatever the actual causes may be for the fluctuations of the $\delta^{34}S$ values of oceanic sulfate during the geologic past, it is obvious that this behavior of oceanic sulfate deviates strongly from that expected from an extreme concept of a steady state ocean. In such a steady-state view the partition into reduced and oxidized reservoirs would be at a fixed ratio. According to Garrels and Perry (1974) the range of $\delta^{34}S$ from ±30‰ to ±10‰ corresponds to a variation of ±30‰ in the average total amount of sulfate stored in sedimentary rocks and in the ocean. In addition, it is obvious that fluctuations in the three major mechanisms for sulfate removal (a) evaporite formation, b) bacterial reduction and c) cycling through midocean ridges) all have occurred largely in response to changes in the geography of the ocean basins and/or of the adjacent seas.

While the partial cycle between ocean and evaporites only involves sulfate transfer from one reservoir to the other, bacterial sulfate reduction as well as weathering of sulfides from argillaceous sediments change the valence state of the sulfur. Therefore, during a period with increased rate of one of these two processes, appreciable amounts either of organic compounds or of free atmospheric oxygen are needed. Especially in the latter case, oxygen consumption during weathering is appreciable. Holland (1978) calculated that oxygen consumption during complete oxidation of 1 kg of average sediment due to the oxidation of C^0, S^{2-} and "FeO" is approximately 12, 6 and 2g O_2/kg respectively. Whatever the actual number is, the amount of oxygen needed is considerable. According to Li (1972) about 50% of the total amount of photosynthetically produced oxygen during the earth's history is stored in sulfate. Since we have shown that appreciable fluctuations obviously exist between the oxidized and reduced sulfur species during the earth's history, atmospheric oxygen concentrations might have fluctuated within small limits.

In this connection the oxygen isotope composition of oceanic sulfate is briefly discussed. Data on $\delta^{18}O$ (sulfate) of various evaporites have been reported much less than those on sulfur, notable exceptions are besides a few other the data by Sakai (1972) and Claypool et al. (1972). These data allow the construction of a tentative oxygen isotope curve, which shows much less fluctuations than the sulfur isotope curve. Holser et al. (1979) have demonstrated that the oxygen isotope composition of oceanic sulfate is controlled by a dynamic balance of sulfate inputs (mainly from weathering of sulfides and sulfates) and outputs (mainly through evaporite formation and sulfate reduction). One of the

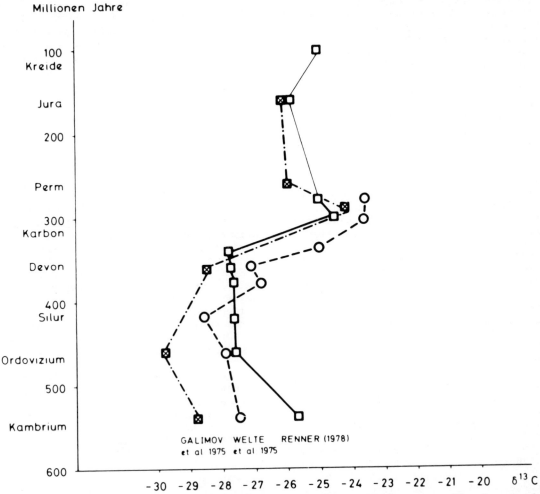

Fig. 2. $\delta^{13}C$ age curves of kerogen from sediments of three different areas (Russian platform, Galimov et al., 1975; Rheinisches Schiefergebirge, Germany, Welte et al., 1975; Williston Basin, Canada, Renner, 1978).

chief unknowns is the oxygen isotope fractionation during sulfate reduction which has been measured in laboratory experiments to be one fourth of that of the sulfur isotope fractionation (Mizutani and Rafter, 1969). However, the question remains if this ratio is relevant for all geological environments (Holser et al., 1979).

An alternative interpretation would be that the $\delta^{18}O$ value of oceanic sulfate is chiefly controlled by reaction and isotopic exchange of sea water sulfate with hot basaltic material, mainly at midocean ridges.

Carbon

The two main exogenic carbon reservoirs a) oxidized inorganic carbon (mostly carbonate sediments) and b) reduced organic carbon (mostly fine dispersed sedimentary organic matter) are

isotopically separated from each other, the former concentrating the heavy isotope ^{13}C, the latter concentrating the light isotope ^{12}C. If the total exogenic carbon remains constant, growth of one reservoir must be compensated for by shrinking of the other reservoir with a resulting shift in the ^{13}C concentration. Although there have been some speculations about the systematic changes in the isotopic composition of inorganic and organic carbon during the earth's history (Jeffery et al., 1955; Compston, 1960; Weber, 1967) most workers agreed until a few years ago that the carbon isotopic composition of carbonates did not show any definite age trend (for instance Keith and Weber, 1964; Veizer and Hoefs, 1976).

However, as more and more carbon isotope analyses - both in carbonates and in organic matter - become available and as the stratigraphic derivation of samples has become refined, a pattern of

^{13}C variation with time is beginning to emerge.

The first piece of evidence stems from the analyses of sedimentary organic matter mainly from the Paleozoic. From three different areas in the world a consistent trend towards heavier $\delta^{13}C$ values in the kerogen from Devonian towards Carboniferous and Permian time has been reported (Galimov et al., 1975, for the Russian platform, Welte et al., 1975, for the Rheinisches Schiefergebirge, Germany, and Renner, 1978, for the Williston Basin, Canada). This relationship is shown in Figure 2: although there are some differences in the absolute values between the 3 different areas, the overall generalized trend is the same.

Secondly, Renner (1978) analyzed besides the kerogen the carbonate carbon and found as in the kerogen the same parallel trend in the carbonate (see Figure 3). For a totally different geologic period (Tertiary through Recent) Fischer and Arthur (1977) observed a similar parallel trend for both carbonate and organic carbon. Veizer et al. (1979) evaluated statistically the compilation of $\delta^{13}C$ values of marine carbonates of Veizer and Hoefs (1976) and also found such a shift towards heavier $\delta^{13}C$ values in the late Paleozoic. It is noteworthy that especially in Permian time a relatively strong enrichment of ^{13}C may have occurred (Magaritz and Schulze, 1979).

The secular variation in carbon isotope ratios can be explained by at least two different factors.

1) The withdrawal of photosynthetically fixed carbon into sediments may fluctuate, since, during photosynthesis, ^{12}C is preferentially fixed, the remaining CO_2 or HCO_3 becomes enriched in ^{13}C. Fluctuation in the fixation of organic matter may result from two different processes:

   a) increase in the rate of carbon burial related to changes in the oxidizing potential of the oceanic system which has been proposed for instance by Fischer and Arthur (1977), Ryan and Cita (1977)

   b) increase in photosynthetic activity; this has been favored by Tappan (1968), Welte (1970), Welte et al. (1975) and others. However, this increase can be only observed if the rate of carbon burial remains constant.

Both mechanisms are equally geologically plausible. An increase in photosynthesis could occur from Devonian towards Carboniferous time due to the appearance and increased productivity of land plants (Welte et al., 1975). An increase in the rate of carbon burial could occur in the Cretaceous for instance, where probably large parts of the ocean floor were poorly supplied with oxygen, e.g. Ryan and Cita (1977). However, the determination of carbon isotope ratios in sedimentary organic matter cannot differentiate between both possibilities.

2) Temporal variations in the rates of carbonate deposition will also change the carbon isotope composition since during carbonate precipitation preferentially ^{13}C is fixed. This implies that during periods of maximal carbonate withdrawal the remaining CO_2 or HCO_3^- should be enriched in ^{12}C. However, as far as I know, no well-founded estimates exist about varying rates of carbonate precipitation and preservation, so nothing can be said at the moment about possible variation in the carbonate reservoir during the last 600 million years.

In summary, these findings do not favor the idea of constant global carbon reservoirs and, similar to sulfur isotope variation, imply that the oxidized and reduced reservoirs have changed with time. Furthermore, the parallel trend observed in both, carbonates and kerogen, argues against a temperature-dependent fractionation as the main reason for these changes.

Similar conclusions have been drawn from model calculations by Garrels and Perry (1974) and by Veizer et al. (1979), who found a correlation between the $^{13}C/^{12}C$ ratios in sedimentary carbonates and the $^{34}S/^{32}S$ ratios in contemporaneous sulfates. These model calculations showed that the slope of the regression line is about 0.1, meaning that if the $\delta^{34}S$ of oceanic sulfate changes by 10‰, the $\delta^{13}C$ of the carbonate changes by about 1‰. Since the ratio of carbonate carbon to organic carbon is about 4:1, this equally means a change in $\delta^{13}C$ of organic carbon by about 4‰. So the changes in the reservoir of organic carbon is much more sensitive than that of carbonate carbon.

Strontium

Gast (1955) and later on Hurley et al. (1965) clearly recognized that $^{87}Sr/^{86}Sr$ ratios in limestones show some variations through geological time. This more qualitative statement was put in quantitative numbers by Peterman et al. (1970), Dasch and Biscaye (1971), Veizer and Compston (1974) and Brass (1976). Although the strontium isotopic composition of carbonates does not only depend upon the isotopic composition of the solution from which they were precipitated (they also depend upon diagenetic recrystallization effects) Veizer and Compston (1974) made probable that the lowest $^{87}Sr/^{86}Sr$ ratio observed for any suite of carbonates may be taken as the value of the contemporary seawater. The above mentioned papers clearly demonstrated that the $^{87}Sr/^{86}Sr$ ratio of ocean water has varied between values as low as about 0.7070 in the Jurassic period to values as high as 0.7090 at present. Furthermore, similar high ratios may have been in early Cambrian time (Veizer and Compston, 1974). Figure 4 summarizes the observed fluctuation in $^{87}Sr/^{86}Sr$ (from Faure et al., 1978).

Three sources are considered for strontium in ocean water

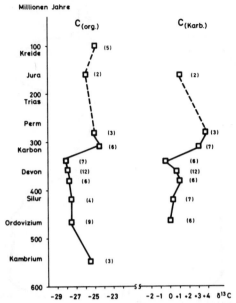

Fig. 3. $\delta^{13}C$ age curves of carbonates and kerogen for sediments from the Williston Basin, Canada (after Renner, 1978) (number in brackets represents number of samples analyzed).

1) the continental crust which through erosion and river transport supplies the ocean with a continuous input of material relatively enriched in radiogenic Sr. Although the average $^{87}Sr/^{86}Sr$ ratio of today's continental run-off strontium cannot be given precisely at the moment, we take a value of 0.716 as a reasonable estimate, given by Veizer and Compston (1974) for freshwaters from the Canadian shield.

2) the mantle strontium is released by oceanic and continental volcanism, notably at ridge systems during recent times. The mean $^{87}Sr/^{86}Sr$ ratio of fresh deep-sea basaltic rock may be near 0.7030, while the mean modified ratio of the oceanic crust may be taken as about 0.7040 (modified by exchange with seawater near the spreading oceanic ridges (Spooner, 1976)

3) the limestone recycling process which seems to be an important source of strontium in seawater (Garrels and MacKenzie, 1971). Since the $^{87}Sr/^{86}Sr$ ratio of marine limestones is not very different from ocean water this source does not have a large effect on the strontium isotopic composition.

These three sources maintain the $^{87}Sr/^{86}Sr$ ratio of the ocean by balancing the continental run-off flux against the hydrothermal basaltic flux near the oceanic ridges. For the last 200 million years the proportions of these two sources appear to reflect the history of global tectonics. The post Cretaceous increase in the strontium isotope composition may be explained by an increase in land area. Variations in land area may themselves have been partly a consequence of variations in global mean sea-floor spreading rate.

Oxygen

The amount of oxygen bound in ocean water is huge compared to the other elements discussed in this paper like sulfur, carbon and strontium. So, to alter the oxygen isotopic composition of ocean water demands huge amounts of oxygen being isotopically very different from the ocean water composition. It is generally agreed that relatively short-term variations of $\delta^{18}O$ occurs during the ice ages. If all the ice sheets in the world were melted containing fresh water enriched in ^{16}O, the present $\delta^{18}O$ value of ocean water would be lowered by about 1‰. Estimation for the maximum enrichment of ^{18}O during the Pleistocene glaciations range from 0.5‰ (Emiliani, 1965) to 1.5‰ (Shackleton, 1967).

Throughout Phanerozoic times the oxygen isotopic composition of ocean water has probably fluctuated within those limits. Muehlenbachs and Clayton (1976) presented a model in which the isotopic composition of ocean water is held constant by two different processes: a) the low temperature weathering of oceanic crust depletes ocean water in ^{18}O, because ^{18}O is preferentially bound in the weathering products whereas b) the high-temperature hydrothermal alteration of ocean ridge basalts enriches ocean water in ^{18}O, because ^{16}O is consumed by the hydrothermal alteration of oceanic crust. If sea floor-spreading ceased, the $\delta^{18}O$ of the oceans would slowly change to lighter values because of continued continental weathering, but at a rate measured in billions of years.

There are major uncertainties for the Precambrian time. Based on the well-documented fact that cherts show a secular variation in $^{18}O/^{16}O$

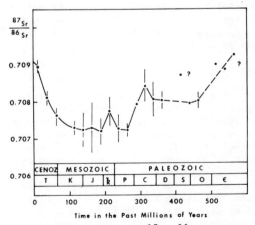

Fig. 4. Changes in the $^{87}Sr/^{86}Sr$ ratio of the oceans of Phanerozoic age (from Faure et al., 1978).

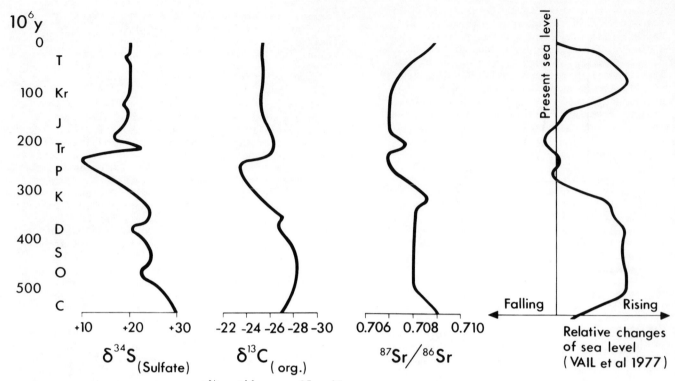

Fig. 5. Age curves of δ^{34}S, δ^{13}C and ^{87}Sr/^{86}Sr ratio together with the sea level curve of Vail et al. (1977).

with geologic time, Becker and Clayton (1976) argued that the Precambrian ocean had a δ^{18}O values of at least as light as -3.5‰, whereas Perry (1976) and Perry et al. (1978) postulate even lighter δ^{18}O values between -8 and -12‰. However, the same trend can also be attributed to post-depositional isotopic exchange with low ^{18}O meteoric waters (Kolodny and Epstein, 1976) or to changes in ocean water temperatures (Knauth and Epstein, 1976, Knauth and Lowe, 1978). If temperature is the main factor, the data of Knauth and Lowe (1978) imply that the ocean surface temperature should have been about 70°C 3.4 billion years ago.

Atmospheric Oxygen

Concerning the oxygen isotopic composition of atmospheric oxygen, the following speculation can be performed: It is well known that atmospheric oxygen is not in isotopic equilibrium with ocean water (Dole effect). The observed isotopic enrichment in ^{18}O seems to have been caused by a preferential uptake of ^{16}O during respiration. Therefore, the δ^{18}O value of atmospheric oxygen is balanced between input from photosynthesis and output by respiration. If it would be possible to determine the δ^{18}O value of atmospheric oxygen in the geologic past, an important parameter reflecting possible ecological changes in the terrestrial biosphere would be available.

One interesting attempt for such a determination has been carried out by Heinzinger et al. (1974). They measured the ^{18}O/^{16}O ratios in oxide phases of cosmic spherules, i.e. ablation droplets of iron meteorites, oxidized in the atmosphere at high temperatures. With this technique, they were able to show that Tertiary spherules were not different from today's ones, but the situation was different during the Devonian, which is geologically not unreasonable. Pre-Devonian air oxygen should lack the contributions of land plants and should be isotopically lighter than today (δ^{18}O air oxygen) $\sim$ +10‰ as Wagener, 1973, has estimated).

Conclusions

As has been demonstrated in the foregoing discussion, fluctuations in the isotopic composition of sulfur, carbon, strontium and oxygen during the last 500 million years obviously occurred. Some of the possible parameters influencing the changing isotopic composition have been mentioned, such as varying rates of weathering, bacterial sulfate reduction, photosynthesis and carbon burial etc. However, some of these parameters are not totally independent from each other. One principal factor affecting many of these parameters - and not mentioned up till now - may be a varying sea level.

Today, the world's ocean covers 71% of the

earth's surface, but it has been recognized for a long time that this figure is nearly a minimum number and that during long periods of the earth's history it might have increased to 80% and more. In Figure 5 the isotopic variations of the elements sulfur, carbon and strontium are schematically drawn together with the curve of relative changes of sea level after Vail et al. (1977). Global highstands of sea-level are characterized by wide-spaced shallow seas on the continental shelves, whereas global lowstands are characterized by reduction of the area of shallow seas and an increased rate of erosion.

Looking at the various curves, one interesting feature is quite obvious. Very pronounced changes seem to occur during the Permian and Permian/Triassic boundary. This period is especially noteworthy because in the opinion of some paleontologists the mass extinction during this period appears to be the most devastating event in the history of life and of the ocean. Two of the more accepted explanations for this worldwide observed mass extinction are a) a strong reduction in the area of shallow seas (e.g. Schopf, 1974) or b) a reduction of ocean salinity due to either a formation of deep brines or huge amounts of evaporites (Fischer, 1964; Stevens, 1977). Whatever the actual cause may have been, both processes might have an influence on the chemistry of the ocean. The regression theory could explain the low $\delta^{34}S$ values of oceanic sulfate (increased erosion of light sedimentary sulfide) and the relatively heavy $\delta^{13}C$ (org. matter) (higher photosynthetic rates on the continents). However, the decrease in the $^{87}Sr/^{86}Sr$ ratio during that period cannot be explained by a lowstand of sea level, on the contrary, the increased erosion rate should result in higher $^{87}Sr/^{86}Sr$ ratios.

If we ask for the cause of sea level changes, Pitman (1978) has argued that except for glacial effects, volume change in the midocean ridge system is the fastest and volumetrically the most significant way to change the sea level. Changes in the volume or elevation of midocean ridges which are related to changes in the rate of sea-floor spreading, appear to be of sufficient magnitude to account for the drastic changes in sea level. Glaciation and deglaciation are also responsible for sea level changes, but seem to be subordinate relative to geotectonic mechanisms. The ultimate source of the changes in ocean water chemistry seems to be, therefore, varying geotectonic forces.

Summary

The observed variations in the isotopic composition of sulfur, carbon, strontium and oxygen argue against a constant chemical composition of ocean water during the earth's history and favor sizable temporal variations in the chemical compsotion, a view which has already been taken earlier by Broecker (1971).

References

Becker, R.H., and R.N. Claton, Oxygen isotope study of a Precambrian banded iron formation, Hammersley Range, Western Australia, Geochim. Cosmochim. Acta, 40, 1153, 1976.

Brass, G.W., The variation of the marine $^{87}Sr/^{86}Sr$ ratio during Phanerozoic time: interpretation using a flux model, Geochim. Cosmochim. Acta, 40, 721, 1976.

Broecker, W.S., A boundary condition of the evolution of atmospheric oxygen, J. Geophys. Res., 75, 3553, 1970.

Broecker, W.S., A kinetic model for the chemical composition of sea water, Quaternary Res., 1, 188, 1971.

Chave, K.E., Evidence on history of sea water from chemistry of deeper subsurface waters of ancient basins, Bull. Amer. Ass. Petrol. Geol., 44, 357, 1960.

Claypool, G.E., W.T. Holser, I.R. Kaplan, H. Sakai, and I. Zak, Sulfur and oxygen isotope geochemistry of evaporite sulfate (Abstr.), Geol. Soc. Am. Programs Abstr., 4, 473, 1972.

Compston, W., The carbon isotopic composition of certain marine invertebrates and coals from the Australian Permian, Geochim. Cosmochim. Acta, 18, 1, 1960.

Dasch, E.J., and P.E. Biscaye, Isotopic composition of strontium in Cretaceous to Recent pelagic foraminifera, Earth Planet. Sci. Letters, 11, 201, 1971.

Emiliani, C., Isotopic paleotemperatures, Science, 154, 851, 1966.

Faure, G., R. Assereto, and E.L. Tremba, Strontium isotope composition of marine carbonates of Middle Triassic to Early Jurassic age, Lombardic Alps, Italy, Sedimentology. 25, 523, 1978.

Fischer, A.G., Brackish oceans as the cause of the Permo-Triassic marine faunal crisis, In: Problems in paleoclimatology, Ed. by A.E. Nairn, Interscience, New York, 1964.

Fischer, A.G., and M.A. Artur, Secular variations in the lepagic realm, Soc. Econ. Paleont. Mineral., Sec. Publ., 25, 19, 1977.

Galimov, E.M., A.A. Migdisov, and A.B. Ronov, Variations in the isotopic composition of carbonate and organic carbon in sedimentary rocks during the earth's history, Geochim. Inter. 12, 1, 1975.

Gast, P.W., Abundance of Sr^{87} during geologic time, Geol. Soc. Am., Bull. 66, 1449, 1955.

Garrels, R.M., and F.T. MacKenzie, Evolution of sedimentary rocks, W.W. Norton Publ., New York, 1971.

Garrels, R.M., and E.A. Perry, Cycling of carbon, sulfur and oxygen through geologic time, In: The Sea, ed. E.D. Goldberg, vol. 5, 303, 1974.

Heinzicker, K. M. Schidlowski, and C. Junge, Isotopic composition of atmospheric oxygen during the geological past, Z. Naturforsch. 29a, 964, 1974.

Holland, H.D., The geologic history of sea water –

an attempt to solve the problems, Geochim. Cosmochim. Acta, 36, 637, 1972.

Holland, H.D., The chemistry of the atmosphere and oceans, Wiley Interscience, New York, 1978.

Holser, W.T., Catastrophic chemical events in the history of the ocean, Nature, 267, 403, 1977.

Holser, W.T., and I.R. Kaplan, Isotope geochemistry of sedimentary sulfates, Chem. Geol., 1, 93, 1966.

Holser, W.T., I.R. Kaplan, H. Sakai, and I. Zak, Isotope geochemistry of oxygen in the sedimentary sulfate cycle, Chem. Geol., 25, 1, 1979.

Hurley, P.M., H.W. Fairbairn, and W.H. Pinson, Evidence from western Ontario of the isotopic composition of Sr in Archean seas, Geol. Soc. Am. Spec. Paper 87, 184, 1965.

Jeffery, P.M., W. Compston, D. Greenhalch, and Y. Delaeter, On the C abundance of limestones and coals, Geochim. Cosmochim. Acta, 7, 255, 1955.

Junge, C.E., M. Schidlowski, R. Eichmann, and H. Pietrek, Model calculations for the terrestrial carbon cycle: Carbon isotope geochemistry and evolution of photosynthetic oxygen, J. Geophys. Res., 80, 4543, 1975.

Keith, M.L., and J.N. Weber, Carbon and oxygen isotopic composition of selected limestones and fossils, Geochim. Cosmochim. Acta, 28, 1787, 1964.

Knauth, L.P., and Epstein, S., Hydrogen and oxygen isotopic ratios in nodular and bedded cherts, Geochim. Cosmochim. Acta, 40, 1095, 1976.

Knauth, L.P., and D.R. Lowe, Oxygen isotope geochemistry of cherts from the Onverwacht group (3.4 billion years), Transvaal, South Africa, with implications for secular variations in the isotopic composition of cherts, Earth Planet. Sci. Letters, 41, 209, 1978.

Kolodny, J., and S. Epstein, Stable isotope geochemistry of deep-sea cherts, Geochim. Cosmochim. Acta, 40, 1195, 1976.

Kramer, J.R., History of sea water: Constant temperature pressure equilibrium models compared to fluid inclusion analyses, Geochim. Cosmochim. Acta, 29, 921, 1965.

Li, Y.-H., Geochemical mass balance among lithosphere, hydrosphere and atmosphere, Am. J. Sci., 272, 119, 1972.

Magaritz, M., and K.H. Schulze, Carbon isotope anomaly of the Permian, (to be published).

Mizutani, Y., and T.A. Rafter, Bacterial fractionation of oxygen isotopes in the reduction of sulphate and in the oxidation of sulphur, N.Z.J. Sci., 12, 60, 1969.

Mottl, M.J., and H.D. Holland, Chemical exchange during hydrothermal alteration by seawater. I. Experimental results for major and minor components of seawater, Geochim. Cosmochim. Acta, 42, 1103, 1978.

Muehlenbachs, K., and R.N. Clayton, Oxygen isotope composition of the oceanic crust and its bearing on seawater, J. Geophys. Res., 81, 4365, 1976.

Nielsen, H., Schwefelisotope im marinen Kreislauf und das δ^{34}S der früheren Meere, Geol. Rundschau, 55, 160, 1965.

Nielsen, H., Model evaluations of the sulfur isotope budget of ancient oceans, Mezh'dun. Geokhim. Kongress Dokl., 4/1, 129, 1973.

Nielsen, H., and W. Ricke, Schwefel-Isotopen-Verhältnisse von Evaporiten aus Deutschland, ein Beitrag zur Kenntnis von ^{34}S im Meerwasser-Sulfat, Geochim. Cosmochim. Acta, 28, 577, 1964.

Ohmoto, H., D.R. Cole, and M.J. Mottl, Experimental basalt-seawater interaction: sulfur and oxygen isotope studies, Eos, 57, 342, 1976.

Perry, E.C.R., The oxygen chemistry of ancient chert, Earth Planet. Sci. Letters, 3, 62, 1967.

Perry, E.C., Jr., S.N. Ahmad, and T.M. Swulius, The oxygen isotope composition of 3,800 m.y. old metamorphosed chert and iron formation from Lsukasia, West Greenland, J. Geol., 86, 223, 1978.

Peterman, Z.E., C.E. Hedge, and H.A. Tourtelot, Iostopic composition of Sr in seawater throughout Phanerozoic time, Geochim. Cosmochim. Acta, 34, 105, 1970.

Pitman, W.C., Relationship between eustacy and stratigraphic sequences of passive margins, Bull. Geol. Soc.

Rees, C.E., The sulphur isotope balance of the ocean: an improved model, Earth Planet. Sci. Letters, 7, 366, 1970.

Renner, R., Geochemische und isotopengeochemische Unterscuhungen an marinen Sedimentgesteinen des Williston Basin (Südkanada), Diplomarbeit Universität Göttingen, 1978.

Rubey, W.W., Geologic history of sea-water, Geol. Soc. Am., Bull. 62, 1111, 1951.

Ryan, W.B.F., M.B. Cita, Ignorance concerning episodes of ocean-wide stagnation, Marine Geology, 23, 197, 1973.

Sakai, H., Oxygen isotopic ratios of some evaporites from Precambrian to Recent ages, Earth Planet. Sci. Letters, 15, 201, 1972.

Schidlowski, M., C.E. Junge, and H. Pietrek, Sulfur isotope variations in marine sulfate evaporites and the Phanerozoic oxygen budget, J. Geophys. Res., 82, 2557, 1977.

Schopf, T.J.H., Permo-Triassic extinctions: Relations to sea floor spreading, J. Geol., 82, 129, 1974.

Shackleton, N.J., Oxygen isotope analyses and Pleistocene temperatures re-assessed, Nature, 215, 15, 1967.

Spooner, E.T.C., The Sr-isotopic composition of seawater and seawater-oceanic crust interaction, Earth Planet. Sci. Letters, 31, 167, 1976.

Stevens, C.H., Was development of brackish oceans a factor in Permian extinctions?, Geol. Soc. Am., Bull. 88, 133, 1977.

Tappan, H., Primary production, isotopes, extinctions and the atmosphere, Paleogeography, Paleoclimatology, Paleoecology, 4, 187, 1968.

Thode, H.G., and J. Monster, Sulfur isotope geochemistry of petroleum evaporites and ancient seas, <u>Am. Ass. Petrol. Geol. Bull., 42</u>, 2619, 1965.

Vail, P.R., R.M. Mitchum, and S. Thompson, Seismic stratigraphy and global changes of sea level, Part IV: Global cycles of relative changes of sea-level, In: Seismic stratigraphy - applications to hydrocarbon exploration, <u>Am. Assoc. Petrol. Geol., Memoir 26</u>, 83, 1977.

Veizer, J., and W. Compston, $^{87}Sr/^{86}Sr$ composition of seawater during the Phanerozoic, <u>Geochim. Cosmochim. Acta, 38</u>, 1461, 1974.

Veizer, J. and J. Hoefs, The nature of O^{18}/O^{16} and C^{13}/C^{12} secular trends in sedimentary carbonate rocks, <u>Geochim. Cosmochim. Acta, 40</u>, 1387, 1976.

Veizer, J., W.T. Holser, and C.K. Wilgus, Correlation of $^{13}C/^{12}C$ and $^{34}S/^{32}S$ secular variations, <u>Geochim. Cosmochim. Acta</u>, (in press).

Wagener, K., Entwicklung der irdischen Atmosphäre durch Evolution der Biosphäre, Rheinisch-Westfäl. Akad. Wiss. Natur., Ing. Wirtschaftswiss. Vortrag, <u>233</u>, 36, 1973.

Weber, J.N., Possible changes in the isotopic composition of the oceanic and atmospheric carbon reservoir over geologic time, <u>Geochim. Cosmochim. Acta, 31</u>, 2343, 1967.

Welte, D.H., Organisher Kohlenstoff und die Entwicklung der Photosynthese auf der Erde, <u>Naturwissenschaften, 57</u>, 17, 1970.

Welte, D.H., W. Kalkreuth, and J. Hoefs, Age-trend in carbon isotopic composition in Paleozoic sediments, <u>Naturwissenschaften, 62</u>, 482, 1975.

MINERAL DEPOSITS AS GUIDES TO SUPRACRUSTAL EVOLUTION

R. W. Hutchinson

Department of Geology, The University of Western Ontario, London, Ontario N6A 5B7

Abstract. Evolutionary changes in major types of mineral deposits from Archean to Recent time provide guides to the main events and sequence of earth's supracrustal tectonic evolution and related processes. Evolutionary changes through time in ferruginous sedimentary rocks reflect a changing source of iron through geologic time and also oxygenation of the oceans about 2.2 b.y. ago. Differences between nickel deposits in Archean, Proterozoic and Phanerozoic rocks may be due to evolutionary changes in crust-mantle thickness and mantle composition. Early-formed lode gold deposits in eugeosynclinal volcano-sedimentary sequences of Archean and younger age were uplifted and metamorphosed by subsequent orogeny and became sources for paleoplacer and placer gold deposits. Evolutionary changes produced different types of massive base-metal sulphide deposits and these reflect changes and diversification through time in the tectonic setting of sea floor exhalative hydrothermal activity. Distinct evolutionary changes in uranium deposits are due to evolution of favorable granitoid crustal source rocks, to oxygenation of the atmosphere-hydrosphere and to biogenic evolution, with consequent changes in the mechanisms of uranium derivation, transportation and deposition.

Deposition of the Archean greenstone belt supracrustal sequences took place under reducing conditions and was dominated by volcanism, with which all the Archean mineral deposit types are closely affiliated. Two stages of volcanism are recognizable. The earlier was of ultramafic-komatiitic nature, but subsidence of these rocks with partial melting produced later, differentiated tholeiitic-calc-alkaline volcanism. The tectonic style involved predominantly vertical movement so that the later volcanic rocks were superposed on the earlier ones. Eparchean orogeny produced an extensive, thick ensialic crust and Proterozoic tectonism involved major rifting of this crust. Initial rift fault displacement was mainly vertical, forming large yoked basins. Proterozoic mineral deposits were generated in various parts, or at various stages of development of these basins, and mineral deposit genesis was also materially affected by oxygenation of the atmosphere-hydrosphere in Early Proterozoic time. In Mid-Proterozoic time rift fault displacement changed to major lateral separation, perhaps due to a period of global expansion. This may have marked the first appearance of deep ocean basins separating major continental blocks. Continued lateral movement, with interaction of oceanic and continental crustal blocks, evolved in Late Precambrian time to produce various types and configurations of plate tectonic boundaries. Earlier Precambrian tectonic environments and mineral deposit types occur again in some of these configurations; other new configurations give rise to new variant deposit types, whereas some older tectonic environments and their deposit types become extinct.

Introduction

Most studies of ore deposits are relatively restricted in scope, concentrating mainly on descriptions of the ores, considering problems of their genesis or evaluating their profitability. The considerable scientific importance of ore deposits as unique and distinctive rock types has perhaps not been adequately considered, although the recent paper by Watson (1975) and an earlier one by Pereira and Dixon (1965, 1966) are notable exceptions. Broad comparative studies of important types of mineral deposits reveal distinctive evolutionary changes through geological time. These changes are undoubtedly responses to evolutionary changes in the conditions of the earth's crust, mantle, hydrosphere, atmosphere, or biosphere under which the deposits were formed. Thus they provide useful information about, and a different approach to the general evolution of the earth. This paper outlines some examples of evolutionary changes in a few major ore types, and attempts to interpret their significance in terms of earth's evolution.

Evidence from Iron-bearing Sedimentary Rocks

Description

A substantial proportion of the world's iron ore is derived from iron-rich sedimentary rocks which

METALLOGENIC EVOLUTION – IRON FORMATIONS

	ARCHEAN	PROTEROZOIC	PHANEROZOIC	
Age (b.y.)	>2.5	~2.0	~0.5	~0.15
Type	Algoman	Superior	Clinton	Minette
Rock Associations				
– volcanic	Major differentiated types	Minor tholeiitic basalt	Nil	Nil
– sedimentary	Volcanoclastic	Epiclastic	Epiclast	Epiclast
Depositional Environment	Deep basin "eugeosynclinal"	Shallow shelf "miogeosynclinal"	Littoral	Littoral
Extent				
– lateral	Few Miles;	Hundreds of miles	Limited	Limited
– vertical	Few hundred ft.	Few thousand ft.	30 feet	30 feet
– continuity	Limited	Great	Discont.	Discont.
Texture	Fine grained	Fine grained-oolitic	Oolitic	Oolitic
Facies	Low Eh	High Eh	High Eh	High Eh
– changes	Sudden	Gradual	Minor	Minor
– oxide	Abundant Magnetite>hematite	Predominant Hematite>magnetite	Hematite	Goethite
– carbonate	Abundant	Rare	Siderite	Siderite
– sulfide	Abundant	Rare	Rare	Rare
– silicate	Abundant	Rare	Rare	Rare
Associated Deposits	Cu Zn	Nil; antipath	Nil	Nil
	Au	Minor	Nil	Nil
	Ni	Nil	Nil	Nil
Repetitions	All later ages	Nil (late Prot.?)		

Figure 1. Metallogenic Evolution of Iron-Rich Sedimentary Rocks

provide an excellent example of evolutionary changes through time. Moreover ferruginous sedimentary rocks are closely related spatially and probably genetically to other types of ores (Hutchinson, 1973, 1980) and are therefore particularly pertinent examples.

The oldest iron-bearing sedimentary rocks are termed Algoman-type Iron Formation (Gross, 1966) and are widespread in Archean greenstone belts of all the world's ancient cratons. Examples are found as widely separated as Adams, Sherman, and Moose Mountain mines in the Superior structural province of northeastern Ontario (Dubuc, 1966; Boyum and Hartviksen, 1970; Markland, 1966), the Koolyanobbing deposit of the Yilgarn Block in Australia (B.H.P. staff, 1975), the deposits of Bomvu Ridge in the Kaapvaal Craton of Swaziland in southern Africa (Davies and Urie, 1956), and the ores of the Sydvarangar district in the Archean Bjornevann Group of the Baltic Shield in northern Norway (Frietsch et al., 1979; Bugge, 1978). The general geological characteristics of these iron formations are summarized in Figure 1, where they are also compared with younger iron-rich sedimentary rocks. The Algoman-type iron formations occur in Archean greenstone belts with a distinctive eugeosynclinal lithofacies assemblage that includes a very thick and varied sequence of differentiated, tholeiitic to calc-alkaline volcanic rocks, related fragmental-pyroclastic rocks and volcanoclastic greywackes. These iron formations are relatively thin and discontinuous, traceable along strike for only a few tens of miles. They range up to a few hundred feet thick. They exhibit rapid changes amongst

oxide, carbonate, sulphide and silicate facies, and two or more of these facies are commonly intercalated in the thinly banded and regularly laminated, alternating iron- and silica-rich strata. Soft-sediment slump and deformation structures are common, reflecting their initial deposition as water-saturated chemical sediments in unstable, subsiding basins accompanied by extensive volcanism. These iron formations are very fine-grained and cherty, except where they have been recrystallized by later metamorphism, and in general they are highly reduced. Ferrous sulphide, carbonate and silicate facies are each abundant and in the oxide facies, magnetite greatly predominates over hematite. Where deeply weathered, as in Western Australia and southern Africa, they may be hematitic near surface, but drilling at Koolyanobbing revealed the presence of magnetite and pyrite at depth (pers. comm.). These iron formations have a close spatial and genetic affiliation with other important metalliferous ores, notably with certain nickel sulphide deposits, gold lodes and massive base-metal sulphide bodies. Although they are commonest and largest in Archean rocks, Algoman-type iron formations reappear in rocks of many later geological periods as, for example in the Ordovician of New Brunswick where they are again closely associated with some of the massive base-metal sulphide deposits (Luff, 1977).

Contrasting markedly with the Algoman-type Archean iron formations are the vastly important Superior Type which are of early Proterozoic age. These iron formations and their contained ore deposits are unique and constitute perhaps the best example of a major metallogenic epoch on the earth for they occur in a remarkably narrow time-stratigraphic interval of about 2.2 billion years in age on virtually all of the earth's Precambrian Shields. Examples are numerous and include the great Lake Superior and Labrador iron ranges of North America (Marsden, 1968; Gross, 1968), the deposits of the Hamersley Basin in Western Australia (Trendall, 1973), of the Transvaal System in South Africa (Snyman, 1976; Fourie, 1976), and of Cerro Bolivar in Venezuela (Ruckmick, 1963), the deposits of Serra dos Carajás in Para and of the Quadrilátero Ferífero of Minas Gerais in Brazil (Tolbert et al., 1971; Beisieguel et al., 1973; Dorr II, 1973; Ladeira, 1980), as well as those of Krivoi Rog in the Soviet Union (Alexandrov, 1973). The general geological characteristics of these iron formations are also summarized in Figure 1. Unlike the Algoman-type, the Superior Type iron formations are associated with shallow water, miogeosynclinal and shelf-facies sedimentary rocks. Mafic and tholeiitic volcanic rocks are associated with the iron formations in some places, as in Western Australia and in Para, Brazil (Trendall, 1973; Tolbert, 1971), but are elsewhere lacking. The thickness and continuity of Superior Type iron formations are very much greater than those of their Archean antecedents. Narrow laminae as well as thicker beds are traceable for over 200 miles in the Hamersley Basin of Western Australia and the iron formations range up to several thousand feet in thickness (Trendall, 1973). Although commonly fine-grained and cherty, the Superior Type iron formations in places are prominently oolitic or pisolitic (Dimroth and Kimberley, 1976). The oxide facies is much more abundant than the other three reduced facies, and in the oxide facies hematite predominates over magnetite. Unlike the Algoman-type, Superior Type iron formations are not significantly associated with, in fact may be antipathetic to other metalliferous deposits. Moreover they do not reappear significantly in younger rocks.

Finally the Clinton and Minette types of Paleozoic and Mesozoic ironstones are vastly different from either of the Precambrian types of iron formation. Although there are differences between the Clinton and Minette ores, their general geological characteristics are similar and are summarized together in Figure 1. These Phanerozoic ironstones are prominently oolitic and pisolitic, probably due to the diagenetic replacement of calcareous oolite by hematite or siderite (Kimberley, 1974). Thus the ironstones do not appear to be primary chemical precipitates like the Precambrian types, and they are associated with clastic, littoral and shallow water sedimentary rocks of estuarine-deltaic or lagoonal deposition. They are far smaller than either of the Precambrian types and are notably lenticular and discontinuous (Simpson and Gray, 1969). They have no affiliation whatsoever with volcanism and no association with other metalliferous deposits.

Interpretation

It is suggested that the obvious change in depositional environment of these iron-bearing sedimentary rocks through time reflects two major evolutionary changes on the earth; firstly a change in the source of iron deposited in these rocks, and secondly a change from a reducing to an oxidizing hydrosphere-atmosphere. Widespread, indeed possibly global Archean volcanism initially provided iron which was extracted by seawater from the thick, water-charged, subaqueously deposited and differentiated Archean volcanic sequences. This iron was carried to the sea floor by convective brine circulation (Hutchinson et al., 1980), discharged at sea floor fumaroles and precipitated near the vents due to local supersaturation to form the Algoman-type iron formations. Rapid facies changes in these iron formations reflect local variations in anion supply and fO_2 that accompanied the variations in volcanicity and related exhalative activity. But under reducing conditions of the Archean atmosphere-hydrosphere (Cloud, 1973) the greater solubility of ferrous compared to ferric iron must have caused much iron to go into solution, possibly as ferrous chlorides, carbonates and sulphates, and it is therefore suggested that reduced iron salts were abundant in the Archean oceans. When the hydrosphere-atmosphere became oxygenated about 2.2 b.y. ago (Cloud, 1972, 1973) this seawater provided a vast source of iron which was rapidly oxidized to ferric state, and consequently precipitated in shallow cratonic basins all over the globe in a restricted stratigraphic zone at about this time. Preferential iron precipitation in shallow cratonic basins may have been due to the abundant growth there of oxygen-generating algal organisms, as suggested by the common stromatolytic mats in associated carbonate rocks of this age, as in the Kona dolomite of Michigan, the Denault dolomite of Labrador and the Malmani dolomite of the Transvaal System in South Africa (Gair and Thaden, 1968; Gross, 1968; Erikson, 1972). Although not formed by evaporation of seawater, when considered as wide-spread, cratonic, marine chemical sedimentary strata, the Superior Type iron formations may be regarded as the "evaporites" of early Proterozoic time.

In Phanerozoic time differentiated, island arc-type volcanism, like that of earlier Archean time, occurred along continent margin orogenic belts, causing the Algoman-type iron formations to recur locally in younger rocks, but the Superior Type does not reappear due to virtually complete exhaustion of iron from the world's oceans at the time of oxygenation. Local mafic volcanism apparently accompanied the deposition of some, but not all of the Superior Type iron formations as evidenced by the Fortescu volcanic rocks beneath the Hamersley iron formations in Western Australia (Trendall, 1973) and the basalts of the Grão Para Group beneath the Carajás iron formation at Serra dos Carajás in Para, Brazil (Dorr II, 1973;

METALLOGENIC EVOLUTION – NICKEL DEPOSITS

	ARCHEAN	PROTEROZOIC	PHANEROZOIC
Age (b.y.)	>2.5	2.2 – 1.0	0.6 – 0.06
Type	Volcanogenic sulfide	Magmatic sulfide	Lateritic- silicate
Ig. Rk. Associations			
– type	Extrusive	Intrusive	Intrusive
– composition	Ultramafic	Mafic	Ultramafic
– alteration	Extensive	Slight	Extensive
– differentiation	Poor	Very good	Poor
Formational Environment	Ultramafic volcanism Eugeosynclinal subsidence	Intrusive Miogeosynclinal shelf	Intrusive (Ophiolitic)
Mineralogy	Po–py–pent (cpy),High Ni- sulfides (millerite, etc.)	Po–pent–cpy	Garnierite
Metal Content	Ni > Cu	Ni ÷ Cu	Ni only
Ni	High	High	High
Cu	Low	High	Trace
Ni/Cu	30 – 5	5 – 0.2	n.a.
P G M	Low	High	?
Form	Massive>disseminated Irregular discontinuous Lenticular	Disseminated>massive Regular continuous Layered	Surface Laterite
Associated Deposits	Algoman I. F. Au	Nil Nil	Nil Nil
Repetitions	Nil	Few	–

Figure 2. Metallogenic Evolution of Nickel Deposits

Tolbert, 1971). This volcanism may have supplied additional iron to the local basins of iron form-ation precipitation, again by leaching and marine fumerolic activity, but elsewhere ferrous iron-charged seawater was the primary source of iron.

Phanerozoic ironstones reveal a still later change to iron derivation by continental weathering of iron-bearing rocks in an oxygenated atmosphere. Under these conditions, most iron is concentrated on the continents in laterites and lateritic soils. However, local transport, due to local reduction of iron to ferrous state, as in tropical organic-rich rivers, carries a minor amount to the continent margin where it is precipitated, probably in the zone of mixing of fresh meteoric, with saline ocean waters, to form the Phanerozoic ironstones. As a result these are relatively small, lenticular and discontinuous, are associated with littoral, clastic sedimentary rock and have no affiliation with volcanism or associated metalliferous deposits.

Evidence from Nickel Deposits

Description

The world's oldest nickel deposits are of a type which appears to be exclusively of Archean age. The geological characteristics of these deposits are summarized in Figure 2 where they are also compared with those of younger nickel deposits. The Archean ores have been designated volcanogenic (Naldrett, 1973) because of their unique associa-tion with distinctive ultramafic, komatiitic extrusive volcanic rocks. These ultramafic rocks are very highly altered and poorly differentiated. The orebodies are small, lenticular pods of massive, high-grade nickel-iron sulphides, with comparatively low contents of copper and platinum-group metals. The Ni/Cu ratio of these deposits therefore favors Ni and ranges from about 5 to as much as 50. The ores are mainly composed of pyrrhotite and pentlandite, but pyrite is an important component, and the nickel-rich sulphide minerals such as millerite, violarite, bravoite, heazlewoodite, and even the nickel-iron alloy awaruite, are minor but not uncommon accessory minerals. These nickel deposits may be found associated with Algoman-type iron formations, as at Mount Windara in Western Australia (Roberts, 1975), and in places, as in the Kalgoorlie-Kambalda region of Western Australia, with gold deposits. They occur in the somewhat older or stratigraphically lower, ultramafic portions of the greenstone belt successions, as in the Kambalda district of Western Australia (Ross and Hopkins, 1975), in the Sebakwian rocks at Shangani in Zimbabwe-Rhodesia (Viljoen, M.J. et al., 1976), and at the base of the Tisdale Group of the Porcupine District in northern Ontario (Coad, 1977; Kerrich et al., 1979).

Proterozoic nickel-sulphide deposits (Fig. 2) are markedly different from the Archean ores and are generally considered to be of magmatic origin (Souch et al., 1969; Cousins, 1969; Haapala, 1969) although this view is currently being questioned (Naldrett and Cabri, 1976; Fleet, 1977, 1979; Naldrett, 1979). Regardless of their origin, the Proterozoic nickel sulphide deposits are associ-ated with distinctive ultramafic to mafic, layered intrusive igneous complexes. These are well differentiated, with cryptic and rhythmic lay-ering, and they are composed of relatively fresh, unaltered igneous rocks. The orebodies are larger and more regular, although commonly lower grade, than those of Archean age. Some of the orebodies, like those of the Merensky Reef (Cousins, 1969) or the Great Lakes nickel deposit (Geul, 1970) consist of disseminated sulphide-bearing orthopyroxenite layers within these complexes. Elsewhere, as at Sudbury (Souch et al., 1969), lenses of higher grade, in places massive sulphide occur at or near the contacts of the intrusions. The ores are mainly composed of pyrrhotite and pentlandite, but they are richer in chalcopyrite and minerals of the platinoid elements than their Archean counterparts. The Ni/Cu ratio in these ores therefore ranges near unity, probably from about 5 to 0.2. Pyrite and the nickel-rich sulphide minerals however are less common than in the Archean type. Although many of the most important examples are of Proterozoic age, as the Merensky Reef in the Bushveld Complex of South Africa (Cousins, 1969), the Great Dyke of Rhodesia (Worst, 1960), the Sudbury Intrusion in Ontario (Souch et al., 1969), and the Duluth Complex in Minnesota (Mainwaring and Naldrett, 1977), there are similar examples in younger rocks as at Insizwa in South Africa (Maske, 1966) and Noril'sk in the Soviet Union (Naldrett et al., 1979).

Phanerozoic nickel deposits (Fig. 2) are markedly different from both Precambrian types. Sulphide nickel ores are extremely rare and, with the notable exception of Noril'sk, of minor importance in Phanerozoic rocks. Instead, Phanerozoic nickel ores are mainly lateritic nickel silicates containing garnierite or other hydrated nickel-magnesium silicates derived by the weathering of highly serpentinized ultramafic rocks (Cumberlidge and Chace, 1968; INAL staff, 1975; Trescases, 1973). In most of these ores there is no evidence that a nickel sulphide phase was present before weathering or serpentinization, and the nickel appears to have been derived directly by weathering and lateritic concentration of nickel from serpentine. The importance and abundance of lateritic nickel deposits in Phanerozoic rocks compared with their rarity in Precambrian terrain is the exact opposite to the case of nickel sulphide ores.

Interpretation

These differences suggest evolutionary changes in crust-mantle relationships, thicknesses and compositions through time. Perhaps in earliest Archean timea thin or discontinuous sialic crust rested on a relatively thick, near-surface, and poorly differentiated upper mantle containing abundant reduced sulphur gas species similar to $(\bar{\bar{S}})$ or $(\bar{HS})$, possibly because mantle degassing had not yet progressed beyond its initial stages. Under these conditions, deep partial melting of mantle (Naldrett and Cabri, 1976), with high heat flow, led to the extrusion in the greenstone belts of the ultramafic, komatiitic flow sequences and readily also to relatively early, high temperature partitioning of nickel, either into a metallic nickel-iron-sulphide liquid or perhaps into a

hydrothermal aqueous fluid. By early Proterozoic time the sialic crust had been substantially thickened due both to greenstone belt evolution and late Archean granitic plutonism. The mantle had been largely depleted of its reduced sulphur gases which had formed the various types of Archean pyritic massive sulphide bodies and were also dispersed in the reduced Archean atmosphere-hydrosphere. At about 2.2 b.y. ago the mantle too, may have become oxygenated. Under these conditions, nickel sulphides occurred only as much sparser, disseminated layers in, or as pods near the intrusions, probably where these had been locally sulphurized by encounter with an external and surficial sulphur source (Naldrett, 1966; Naldrett and Macdonald, 1980). By Phanerozoic time these same changes had progressed even further, and the mantle was essentially devoid of sulphur which had been degassed either into the massive pyritic sulphides, seawater sulphate and saline marine evaporites of the upper crust-atmosphere-hydrosphere, or perhaps concentrated downward into a troilitic, outer nickel-iron core. Under these conditions, partial melting of mantle produced ultramafic melts, during the crystallization of which nickel was normally combined into the silicate structures, initially into olivine orthopyroxene and, after hydrous alteration, into serpentine. Here it was directly available in silicate form during later lateritic weathering for concentration and formation of the Phanerozoic nickeliferous laterites. Locally however, where intrusions encountered an external source of sulphur, perhaps as at Noril'sk (Naldrett et al., 1979), the Proterozoic type of nickel sulphide deposit reappears in younger Phanerozoic rocks. However, the major, permanent evolutionary changes in mantle composition, oxidation-reduction state, and perhaps in crust-mantle thicknesses, prevented recurrence of the more primitive and earlier Archean type of volcanogenic nickel sulphide deposit in these younger rocks.

Evidence from Gold Ores

Description

The world's oldest important gold ores are the lode deposits which occur in greenstone belts of all the world's Archean cratons. Examples are numerous and include the Yellowknife, Porcupine and Kirkland Lake districts in the Slave and Superior Structural Provinces of Canada (Boyle, 1961; Ferguson et al., 1968; Ridler, 1970), the Kalgoorlie district of the Yilgarn Block in West Australia (Woodall, 1975), the Barberton and Steynsdorp districts in the Kaapvaal Craton of southern Africa (Anhaeusser, 1976; Viljoen, R.P. et al., 1969), the various gold deposits of Rhodesia (Fripp, 1976), the famous Morro Velho mine in the Archean Rio des Velhas rocks of Minas Gerais in Brazil (Dorr II, 1973; Ladeira, 1980), and the Ashanti Goldfields in the Tarkwaian rocks of Ghana in west Africa (Ntiamoah-Adjaquah, 1974).

METALLOGENIC EVOLUTION – GOLD DEPOSITS

	ARCHEAN	PROTEROZOIC	PHANEROZOIC
	Early ←→ late MM	Post Archean	
Age (b.y.)	>2.5	>2.2	0.6 – 0.06
Type	Concordant lodes - →Lodes and		
	↘ Disc. lodes→ -Paleoplacers - - - - - - - - - →↘ placers		
Rock Associations			
– volcanic	Major differentiated	Minor	
– sedimentary	Volcanoclastic-chert	Conglomerate quartzite	Repeat
Formational Environment	Mafic volcanism	Continental deltaic	
	Eugeosynclinal subsid.	High "E"	
	Chemical sedimentation	Clastic sedimentation	geological
Metal Association			
Ag	Relatively high	Very low	
Cr – Pt	Present	Minor	environments of
U	Absent	Important	
Form	Stratiform	Concordant reef	earlier types
	↘ Stratabound	channels	
Associated Deposits	Algoman I. F.	U.,	
	Cu Zn	others insignificant	
	Ni		
Repetitions	In later ages	In later ages - - - - - - - - - - -⌐	

Figure 3. Metallogenic Evolution of Gold Deposits

The geological setting of these deposits in the Archean greenstone belts is summarized in Figure 3, where it is also compared with those of younger gold deposits, and is very similar to that of the Algoman-type iron formations. These gold ores have a wider stratigraphic range than the Archean massive nickel sulphide deposits in the greenstone belts, as they are not only associated with ultramafic-mafic volcanic rocks low in the successions, as in the Kalgoorlie or Porcupine districts, but also with felsic volcanic rocks higher in the successions, as at Agnico-Eagle in the Joutel-Poirier district of Quebec (Barnett et al., in press). Thus they are associated with both Archean volcanogenic nickel sulphide deposits and Archean volcanogenic massive copper-zinc sulphide deposits. These gold lodes have a significant silver content, with a fineness ranging from about 600 to 900, and also an association with ultramafic elements other than nickel, notably with chromium and platinum. However they lack any association with uranium. Some of the lodes are concordant and were apparently deposited originally as gold-rich chemical sedimentary strata intercalated with volcanic, volcanoclastic and pyroclastic rocks (Fryer et al., 1978). Other lodes are definitely discordant and transect the former, clearly reflecting a later redistribution of gold by some intrusive or metamorphic event. These gold lodes reappear in the rocks of many younger geologic ages as for example in the early Paleozoic rocks of the Ballarat-Bendigo district in Victoria, Australia and in the Jurassic rocks of the Mother Lode in California (Bowen and Whiting, 1975; Knopf, 1929; Evans and Bowen, 1977).

The distinctive Proterozoic paleoplacers of the Witwatersrand Basin in South Africa are the world's most important gold deposits (Brock and Pretorius, 1964; Fells and Glynn, 1976). Although of lesser importance, similar deposits occur in the Jacobina district of Brazil (Gross, 1968), in the Tarkwaian rocks of Ghana in West Africa (Sestini, 1971), and in the Nullagine Formation of West Australia (W. White, Sydney, pers. comm.). These ores are markedly different from the Archean lodes. They are associated with coarse-grained, fluviatile-deltaic clastic sedimentary rocks. They have no significant affiliation with nickel, chrome, platinum or copper deposits. They have much lower silver content than the Archean lodes, with a fineness of 900 or more, and they have a very important association with uranium which is economically recoverable from many of the gold ores as by-product. The gold deposits themselves have no direct or significant association with volcanic rocks, although mafic tholeiitic extrusive rocks or sill-like intrusions are commonly present in the succession, as the Ventersdorp lavas of the Witwatersrand or the Sudbury gabbros of Blind River, Ontario. The orebodies are concordant quartz pebble conglomerate strata that have a distinctive pyritic matrix as well as a distinctive detrital heavy mineral suite (Coetzee, 1965), that testify to their placer origin.

Interpretation

This comparison suggests that gold, like iron and nickel, was originally derived from the volcanic host rocks of the lode gold ores by convective hydrothermal leaching that accompanied the widespread Archean subaqueous volcanism. The concordant ankeritic veins were initially deposited as auriferous chemical sedimentary carbonate-chert layers intercalated with volcanic and volcanoclastic strata. They were later metamorphosed, perhaps by subvolcanic-porphyritic stocks coeval with later-stage Archean felsic volcanism, by ep-Archean granodioritic plutonism, or both.

METALLOGENIC EVOLUTION – MASSIVE BASE METAL SULFIDE FAMILY

	VOLCANIC ROCKS	CLASTIC SEDIMENTARY ROCKS	APPROXIMATE AGE RANGE		EXAMPLES
I. EXHALATIVE VOLCANOGENIC GROUP					
1. PRIMITIVE TYPE ZN-CU: AG-AU	FULLY DIFFERENTIATED SUITES BASALTIC TO RHYODACITIC	VOLCANOCLASTICS GREYWACKES	ARCHEAN EARLY PROTEROZOIC EARLY PHANEROZOIC	>2.5 >1.8 Є-DEV.	NORANDA, KIDD CREEK JEROME, FLIN FLON, CRANDON WEST SHASTA
2. POLYMETALLIC TYPE PB-ZN-CU: AG-AU	BIMODAL ? SUITES THOLEIITIC BASALTS CALC-ALKALIC LAVAS, PYROCLASTICS	VOLCANOCLASTICS INCREASING CLASTICS MINIMUM CARBONATES	EARLY PROTEROZOIC PHANEROZOIC	>1.8 ORD. MESO. TER.	PRESCOTT, SUDBURY BASIN, MT. ISA NEW BRUNSWICK EAST SHASTA JAPAN
3. CUPREOUS PYRITE TYPE CU: AU	OPHIOLITIC SUITES THOLEIITIC BASALTS	MINOR TO LACKING	PHANEROZOIC	Є-ORD. MESO.	NEWFOUNDLAND CYPRUS, TURKEY, OMAN
4. KIESLAGER TYPE CU-ZN: AU	MAFIC; THOLEIITIC(?) (AMPHIBOLITE)	GREYWACKE, SHALE(?) (BIOT.-AMPHIB. SCHIST)	LATE PROTEROZOIC PALEOZOIC	1.2–.8	MATCHLESS-OTJIHASE S.W. AFRICA; DUCKTOWN, TENN; BESSHI, JAPAN; GOLDSTREAM, B.C.
II. EXHALATIVE SEDIMENTARY GROUP					
5. CLASTIC HOSTED TYPE PB-ZN: AG	MINOR, BASALTIC (GABBROIC-AMPHIBOLITIC INTRUSIVE SHEETS)	ARGILLITE, TURBIDITE	MID PROTEROZOIC PALEOZOIC	1.7–1.0 Є-DEV.	SULLIVAN, BROKEN HILL, McARTHUR RIVER ANVIL, MEGGEN, RAMMELS-BERG, HOWARDS PASS
6. CARBONATE HOSTED TYPE ZN-PB: (AG)	MINOR TO LACKING (TUFFACEOUS BEDS)	CARBONATE LST.-DOL. SANDSTONE, SHALE	LATE PROTEROZOIC PHANEROZOIC	~1.0 MISS.	BALMAT NAVAN, SILVERMINES, TYNAGH, IRELAND

Figure 4a. Metallogenic Evolution of Massive Base Metal Sulphide Deposits

Th⁀s caused local remobilization of gold, and generated the discordant veins and stockworks. Finally, uplift accompanying Kenoran orogeny at the close of Archean time exposed both concordant and discordant lodes to weathering and erosion, still under the reducing atmospheric conditions of earliest Proterozoic time. Thus they became sources for the important and distinctive gold-uranium paleoplacer deposits of this age around the world. Repetitions in younger rocks of the same sequence of concordant lodes, discordant lodes and placer deposits, occur along continent-margin orogenic belts. Here marine volcanism, like that of Archean time, generated first concordant lodes, and these were subsequently intruded and metamorphosed to form discordant veins and stockworks. The deposits of the Mother

Lode Belt in California are examples. Subsequent Nevadan Orogeny, like the Kenoran orogeny at the close of Archean time, caused uplift of the lodes, weathering, erosion, and formation of the famous high- and low-level California gold placers.

Evidence from Massive Base Metal Sulphide Deposits

Description

Massive base-metal sulphide deposits also exhibit marked evolutionary changes in their geological characteristics through time. These are extremely complex and have been discussed in detail by Hutchinson (1973, 1980). They are summarized on Figures 4a and 4b where various geological characteristics of the different types

METALLOGENIC EVOLUTION – MASSIVE BASE METAL SULFIDE FAMILY

	DEPOSITIONAL ENVIRONMENT	TECTONIC ENVIRONMENT	
		GENERAL CONDITIONS	PLATE TECTONIC SETTING
I. EXHALATIVE VOLCANOGENIC GROUP			
1. PRIMITIVE TYPE ZN-CU: AG-AU	EXTENSIVE, EVOLVING, DEEP TO SHALLOW WATER, THOLEIITIC TO CALC-ALKALINE MARINE VOLCANISM	MAJOR SUBSIDENCE COMPRESSION	SUBDUCTION AT CONSUMING MARGIN, ISLAND ARC
2. POLYMETALLIC TYPE PB-ZN-CU: AG-AU	EXPLOSIVE SHALLOW CALC-ALKALINE-ALKALINE MARINE-CONTINENTAL VOLCANISM	SUBSIDENCE REGIONAL COMPRESSION BUT LOCAL TENSION	BACK-ARC OR POST-ARC SPREADING; CRUSTAL RIFTING AT CONSUMING MARGIN
3. CUPREOUS PYRITE TYPE CU: AU	DEEP THOLEIITIC MARINE VOLCANISM	MINOR SUBSIDENCE TENSION	OCEANIC RIFTING AT ACCRETING MARGIN
4. KIESLAGER TYPE CU-ZN: AU	DEEP MARINE SEDIMENTATION AND THOLEIITIC VOLCANISM	MAJOR SUBSIDENCE COMPRESSION	FORE-ARC TROUGH OR TRENCH
II. EXHALATIVE SEDIMENTARY GROUP			
5. CLASTIC HOSTED TYPE PB-ZN: AG	DEEP MARINE SEDIMENTATION; MINOR THOLEIITIC ACTIVITY	MAJOR SUBSIDENCE TENSION, RIFTING	SEPARATION AT CONTINENTAL RIFT; AULACOGENIC TROUGH OR TRENCH
6. CARBONATE HOSTED TYPE ZN-PB: (AG)	SHALLOW MARINE-SHELF SEDIMENTATION	MINOR SUBSIDENCE, TENSION	SHELF; LOCAL FAULT-CONTROLLED BASIN

Figure 4b. Metallogenic Evolution of Massive Base Metal Sulphide Deposits

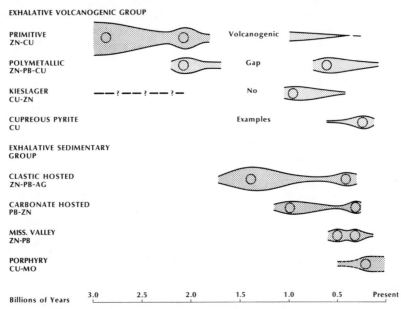

MASSIVE BASE METAL SULFIDE FAMILY
TIME RANGES AND PEAKS OF TYPES

EXHALATIVE VOLCANOGENIC GROUP

PRIMITIVE ZN-CU   Volcanogenic

POLYMETALLIC ZN-PB-CU   Gap

KIESLAGER CU-ZN   No

CUPREOUS PYRITE CU   Examples

EXHALATIVE SEDIMENTARY GROUP

CLASTIC HOSTED ZN-PB-AG

CARBONATE HOSTED PB-ZN

MISS. VALLEY ZN-PB

PORPHYRY CU-MO

Billions of Years   3.0   2.5   2.0   1.5   1.0   0.5   Present

Figure 5.   Time Distribution of Massive and Other Sulphide Deposit Types

and ages of these deposits are also compared. Broad aspects of this evolution include firstly, a change from lenticular, stratabound, copper-gold-rich, volcanic-hosted types in Archean and early Proterozoic time to increasingly stratiform and regular, lead-silver rich, sediment-hosted types in mid-Proterozoic time. Secondly, there is a marked absence of all volcanic-hosted types in mid-Proterozoic time, from about 1.8 b.y. ago to about 1.0 b.y. ago, which coincides with the major appearance of the important sediment-hosted types. This is shown by a plot of the approximate time ranges of the various types in Figure 5. Thirdly, as in the case of Algoman-type iron formations and the lode gold deposits, the older, more primitive copper-gold-rich types of massive base-metal sulphide ores recur first in the earliest stages of Phanerozoic orogenic belt evolution and are followed by reappearance of the younger, lead-silver-rich types. Thus the same relative age progression in the deposit types seen throughout geologic time on a worldwide basis is preserved and recapitulated on the much shorter time span and space scale represented in Phanerozoic orogenic belts. The relationship is analogous to the expression "ontogeny duplicates phylogeny" in biological science. Fourthly, there is clearly a diversification of types through time, with the appearance of both new types of massive sulphides and of other related types such as Mississippi Valley type zinc-lead ores (Hutchinson, 1980) and porphyry copper deposits (Hutchinson and Hodder, 1972). This diversification is diagrammatically illustrated in Figure 5 which shows the approximate main time distribution and "peaks" of these

various deposit types. The diversification is also comparable to that seen in faunal evolution.

Interpretation

These changes are believed to reflect evolutionary changes in Earth's crustal tectonic processes, particularly an evolution from earlier ensimatic to later ensialic supracrustal environments. It is suggested, as for iron, nickel and gold, that the earliest primitive type (Figs. 4a, b) of volcanogenic massive sulphide deposits was formed by seafloor exhalative hydrothermal activity accompanying the late felsic stages of prolonged differentiated volcanism in the Archean greenstone belts. Similar volcanism and activity recurred in places in early Proterozoic time generating additional primitive type deposits. But there was also a tendency to bimodal volcanism of increasingly ensialic nature, with more explosive, felsic activity, perhaps in shallower, more cratonic basins, and sea floor exhalative hydrothermal activity accompanying this environment formed the increasingly Pb-Ag rich polymetallic ores for the first time. Subsequently in mid-Proterozoic time, about 1.8 b.y. ago, there was a major change in crustal tectonics to initiate environments dominated by the deposition of deep water, turbiditic siltites-argillites. Exhalative hydrothermal activity accompanying this tectonic-depositional environment formed many important deposits of the clastic-hosted exhalative sedimentary massive sulphide type (Figs. 4a, b) that are so important in rocks of mid-Proterozoic age (Fig. 5), particularly in Australia, but also elsewhere as at Kimberley in British Columbia. Mafic magmatism, forming tholeiitic basalts, or intrusive diabase-gabbro (amphibolite) sills also occurred commonly in this environment. Shallow cratonic basins, presumably flanking these deeper sedimentary troughs, were sites of shelf carbonate and coarser clastic sedimentation, and sea floor exhalative hydrothermal activity here in mid- to late-Proterozoic time formed massive sulphides of the carbonate-hosted type. Finally, in Phanerozoic time, all of the same tectonic and resulting supracrustal depositional environments reappear, in more or less the same relative age sequence, along continent-margin orogenic belts over consuming plate boundaries as these belts go through their own tectonic evolution. The formation of the various massive sulphide deposit types accompanies this repetitive and cyclic tectonic evolution and results in the reappearance of the various types in Phanerozoic rocks. A single new type, the cupreous pyrite deposits (Figs. 4a, b) appears however in the unique ophiolitic environment of Phanerozoic accreting plate boundaries. Most such deposits are presumably destroyed during subduction of this oceanic crust resulting from sea floor spreading, but a few are preserved and incorporated into the continental crust by obduction.

Evidence from Uranium Deposits

Description

The evolution of uranium deposits through time has been much discussed in recent years (McMillan, 1978; Robertson et al., 1978). Its earliest important aspect is the absence of significant uranium deposits in Archean rocks. Uranium is like lead in this regard; there are no major uranium- or lead-producing deposits in Archean rocks throughout the world. The oldest significant uranium deposits are paleoplacers of earliest Proterozoic age, as at Elliot Lake in the Aphebian rocks of Ontario (Theis, 1978). These deposits are essentially identical to the paleoplacer gold ores, and the same examples may be cited, as the Witwatersrand of South Africa where uranium is an important by-product of gold mining. In later Proterozoic time, after oxygenation of the hydrosphere-atmosphere about 2.2-2.1 b.y. ago, the important, high-grade, unconformity- related vein or lens type of uranium deposits appears. This type is also commonly rich in other metals, silver, nickel, copper and others. Examples are the Rabbit Lake, Key Lake, Cluff Lake deposits of northern Saskatchewan (Hoeve and Sibbald, 1978) and the Nabarlek-Jabaluka-Ranger deposits of Northern Territory in Australia (Anthony, 1975; Rowntree and Mosher, 1975; Eupene et al., 1975).

Subsequent Proterozoic orogeny, such as the Hudsonian (1.7 b.y.) of the Churchill Province in Canada, or the Grenvillian (0.1 b.y.) of the Grenville Province in eastern Canada (Stockwell et al., 1970) and the Namaqualand orogeny of south-

western Africa was important in generating the significant syenite-, granite-, and pegmatite-affiliated type of uranium deposits. Examples are the Faraday and Bicroft deposits (Hewitt, 1967) of southeastern Ontario, possibly the Gunnar Mine ores of northern Saskatchewan and the important Rössing deposit of Namibia (Berning et al., 1976). These may have been generated in part by orogeny, with accompanying plutonic intrusions which remobilized and redistributed uranium from the two earlier types of deposits.

Although not large producers, uraniferous carbonaceous marine shales, like the Chattanooga of the southeastern United States (Bates and Strahl, 1958) and the Upper Cambrian Billingen deposits of the Ranstad area in Sweden (Ruznicka, 1978) are important potential uranium sources in early Paleozoic rocks. They clearly indicate incorporation of uranium into these reduced sediments either during or shortly after marine sedimentation.

In still younger, late Paleozoic rocks the important continental sandstone-affiliated type of uranium deposit makes its first appearance and forms widespread and important orebodies in similar but later sandstones of Mesozoic-Tertiary age. The best examples are the many deposits of the Colorado plateau (Fischer, 1968), and those of the Wyoming intermontane basins (Rosholt et al., 1964; Melin, 1964). It is interesting that this type of uranium deposit has many similarities to the earliest Proterozoic paleoplacers. Both types occur in rapidly deposited, coarse clastic sedimentary rocks. These were laid down in shallow water, cratonic or craton-margin basins, following uplift that was accompanied by granitic plutonism or extrusive rhyolitic volcanism. Both contain significant amounts of pyrite. Probably however, the mode of uranium transportation in the two was different; clastic transport of earlier detrital uraninite grains in the case of the paleoplacers and chemical transport of hexavalent, oxidized uranium in solution in the case of the sandstone type. This is suggested by their differing minor element contents and associations; notably high gold and thorium with other detrital heavy minerals in the paleoplacers as compared to low gold-thorium but high vanadium, base metals, and molybdenum in the sandstone type. Considering these many similarities however, the sandstone deposits may be considered as younger analogues of the older paleoplacers.

Interpretation

These changes in the nature of uranium deposits through time reflect important evolutionary global changes of two broad types. Firstly, they are partly due to the development of extensive, thick ensialic crust. Felsic igneous rocks are considerably richer in uranium than more mafic igneous rocks (Dyck, 1978). Granites and pegmatites are considered a likely source for detrital uraninite in paleoplacer deposits

(Robertson, 1976) and continental felsic volcanic rocks are a possible source of uranium for the sandstone type deposits (Mrak, 1968). These relationships suggest that felsic igneous rocks and sialic, continental crust represent an essential stage of uranium preconcentration, and a necessary source of uranium for the different types of deposits which formed subsequently by various processes of additional uranium concentration and enrichment. The first major and worldwide period of granitic plutonism and orogeny occurred at the end of Archean time and if, as this suggests, granitic-ensialic crust was thin and poorly developed during Archean time the paucity of this preconcentrated sialic source may be one factor responsible for the worldwide absence of significant uranium deposits in Archean rocks.

Another factor may be the broad evolutionary changes in the earth's atmosphere-hydrosphere-biosphere system, particularly inversion from a reduced to an oxidized condition at about 2.2-2.1 b.y. ago (Cloud, 1973, 1972). Initially, reducing conditions in Archean time would have prevented aqueous solution transport of insoluble, tetravalent, reduced uranium. This aspect, combined with sparse granitic sources for detrital uranium minerals, may explain the lack of uranium deposits in Archean rocks. In earliest Proterozoic time, before oxidation but following the worldwide granitic plutonism of latest Archean time, uranium transport in aqueous solution remained impossible. However abundant detrital uranium-thorium and other accessory heavy minerals were available from uplift and weathering of granitic terrains. Abundant pyrite, also stable under reducing weathering conditions, was also available as a detrital mineral from weathering of massive sulphide bodies in the uplifted and deformed greenstone belts. These conditions provide a reasonable explanation for the worldwide occurrence of the large, but low-grade, thorium-rich uranium paleoplacers in early Proterozoic rocks.

Oxidation of the atmosphere-hydrosphere permits aqueous solution transport of soluble, oxidized, hexavalent uranium, and its separation during weathering from insoluble thorium, with which it is closely associated in granitic igneous rocks (Dyck, 1978). This appears to have been vital to formation of the rich, unconformity-related vein or lens uranium deposits of mid-Proterozoic rocks. Uranium was apparently taken into solution in oxidized meteoric waters during weathering, probably particularly following episodic orogenic granitic plutonism and uplift, when favourable source rocks were exposed to rapid erosion. The soluble uranium was transported in solution, with minor amounts of other oxide- and acid-soluble metals, along or near the unconformity surface between the older uplifted terrain and its cover of younger sediment, commonly in coarse clastic sedimentary rocks deposited on this surface. Precipitation occurred near, commonly just below this unconformity where meteoric waters encountered any

reducing agent. This caused reduction of the water, precipitation of reduced, insoluble tetravalent uranium and of most other transported metals, particularly where sulphur may have been generated from the same waters by sulphate reduction. Examples are known of precipitation by such reducing agents as mafic dykes, which intruded the basal coarse clastic sediments during meteoric water migration (Stanton, 1979) and by graphitic rocks in the older basement beneath the unconformity (Hoeve and Sibbald, 1978). Presumably simple reducing conditions below the paleo-water table may also have caused uranium deposition, and earlier pyritic bodies were also likely causes. Corresponding oxidation of the rocks near the depositional site caused generation of the hematitic alteration so commonly associated with these deposits (Dawson, 1956; Tremblay, 1978). These conditions offer a reasonable broad explanation for the appearance in mid-Proterozoic time of the unconformity related, high-grade uranium deposits. They also apply in all later geological periods, explaining many similar deposits in rocks of younger age such as the Schwartzwalder mine of Colorado (Carpenter et al., 1979).

Despite its local precipitation by reducing agents to form the high grade veins or lenses, most of the uranium leached from continental rocks during weathering by circulating, oxidizing meteoric waters must have been carried to the oceans. Here much of it was again deposited in reducing marine environments due to the conversion of soluble hexavalent to insoluble tetravalent uranium. One very extensive reducing marine environment is that of black, carbonaceous-pyritic shale, and this may explain the high uranium content in many of these rocks (Bell, 1978), particularly in early Paleozoic strata.

Finally the mid-Paleozoic evolution of terrestrial flora and fauna, which in places were buried before decay in rapidly deposited, coarse, continental clastic sediments, provided a new reducing, and potential uranium-precipitating agent. Oxygenated, circulating meteoric groundwaters carrying soluble hexavalent uranium and flowing through these porous strata deposited uranium where they encountered the buried organic debris. These relationships explain the appearance in late Paleozoic time of the important sandstone type of uranium deposit, and its continuing importance in rocks of younger geologic age.

Summary of Metallogenic Evolution

From the foregoing it is clear, whatever the explanation, that these and probably other types of mineral deposits, have undergone major evolutionary changes from earliest Archean to Recent time. It is also clear that there are close genetic and metallogenic relationships between many types of deposits that have generally been considered as entirely independent of one another. Examples are the close spatial

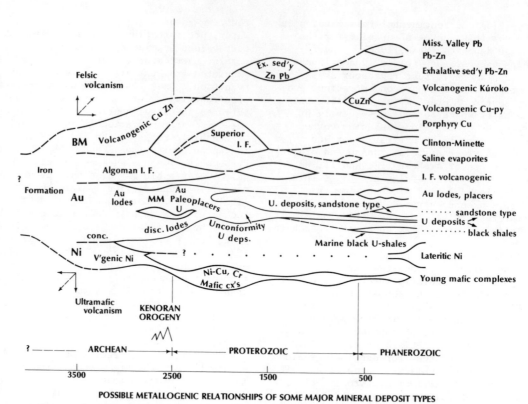

POSSIBLE METALLOGENIC RELATIONSHIPS OF SOME MAJOR MINERAL DEPOSIT TYPES

Figure 6.   Metallogenic Relationships of Major Mineral Deposit Types

and genetic relations between nickel sulphide, gold, and copper-zinc sulphide deposits on the one hand, and Algoman-type iron formations on the other. Mississippi Valley type zinc-lead ores are very similar in geological environment and time distribution to exhalative sedimentary, carbonate-hosted massive sulphide bodies (Hutchinson, 1980). Porphyry copper deposits have many close geological similarities and metallogenic relationships to massive base metal sulphide deposits (Hutchinson and Hodder, 1972). The metallogenic evolution and relationships of all these deposit types are represented diagrammatically in Figure 6.

Synthesis

Considered collectively and in relationship to one another, because all are interrelated and none independent, these relationships provide additional information about the course of earth's supra-crustal evolution. It is emphasized however, that because there are no significant mineral deposits in the ancient Archean gneiss terrains (Hunter, 1970, 1974) this approach yields no information about the nature and evolution of these terrains or about their relationships to the greenstone belts where the important mineral deposit types are found, and to which this synthesis mainly applies.

The geological environment of the Archean nickel sulphide deposits suggests that in the greenstone belts the earliest Archean volcanism took place through rifts, either in a thin crust, or perhaps directly from and on the Archean protomantle (Fig. 7). This process may have been analogous to the initial differentiation and production of new oceanic crust from the mantle at modern accreting plate boundaries. In this manner the ultramafic, komatiitic volcanic sequences of earliest Archean age were formed. They contain the significant volcanogenic type of Archean nickel deposit and associated Algoman-type iron formation. Modern analogues of these deposits may be the nickel-iron-manganese nodules of the deep ocean floor (Bonatti, 1978). Prolonged ultramafic volcanism of this type resulted in gravitational instability and major subsidence, or vertical subduction (Fig. 8). Partial melting of this vertically subducted "proto-oceanic lithosphere" in the subsiding zone may subsequently have generated the differentiated tholeiitic to calc-alkaline vol-canic sequences of the younger Archean greenstone belts. The latter were superposed on the former as observed in the Abitibi greenstone belt of northern Ontario-Quebec (Fig. 9) (Pyke, 1975; Pyke et al., 1978; Alsac et Latulippe, 1979). These younger sequences contain the important gold lodes and copper-zinc massive sulphide deposits, as well as additional Algoman-type iron formations. This second crustal differentiation may be compared to that occurring in volcanic arcs over slabs of subducted oceanic crust along Phanerozoic

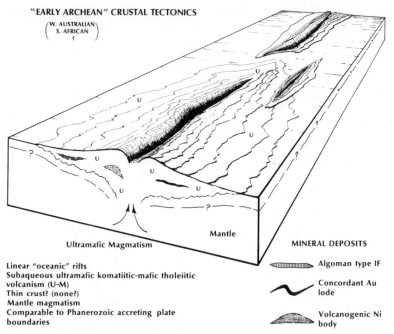

"EARLY ARCHEAN" CRUSTAL TECTONICS

$\left(\begin{array}{l}\text{W. AUSTRALIAN}\\\text{S. AFRICAN}\\?\end{array}\right)$

Ultramafic Magmatism

Mantle

Linear "oceanic" rifts
Subaqueous ultramafic komatiitic-mafic tholeiitic
volcanism (U-M)
Thin crust? (none?)
Mantle magmatism
Comparable to Phanerozoic accreting plate
boundaries

MINERAL DEPOSITS

Algoman type IF

Concordant Au
lode

Volcanogenic Ni
body

Figure 7. Diagrammatic Illustration of Early Archean Crustal Tectonics

consuming plate boundaries. The major difference appears to lie in the vertical tectonism that produced the Archean successions vs. the lateral translation of crustal blocks that is so important in more recent plate tectonics.

Although its sources and mechanisms are not clear, culmination of the prolonged Archean volcanism appears to have been worldwide granodior-

itic plutonism of the Kenoran or Eparchean orogeny (Fig. 10; left). The volcanic sequences of the greenstone belts were deformed and uplifted by this activity which appears to have formed, perhaps for the first time, a thick stable continental crust. It thereby provided the possibility for a new tectonic style in the form of rifting of continental crust. Major conti-

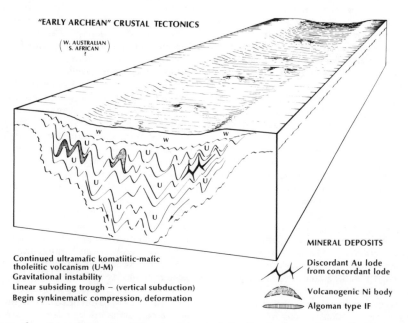

"EARLY ARCHEAN" CRUSTAL TECTONICS

$\left(\begin{array}{l}\text{W. AUSTRALIAN}\\\text{S. AFRICAN}\\?\end{array}\right)$

Continued ultramafic komatiitic-mafic
tholeiitic volcanism (U-M)
Gravitational instability
Linear subsiding trough – (vertical subduction)
Begin synkinematic compression, deformation

MINERAL DEPOSITS

Discordant Au lode
from concordant lode

Volcanogenic Ni body

Algoman type IF

Figure 8. Diagrammatic Illustration of Early Archean Crustal Tectonics

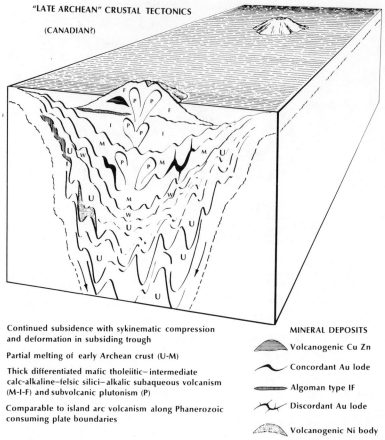

"LATE ARCHEAN" CRUSTAL TECTONICS

(CANADIAN?)

Continued subsidence with sykinematic compression
and deformation in subsiding trough

Partial melting of early Archean crust (U-M)

Thick differentiated mafic tholeiitic–intermediate
calc-alkaline–felsic silici–alkalic subaqueous volcanism
(M-I-F) and subvolcanic plutonism (P)

Comparable to island arc volcanism along Phanerozoic
consuming plate boundaries

MINERAL DEPOSITS

Volcanogenic Cu Zn

Concordant Au lode

Algoman type IF

Discordant Au lode

Volcanogenic Ni body

Figure 9.   Diagrammatic Illustration of Late Archean Crustal Tectonics

nental rift systems appear to dominate Proterozoic tectonics. They influence the location and development of major Proterozoic sedimentary basins and are closely related to many Proterozoic mineral deposits. Early Proterozoic displacement on these rift systems may have been mainly vertical, forming great yoked basins of asymmetric subsidence like that of the Witwatersrand (Brock and Pretorius, 1964) or the Huronian of northern Ontario (Card and Hutchinson, 1972; Kumarepeli and Saul, 1966). Here the important paleoplacer gold-uranium ores were deposited in thick clastic wedges, perhaps fringed by stromatolytic or algal reefs, along the vertically subsiding margin of the rift (Fig. 10; right). Major mafic intrusions accompanied this rifting, like the Nipissing diabase of Ontario, or the widespread, remarkably continuous, early to late Proterozoic diabase dykes that are present in virtually all Archean cratons. Elsewhere mafic magmatism took the form of major extrusions like the Ventersdorp lavas of the Witwatersrand, but major mafic igneous activity was extensive throughout Proterozoic time (Card and Hutchinson, 1972), and was presumably a consequence of the extensive cratonic rifting which penetrated the

mantle. In some of the layered and differentiated intrusions magmatic nickel-copper-platinum deposits, for example the Merensky Reef of the Bushveld Complex or the podiform Sudbury ores were formed, perhaps where local sulphurization of the intrusions took place.

Slightly later in Proterozoic time oxygenation of the oceans, probably due to biogenic evolution, resulted in the widespread deposition of Superior Type iron formation. This occurred in broad shallow parts of the same yoked basins, or their successors, and perhaps in pools of evaporite-like deposition isolated behind extensive algal banks (Fig. 11). Here oxygen supply from the algal organisms would have been direct and prolific, causing ferric iron precipitation from the ferrous iron-charged seawater. In mid-Proterozoic time however, it is suggested that major lateral displacement began on the same, or new continental rift systems. The important exhalative sedimentary, clastic-hosted type of massive sulphide deposit was formed in these widening, deepening and subsiding rifts (Fig. 12), or perhaps in related but narrower, intracratonic aulacogenic rifts where separation was minor. An example may be the deeply buried rift that coincides with the southern

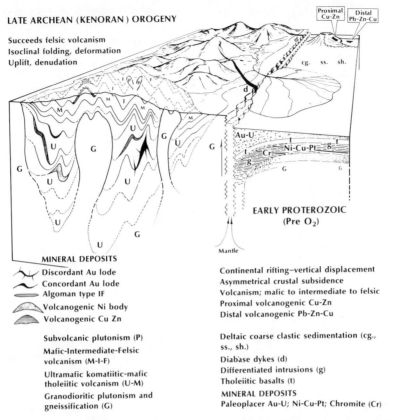

LATE ARCHEAN (KENORAN) OROGENY

Succeeds felsic volcanism
Isoclinal folding, deformation
Uplift, denudation

EARLY PROTEROZOIC
(Pre O₂)

MINERAL DEPOSITS

Discordant Au lode
Concordant Au lode
Algoman type IF
Volcanogenic Ni body
Volcanogenic Cu Zn

Subvolcanic plutonism (P)
Mafic-Intermediate-Felsic
volcanism (M-I-F)
Ultramafic komatiitic-mafic
tholeiitic volcanism (U-M)
Granodioritic plutonism and
gneissification (G)

Continental rifting—vertical displacement
Asymmetrical crustal subsidence
Volcanism; mafic to intermediate to felsic
Proximal volcanogenic Cu-Zn
Distal volcanogenic Pb-Zn-Cu

Deltaic coarse clastic sedimentation (cg.,
ss., sh.)
Diabase dykes (d)
Differentiated intrusions (g)
Tholeiitic basalts (t)
MINERAL DEPOSITS
Paleoplacer Au-U; Ni-Cu-Pt; Chromite (Cr)

Figure 10. Diagrammatic Illustration of Kenoran Orogeny and early Proterozoic Crustal Tectonics

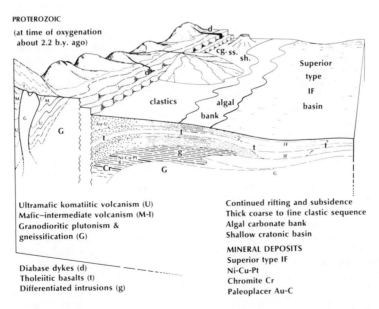

PROTEROZOIC
(at time of oxygenation
about 2.2 b.y. ago)

Ultramafic komatiitic volcanism (U)
Mafic–intermediate volcanism (M-I)
Granodioritic plutonism &
gneissification (G)

Diabase dykes (d)
Tholeiitic basalts (t)
Differentiated intrusions (g)

Continued rifting and subsidence
Thick coarse to fine clastic sequence
Algal carbonate bank
Shallow cratonic basin

MINERAL DEPOSITS
Superior type IF
Ni-Cu-Pt
Chromite Cr
Paleoplacer Au-C

Figure 11. Diagrammatic Illustration of early Proterozoic Crustal Tectonics

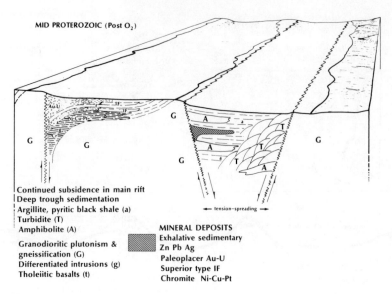

MID PROTEROZOIC (Post O₂)

Continued subsidence in main rift
Deep trough sedimentation
Argillite, pyritic black shale (a)
Turbidite (T)
Amphibolite (A)

Granodioritic plutonism &
gneissification (G)
Differentiated intrusions (g)
Tholeiitic basalts (t)

← tension-spreading →

MINERAL DEPOSITS
Exhalative sedimentary
Zn Pb Ag
Paleoplacer Au-U
Superior type IF
Chromite Ni-Cu-Pt

Figure 12.   Diagrammatic Illustration of mid-Proterozoic Crustal Tectonics

extension of the Rocky Mountain trench, and with the important Sullivan orebody in southern British Columbia (Kanasewich, 1968).

Major separation along these rifts may have been due to a mid-Proterozoic period of global expansion and may have marked the first appearance of the deep ocean basins and major continental blocks more or less as they are known today. Global expansion, with resulting tensional tectonics might also explain the rarity in rocks of this age of differentiated, compression-generated calc-alkaline volcanic rocks (Miyashiro, 1975; Christiansen and Lipman, 1972), and the corresponding absence of exhalative volcanogenic massive sulphide deposits. Moreover lode gold deposits of the older Archean type, which reappear in the subduction zone-generated volcanic rocks of Phanerozoic orogenic belts are also notably lacking in rocks of mid-Proterozoic age. Finally, there is an absence of oceanic crust in the form of obducted ophiolites in rocks of mid-Proterozoic age (Glikson, 1979). These features too, might result from major global expansion.

Finally, the major lateral separations of mid-Proterozoic time evolved, by late Precambrian-early Phanerozoic time, into the mechanisms of plate tectonics. These are marked by important lateral translation and interaction of oceanic and continental crustal plates. The various configur-ations and types of plate boundaries however, result in duplications and reappearance of all the earlier mineral deposit types and their geological environments, as well as in the appearance of new, diversified types. These relationships are particularly well illustrated by the plate tectonic positioning of the various massive base metal sulphide types and certain related ores (Fig. 13). The cupreous pyrite type is generated

in ophiolitic rocks at accreting plate margins and, along with iron-manganese-nickel nodules, this activity may be analogous to the komatiitic volcanism and mineral deposit generation of earliest Archean time. The primitive, copper-zinc-rich, volcanogenic massive sulphide types are formed in fore-arc trench or island arc environ-ments over consuming plate boundaries. They are also accompanied in this environment by the lode gold deposits and Algoman-type iron formations. Finally, the later polymetallic and sedimentary exhalative massive base metal sulphide types are generated in back- or post-arc, craton margin or intracratonic basins. Like those of early and mid-Proterozoic time, these are increasingly ensialic and probably related to back-arc spreading, tension and crustal rifting. Along consuming plate margins, as subduction continued through time, two important new base metal deposit types appeared (Hutchinson, 1980). The Mississippi Valley type zinc-lead ores formed in the same environment as the carbonate-hosted, exhalative sedimentary massive sulphides, which they closely resemble but probably by diagenetic reactions between already deposited, unlithified carbonate sediments and exhalative metalliferous fluids that did not reach the sea floor. The many similarities between porphyry copper deposits and the deep, copper-rich altered zones below volcanogenic massive sulphide bodies suggest that the former are subvolcanic equivalents of the latter. Perhaps the porphyry-type deposits were generated when and where prolonged subduction, with overriding of oceanic by continental crust, prevented exhalative fluids from reaching the sea floor (Hutchinson, 1980). This might have been accompanied by elimination of back-arc spreading and back-arc basins, and by a change from submarine to subaerial

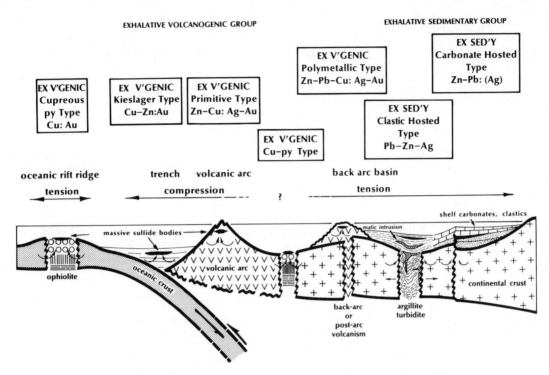

EXHALATIVE VOLCANOGENIC GROUP   EXHALATIVE SEDIMENTARY GROUP

Figure 13. Diagrammatic Illustration of Phanerozoic Plate Tectonics showing Tectonic Settings of Various Massive Base Metal Sulphide Deposits

volcanism. These conditions currently apply along the Andean margin of South America, one of the world's most important porphyry copper regions.

Conclusions

Consideration of evolutionary changes through time in various types of mineral deposits and their depositional environments reveal close genetic and metallogenetic links among these types, whereas the types have previously been considered as largely independent of one another, and unrelated. These evolutionary changes are responses to changing supra-crustal tectonic processes through time, and hence indicate the events and sequence of earth's supracrustal evolution.

The development of Archean greenstone belts was dominated by prolonged volcanism, and the formation of important mineral deposits in Archean rocks was closely linked to this process. Initially, Archean volcanism took the form of ultramafic extrusive activity, and in this environment important volcanogenic nickel sulphide deposits, reduced ferrous iron-rich, Algoman-type iron formations, and some gold lodes were formed. Later in Archean time, probably due to subsidence and partial melting of the earlier ultramafic sequences, volcanism changed to the generation of thick, differentiated, tholeiitic to calc-alkaline sequences. Deposits generated in this environment include additional important gold lodes, a primitive type of massive

base metal sulphide and additional Algoman-type iron formations. No significant uranium deposits were formed in Archean time because there was no extensive granitic-sialic source for detrital uranium minerals and because aqueous solution transport of uranium was generally not possible under reducing atmosphere-hydrosphere conditions.

Proterozoic tectonism was dominated by rifting of a thickened, continental crust largely formed by granodioritic plutonism of the Kenoran orogeny at the end of Archean time. Initial displacement on the rift faults was mainly vertical, forming extensive yoked basins in which asymmetric subsidence and deposition of coarse, fluviatile clastic sediments occurred along the down-thrown rift margins. In these basins the paleoplacer gold-uranium deposits were formed, with rocks which were derived from weathering of the surrounding uplifted Archean terrain. Extensive mafic magmatism accompanied the rifting, forming thick basalt sequences or extensive thick sheets of gabbro or diabase. Some of the latter were locally sulphurized and differentiated to form magmatic nickel-copper-platinoid deposits. The Proterozoic, Superior-type iron formations reflect the relatively sudden oxygenation of the atmosphere-hydrosphere about 2.2 b.y. ago, with the resulting worldwide deposition of these highly hematitic iron formations as evaporite-like chemical sediments in shallow, perhaps algal reef-fringed portions of the same yoked basins. Atmospheric oxygenation permitted aqueous solution transport of uranium and the rich, unconformity-related uranium deposits were formed subsequently, particularly following periods of granitic orogeny, uplift and weathering.

Somewhat later in Proterozoic time major lateral separation occurred along the rifts, perhaps due to a period of global expansion at this stage of earth history. The appearance of deep ocean basins separating major continental blocks may have been due to this expansion and separation along the rifts. The important lead-zinc-silver-rich, clastic sediment-hosted type of massive sulphide deposit was formed in these opening, deepening troughs, whereas the carbonate-hosted type was formed in flanking, shallow, carton-margin shelf environments.

In late Precambrian time lateral movement of oceanic and continental blocks led to evolution of the differing types and configurations of plate boundaries as envisaged in current plate tectonic concepts. These are more diverse and complex than their Precambrian antecedents, with more possible variations in tectonic environment. Some of these environments are replicas of earlier Precambrian ones, although they persist for shorter time periods than the latter, and their spatial extent is more limited within narrow, linear Phanerozoic orogenic belts. This is the case particularly in the early stages of subduction along consuming plate boundaries where "Late Archean-like" volcanism produces thick, differentiated subaqueous flow sequences. Here Archean-like mineral deposits also reappear in much younger rocks, and these include Algoman-type iron formations, gold lodes and primitive, copper-zinc-rich volcanogenic massive sulphides. Other Phanerozoic plate tectonic environments are new variants or diversifications, where new forms of mineral deposits appear. Examples are the cupreous pyrite type of volcanogenic massive sulphide along accreting plate boundaries, the appearance of extensive porphyry copper deposits along mature, long-continued subductive margins, or of sandstone-type uranium deposits in continental clastic sedimentary rocks after the evolution of terrestrial organisms. Nevertheless, some geological characteristics seen in older deposit types may be repeated in the new variant types, as for example in the depositional environment of host rocks for the paleoplacer and sandstone-type uranium deposits. Some important Precambrian deposit types disappear however in younger rocks, presumably because evolutionary changes in broad tectonic processes or supracrustal conditions have eliminated their essential generative environments. Examples are the Early Archean volcanogenic nickel sulphide ores and the Proterozoic Superior-type iron formations. These various continuations or repetitions, diversifications and extinctions of mineral deposit types and environments through time are analogous to those known in faunal evolution.

Acknowledgements. The broad comparisons between mineral deposits of various types and ages in many parts of the world that are the basis of this paper were made possible in large part by field visits during a year's sabbatical leave in 1973-74. This leave was partly funded by the University of Western Ontario, the National Research Council of Canada, the Economic Geology Research Unit of the University of the Witwatersrand, and by Chevron Resources Co. In addition many geologists, mining companies and governmental geological agencies in many countries provided invaluable assistance in the field, made available their goelogical information and gave helpful advice and suggestions. The writer is deeply indebted to all these organizations and geologists, and their considerable contributions to this paper are gratefully acknowledged.

During final manuscript preparation my friend and colleague Dr. Gordon Suffel has collated the references cited in the text and has prepared the essential supporting bibliography. He and a second friend and colleague, Dr. R.W. Hodder, have also provided continued and valued comment, criticism and discussion of the ideas presented in this paper. Their considerable resulting contributions to it are very much appreciated.

References

Alexandrov, E.A., The Precambrian banded iron-formations of the Soviet Union, Econ. Geology, 68(7), 1039-1062, 1973.

Alsac, C. et M. Latulippe, Quelques aspects pétrographiques et géochimiques du volcanisme archéen du Malartic en Abitibi (Province du Quebec, Canada), Can. Jour. Earth Sciences, 16(5), 1041-1059, 1979.

Anhaeusser, C.R., The nature and distribution of Archaean gold mineralization in Southern Africa, Mineral Science and Engineering, 8(1), 46-84, 1976.

Anthony, P.J., Nabarlek uranium deposit, in Knight, C.L., ed., Economic Geology of Australia and Papua New Guinea, 1. Metals. The Australian Inst. Min. Met., Mon. Series No. 5, 304-308, 1975.

Barnett, E.S., R.W. Hutchinson, A. Adamcik, and R. Barnett, Geology of the Agnico-Eagle gold deposit, Quebec, in Hutchinson, R.W., ed., Precambrian sulphide deposits, Robinson Volume, Geol. Assoc. Canada (in press).

Bates, T.F., and E.O. Strahl, Mineralogy and chemistry of uranium-bearing black shales, in Second UN Int. Conf. Peaceful Uses Atomic Energy, Geneva 1958, 2, 407-411, 1958.

Beisieguel, V.R., A.L. Bernadelli, N.F. Drummond, A.W. Ruff, and J.W. Tremaine, Geologia e recursos minerais da Serra dos Carajás, Revista Brasileira de Geociencias, 3(4), 215-242, 1973.

Bell, R.T., Uranium in black shales - a review, in Kimberley, M.M., ed., Short course in uranium deposits: their mineralogy and origin, Miner. Assoc. Canada Short Course Handbook, 3, 307-329, 1978.

Berning, J., R. Cooke, S.A. Hiemstra, and U.

Hoffman, The Rossing uranium deposit, South West Africa, Econ. Geology, 71(1), 351-368, 1976.

Bonatti, E., The origin of metal deposits in the oceanic lithosphere, Scientific American, 238, 54-61, 1978.

Bowen, K.G., and R.G. Whiting, Gold in the Tasman geosyncline, Victoria, in Knight, C.L., ed., Economic geology of Australia and Papua New Guinea, 1. Metals, Australasian Inst. Min. Met., Mon. Series No. 5, 647-659, 1975.

Boyle, R.W., The geology, geochemistry and origin of the gold deposits of the Yellowknife District, Geol. Surv. Canada, Memoir 310, 193 p., 1961.

Boyum, B.H., and R.C. Hartviksen, General geology and ore grade control at the Sherman mine, Temagami, Ontario, Can. Inst. Min. Met. Bull., 63(701), 1059-1068, 1970.

Brock, B.B., and D.A. Pretorius, Rand Basin sedimentation and tectonics, in Haughton, S.H., ed., The geology of some ore deposits in southern Africa, Geol. Soc. South Africa, Johannesburg, 1, 549-600, 1964.

Broken Hill Proprietary (B.H.P.) Staff, Koolyanobbing iron ore deposits, W.A., in Knight, C.L., ed., Economic geology of Australia and Papua New Guinea, 1. Metals. The Australasian Inst. Min. Met., Mon. Series No. 5, 940-942, 1975.

Bugge, J.A.W., The Sydvarangar type of quartz-banded iron ore, with a synopsis of Precambrian geology and ore deposits of Finnmark (abstract), Metallogeny of the Baltic Shield, Helsinki Symposium 1978, Abstracts, Academy of Finland, 17-19, 1978.

Card, K.D., and R.W. Hutchinson, The Sudbury structure: its regional geological setting, in Guy-Bray, J.V., ed., New developments in Sudbury geology, Geol. Assoc. Canada Special Paper No. 10, 67-78, 1972.

Carpenter, R.H., J.R.L. Gallagher, and G.C. Huber, Modes of uranium occurrences of Colorado Front Range, Colorado School of Mines Quarterly, 74(3), 50-55, 1979.

Christiansen, R.L., and P.W. Lipman, Cenozoic volcanism and plate-tectonic evolution of the western United States. II. Late Cenozoic, Phil. Trans. Roy. Soc. London, A, 271, 249-284, 1972.

Cloud, P., A working model of the primitive earth, Am. Jour. Science, 272, 537-548, 1972.

Cloud, P., Paleoecological significance of the banded iron-formation, Econ. Geology, 68(7), 1135-1143, 1973.

Coad, P.R., Nickel sulphide deposits associated with ultramafic rocks of the Abitibi Belt and economic potential of mafic-ultramafic intrusions, Ont. Geol. Survey Open File Rept. 5232, 105 p., 1977.

Coetzee, F., Distribution and grain size of gold, uraninite, pyrite and certain other heavy minerals in gold-bearing reefs of the Witwatersrand basin, Trans. Geol. Soc. South Africa, 68, 61-68, 1965.

Cousins, C.A., The Merensky Reef of the Bushveld igneous complex, in Wilson, H.D.B., ed., Magmatic ore deposits, Econ. Geology Monograph 4, 239-251, 1969.

Cumberlidge, J.T., and F.M. Chace, Geology of the Nickel Mountain mine, Riddle, Oregon, in Ridge, John D., ed., Ore deposits in the United States, 1933-1967 (Graton-Sales Volume), Am. Inst. Min. Met. and Pet. Engineers, New York, 2, 1650-1672, 1968.

Davies, D.N., and J.G. Urie, The Bomvu Ridge haematite deposits, Swaziland Geol. Survey Special Rept. 3, 23 p., 1956.

Dawson, K.R., Petrology and red coloration of wallrocks, radioactive deposits, Goldfields region, Saskatchewan, Geol. Survey Canada Bull. 33, 1956.

Dimroth, Erich, and M.M. Kimberley, Precambrian atmospheric oxygen: evidence in the sedimentary distributions of carbon, sulfur, uranium and iron, Can. Jour. Earth Sciences, 12(9), 1161-1185, 1976.

Dorr II, J. Van N., Iron-formation in South America, Econ. Geology, 68(7), 1005-1022, 1973.

Dubuc, F., Geology of the Adams mine, Can. Inst. Min. Met. Bull., 59(646), 176-181, 1966.

Dyck, W., The mobility and concentration of uranium and its decay products in temperate surficial environments, in Kimberley, M.M., ed., Short course in uranium deposits: their mineralogy and origin, Miner. Assoc. Canada Short Course Handbook, 3, 57-100 (see appendices), 1978.

Eriksson, K.A., Cyclic sedimentation in the Malmani Dolomite, Potchefstroom Synclinorium, Geol. Soc. South Africa Trans., 75, 85-95, 1972.

Eupene, G.S., P.H. Fee, and R.G. Colville, Ranger One uranium deposits, in Knight, C.L., ed., Economic geology of Australia and Papua New Guinea, 1. Metals. The Australasian Inst. Min. Met., Mon. Series No. 5, 308-317, 1975.

Evans, J.R., and O.E. Bowen, Geologic map and sections of the southern Mother Lode, Tuolumne and Mariposa counties, California, Ca. Div. Mines and Geology Map Sheet No. 36, 1977.

Fells, P.D., and Christopher Glynn, Gold 1976. Mining Survey, No. 82, Johannesburg, 2-16 (Table world gold production, 1970-1975), 1976.

Ferguson, S.A., B.S.W. Buffam, O.F. Carter, A.T. Griffis, T.C. Holmes, M.E. Hurst, W.A. Jones, H.C. Lane, and C.S. Longley, Geology and ore deposits of Tisdale township, District of Cochrane, Ont. Dept. Mines, GR 58, 117 p. (accompanied by Map 2075, scale 1 inch to 1,000 feet), 1968.

Fischer, R.P., The uranium and vanadium deposits of the Colorado Plateau region, in Ridge, J.D., ed., Ore deposits of the United States, 1933-1967 (Graton-Sales Volume), Am. Inst. Min. Met. and Pet. Engineers, New York, 1, 735-746, 1968.

Fleet, M.E., Partitioning of Fe, Co, Ni and Cu between sulfide liquid and basaltic melts and the composition of Ni-Cu sulfide deposits, Econ. Geology, 74(6), 1517-1519 (discussion), 1979.

Fleet, M.E., Origin of disseminated copper-nickel sulfide ore at Frood, Sudbury, Ontario, Econ.

Geology, 72(8), 1449-1456, 1977.

Fourie, G.P., Sischen Mine, northern Cape Province, in Coetzee, C.B., ed., Mineral resources of South Africa, Geol. Survey, Dept. of Mines, Pretoria, 150-152, 1976.

Frietsch, R., H. Papunen, and F.M. Vokes, The ore deposits in Finland, Norway and Sweden - a review, Econ. Geology, 74(5), 975-1001, 1979.

Fripp, R.E.P., Stratabound gold deposits in Archean banded iron-formation, Rhodesia, Econ. Geology, 71(1), 58-75, 1976.

Fryer, B.J., R. Kerrich, R.W. Hutchinson, M.G. Peirce, and D.S. Rogers, Archaean precious-metal hydrothermal systems, Dome Mine, Abitibi Greenstone belt. I, Patterns of alteration and metal distribution, Can. Jour. Earth Sciences, 16(3), 421-439, 1978.

Gair, J.E., and R.E. Thoden, Geology of the Marquette and Sands quadrangle, Marquette county, Michigan, U.S. Geol. Survey Prof. Paper 397, 1968.

Geul, J.J.C., Devon and Pardee townships and the Stuart location, Ont. Dept. Mines Geol. Rept. 87, 31-36, 1970.

Glikson, A.Y., The missing Precambrian crust, Geology, 7, 449-454, 1979.

Gross, G.A., Geology of iron deposits in Canada, v. 3. Iron ranges of the Labrador geosyncline, Geol. Survey Canada, Econ. Geol. Rept. 22, 22, (1968).

Gross, G.A., Principal types of iron-formation and derived ores, Can. Min. Met. Bull., 59(646), 150-153, 1966.

Gross, W.H., Evidence for a modified placer origin for auriferous conglomerates, Canavieiras mine, Jacobina, Brazil, Econ. Geology 63(3), 271-276, 1968.

Haapala, P.S., Fennoscandian nickel deposits, in Wilson, H.D.B., ed., Magmatic ore deposits, Econ. Geology Monograph 4, 262-275, 1969.

Hewitt, D.F., Uranium and thorium deposits of southern Ontario, Ont. Dept. Mines, Mineral Resources Circ. 4, 1967.

Hoeve, J., and T.I.I. Sibbald, Uranium metallogenesis and its significance to exploration in the Athabasca Basin, in Parslow, G.R., ed., Uranium exploration techniques, Sask. Geol. Soc. Special Pub. No. 4, 161-168, 1978.

Hunter, D.F., Crustal development in the Kaapvall craton. I, The Archean, Precambrian research, 1, 259-294, 1974.

Hunter, D.F., The ancient gneiss complex in Swaziland, Trans. Geol. Soc. South Africa, 73, 107-150, 1970.

Hutchinson, R.W., Massive base metal sulfide deposits as guides to tectonic evolution, in Strangway, D.W., ed., Continental crust and its mineral deposits, Geol. Assoc. Canada, J. Tuzo Wilson Volume, Special Paper 20, 659-684, 1980.

Hutchinson, R.W., Volcanogenic sulfide deposits and their metallogenic significance, Econ. Geology, 68(8), 1223-1246, 1973.

Hutchinson, R.W., and R.W. Hodder, Possible tectonic and metallogenic relationships between porphyry copper and massive sulphide deposits, Can. Min. Met. Bull., 65(718), 34-40, 1972.

Hutchinson, R.W., W.S. Fyfe, and R. Kerrich, Deep fluid penetration and ore deposition, Min. Sci. and Eng. 12(3), 107-120, 1980.

INAL Staff, Nickeliferous laterite deposits of the Rockhampton area, Q., in Knight, C.L., ed., Economic geology of Australia and Papua New Guinea, 1. Metals. The Australasian Inst. Min. Met., 1001-1006, 1975.

Kanasewich, E.R., Precambrian rift: genesis of stratabound ore deposits, Science, 161, 1002-1005, 1968.

Kerrich, R., D. Robinson, R.W. Hodder, and R.W. Hutchinson, Field relations and geochemistry of Au, Ni and Cr deposits in ultramafic-mafic volcanic rocks, in Pye, E.G., ed., Geoscience Research Seminar Abstracts, Ont. Geol. Survey, 12, 1979.

Kimberley, M.M., Origin of iron ore by diagenetic replacement of calcareous oolite, Unpub. Ph.D. Thesis, Princeton Univ., Princeton, N.J., 1, 345 p; 2, 386 p., 1974.

Knopf, Adoph, The Mother Lode system of California, U.S. Geol. Survey Prof. Paper 157, 87 p., 1929.

Kumarepeli, P.S., and V.A. Saull, The St. Lawrence valley system; a North American equivalent of the East African rift valley system, Can. Jour. Earth Sciences, 3(5), 639-658, 1966.

Ladeira, E.A., Metallogenesis of gold at the Morro Velho mine and in the Nova Lima district, Quadrilatero Ferrifero, Minas Gerais, Brazil, Ph.D. Thesis, Univ. of Western Ontario, London, Ontario, Canada, 272 p., 1980.

Lipman, P.W., H.J. Prestka, and R.L. Christiansen, Cenozoic volcanism and plate-tectonic evolution of the western United States. I. Early and Middle Cenozoic, Phil. Trans. Roy. Soc. London, A, 271, 217-248, 1972.

Luff, W.M., Geology of Brunswick No. 12 mine, Can. Inst. Min. Met. Bull., 70(782), 109-119, 1977.

Mainwaring, P.R., and A.J. Naldrett, Country-rock assimilation and the genesis of Cu-Ni sulfides in the Water Hen intrusion, Duluth Complex, Minnesota, Econ. Geology, 72(7), 1269-1284, 1977.

Markland, G.D., Geology of the Moose Mountain Mine and its application to mining and milling, Can. Min. Met. Bull., 59(646), 159-170, 1966.

Marsden, R.W., Geology of the iron ores of the Lake Superior region in the United States, in Ridge, J.D., ed., Ore deposits of the United States, 1933-1967 (Graton-Sales Volume), Am. Inst. Min. Met. and Pet. Engineers, New York, 1, 489-507, 1968.

Maske, S., The petrography of the Ingeli mountain range, Annals Univ. Stellenbosch, 41, Series A, No. 1, 109 p., 1966.

Maslyn, R.M., An epigenetic model for the formation of the Schwartzwalder uranium deposit, Econ. Geology, 73(4), 552-557, 1978.

McMillan, R.H., Genetic aspects and classification of important Canadian uranium deposits, in Kimberley, M.M., ed., Short course in uranium deposits: their mineralogy and origin, Miner.

Assoc. Canada Short Course Handbook, 3, 187-204, 1978.

McMillan, R.H., Metallogenesis of Canadian uranium deposits, in Geology, mining and extractive processes of uranium: A symposium co-sponsored by the Inst'n. Min. Met. and The Commission of European Communities, London, 145-157, Jan. 1977.

Melin, R.E., Description and origin of uranium deposits in Shirley Basin, Wyoming, Econ. Geology, 59(5), 835-849, 1964.

Miyashiro, A., Volcanic rock series and tectonic setting, in Annual Review of Earth and Planetary Sciences, 3, 251-270, 1975.

Mrak, V.A., Uranium deposits in the Eocene sandstones of the Powder River basin, Wyoming, in Ridge, J.D., ed., Ore deposits of the United States, 1933-1967 (Graton-Sales Volume), Am. Inst. Min. Met. and Pet. Engineers, New York, 1, 838-848, 1968.

Naldrett, A.J., and A.J. Macdonald, Tectonic settings of Ni-Cu sulfide ores: their importance in genesis and exploration, in Strangway, D.W., ed., Continental crust and its mineral deposits, Geol. Assoc. Canada, J. Tuzo Wilson Volume, Special Paper 20, 633-658, 1980.

Naldrett, A.J., Partitioning of Fe, Co, Ni and Cu between sulfide liquid and basaltic melts and the composition of Ni-Cu sulfide deposits - a reply and further discussion, Econ. Geology 72(6), 1520-1528, 1979.

Naldrett, A.J., Nickel sulphide deposits - their classification and genesis, with special emphasis on deposits of volcanic association, Can. Min. Met. Bull., 66(739), 45-63, 1973.

Naldrett, A.J., The role of sulphurization in the genesis of iron-nickel sulphide deposits of the Porcupine District, Ontario, Can. Min. Met. Bull., 59(648), 489-497, 1966.

Naldrett, A.J., et al., The composition of Ni-sulfide ores, with particular reference to their content of PGE and Au, Can, Mineralogist, 17, Pt. 2, 403-415 (see Noril'sk, p. 405), 1979.

Naldrett, A.J., and L.J. Cabri, Ultramafic and related rocks: their classification and genesis with special reference to the concentration of nickel sulfides and platinum group elements, Econ. Geology, 71(7), 1131-1158, 1976.

Ntiamoah-Adjaquah, R.J., Obuasi gold deposits of Ghana: their genesis in light of comparison to selected deposits, M.Sc. Thesis, Univ. of Western Ontario, London, Ontario, Canada, 177 p., 1974.

Pereira, J., and C.J. Dixon, Evolutionary trends in ore deposition, Trans. Inst'n. Min. Met., 75, B92-B96, 1966.

Pereira, J., and C.J. Dixon, Evolutionary trends in ore deposition, Trans. Inst'n. Min. Met., 74, B506-B527, 1965.

Pyke, D.R., On the relationship of gold mineralization and ultramafic volcanic rocks in the Timmins area, Ont. Div. Mines, Misc. Paper 62, 23 p., 1975.

Pyke, D.R., J.G. MacVeigh, and R.S. Middleton, Volcanic stratigraphy and geochemistry in the Timmins mining area, Geol. Assoc. Canada, Field trip guidebook, 1978.

Rajman, V., and A.J. Naldrett, Partitioning of Fe, Co, Ni and Cu between sulfide liquid and basaltic melts and the composition of Ni-Cu sulfide deposits, Econ. Geology, 73(1), 82-93, 1978.

Ridler, R.H., Relationship of mineralization to volcanic stratigraphy in the Kirkland-Larder Lakes area, Ontario, Geol. Assoc. Canada, Proceedings, 21, 33-42, 1970.

Roberts, J.B., Windarra nickel deposits, in Knight, C.L., ed., Economic geology of Australia and Papua New Guinea, 1. Metals. The Australasian Inst. Min. Met., Mon. Series No. 5, 129-143, 1975.

Robertson, D.S., J.E. Tilsley, and G.M. Hogg, The time-bound character of uranium deposits, Econ. Geology, 73(8), 1400-1419, 1978.

Robertson, J.A., The Blind River uranium deposits: the ores and their setting, Ont. Div. Mines, Misc. Paper 65, 45 p., 1976.

Rosholt, J.N., E.N. Harshman, W.R. Shields, and E.L. Garner, Isotopic fractionation of uranium related to roll features in sandstone, Shirley Basin, Wyoming, Econ. Geology, 59(4), 570-585, 1964.

Ross, J.R., and G.M.F. Hopkins, Kambalda nickel sulphide deposits, in Knight, C.L., ed., Economic geology of Australia and Papua New Guinea, 1. Metals. The Australasian Inst. Min. Met., Mon. Series No. 5, 100-121, 1975.

Rowntree, J.C., and D.V. Mosher, Jabiluka uranium deposits, in Knight, C.L., ed., Economic geology of Australia and Papua New Guinea, 1. Metals. The Australasian Inst. Min. Met., Mon. Series No. 5, 321-326, 1975.

Ruckmick, J.C., The iron ores of Cerro Bolivar, Venezuela, Econ. Geology, 58(2), 218-236, 1963.

Ruzicka, V., Phanerozoic uranium deposits and occurrences in Europe and eastern North America, in Kimberley, M.M., ed., Short course in uranium deposits: their mineralogy and origin, Miner. Assoc. Canada Short Course Handbook, 3, 217-228, 1978.

Sestini, G., Sedimentology of a paleoplacer: the gold-bearing Tarkwaian of Ghana, in Amstutz, G.C. and Bernard, A.J., eds., Ores in sediments, Int. Union of Geol. Sciences, Ser. A., No. 3, 275-305, 1971.

Simpson, T.A., and T.R. Gray, The Birmingham red-ore district, Alabama, in Ridge, J.D., ed., Ore deposits of the United States, 1933-1967 (Graton-Sales Volume), Am. Inst. Min. Met. and Pet. Engineers, New York, 1, 187-206, 1969.

Smart, P.G., P.G. Wilkes, R.S. Needham, and A.L. Watchman, Geology and geophysics of the Alligator Rivers region, in Knight, C.L., ed., Economic geology of Australia and Papua New Guinea, 1, Metals. The Australasian Inst. Min. Met., Mon. Series No. 5, 285-301, 1975.

Snyman, A.A., Thabazimbi Mine, central Transvaal Basin, in Coetzee, C.B., ed., Mineral resources of South Africa, Geol. Survey, Dept. Mines, Pretoria, 152-155, 1976.

Souch, B.E., T. Podolsky, and Geological Staff, The sulfide ores of Sudbury: their particular relationship to a distinctive inclusion-bearing facies of the Nickel Irruptive, in Wilson, H.D.B. ed., Magmatic ore deposits, Econ. Geology Monograph 4, 252-261, 1969.

Stanton, R.A., Genesis of uranium-copper occurrence 74-1E, Baker Lake, Northwest Territories, M.Sc. Thesis, Univ. of Western Ontario, London, Ontario, Canada, 167 p., 1979.

Stockwell, C.H., et al., Geology of the Canadian Shield, in Douglas, R.J.W., ed., Geology and economic minerals of Canada, Geol. Survey Canada, Econ. Geology Rept. No. 1 (5th ed.), 44-150 and Fig. iv-3, p. 49), 1970.

Theis, N.J., Mineralogy and setting of Elliot Lake deposits, in Kimberley, M.M., ed., Short course in uranium deposits: their mineralogy and origin, Miner. Assoc. Canada Short Course Handbook, 3, 331-338, 1978.

Tolbert, G.E., J.W. Tremaine, G.C. Melcher, and C.B. Gomez, The recently discovered Serra dos Carajás iron deposits, Northern Brazil, Econ. Geology, 66(7), 985-994, 1971.

Tremblay, L.P., Geologic setting of the Beaverlodge-type of vein - uranium deposit and its comparison to that of the unconformity-type, in Kimberley, M.M., ed., Short course in uranium deposits: their mineralogy and origin. Miner. Assoc. Canada Short Course Handbook, 3, 431-456, 1978.

Trendall, A.F., Precambrian iron-formations of Australia, Econ. Geology, 68(7), 1023-1034, 1973.

Trescases, J.J., Weathering and geochemical behaviour of the elements of ultramafic rocks in New Caledonia, in Fisher, N.H., ed., Metallogenic provinces and mineral deposits in the Southwestern Pacific, Bur. Mineral Resources, Geology and Geophysics (B.M.R.), Canberra, 149-161, 1973.

Viljoen, M.J., et al., The geology of the Shangani nickel deposit, Rhodesia, Econ. Geology, 71(1), 76-95, 1976.

Watson, Janet, Mineralization in Archaean provinces, in Windley, B.F., ed., The early history of the earth, John Wiley and Sons, 443-466, 1975.

Woodall, R., Gold in the Precambrian Shield of Western Australia, in Knight, C.L., ed., Economic geology of Australia and Papua New Guinea, 1. Metals. The Australasian Inst. Min. Met., Mon. Series No. 5, 175-184, 1975.

Worst, B.G., The Great Dyke of Southern Rhodesia, Southern Rhodesia Geol. Survey Bull. No. 47, 122-125, 1960.

INFERENCES FROM OTHER BODIES FOR THE EARTH'S COMPOSITION AND EVOLUTION

William M. Kaula

Department of Earth and Space Sciences
University of California, Los Angeles, California 90024

Abstract. The compositions of the terrestrial planets depend primarily on distance from the sun, secondarily on planet size, with positive correlations of both with volatile content. The major uncertainties pertain to volatiles, in particular, the great differences in inert gas contents of Venus, the Earth, and Mars are baffling. The main inference for the Earth dependent on its comparison with other bodies is that it is depleted in elements of intermediate volatility, such as S, K, Na.

The evolution of a terrestrial planet has, until recently, appeared to depend mainly on planet size: the larger the planet, and hence the lower its area/mass ratio, the longer persists its tectonic activity. However, the Pioneer Venus altimetric and gravimetric data indicate that Venus is more like Mars than the Earth in having a thicker crust and no indications of plate tectonic activity. An obvious immediate cause is the higher surface temperature, leading to a greater depth for the basalt-eclogite phase transition which in turn inhibits recycling of crustal material. A plausible ultimate cause of the differences in evolution of the Earth and Venus is that the Earth had a major initial energetic input from a great impact, which also created the moon.

Introduction

Planetary exploration in the last 15 years has greatly increased our knowledge of the other planets. This knowledge has made cosmogony and comparative planetology more meaningful disciplines. The usual procedure is to utilize our knowledge of the Earth to help understand the origin, evolution, and structure of the other terrestrial bodies. In this paper we shall attempt to reverse the procedure, enquiring whether certain properties of the Earth are what should be plausibly expected for the end member of the class of predominantly silicate solar system bodies.

Such an enquiry is, of course, strongly influenced by accessibility of data: for volatile composition, comparisons with Mars, Venus, & stony meteorites; for refractory composition, compari- sons with the moon & stony meteorites; and for dynamical & tectonic behavior, comparisons with the moon, Mars, and Venus. The principal data for bodies defined as being "terrestrial" by being largely iron and silicate in composition are given in Table 1. For study of the Earth, the smaller bodies listed below Mars are relevant almost entirely for inferring bulk composition. In particular, chondrites are important as giving the best estimates of the involatile part of the primordial abundance, while achondrites, in particular eucrites, are samples of the smallest body, or bodies, known to have differentiated.

Comparisons among planets fall naturally into the two categories of bulk composition, dependent almost entirely on origin circumstances, and evolution, dependent largely on the properties of the planets in isolation from one another.

Bulk Composition

The decrease in mean density with distance from the sun, Mercury-Earth-Mars, is generally taken to indicate a decreasing ratio of refractory to volatile constituents. This trend is most evident in elements of intermediate volatility: i.e., with condensation temperatures in the range $\sim$600-1200°K. Figure 1, from Morgan & Anders [1979], is a summary of these data, in the form of depletion with respect to cosmic abundances based mainly on carbonaceous chondrite meteorites. While the curves are labeled by "index" elements, it is important to emphasize that elements of similar condensation temperature appear to have similar depletions, regardless of differences in their geochemical behavior after condensation. Hence, explanations of depletions by intra-planetary processes (such as putting potassium or sulfur in the Earth's core) should also discuss why these elements are enriched in bulk composition relative to others of similar volatility.

Figure 1 shows another trend besides that associated with solar distance pertaining to elements of moderate volatility: $400°K \lesssim T_c \lesssim 1200°K$, where T_c is condensation temperature. The retention of such elements appears to be positively correlated with planet size, a contradiction of

TABLE 1. Terrestrial Body Properties

Body	Distance from Sun AU	Mass Earth Masses $M_\oplus$	Mean Density g cm⁻³	Density reduced to 10 kb g cm⁻³	Moment of Inertia C/MR²	Metal Metal+Silicate	Crustal Thickness km	Ocean & Atmosphere Total Mass
Mercury	0.39	0.055	5.44	5.31		0.68	≥7	<10^{-14}
Venus	0.72	0.815	5.24	3.95		~0.26	≳30	6.5×10^{-5}
Earth	1.00	1.00	5.52	4.03	0.332	~0.31	14	2.3×10^{-4}
Moon	1.00	0.012	3.34	3.34	0.393	<0.01	75	<10^{-14}
Mars	1.52	0.107	3.93	3.71	0.365	0.19	>28	7×10^{-10}
Asteroids C	~2.4–3.6	<2×10^{-4}	~2.2	2.4				
S	~2.3–3.1	<5×10^{-6}	2.5–3.5	2.7–3.7				
O	~2.3–4.0	<4×10^{-5}						
Meteorite Parent Bodies	2.2–2.5?	<4×10^{-5}?	2.0–3.7	2.2–4.0				
Io	5.2	0.015	3.52	3.52				
Europa	5.2	0.008	3.45	3.45				

Based on: Wood et al. [1981], Chapman [1979], Ringwood & Anderson [1977].
Note: Asteroid types are C: carbonaceous; S: stony; O: other.

the correlation with solar distance in the cases of Mars and probably the eucrite parent body. However, this correlation with size is not uniform; some meteorites which appear to have come from small bodies, most notably carbonaceous chondrites, are volatile rich.

The data for gases, given in Table 2, seems even more unsystematic, particularly since the Viking and Pioneer Venus findings. While both Mars and Venus confirm the general trends of positive correlation of volatile retention with planet size, they do so in a very uneven way. Most remarkable are the tremendous differences in primordial argon. These have not been explained convincingly; attempts to ascribe them to either a pressure gradient [Pollack & Black, 1979] or solar wind implantation [Wetherill, 1980] require rather ad hoc modeling assumptions. Meanwhile, the relatively high inert gas abundances in carbonaceous chondrites remain to remind us that the losses of primordial inert gases probably took place when most of the protoplanetary material was in very small bodies.

Processes which have been suggested to account for compositional variations in inner solar system include: (1) heating of the nebula gas by the early sun, leading to temperature decreasing with distance from the sun, and thus an increasing content of volatiles in condensed material; (2) a decrease of this heating with time, thus leading to an increase of volatile content with the time of condensation; (3) collection of condensed materials within planetesimals, cutting off chemical interaction with the gas; (4) a sweeping away of gas and particulate matter by the early sun, leading to a decrease in volatile content with time of collection in a planetesimal; (5) heating of planetesimals, by the short-lived radioisotope ^{26}Al or electromagnetic induction or collisions, leading to planet-like differentiations; (6) breakup of differentiated planetesimals by collisions, leading to separation of their silicate and metallic parts; (7) differential gas drag on planetesimal fragments, leading to loss of the smaller silicate fragments relative to the metallic; (8) gravitational scattering by the Earth and Venus as they got larger, leading to a blurring of compositional gradients determined mainly by temperature; (9) gravitational scattering by Jupiter, leading to an admixture of volatile material to the inner solar system and an inhibition of the growth of Mars; (10) major impacts into the Earth and/or Venus, leading to breaking off of protosatellite material; and (11) gravitational capture of gas, as well as solids, by planets as they got large, leading to a larger volatile content in the outer parts of the larger planets, Earth and Venus.

The above list includes most of the processes considered as plausibly significant by most people working in solar system origin, but it is far from including all those possibly significant, let alone those suggested as such. Since models incorporating these processes are the only way

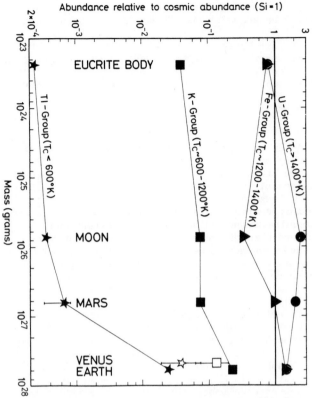

Abundance relative to cosmic abundance (Si=1)

EUCRITE BODY

Tl-Group (Tc < 600°K)

K-Group (Tc~600-1200°K)

Fe-Group (Tc~1200-1400°K)

U-Group (Tc~1400°K)

MOON

MARS

VENUS
EARTH

Mass (grams)

Fig. 1. Compositions of five terrestrial bodies, represented by index elements for four cosmochemical groups. From Morgan & Anders [1979].

to connect the small number of disparate objects available to us as products of the formation process, it is not feasible to infer a bulk composition of the Earth which is a well-constrained consequence of (1) a starting cosmic composition from the sun & carbonaceous chondrites; (2) the Earth's bulk properties (mean density, moment-of-inertia, etc.); and (3) the differences in properties of the Earth from other terrestrial bodies, particularly its nearest neighbors: the moon, Venus, and Mars. Table 3 is the consequence for the Earth of a recent group effort to construct bulk compositions of terrestrial mantles

[Wood et al., 1981]. The principal variations among the compositions dependent on solar system origin models (implicit as well as explicit) are among the volatiles O, H_2O, S, K, Na. Some of these variations are ad hoc--i.e., invoked to explain core density or core heating--while some are rather arbitrary: e.g., the "equilibrium condensation" models depended on a particular assumed nebula temperature gradient. If the core and mantle densities of the Earth were not known from seismology (i.e., if we had only the mean density and moment-of-inertia of the whole planet), then the uncertainties arising from origin circumstances would extend to the Fe:Si and Mg:Si ratios. Hence the solar system origin constraints on Earth composition really are significant only for <u>differences</u> of behavior among elements of similar characteristics, such as condensation temperature. For this reason, estimates based on empirical extrapolation from meteorites, such as E3 and E5, are probably more realistic than the more ideal, but still very incomplete, equilibrium condensation models, E1 and E2.

In the converse mode of using what we know about the Earth to infer the composition of the other terrestrial planets, it has long been accepted that Mercury & the moon are radically different, most obviously because of their much smaller size and differing dynamic circumstances; that Mars has a plausibly higher oxidation level; and that Venus required some fine tuning to account for its lower density--more oxidation or less iron or less eclogite, plus possibly less sulfur [Ringwood & Anderson, 1977; Anderson, 1979].

In the aforestated context, the Viking findings about the Martian atmosphere (see Table 2) showed it to be surprisingly deficient in volatiles. The principal constraint requiring that the sparsity of Martian atmosphere (compared to the atmosphere <u>plus</u> ocean on Earth) arises from a deficiency in bulk composition, rather than inefficiency of outgassing, is the low $^{36+38}Ar$:^{40}Ar ratio [Anders & Owen, 1977]. The outgassing efficiency of these isotopes should be negligibly different; it is implausible that the primordial argon would be much deeper than the primordial potassium; and more of the primordial argon should be outgassed than the radiogenic argon, since it has been in the planet longer. The deficiencies

TABLE 2. Gaseous Abundances of the Planets
$Log_{10}(g/g)$

	$^{36+38}Ar$	^{40}Ar	Xe	N_2	CO_2	H_2O	$^{38}Ar/^{36}Ar$	$^{20}Ne/^{36}Ar$
Sun	-4.0		-7.8	-2.9	-1.9	-2.0		31.
Venus*	-8.4	-8.5		-5.4	-4.0	-7.0	0.18	0.4
Earth	-10.3	-8.0	-10.9	-5.6	-3.8	-3.6	0.19	0.6
Mars*	-12.7	-9.3	-14.0	-6.9	-7.5	-5.3	0.2	
C-Chondrite	-9.1		-10.3				0.2	0.1-15

Based on: Pollack & Black [1979] and Phillips et al. [1981].
*Assuming total outgassing.

TABLE 3. Model Compositions (Major and Heat-Producing Elements) of the Earth

		E1	E2	E3	E6	E7
Mantle + Crust	SiO_2	48.5	47.2	48.0	45.9	45.7
	TiO_2	0.18	0.17	0.27	0.3	0.1
	Al_2O_3	3.5	3.4	5.2	3.9	3.4
	Cr_2O_3			1.1	0.3	0.4
	MgO	34.5	33.6	34.3	38.0	38.4
	FeO	8.3	10.6	7.9	8.1	8.0
	MnO			0.12	0.1	0.1
	CaO	3.3	3.2	4.2	3.2	3.1
	Na_2O	1.5	1.5	0.33	0.1	0.4
	H_2O	0.13	0.026	0.11		0.3
	K(ppm)	1324	1289	262	129	387
	U(ppm)	0.019	0.019	0.029	0.021	0.026
	Th(ppm)	0.086	0.066	0.100	0.075	0.108
Core	Fe	68.9	71.8	89.2	86.2	77.7
	Ni	5.0	5.5	5.7	4.8	6.0
	S	26.1	22.7	5.1	1.0	16.4
	O				8.0	
Relative Masses	Mantle + Crust	69.2	71.9	63.9	68.9	68.3
	Core	30.8	28.1	36.1	31.2	31.7

E1: Equilibrium condensation, assuming T = $600°K/a^{1.1}$, a = solar distance in AU.

E2: Model E1 modified by a weighting function to allow for dynamical scattering.

E3: Ganapathy and Anders [1974].

E6: Model of Ringwood [1979], 15% component A + 85% component B.

E7: Empirical model: mantle and core from Ringwood [1979], crust from Smith [1977]. From Wood et al. [1981].

of Ar, Xe, (and perhaps N_2) in Mars, together with its higher oxidation level, indicate that inert gases must be considered as a different class from chemically active gases in solar system origin studies [Pollack & Black, 1979].

The Pioneer Venus findings about the Venerean atmosphere (see Table 2) are even more of a wrench in the opposite direction. The high $^{36+38}$Ar abundance, by similar arguments, must reflect bulk composition dependent on origin circumstances. The enrichment by a factor of 100 over the Earth, while N_2 and CO_2 are comparable, is, as mentioned above, difficult to ascribe to nebula circumstances. The least implausible hypothesis is an admixture of a "solar" component to a "planetary" component [Wetherill, 1980], similar to that long used to explain differences of volatile composition between grains and matrix in meteorites [Wasson, 1974, 97-109]. For example, if one part of "solar" component, as given in Table II, is added to 20,000 parts "Earth" component, the $^{36+38}$Ar proportion is changed from $10^{-10.3}$ to $10^{-8.3}$, but the N_2, CO_2, and H_2O proportions are negligibly changed. The difficulty is imaging how in a continuous soup like the

nebula, at least 100 times as much of anything could be added to the Venus zone as to the Earth zone.

While quantitatively the volatiles are a very minor part of planetary constitution, we cannot feel confident about any estimates of composition based on solar system origin models until their variations are satisfyingly explained.

Planetary Evolution

After the Apollo and Mariner projects, it seemed fairly evident that the overwhelming factor in terrestrial planet evolution is size. All planets apparently were hot enough in their outer parts as a consequence of formation to differentiate sizeable crusts, but subsequent evolution has been mainly governed by planetary area/mass ratio: the smaller the planet, the lower its level of convective activity, and the quicker its outer layers cooled to form a lithosphere, which now appears to be one continuous plate on all the terrestrial planets except the earth [Kaula, 1975].

An evident consequence of the aforegoing evolutionary pattern is an inverse correlation of lithospheric thickness with planet size. The Apollo seismometry established that the lunar lithosphere is ∿800 km thick, while the large variations in Martian topography and gravity, together with their high correlation, are consistent with a lithosphere ∿300 km thick [Sleep & Phillips, 1979]. The one plate planets, Mars, Mercury, and the moon, also evidence the counteraction of heating of the inner parts from radioactivity against the cooling of the outer parts. The consequences are rough balances of overall thermal expansion and contraction in planetary evolution, as evidenced by the absence of marked net tension or compression in the surface features. Arising from the evolution of this expansion versus contraction is a direct correlation of the time of shutting off of volcanism with planet size: the smaller the planet, the earlier compression became dominant over tension. The principal variations in this pattern determined by planet size arose mainly from composition. Mercury's surface evidences a net compressive regime, as would be the result of early core formation, resulting from a large metal:silicate ratio, hence a large energy release, self-accelerating to early conclusion, with predominant cooling ever since. The surface of Mars evidences a net tensional regime, consistent with late core formation, resulting from a small metal:silicate ratio, hence a small energy release, requiring appreciable radioactive heating to both start and bring to completion [Solomon & Chaiken, 1976; Solomon, 1977].

Another inverse correlation with planetary size shown in Table I is crustal thickness. The mean crustal thicknesses of the Earth, moon, and Mars are all about one fourth the depth at which the basalt-eclogite phase transition occurs, suggesting that crustal formation is dependent on the tendency of basaltic differentiate to float or sink.

The principal evolutionary puzzle raised by the more detailed Viking examination of Mars is the evidence of erosive activity by, most likely, water which appeared to have been released sporadically to create "scablands". Apparently the cold surface of Mars leads to its volatiles not being outgassed but to being trapped in a permafrost layer, from which the sporadic releases were triggered by impact or volcanic events [Baker, 1978]. The subsequent loss of the water is a problem of atmospheric evolution, but on the whole the Viking findings have not required a drastic alteration of ideas of solid planet evolution.

The new findings by the Pioneer Venus project are a different story, however. The most prominent topographic features on Venus mapped by the radar altimeter are plateaus and mountain ranges 5-10 kilometers above the general terrain [Pettengill et al., 1980]. These features suggest that Venerean tectonics is internally generated. There are some craters, indicating that the Venerean surface is rather old, but nothing like the great impact basins which dominate the lunar surface. However, although the most prominent features appear to be endogenic, there is little to suggest plate tectonics on Venus: no long ridges with concave slopes like terrestrial ocean rises, no arcuate sharp ridges suggesting subduction zones. The gravity field inferred from perturbations of the spacecraft appears to be milder but of higher correlation with the topography than the Earth's [Ananda et al., 1980; Sjogren et al., 1980], further suggesting a one-plate planet with features supported passively by a thick lithosphere. Studies of the gravity & topography suggest compensation depths of 100 km or more; hence, the support may be dynamic [Phillips et al., 1981].

In short, the thermotectonics of Venus resembles that of Mars much more than that of the Earth, which is rather inconsistent with the notion that size is the dominant factor in planetary evolution. The most evident differences of Venus from the Earth are (1) a surface temperature $\sim$500°C higher and (2) a much lower water content in the atmosphere (see Table 2).

The obvious inference from the higher surface temperature is that temperatures high enough for convection to dominate would be reached at a shallower depth, with the result that tectonic plates would be smaller, etc. However, for thermal convection in a highly viscous medium the thickness of the boundary layer--the region in which conductive transport dominates over convective--is determined by the thermal diffusivity, rather than the kinematic diffusivity [Peltier, 1980]. For a given heat transport, the moderate decrease in conductivity resulting from the higher temperatures would thus act to only moderately decrease the boundary layer

thickness, and corresponding to increase the temperature gradient in the layer. Anderson [1979] suggests that a consequence of the higher temperatures would be partial melting of eclogitic material, at depths of $\sim$40-80 km, resulting in the removal of the garnet and retention of a much greater proportion of the basaltic component in the low density phase. Anderson [1979] utilizes this phenomenon to explain why the mean density of Venus is lower than that of the Earth (reduced to the same pressure). A more important inference may be that a much higher proportion of crustal material, and hence radioactive heat sources, is retained near the surface, rather than recycled to the mantle, as in the Earth. Consequently Venerean evolution might be more Mars-like because of the lesser heat sources at depth. Consistent with this hypothesis are the greater crustal thickness implied by the greater topographic range and the lesser energy sources at depth implied by the lack of a magnetic field.

The low H_2O content of the Venerean atmosphere is generally explained as the consequence of photodissociation followed by loss of H_2 off the top and recombination of O_2 with surface rock [Pollack, 1979; Walker, 1975]. The level of tectonic activity required to expose sufficient oxidizable rocks is not excessive. The effects of lower H_2O content and higher oxidation level would be to increase the viscosity, thus inhibiting recycling of crustal material and stiffening the lithosphere. However, it is difficult to see how the absence of water would have significant effect on lithospheric spreading: if plate tectonics is occurring on Venus, one would expect to see the topography characteristic of the cooling boundary layer. More likely to be significant are the effects of the lack of water at subduction zones: if the breakdown of subducted hydrates is important to flux island arc volcanism, then compressional zones on Venus would tend to resemble collisional belts on the Earth, such as the Himalayas, rather than the island arcs characteristic of the subduction of oceanic lithosphere. But it is more likely that because of the greater thickness of buoyant crustal material, plate tectonics has ceased entirely on Venus [cf. McGill, 1979].

The radically different tectonic style on Venus should provoke thought about the implications of the obvious differences of the Earth from Venus: the lower surface temperature and the greater water content of, at least, crustal rocks. Foremost is that to maintain a thin crust and appreciable energy sources at depth, the Earth must have recycled an appreciable volume of crustal material. Isotopic data do not preclude such recycling prior to $\sim$2.5 Ga ago [Wasserburg & DePaolo, 1979]. Secondly the resistance to subduction of crustal material must depend on some minimal ratio of crustal thickness to lithospheric thickness which is exceeded on Venus but not on the Earth. How Venus evolved to this state appears to depend on its surface temperature, which

in turn is the result of its atmospheric evolution
--suggesting that the volatile tail wags the solid
dog. Intuitively one would expect events of
greater energetic order-of-magnitude, core differentiation or planetary formation, itself, to
be more important. It seems unlikely that Venus
escaped core differentiation. More plausible is
a difference in primordial energy suggested by
Venus's slower spin and lack of a satellite. The
almost-twin planets may have been jolted onto
different evolutionary tracks just by one having
been hit by a major initial impact--say, Mars-
sized--while the other wasn't.

Acknowledgments. This work has been supported
by NASA grants NAS 2-9128 and NGL 05-007-002.

References

Ananda, M.P., W.L. Sjogren, R.J. Phillips, R.N.
Wimberly, and B.G. Bills, A low-order global
gravity field of Venus and dynamical implications, J. Geophys. Res., 85, 8303-8318, 1980.

Anders, E., and T. Owen, Mars and Earth: origin
and abundance of volatiles, Science, 198, 453-
465, 1977.

Anderson, D.L., Tectonics and composition of
Venus, Geophys. Res. Let., 7, 101-104, 1980.

Chapman, C.R., The asteroids: nature, inter-
relations, origin and evolution, in Gehrels, T.,
ed., Asteroids, Univ. Arizona Press, Tucson,
25-60, 1979.

Ganapathy, R., and E. Anders, Bulk compositions
of the moon and Earth estimated from meteorites,
Proc. Lunar Sci. Conf. 5th, 1181-1206, 1974.

Kaula, W.M., The seven ages of a planet, Icarus,
26, 1-15, 1975.

McGill, G.E., Venus tectonics: another Earth or
another Mars?, Geophys. Res. Let., 6, 739-742,
1979.

Morgan, J.W., and E. Anders, Chemical composition
of Mars, Geochim. Cosmochim. Acta, 43, 1601-
1610, 1979.

Peltier, W.R., Mantle convection and viscosity,
in A. Dziewonski & E. Boschi, eds., Physics of
the Earth and Planetary Interiors, Academic
Press, 1980.

Pettengill, G.H., E. Eliason, P.G. Ford, G.B.
Loriot, H. Masursky, and G.E. McGill, Pioneer
Venus radar results, altimetry and surface

properties, J. Geophys. Res., 85, 8261-8270,
1980.

Phillips, R.J., W.M. Kaula, G.E. McGill, and M.C.
Malin, Tectonics and evolution of Venus,
Science, in press, 1981.

Pollack, J.B., Climatic change on the terrestrial
planets, Icarus, 37, 479-553, 1979.

Pollack, J.B., and D.C. Black, Implications of
the gas compositional measurements of Pioneer
Venus for the origin of planetary atmospheres,
Science, 205, 56-59, 1979.

Ringwood, A.E., Origin of the Earth and Moon,
Springer-Verlag, New York, 295 pp., 1979.

Ringwood, A.E., and D.L. Anderson, Earth and
Venus: a comparative study, Icarus, 30, 243-
253, 1977.

Sjogren, W.L., R.J. Phillips, P.W. Birkeland, and
R.N. Wimberly, Gravity anomalies on Venus, J.
Geophys. Res., 85, 8295-8302, 1980.

Sleep, N.H., and R.J. Phillips, An isostatic
model for the Tharsis Province, Mars, Geophys.
Res. Let., 6, 803-806, 1979.

Smith, J.V., Possible controls on the bulk composition of the Earth: implications for the
origin of the Earth and moon, Proc. Lunar Sci.
Conf. 8th, 333-369, 1977.

Solomon, S.C., The relationship between crustal
tectonics and internal evolution in the moon
and Mercury, Phys. Earth Planet. Inter., 15,
135-145, 1977.

Solomon, S.C., and J. Chaiken, Thermal expansion
and thermal stress in the moon and terrestrial
planets: clues to early thermal history,
Proc. Lunar Sci. Conf. 7th, 3229-3243, 1976.

Walker, J.C.G., Evolution of the atmosphere of
Venus, J. Atmos. Sci., 32, 1248-1256, 1975.

Wasserburg, G.J., and D.J. DePaolo, Models of
Earth structure inferred from neodymium and
strontium isotopic abundances, Proc. Nat. Acad.
Sci., 76, 3594-3598, 1979.

Wasson, J.T., Meteorites, Springer-Verlag,
New York, 316 pp., 1974.

Wetherill, G.W., Could the solar wind have been
the source of the high concentration of ^{36}Ar in
the atmosphere of Venus, Lunar and Planetary
Science XIX, 1239-1241, 1980.

Wood, J.A., and 12 others, Mantles of the terrestrial planets, in Balsaltic Volcanism on the
Terrestrial Planets, in press, 1981.

PLATE TECTONIC PATTERNS AND CONVECTION IN THE PHANEROZOIC

E.R. Kanasewich, M.E. Evans and J. Havskov*

Department of Physics, University of Alberta, Edmonton, Canada T6G 2J1

*Instituto de Ingenieria, Universidad de Mexico, Mexico

Abstract. Paleomagnetic observations, oceanic
magnetic lineations, and the present continental
margins are combined in an interactive computer
program to generate maps for seven periods in
the Phanerozoic Era. By using geological evidence on the position of geosynclines, volcanic
assemblages, and zones of diastrophism an attempt
is made to model a reconstruction of major plate
boundaries in the past. During the Cambrian and
Ordovician Periods the continental segments were
as widely dispersed as at present and formed a
ring of plates on the paleoequator but with
North Africa and South America contiguous and
close to the south pole. In these, as in other
periods, there is symmetry about the spin axis.
Extensive plate motion occurred at about the
time of the Caledonian Orogeny when the continental segments evolved toward the formation of
a single large group called Pangaea by Wegener.
This evolution occupied much of the late Paleozoic and Mesozoic Eras. Toward the end of the
Cretaceous Period a more dispersed form began to
develop yielding the present pattern of two
antipodal quasi-circular plates separated by a
ring of more irregular quasi-elliptical plates.
The results suggest the presence of a slowly
evolving mantle-wide convection system that is
symmetric about the Earth's spin axis. The
dominant pattern during the late Paleozoic and
the Cenozoic was of plates with dimensions of
60-120° in which the convective pattern may be
described in terms of spherical harmonics of the
third order. In the Mesozoic and late Paleozoic
the plate sizes were of order of 90-180° and
physical parameters describing convection were
presumably dominated by second order terms.

Introduction

A series of azimuthal-equidistant map projections, centered on each of the lithospheric
plates were used by Kanasewich (1976) to demonstrate a high degree of ordering and symmetry
in the major plates at the present time. Major
plates cover 9-20% of the earth's surface and
are clearly differentiated in size and shape

from mini-plates or splinters that cover area
of 3% or less (Nasca, 3%; Philippines, 1.6%;
Arabia, 1%; Cocos, 0.6%). The lithosphere was
shown to be highly organized with two antipodal
quasi-circular plates (African and Pacific),
120° in diameter, separated by a ring of quasi-
elliptical plates. The semi-major and semi-
minor axes are defined by triple junctions. The
normal Mercator projection distorts the pattern
in high latitudes but if the continental outlines are rotated so that the great circle path
through the north and south poles and the centers
of the ring plates becomes the equator in an
Eckert projection, the distortion becomes
minimal (Fig. 1). The Eckert (Ortelius) projection has equal spacing of parallels and
displays the entire earth with an approximation
to an equal-area projection. The quasi-elliptical plates have their major axes all aligned
at about the same angle to the 'pseudo-equator'.
The symmetry inherent in the present pattern and
the large dimensions of the major plates (120°
as a diameter or semi-major axis) is strong
evidence against dynamic systems that produce
random plate distributions. The present
evidence suggests that the physical properties
of the earth may be described in terms of
spherical harmonics with dominant terms of
order three.

Theoretical studies by Chandrasekhar (1961)
investigated three-dimensional convection in
spherical shells of a uniform Newtonian incom-
pressible fluid. A variational principle was
used to determine the Rayleigh number for the
onset of convection in cells with various sizes.
Since the core is liquid with a small viscosity,
compared to the mantle, the lower boundary
condition must have zero shear stress (free
slip). The boundary condition at the lithosphere-asthenosphere contact is more complex
but it is reasonable to assume that the lithosphere is free to move with the underlying
mantle. More complex and irregular plate
boundaries will result because this condition
is only approximated when considering the broad
pattern of convective motion. With upper and

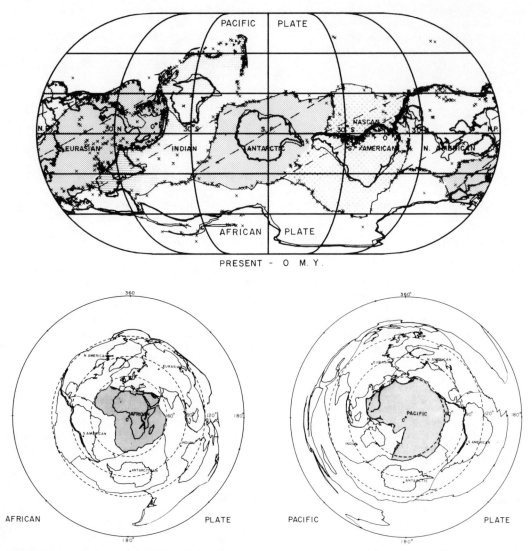

Fig. 1. The ring of quasi-elliptical plates at the present time displayed on an Eckert projection. The great circle path passing through the north and south poles and the center of the ring plates was rotated to form a pseudo-equator in this projection. The orientation of the quasi-elliptical plates is defined by the dashed lines from an African triple junction to a Pacific triple junction. Shallow earthquakes are shown as crosses. The smaller diagrams are azimuthal-equidistant projections of the present Earth centered on the African and Pacific plates respectively.

lower boundaries being free-slip and using the present size of the mantle it is found that a flow generated by internal sources distributed as a spherical harmonic of degree three is most readily excited (Rayleigh number, R = 19,000). Those with degree 2 (R = 22,000) and degree 4 (R = 21,000) are only slightly larger and are probably easily excited. That with degree 6 has a Rayleigh number of 35,000. It has been found by Chamalaun and Roberts (1962) that axially symmetrical modes are excited more readily than unsymmetrical modes. A very detailed knowledge of the boundary conditions

and the physical properties of the mantle would be necessary to make a theoretical prediction of the combination of modes excited and maintained in the flow pattern. However, it seems reasonable to assume that, if mantle-wide convection occurs, the low order spherical harmonics will have the dominant amplitudes.

Plate tectonic models of oceanic lithosphere for the Tertiary and Cretaceous Periods have been made for the Atlantic (Pitman and Talwani 1972), Pacific (Herron 1972; Larson and Pitman 1972; Atwater and Molnar 1973; Molnar et al. 1975; Cooper et al. 1976), Indian (Fisher et al.

1971; McKenzie and Sclater 1971), and Arctic
(Lambert 1974) oceanic areas. These reconstructions using magnetic lineations cannot be
continued beyond the Cretaceous-Jurassic boundary
because of the youthfulness of oceanic basins and
their destruction by subduction. Any further
attempts at modelling in the Mesozoic and
Paleozoic eras must rely on data from continental
crust. However, purely geological evidence does
not, in general, permit a unique solution. Some
of the ambiguity would be reduced if it were
possible to establish the properties of the
plates as a function of time and to determine
the global dynamic system, often referred to as
the driving mechanism. A few properties and
principles that may be of value in reconstructing
a plate tectonic model at any time in the past
have been pointed out by Kanasewich (1976) and
these will be tested and explored in this paper
using paleomagnetic evidence.

In reconstructing a plate tectonic model for
any period some rules or principles should be
established. Some of these are absolute and
result from restrictions imposed by the geometry
of plate motions on a spherical surface. Others
are in the nature of postulates made from
observations, particularly over the sea floor,
at the present time and may not be universally
applicable throughout the earth's history. For
instance, it is assumed in this paper that the
plates remain rigid when they interact even
though it is known that considerable deformation
occurs, particularly when two continental plates
collide as along the Alpine and Himalayan
mountain chains. These continental interactions
and others like the transcurrent motion along
the San Andreas or Great Glen fault systems
involve minor amounts of crust and are assumed
to be second order effects when considering
the global results of continental drift. The
following principles will be observed as
closely as possible in obtaining the models
for each period.

1. The major tectonic features of the earth
may be described by the interaction of six to
nine uncoupled rigid plates of lithosphere with
dimensions 60-120° (Elsasser 1969; Le Pichon
1968; Kanasewich 1976).
2. The plates are created and destroyed along
ridges and trenches, respectively (Vine and
Matthews 1963; Hess 1962; Oliver and Isacks
1967; Isacks et al. 1968).
3. Transform faults conserve lithosphere and
are lines of slip between two plates. They lie
on small circles centered on the pole of relative
motion between two plates (Wilson 1965; McKenzie
and Parker 1967).
4. The poles of rotation between pairs of
plates are relatively stable for long periods of
time (Morgan 1968; McKenzie and Parker 1967).
5. An absolute reference frame for plate
kinematics is defined, to a good approximation,
by minimizing the translational motion of plate

boundaries. The velocity of plates is proportional to the amount of continental lithosphere
they contain. Purely oceanic plates move about
five times faster than purely continental
plates, which, at the present time, move at
about 1.5 cm/year (Minster et al. 1976; Kaula
1975).
6. The continental lithosphere extends to the
500 fathom contour along coast lines (Bullard
et al. 1965).
7. Continental crust, by virtue of its
buoyancy, is not destroyed in any significant
amount along plate margins (McKenzie 1969).
8. A eugeosyncline is direct evidence for a
trench and a subduction zone. Miogeosynclines
and zones of diastrophism are secondary lines
of evidence for the nearby presence of a subduction zone (Dewey and Bird 1970).
9. The continuing presence of a seaway or an
ocean throughout more than one geological period
is taken as direct evidence for the occurrence
of sea floor spreading and the presence of a
ridge system.
10. The geomagnetic field has always been
dominated by a dipole component and, when
averaged over a period of the order of 10^6
years, the axis of the dipole coincides with the
rotation axis of the earth (Torreson et al. 1969;
Hospers 1954; McElhinny and Merrill 1975; Evans
1976).

An interactive computer program was developed
to rotate the continental segments about any
pole of rotation. The paleomagnetic results
were used initially to position each continental
segment on the appropriate latitude and in the
correct orientation so that all averages of
measured poles were exactly on the south pole.
It was found to be most convenient, at this
stage, to initiate an interactive routine that
modified the positions to eliminate overlap of
continental margins while monitoring the results
on a display device. For the Tertiary and
Cretaceous periods the magnetic lineations were
used to establish relative longitude. For all
periods the absolute longitude was obtained from
an application of the fifth principle (see above).
The sum of the squares of the velocities of
equal area portions of all continental blocks
was minimized in a least squares sense to determine longitude. When the relative longitude
could not be obtained from magnetic lineations,
the largest contiguous continental group was
given priority since present evidence indicates
that purely continental plates have the lowest
velocity. The velocity was determined along a
small circle, centered on the pole of relative
motion from one period to the next. This procedure was applied, in order of area, to the
remaining group of continental segments. The
solution is not unique but is the most conservative estimate and is valuable in giving a
quantitative estimate of the minimum velocity
that satisfies the paleomagnetic observations.

Basic Data and Plate Tectonic Models

Various reconstructions of the continents have been made previously using Euler's theorem. Notable examples are by Bullard et al. (1965) and Smith and Hallam (1970) who used the fit of the continental margins to obtain a Triassic model. Smith et al. (1973) made extensive use of paleomagnetic data to model continental positions during several periods. We are able to draw upon much new paleomagnetic data and also the dating of magnetic lineations on the sea floor in obtaining new computer assisted reconstructions of the continental margins for seven periods. The paleomagnetic data are those compiled by Irving (1960a,b, 1961, 1962a,b, 1965), Irving and Stott (1963), and McElhinny (1968a,b, 1969, 1970, 1972a,b) and summarized by McElhinny (1973). To be included in McElhinny's summary a result must satisfy certain minimum acceptability criteria designed to eliminate inadequate data (for details see McElhinny (1973, p. 106). In addition McElhinny excludes many results from tectonically disturbed mobile belts on the grounds that they are probably not representative of the adjoining continental crust. We also had at our disposal a computer file, referred to here as the Ottawa list, compiled under the direction of E. Irving at the Earth Physics Branch, Department of Energy, Mines and Resources, Ottawa, Ontario (Irving et al. 1976). Data were also obtained from a recent compilation by McElhinny and Cowley (1977). Finally, in some cases, very recently reported data were taken directly from the publications involved. Details of the actual observations used may be found in Kanasewich et al. (1978).

Tertiary Period, Anomaly 13–38 Ma

Magnetic lineations for anomaly 13 (Pitman et al. 1974) were matched to give the relative positions of the continental blocks for the Oligocene-Eocene boundary 38 Ma ago (Heirtzler et al. 1968; Anonymous 1964) (Fig. 2). Paleomagnetic observations were not used to determine any of the relative rotations but the mean pole position was used to obtain the absolute position of the spin axis in the Tertiary. The absolute longitude for all continents was obtained by minimizing the velocity of $3 \times 3°$ equal area continental segments between the time of anomaly 13 and the present. This simple procedure gives a solution that compares very well with reconstructions of the east-central Indian Ocean by Sclater and Fisher (1974), the North Pacific by Atwater (1970) and Atwater and Molnar (1973), and the North Atlantic by Pitman and Talwani (1972). Note that the position of magnetic anomaly 13 off the coast of North America relies heavily on the superposition of a very short segment east of Cape Horn in South America.

For purposes of interpretation an azimuthal-

equidistant projection with the origin approximately on the centroid of the continental masses is more useful. This projection has the property that great circle paths from the origin to any point on the sphere transform into radii. Regions at epicentral distances less than 90° have minimal distortion. At greater distances the azimuthal distortion becomes serious, reaching a maximum at 180°, where a point on the opposite side of the earth from the origin is transformed into the bounding circumference. The distorted portion is, of course, conveniently placed on the ocean-dominated portion of the Earth.

Eocene and Oligocene global geology is summarized in Fig. 2 which shows the mid-Tertiary Earth on an azimuthal-equidistant projection centered on the mean of the continental segments and also centered on the opposite side in the Pacific Ocean. The summary maps and world-wide correlation charts of rock formations by Kummel (1970) were the primary source of information but many other maps were consulted throughout this study. The mid-oceanic ridges were obtained from the position of anomaly 13, 38 Ma ago. The position of the subduction zones was inferred from the geologic evidence. Episodes of volcanism in western South America occurred in the Miocene and Pliocene (Harrington 1962) and it is possible that the Phoenix (Larson and Chase 1972) and South American plates were not separated by a subduction zone prior to the Late Eocene. In post-Eocene times this plate separated into the slow moving, dominantly continental, South American plate and the fast moving, dominantly oceanic Nascan plate. The present remnants of the Farallon plate (McKenzie and Morgan 1969) are the Cocos and Juan de Fuca plates. The parts of continents having the maximum and minimum velocities for their small circle paths between the Tertiary and the present time are shown on the figures. It must be emphasized that the velocities and paths are not unique and represent minimum estimates.

It was noted in Fig. 1 that the African and Pacific 'quasi-circular' plates are symmetrically located with respect to the geographic pole at the present time. This same symmetry of the African and Pacific plates is evident in the Tertiary period. The equator passes close to the center of these two plates and also the center C of the entire continental lithosphere. The diameters of the African and Pacific plates are 100 and 120°, respectively. The group of 'ring' plates consists of (1) South America and Phoenix, (2) Antarctica, (3) Australia, (4) India, (5) Eurasia, and (6) North America. The maximum velocity of the continental plates, under the assumption that simultaneous minimization of all segments yields absolute longitude, varies from 8.6 to 0.4 cm/year. The Indian plate moves northward with a velocity of 5.9-8.6 cm/year and the next most rapid plate is the Australian with a northward velocity of 4.5-6.6 cm/year. This

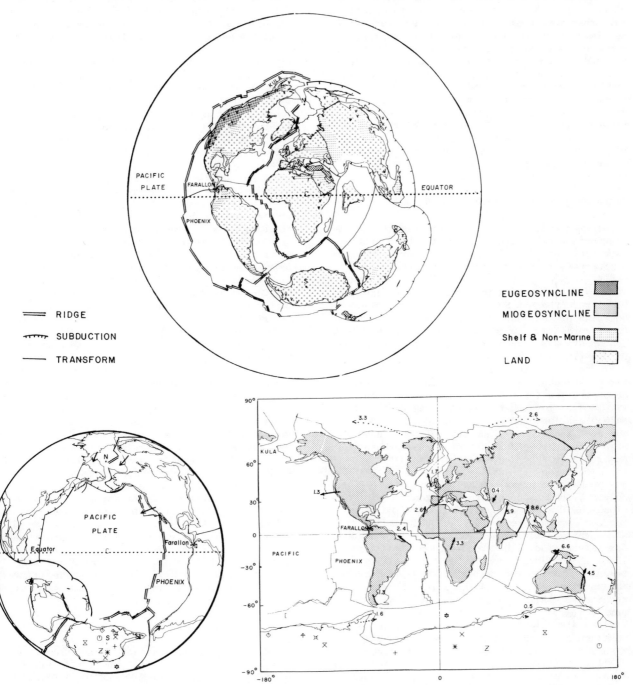

Fig. 2. Tertiary (38 Ma) geology and postulated plate boundaries on an azimuthal equidistant projection. The shape of the Pacific plate is depicted in the lower left diagram, which is antipodal to the map above and contains all areas within 120° of the center. The symbols v and r indicate zones of Tertiary volcanism and reef formation. The modified Mercator map in the lower right shows the longitude determination using a least squares minimization of continental velocities from the Tertiary to the present. The various symbols near the south pole (S) represent the paleomagnetic poles obtained from all the continental segments. The symbols are explained in Fig. 8.

is in accord with present day observations that plates with a high ratio of oceanic to continental crust move most rapidly. Since Antarctica is almost surrounded by spreading centers its movement is geometrically restricted to between 0.5 and 1.6 cm/year. Associated with the 'ring' plates are minor segments in the Mediterranean and the Caribbean Seas and the Kula and Juan de Fuca plates. Such fragments are of great interest to studies of local geology but are unlikely to be a significant part of the boundary conditions that determine the dynamics of the global system. In summary, the plate tectonic pattern in the Tertiary Period was similar to what is seen at the present time.

Cretaceous Period, Anomaly M1-110 Ma

Paleomagnetic observations were used to model the continental arrangement at the base of the Late Cretaceous, 110 Ma ago. A minor separation of South America and Africa is necessary to satisfy the available magnetic lineations for anomaly M1. The solution in Fig. 3 is very similar to that of Smith et al. (1973). The position of the magnetic lineations in the Pacific relative to the North and South American continents were taken from Fig. 7 of Larson and Pitman (1972). Their model used both paleomagnetic poles and the position of anomaly M1 from Larson and Chase (1972) to obtain a convincing demonstration for the existence of two stable triple junctions that separated the Pacific, Kula, Farallon, and Phoenix plates. New Zealand, the Auckland Plateau, and New Caledonia have been rotated toward Antarctica and southeast Australia in accordance with the magnetic lineations that show the area starting to separate at the base of anomaly 32, 76 Ma ago. The Kolyma block has been detached from Siberia along the Chersky foldbelt following the geological evidence reported by Churkin (1969,1972), although the paleomagnetic evidence is ambiguous for this period. The preservation of magnetic lineations M1-M13 (about 110-130 Ma ago) in the Bering Sea basin (Cooper et al. 1976) is strong evidence that Alaska and the Kolyma block have remained a single block throughout this critical period.

Fifty-six paleomagnetic poles from 12 separate continental blocks were used and values of Fisher's precision parameter, K, vary from 15 to 114. Fifty-two of these poles are Cretaceous in age, three are listed as Lower Cretaceous to Upper Jurassic and one has a quoted age range of Upper Jurassic to Paleocene. Most of the data are summarized by McElhinny (1973) with additions from lists XIII and XIV. Newly reported poles permit tighter temporal constraints to be placed on the African data, and the Australian data are those reported by Schmidt (1976).

Fisher's precision parameter increases from 7, for no continental drift between the present and the Cretaceous, to 21 for the model shown in Fig. 3. This increase is significant at the three standard deviation (99%) level. The distribution is not as concentrated as in the Tertiary and, in fact, the poles for the northern group of continents (Laurasia) are clearly separated from the poles of the southern group (Gondwanaland). Any attempt to superimpose the two sets of pole positions leads to substantial overlap of Africa and Eurasia. This pattern of pole clustering is found to be present throughout the Mesozoic and late Paleozoic Eras. Its existence in the Permo-Triassic was reported by Briden et al. (1970). It can be accounted for by the presence of a small non-axial multipole component in the earth's magnetic field in addition to the dominant axial dipole component.

On an azimuthal-equidistant projection the continental lithosphere for the Cretaceous Period displays a most remarkable symmetry. All the continental segments are within 90° of their centroid that lies on the paleo-equator. The Tethys Sea lies opposite the opening in the North Atlantic Ocean forming the North American and Canary basins. Eugeosynclines form a rim nearly all the way around the block of continental lithosphere in Fig. 3. It is apparent that four oceanic plates (Farallon, Pheonix, Kula, and Pacific) generated from the twin sets of ridge-type triple junctions produced active subduction zones around Pangaea. The spreading center that had created the Tethys Sea was weak and beginning to change character towards the end of the Mesozoic Era. The geometric arrangement of the spreading centers and the symmetry of the continental lithosphere is considerably different from those of the Tertiary and the present time. The velocity vectors showing continental velocity along small circle paths are dominantly outward from the African plate. The velocity of all continental segments are more uniform than in the Tertiary, all velocities being less than 5.5 cm/year. A spreading center between South America and Africa is starting to separate these two continents. Although more subdivisions are beginning to be apparent, there are basically only three large continental plates, Gondwanaland, North America, and Eurasia, opposite the four large oceanic plates.

Triassic Period - 190 Ma

A total of 90 paleomagnetic poles from 13 continental segments is available and Fisher's precision parameters range from 31 to 111. The well-known reconstruction of Bullard et al.(1965) has been criticized because it conflicts with geological evidence in the overlapped portion of Central America (McBirney and Bass 1969; King 1970; Ladd 1976). Following the principles established in the introduction, the interactive computer program allows one to arrive at a solution that does not overlap portions of Central America that have outcrops of Triassic or older rocks. The resulting Triassic model is

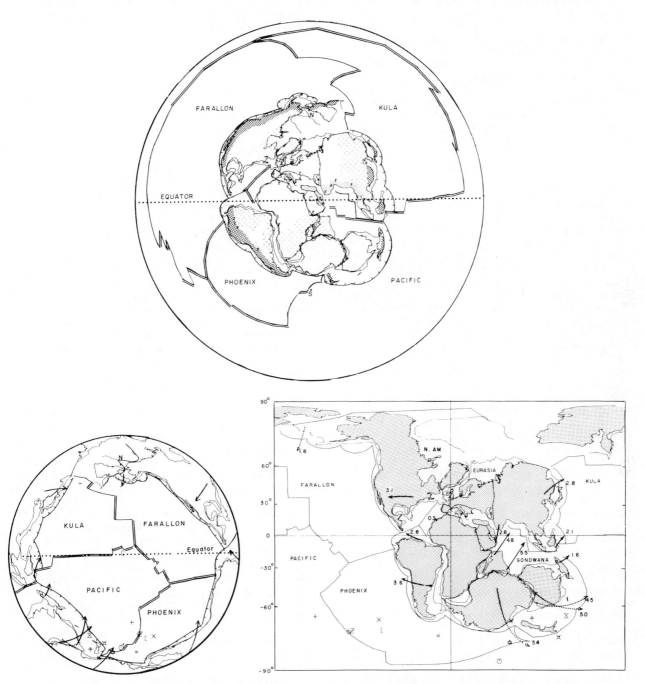

Fig. 3. Cretaceous (110 Ma) geology and postulated plate boundaries on an azimuthal-equidistant projection. The double set of triple junctions in the Pacific is shown with minimal distortion in the lower left diagram, which contains all areas within 120° of the center. The symbol r indicates zones with Cretaceous reef formation. The modified Mercator projections in the lower right show the longitude determination using a least squares minimization of continental velocities from the Cretaceous to the Tertiary. The various symbols near the south pole represent the paleomagnetic poles obtained from the continental segments. The symbols are explained in Fig. 8.

shown in Fig. 4. Fisher's precision parameter increases from 4 for the case of no continental drift to only 21 for this model, but the increase is still significant at the 99.9% level. Our solution is preferred to the one by Bullard et al. (1965) not only because there are no objectionable overlapping portions but also because our reconstruction for the Paleozoic Era indicate that Pangaea did not continue to exist but consisted of two or more continents that moved independently. There are two main changes over the previous solution: (1) North America and Eurasia have been shifted away from Africa to eliminate any overlap of Central America with Precambrian and Paleozoic outcrops; and (2) Kolyma and Alaska have been detached from North America and rotated into a position in accord with the Triassic paleomagnetic data. This second alternative is the more uncertain. Not only is it difficult to know where Alaska and British Columbia should be separated from North America but also the longitude cannot be determined unambiguously. The geological evidence is also ambiguous. Kolyma and Alaska will be kept together with North America in producing models for more ancient periods because the modifications, if required, are easily visualized. Two of the possibilities for the Kolyma group are dotted in Fig. 4. The absolute longitude for Pangaea was obtained using the least squares solution for velocity between 190 and 110 Ma years ago. The paths of finite rotation along small circles were similarly obtained non-unique minimum estimates. From the sparse data on M-type magnetic lineations it is unlikely that there is a large longitudinal shift so the velocities given are probably close to their true values. It is seen that the large continent of Gondwana has a velocity of about 1 cm/year between the Triassic and the Cretaceous Periods. On the other hand, parts of Laurasia have velocities exceeding 4 cm/year in the form of a rotation that opens up the Atlantic Ocean and closes the Tethys Sea. As in the Cretaceous Period, the pole positions form two distinct populations and any attempt at superimposing them leads to a greater degree of continental overlap.

Geological information has been included on the new Triassic model in Fig. 4 and an attempt is made to sketch in the ridges and subduction zones. The continental grouping is not as symmetric as in the Cretaceous Period but the same basic pattern is evident. The equator passes close to the centroid of the two continental plates of Laurasia and Gondwana. Plates dominated by oceanic lithosphere are Kula, Farallon, Pacific, and Tethys. The tectonic activity does not seem to demand more than six or seven large plates. At least one ridge-type triple junction is required in the Pacific and one section of the ridge follows an equatorial path to produce the Tethys Sea. If the Pacific ridge-type triple junction is stable then the

subduction zones that result on the periphery of the continental margins may effectively keep Pangaea as a stable entity for several geological periods.

Permo-Carboniferous Period - 280 Ma

Ninety-four poles are available but these represent only 10 continental fragments. With the exception of India, the K values lie between 25 and 178 and all but one of the poles are Carboniferous or Permian.

The arrangement of North America and Europe relative to Africa and South America must be different from that in the Triassic. Consequently it must be assumed that the 'optimum' fit achieved by Bullard et al. (1965) and Smith and Hallam (1970), insofar as it ever existed, was a transitory phenomenon. As in the Triassic, a limiting case for the absolute longitude was obtained by using a least squares solution on Pangaea as a whole for a minimum velocity between 280 and 190 Ma ago. Much of the velocity is taken up by a general northward drift of Pangaea and is under 5 cm/year for all continents. The Tethys Sea was consistently wider as we proceed to earlier periods. The south pole is centered on the well-known zone of glacial deposits and erosional features of eastern South America, southern Africa, Antarctica, India, and Australia. As in the Mesozoic Era the pole positions for Gondwana and Laurasia form two distinct populations in the late Paleozoic and they cannot be superimposed without a large amount of continental overlap (Fig. 5).

Global geological data are superimposed in Fig. 5 together with postulated subduction zones and ridges. The pattern is similar to that in the Triassic although the symmetry of the continental blocks is distorted by the widening of the Tethyan seaway. Continental collision is the dominant form of interaction and is assumed to be the cause of the final phases of the Appalachian Orogeny in North America, the Hercynian Orogeny in Europe and northwest Africa, and the Uralian Orogeny between the Baltic and the Angaran cratons. The continents of Laurasia and Gondwana must be treated as two or more interacting plates. Subduction of oceanic plates is assumed to be responsible for the Kanimblan Orogeny in Australia and the Antler Orogeny in western North America.

Devonian Period - 370 Ma

Paleomagnetic data are available for only eight continental fragments, and although the total number of poles is 50, four of the eight fragments are represented by less than five poles. Precision parameters range from 20 to 162 and ages cover a considerable span from middle Silurian to Lower Carboniferous.

Paleomagnetic results require a wider dispersion of Laurasia and Gondwanaland in the

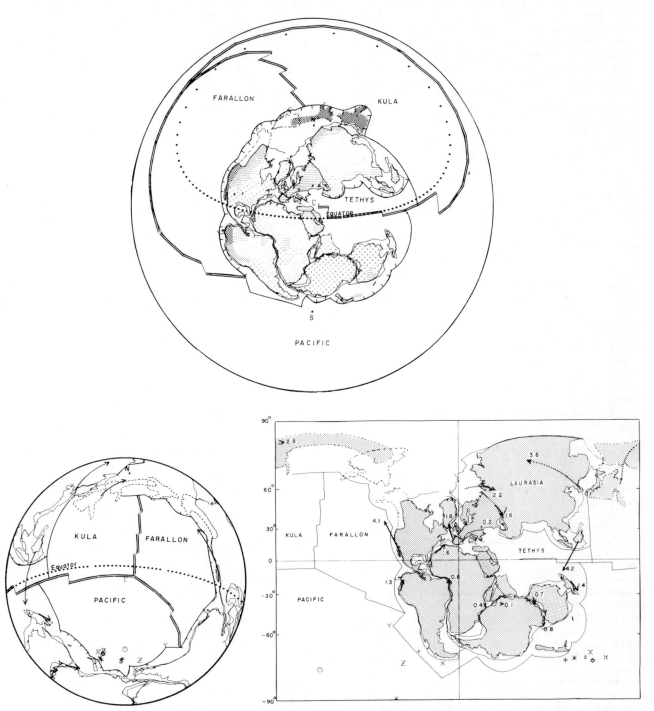

Fig. 4. Triassic (190 Ma) geology and postulated plate boundaries on an azimuthal-equidistant projection. The triple junction in the Pacific is hypothetical but reasonable on the basis of eugeosynclinal deposits, interpreted due to subduction zones, around the periphery of the Pacific Ocean. The symbol r indicates zones of Triassic reef formation. The Kolyma pole (ω) fits well if Kolyma and Alaska are at the same latitude as Japan. Note that Central America does not overlap South America.

Devonian. It is probable that the Tethys seaway was a continuous channel dividing the continental masses into two major parts. The longitude of Siberia, and therefore its position relative to Europe, is ambiguous but the formation of the Ural foldbelt in the Carboniferous places a constraint on the separation. Several continental segments have a northward component of velocity close to 3 cm/year. The distribution of continental lithosphere on an azimuthal-equidistant projection in Fig. 6 shows the same symmetry as in the Mesozoic and upper Paleozoic despite the widening Tethyan gap.

Geological information has been added in Fig. 6. As has been pointed out many times (e.g. Briden and Irving 1964), the distribution of Devonian reefs compares well with the location of the paleomagnetic equator. Note also that the 'Old Red Continent' in Europe straddles the equator, in agreement with the fossil fauna and continental rocks that are interpreted as having been deposited in a tropical and semi-arid climate. Since the distribution of continents and their associated geosynclines is similar to that in the Permo-Carboniferous, the same oceanic ridge-type triple junction is assumed to be active still. One arm of the ridge system forms a near-globe-encircling system around Gondwanaland. The complex pattern of diastrophism cutting across Siberia and China may be indicative of a collision of segments of the Asian landmass but the sparseness of the paleomagnetic observations does not justify any separation.

Ordovician Period - 470 Ma

There is a considerable body of paleomagnetic data for the lower Paleozoic to indicate a major reorganization of continental segments and some large scale continental drift. In an effort to analyze the changes as a function of time an Ordovician reconstruction has been attempted. The quantity and quality of the observational results are lower than for later periods but since the resultant model is similar to an independent one using Cambrian data, it is thought to have some validity. In order to obtain a meaningful Ordovician reconstruction we have restricted the temporal spread of suitable paleomagnetic poles as much as possible. This leads to a compilation of 45 poles from 8 continental segments, although only Siberia is represented by more than 10 poles.

The paleolatitude for all continental segments becomes quite low in the Ordovician and Cambrian Periods. This places all the continental plates on or near the equator (Fig. 7) and to accommodate them the variation in longitude is highly constrained.

Geological data are superimposed in Fig. 7. Africa has undergone a large amount of drift because the south pole now appears in the Tethyan Ocean north of this continent. The continental

segments are arranged symmetrically along the equator. The configuration suggests a plate pattern similar to the one at the present time with the 'ring' plates on the paleo-equator but with dominantly oceanic plates covering the poles. The individual plate boundaries cannot be established with any certainty because of the change in pattern and the imprecision of much of the data. The spreading center in the Tethyan Ocean must have been quite active to have created such a wide seaway and to have carried Gondwana to its indicated position. The 'Pacific' plate, now centered on the north pole, must also have been very active to generate the Caledonian Orogeny, along the periphery of the 'ring' plates. The Caledonian Orogeny was episodic from the Late Cambrian to the Middle Devonian and this appears to have reorganized the plate tectonic pattern drastically. Certainly, the velocity of the continental segments between the Ordovician and the Devonian is rather high, often with a northward component of 5-7 cm/year but much more paleomagnetic data are necessary to document the precise position of the continental segments.

Cambrian Period - 550 Ma

The continental fragments are represented by a total of only 35 poles, some of which are uppermost Precambrian and others Lower Ordovician. With the exception of South America, K values lie between 19 and 99. For North America we have followed Van der Voo et al. (1976) and for India we have used the summary by Wensink (1975). The South American summary is that given by Thompson (1973) who lists five Cambrian poles; individually these poles are of poor quality (only one has an α_{95} under 20°) and as a group they are highly scattered (K = 4, A_{95} = 41°). If taken at face value they imply large amounts of polar wandering within the Cambrian. Although this may in fact be true (see Hailwood (1974), for example) we prefer to adopt the conservative approach of excluding these data until they are corroborated by studies from other continents. The pattern and polar symmetry is very similar to that in the Ordovician.

Geological data and the postulated plate boundaries are shown in Fig. 8. The position of the paleo-equator corresponds well to outcrops built from shells of the reef organism, Archaeocyathus. The proto-Atlantic between Europe and North America that was postulated by Wilson (1966) and by Dewey and Bird (1970) is required here by the paleomagnetic data although the longitudinal change is uncertain. The Tethyan and Australian spreading center must have been fairly intense to generate the rotation of North America, Europe, and Asia with velocities of 3-4 cm/year. The spreading center between Antarctica and Africa must have been dying out as Gondwanaland is a recognizable entity in the Ordovician. All continental

Fig. 5. Permo-Carboniferous (280 Ma) geology and postulated plate boundaries on an azimuthal-equidistant projection. The locations of glacial indicators are denoted by G and reefs by r.

plates require a minimum velocity of between 2 and 4 cm/year between the Cambrian and Ordovician Periods.

Interpretation

Using paleomagnetic observations and a small number of principles based on plate tectonic data and concepts it has been possible to reconstruct continental fragments for six periods between the Cretaceous and the Cambrian in a statistically significant manner. The grouping of the paleomagnetic poles shows an improvement in Fisher's precision parameter at the 99% or three standard deviation significance level. The model for the Tertiary period was made using

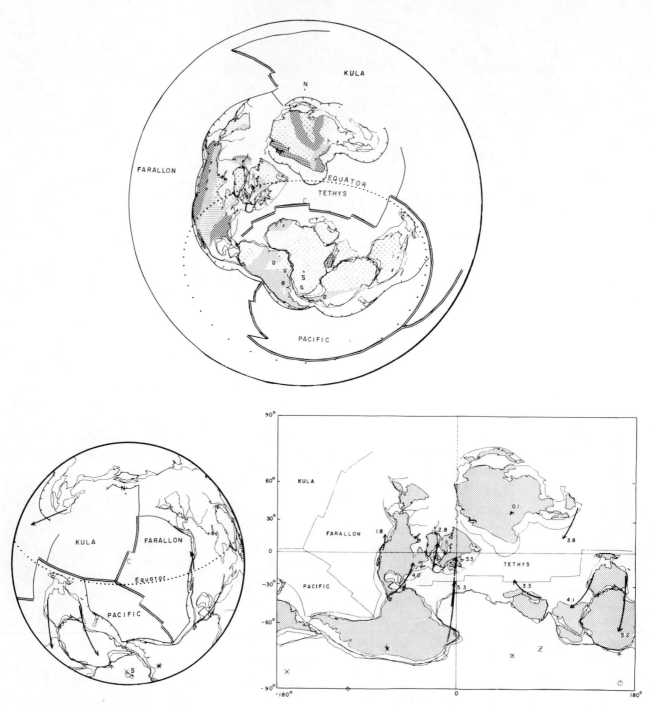

Fig. 6. Devonian (370 Ma) geology and postulated plate boundaries on an azimuthal-equidistant projection. The locations of reefs are indicated by r and glacial indicators by G.

magnetic lineations and is an independent test of the paleomagnetic method.

Models using the paleomagnetic data contain ambiguities due to the uncertainty in the longitude. The minimization with respect to velocity of the plates to define the longitude, although non-unique, has proven to be of great value. Alternate positions of the continental segments can be readily visualized by consulting the Mercator projections and the increase in plate velocity can be estimated by the relative shift from the minimum velocity point. From the

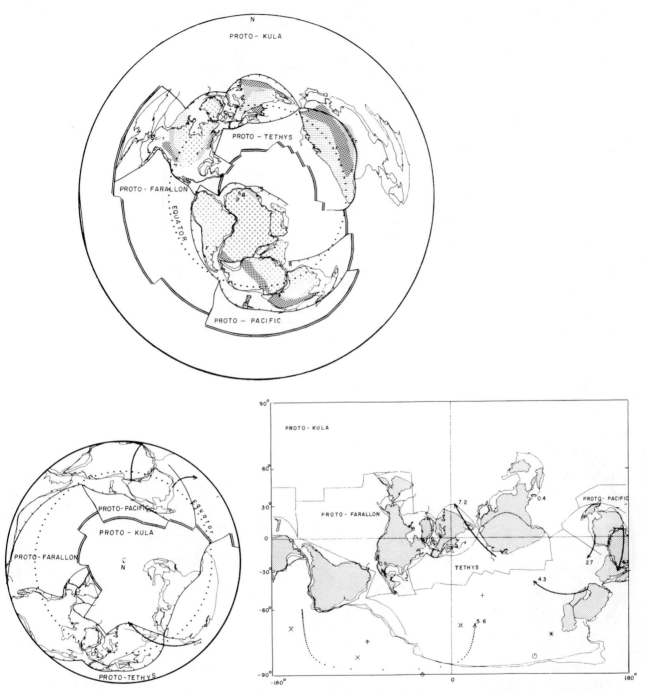

Fig. 7. Ordovician (470 Ma) geology and postulated plate boundaries on an azimuthal-equidistant projection centered on the south pole. In the lower figure the projection is centered on the north pole and shows all areas within 120° of the pole. The locations of glacial indicators are shown by G and reefs by r.

Devonian Period to the present time the dated magnetic lineations and the reconstruction of Gondwana and Laurasia place severe restrictions on the longitude (unless one has reason to believe in a shift of the earth's entire litho-sphere along lines of latitude). In the Devonian Period the North American and Asian continents are restrained to the position shown by the geological evidence for the formation of the Urals. In the Cambrian and Ordovician Periods

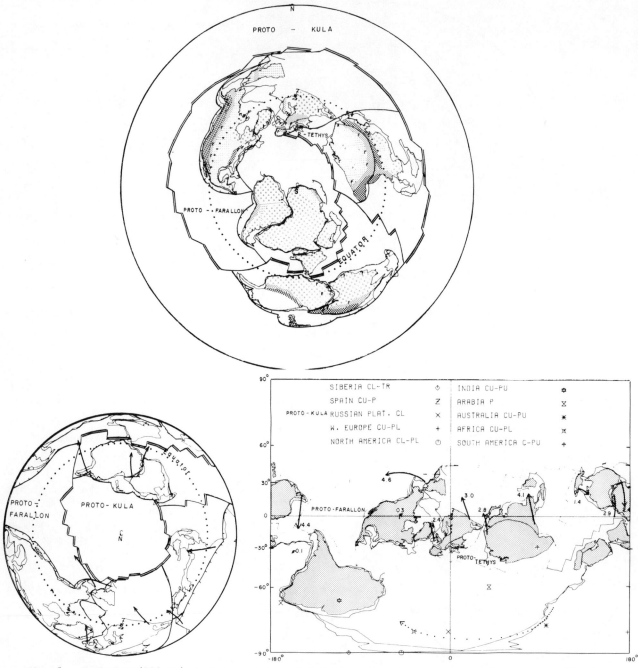

Fig. 8. Cambrian (550 Ma) geology and postulated plate boundaries on an azimuthal-equidistant projection centered on the south pole. Symbol r marks the location of Archaeocyathus fossils. In the lower figure the projection shows all areas within 120° of the north pole.

the shift of all major segments to an equatorial position places very tight constraints on the longitude. In summary, unless the paleomagnetic evidence is missing, as in China, or in error, the relative longitude of all major continental segments is estimated to be constrained within

10°. Their velocities are then within 1 cm/year of their true values.

If we now accept the continental reconstruction as approximately correct, the azimuthal-equidistant maps, with origin on the center of mass of the continental margins, are useful in examining various geometrical properties. In all periods there is present a single large ocean, similar to the present Pacific Ocean. In the early Paleozoic, up to the time of the Caledonian Orogeny, this ocean was centered on the North Pole. Subsequently it occupied an equatorial position antipodal to the Tethys Sea north of Africa. At the end of the Mesozoic Era the center of the Pacific plate shifted to a position antipodal to the center of the African plate. The continental segments are grouped in a cluster that is related to the spin axis. The center of mass C is either on the south pole, as in the Cambrian and Ordovician Periods, or else near the equator. The graph in Figure 9 shows the change in latitude of the mean south paleo-magnetic pole as a function of time. Samples from the Cambrian and Ordovician Systems yield paleomagnetic south poles close to the present equator. A rapid shift in latitude occurred at the time of the Caledonian Orogeny that culmi-nated toward the end of the Ordovician Period. This was concurrent with a major reorganization of the continental segments and relatively rapid continental plate motion. The rest of the Paleozoic is represented by a uniform drift toward the present polar position accompanied by a reorganization of the continents into Gondwana and Laurasia. The Tethys Sea closed uniformly from its oceanic proportions in the lower Paleo-zoic. A high degree of symmetry occurred in the Mesozoic Era with Laurasia and Gondwana locked together to form the supercontinent of Pangaea. The latitude of the pole stabilized near 80° in terms of present day coordinates. The end of the Mesozoic shows a very symmetric arrangement even though Gondwana and Laurasia are separating with the formation of the Atlantic Ocean in a symmetric relation to the Tethys Sea with respect to the map center (Fig. 3). In the Tertiary Period the latitude in Fig. 9 begins to shift, once again, towards 90°. The continental arrangement has the symmetry of the present day with the African and the Pacific quasi-circular plates separated by a ring of quasi-elliptical plates (Fig. 1). It is significant that the variation in the latitude of the pole with time is not random but is a very regular and systema-tic function of time.

The reconstruction of plate boundaries in the past requires further assumptions as outlined in the principles in the introduction. The incom-pleteness of paleomagnetic observations is compounded by gaps in the geological record. However, accepting that the reconstructions have some validity, it is seen that the number of plates varies from six to eight or nine. Small plates, such as the Indian in the Tertiary or the Nascan at the present time, have very rapid motions and disappear quickly as independent entities. There appears to have been seven or eight major plates in the early Paleozoic and Cenozoic Eras. There may have been as few as six in the late Paleozoic and Mesozoic Eras

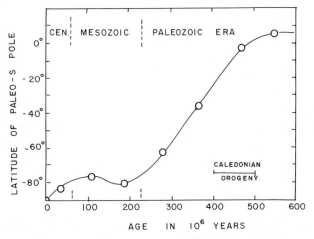

Fig. 9. Variation of the latitude of the mean paleomagnetic south pole in the pre-sent coordinate system as a function of time.

although the number occupying the enlarged Pacific Ocean is speculative. The plate arrange-ment also shows a certain symmetry during each period. The Cambrian and Ordovician Periods have an equatorial ring of dominantly continental plates indicating a close similarity to the present situation. A south oceanic polar plate containing Africa lies opposite the major north polar oceanic plate. The arrangement on an Eckert projection (Fig. 10) should be compared to a similar one of the present plate system (Fig. 1) except that the 'ring' plates are rotated 90° to the spin axis. The upper Paleo-zoic and Mesozoic plate arrangement is one in which the continental blocks have an equatorial position opposite the oceanic group of plates. Following principle 9, there must be a spreading center in the Tethys Sea that is seen to follow the equator. As the Tethys seaway becomes smaller towards the Cretaceous Period the con-tinental and plate arrangement acquires greater symmetry. In conclusion, the arrangement of plates is neither random in space nor time. On the contrary, the plate tectonic pattern appears to have an evolutionary development with a time scale of several hundred million years and a high degree of spatial organization whose physical properties should be described by low order spherical harmonics. Such an ordered kinematic and geometric system on the surface of the earth must be reflected also in the dynamic system within its interior.

The organization of the lithosphere with two antipodal quasi-circular plates separated by a ring of quasi-elliptical plates at the present time was considered by Kanasewich (1976) to be convincing evidence for a mantle-wide convective system. Three dimensional convection models were discussed in the introduction. Two-dimensional models of convective flow can be

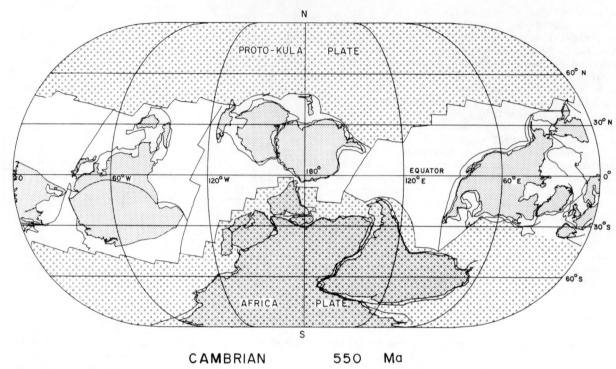

CAMBRIAN   550 Ma

Fig. 10. A model of the plate tectonic pattern in the Cambrian period on an Eckert projection. This figure should be compared to the configuration at the present time in Fig. 1.

used to solve more complex physical states. Thus, Takeuchi and Sakata (1970) have made a theoretical computation on a two-layered Newtonian fluid with two-dimensional recti-linear geometry. The boundary conditions were zero shear stress at the surface and a non-slipping (zero velocity) lower boundary that was being heated. The lower layer occupied 90% of the total thickness and was 1000 times more viscous than the upper (asthenospheric) layer. This model showed that there are larger hori-zontal velocities in the upper layer but that the return flow is in the lower, more viscous, layer. Studies by Peltier (1973) for heating distributed uniformly through a model with the same boundary conditions gave similar results. Davies (1977) studied a model similar to the one by Takeuchi and Sakata (1970) but with both boundary conditions being free-slipping and there was no heating. The upper layer was taken to be 700 km thick and the lower one was 2300 km. Davies (1977) found that five orders of magnitude contrast between the viscosity in the lower and upper layer are necessary to exclude flow from the more viscous layer. The contrast becomes even greater as the upper layer is made thinner. Although the models are simple and the boundary conditions not as complex as those encountered in the real earth, when theoretical studies have been carried out to include the entire mantle, the results indicate that the lower mantle must

be involved in the dynamics of plate tectonics.
    Assuming mantle-wide convection, a series of cross sections have been drawn up in Fig. 11 to illustrate hypothetical flow patterns for the seven periods modelled with paleomagnetic data in Figs. 2-8. Most of the cross sections of the earth cut the azimuthal-equidistant projections along a great circle path between 360, C, and 180 in Figs. 2-8. A polar cross section (Fig. 11h,i) at any longitude for the Cambrian and Ordovician Periods tends to show six cells or a convective pattern that can be described by a third order spherical harmonic. The Devonian Period (Fig. 11g) has a second order convective pattern that is somewhat irregular. The Ordo-vician pattern is also more irregular than the Cambrian one. Comparing the Devonian and Ordo-vician spreading centers on a Mercator projec-tion (Fig. 6,7) it is seen that the change in convective order is evolutionary. The Ordovician Tethys spreading center moved toward the equator and broke through between Asia and Australia to join the Pacific spreading center, which evolved, in the Devonian, into a triple junction. If the Pacific oceanic spreading centers were consis-tently more active than the Tethys one, the rapidly moving Kula, Farallon, and Pacific plates would tend to concentrate the continental segments into the one large continent called Pangaea by Wegener (1929). As the cross sections in Fig. 11d-g show, the convection pattern

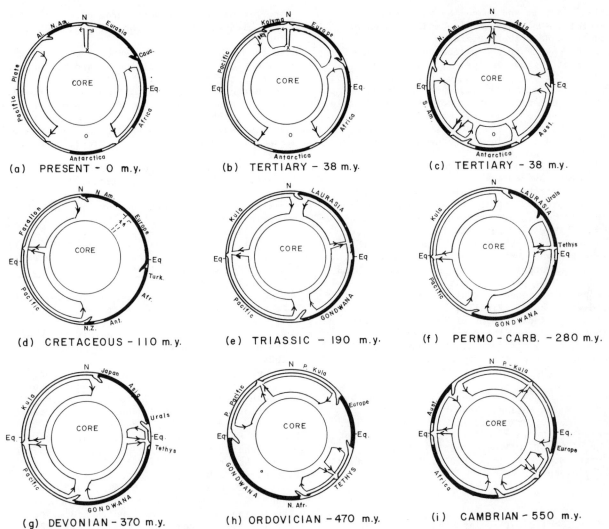

Fig. 11.   Cross sections of the Earth along the great circle path 360-C-180 in the azimuthal-equidistant projections of Figs. 2, 3, 4, 5, 6, 7 and 8.   The dimensions and directions of hypothetical mantle current systems are indicated.

remained stable as a second order system until the Cretaceous Period.   A second order system is still present in the Cretaceous but the Tethys spreading center has decayed away and, as seen in Fig. 3, a transition is occurring to a new order at the beginning of the Cenozoic Era.   This is a more complex system that tended to split Gondwana and Laurasia apart.   The Pacific system developed a double triple junction (Fig. 3) with the four oceanic plates of Pacific, Kula, Farallon, and Phoenix.   In the Tertiary Period (Fig. 2) the spreading center between the Phoenix and Farallon plates began to dominate so that the Kula and Farallon plates were subducted out of existence underneath Asia and North America. The North Atlantic ridge, which first appeared in the Jurassic Period, also became dominant and a spreading center formed three-quarters of the

way around Africa.   The Tertiary spreading centers, like the present ones, are complex, involving many axially symmetric modes of low order spherical harmonics, but the third and first orders predominate.   This produces two main spreading centers that drive the Pacific and Africa plates in a northward direction (Fig. 11a,b).   The two cells have very different velocities and are decoupled by a ring of plates and minor spreading centers.   These show the third order pattern more clearly (Fig. 11c) if another cross section is made through the 'ring' perpendicular to the one in Fig. 11b.

Conclusion

The continental segments, and therefore the lithosphere, have been shown to be highly organized in time and space for the entire Phanerozoic. Only six to nine major plates are necessary to describe satisfactorily the main geological events in the paleomagnetic reconstruction of the continents. It is difficult to conceive of an upper mantle convective system, or other driving mechanism, that will produce plates with dimensions of 60-180° that are stable over a time scale of several hundred million years. Therefore, the evidence seems to be irrefutable that mantle-wide convection was, and is, present. The order of the convective system varied very little during the entire Phanerozoic. A third order system was present during the early Paleozoic Era. The intense world-wide Caledonian Orogeny appears to have resulted from a change in convective pattern to a second order system. This was responsible for the formation of the supercontinent Pangaea. The close of the Mesozoic Era was marked by a second transition from a second order convective system to a third order one. Geologically, this was also accompanied by an increase in tectonic activity on a world-wide basis. Finally, it is concluded that the 10 principles for plate tectonics are a valuable guide for the reconstruction of the geological past. If further observational evidence supports them they may be of assistance in modelling the Precambrian Era.

Acknowledgements. This work was supported by grants from the National Research Council of Canada. The authors would like to express their appreciation for many useful discussions on the subject with J.T. Wilson, R. St. J. Lambert, E. Irving, M.W. McElhinny, B.J.J. Embleton, and R.E. Folinsbee.

References

Anonymous, Geological Society Phanerozoic time scale 1964, Geol. Soc. of Lon. Quart. 120, 260-262, 1964.

Atwater, T., Implications of plate tectonics for the Cenozoic tectonic evolution of Western North America, Geol. Soc. of Am. Bull. 81, 3513-3536, 1970.

Atwater, T., and P. Molnar, Relative motion of the Pacific and North American plates deduced from sea-floor spreading in the Atlantic, Indian and South Pacific Oceans, in Proceedings of a Conference on Tectonic Problems of San Andreas Fault Systems, Ed. by R.L. Kovach and A. Nur, Stanford University, Stanford, Ca., 136-148, 1973.

Briden, J.C., and E. Irving, Palaeoclimatic spectra of sedimentary palaeoclimatic indicators, in Problems in Palaeoclimatology, Ed. by A.E.M. Nairn, Interscience, New York, N.Y., 199-250, 1964.

Briden, J.C., A.G. Smith, and J.T. Sallomy, The geomagnetic field in Permo-Triassic time, Geophys. J.R. astr. Soc., 23, 101-117, 1970.

Bullard, E.C., J.E. Everett, and A.G. Smith,

The fit of the continents around the Atlantic, Phil. Trans. Roy. Soc. Lon. A, 258, 41-51, 1965.

Chamalaun, T., and P.H. Roberts, The theory of convection in spherical shells and its application to the problem of thermal convection in the earth's mantle, in Continental Drift, Ed. by S.K. Runcorn, Academic Press, New York, N.Y., 177-194, 1962.

Chandrasekhar, S., Hydrodynamic and Hydromagnetic Stability, Oxford University Press, London, England, 220-271, 1961.

Churkin, M. Jr., Palaeozoic tectonic history of the Arctic Basin north of Alaska, Science, 165, 549-555, 1969.

Churkin, M. Jr., Western boundary of the North American continental plate in Asia, Geol. Soc. of Am. Bull. 83, 1027-1036, 1972.

Cooper, A.K., D.W. School, and M.S. Marlow, Plate tectonic model for the evolution of the eastern Bering Sea Basin, Geol. Soc. of Am. Bull. 87, 1119-1126, 1976.

Davies, G.F., Whole-mantle convection and plate tectonics, Geophys. J.R. astr. Soc., 49, 459-486, 1977.

Dewey, J.F., and J.M. Bird, Mountain belts and the new global tectonics, J. Geophys. Res., 75, 2625-2647, 1970.

Elsasser, W.M., Convection and stress propagation in the upper mantle, in The Application of Modern Physics to the Earth and Planetary Interiors, Ed. by S.K. Runcorn, John Wiley & Sons, New York, N.Y., 223-246, 1969.

Evans, M.E., Test of the dipolar nature of the geomagnetic field throughout the Phanerozoic time, Nature, 262, 276-277, 1976.

Fisher, R.A., Dispersion on a sphere, in Proc. Roy. Soc. Lon. A, 217, 295-305, 1953.

Fisher, R.L., J.G. Sclater, and D.P. McKenzie, Evolution of the Central Indian Ridge, Western Indian Ocean, Geol. Soc. of Am. Bull. 82, 553-562, 1971.

Hailwood, E.A., Palaeomagnetism of the Msissi Norite (Morocco) and the Palaeozoic reconstruction of Gondwanaland, Earth Planet. Sci. Letters, 23, 376-386, 1974.

Harrington, H.J., Palaeogeographic development of South America, Am. Assoc. Pet. Geol. Bull., 46, 1773-1814, 1962.

Heirtzler, J.R., G.O. Dickson, E.M. Herron, W.C. Pitman, and X. Le Pichon, Marine anomalies, geomagnetic field reversals and motions of the ocean floor and continents, J. Geophys. Res., 73, 2119-2136, 1968.

Herron, E.M., Sea-floor spreading and the Cenozoic history of the east-central Pacific, Geol. Soc. of Am. Bull. 83, 1671-1692, 1972.

Hess, H.H., History of ocean basins, in Petrologic Studies, Ed. by A.J. Engel, H.L. James and B.F. Leonard, Geological Society of America, Buddington Volume, 559-620, 1962.

Hospiers, J., Rock magnetism and polar wandering, Nature, 173, 1183-1184, 1954.

Irving, E., Palaeomagnetic pole positions, Part

I, Geophys. J.R. astr. Soc., 3, 96, 1960a.

Irving, E., Palaeomagnetic pole positions, Part II, Geophys. J.R. astr. Soc., 3, 444-449, 1960b.

Irving, E., Palaeomagnetic pole positions, Part III, Geophys. J.R. astr. Soc., 5, 70-79, 1961.

Irving, E., Palaeomagnetic pole positions, Part IV, Geophys. J.R. astr. Soc., 6, 263-267, 1962a.

Irving, E., Palaeomagnetic pole positions, Part V, Geophys. J.R. astr. Soc., 7, 263-274, 1962b.

Irving, E., Palaeomagnetic pole positions, Part VII, Geophys. J.R. astr. Soc., 9, 185-194, 1965.

Irving, E., and P.M. Stott, Palaeomagnetic directions and pole positions, Part VI, Geophys. J.R. astr. Soc., 8, 249-257, 1963.

Irving, E., E. Tanczyk, and J. Hastie, Catalogue of paleomagnetic directions and poles, publications of the Earth Physics Branch, Energy, Mines and Resources, Ottawa, Ont., Geomagnetic Series, Numbers 5, 6, and 10, 1976.

Isacks, B., J. Oliver, and L.R. Sykes, Seismology and the new global tectonics, J. Geophys. Res., 73, 5855-5899, 1968.

Kanasewich, E.R., Plate tectonics and planetary convection, Can. J. Earth Sci., 13, 331-340, 1976.

Kaula, W.M., Absolute plate motions by boundary velocity minimizations, J. Geophys. Res., 80, 244-248, 1975.

King, P.B., Tectonics and geophysics of eastern North America, in The Megatectonics of Continents and Oceans, Ed. by H. Johnson and B.L. Smith, Rutgers University Press, New Brunswick, N.J., 74-112, 1970.

Kummell, B., History of the Earth (2nd edition), Freeman, San Francisco, Ca. 707 p., 1970.

Ladd, J.W., Relative motion of South America with respect to North America and Caribbean tectonics, Geol. Soc. of Am. Bull., 87, 969-976, 1976.

Lambert, R. St. J., Global tectonics and the Canadian Arctic Continental Shelf, Proceedings of the Symposium on the Geology of the Canadian Arctic, Ed. by J.D. Aitken and D.J. Glass, Canadian Society of Petroleum Geologists, Calgary, Alberta, 5-22, 1974.

Larson, R.L., and C.G. Chase, Late Mesozoic evolution of the western Pacific Ocean, Geol. Soc. of Am. Bull., 83, 3627-3644, 1972.

Larson, R.L., and W.C. Pitman, World-wide correlation of Mesozoic magnetic anomalies and its implications, Geol. Soc. of Am. Bull., 83, 3645-3662, 1972.

Le Pichon, X., Sea-floor spreading and continental drift, J. Geophys. Res., 73, 3661-3697, 1968.

McBirney, A.R., and M.N. Bass, Structural relations of pre-Mesozoic rocks of northern Central America, Am. Assoc. of Petr. and Geol., Memoir II, 269-280, 1969.

McElhinny, M.W., Notes on progress in geophysics. Palaeomagnetic directions and pole positions, Part VIII, Geophys. J.R. astr. Soc., 15, 409-430, 1968a.

McElhinny, M.W., Notes on progress in geophysics. Palaeomagnetic directions and pole positions, Part IX, Geophys. J.R. astr. Soc., 16, 207-224, 1968b.

McElhinny, M.W., Notes on progress in geophysics. Palaeomagnetic directions and pole positions, Part X, Geophys. J.R. astr. Soc., 19, 305-327, 1969.

McElhinny, M.W., Notes on progress in geophysics. Palaeomagnetic directions and pole positions, Part XI, Geophys. J.R. astr. Soc., 20, 417-429, 1970.

McElhinny, M.W., Notes on progress in geophysics. Palaeomagnetic directions and pole positions, Part XII, Geophys. J.R. astr. Soc., 27, 237-257, 1972a.

McElhinny, M.W., Notes on progress in geophysics. Palaeomagnetic directions and pole positions, Part XIII, Geophys. J.R. astr. Soc., 30, 281-293, 1972b.

McElhinny, M.W., Palaeomagnetism and plate tectonics, Cambridge University Press, New York, N.Y., 1973.

McElhinny, M.W., and J.A. Cowley, Palaeomagnetic directions and pole positions, Part XIV, Geophys. J.R. astr. Soc., 49, 313-356, 1977.

McElhinny, M.W., and R.T. Merrill, Geomagnetic secular variations over the past 5 m.y., Rev. of Geophys. and Sp. Phys., 13, 687-708, 1975.

McKenzie, D.P., Speculations on the consequences and causes of plate motions, Geophys. J.R. astr. Soc., 18, 1-32, 1969.

McKenzie, D.P., and W.J. Morgan, The evolution of triple junctions, Nature, 224, 125-133, 1969.

McKenzie, D.P., and R.L. Parker, The north Pacific: an example of tectonics on a sphere, Nature, 216, 1276-1280, 1967.

McKenzie, D.P., and J.G. Sclater, The evolution of the Indian Ocean since the Late Cretaceous, Geophys. J.R. astr. Soc., 24, 437-528, 1971.

Minster, J.B., T.H. Jordan, P. Molnar, and E. Haines, Numerical modelling of instantaneous plate tectonics, Geophys. J.R. astr. Soc., 36, 541-576, 1974.

Molnar, P., T. Atwater, J. Mammerickx, and S.M. Smith, Magnetic anomalies, bathymetry and the tectonic evolution of the South Pacific since the late Cretaceous, Geophys. J.R. astr. Soc., 40, 383-420, 1975.

Morgan, W.J., Rises, trenches, great faults, and crustal blocks, J. Geophys. Res., 73, 1952-1982, 1968.

Oliver, J., and B. Isacks, Deep earthquake zones, anomalous structures in the upper mantle, and the lithosphere, J. Geophys. Res., 72, 4259-4275, 1967.

Peltier, W.R., Penetrative convections in the planetary mantle, Geophysical Fluid Dynamics, 5, 47-88, 1973.

Pitman, W.C., and M. Talwani, Sea-floor spreading in the North Atlantic, Geol. Soc. of Am. Bull., 83, 619-646, 1972.

Pitman, W.C., R.L. Larson, and E.M. Herron, The Age of the Ocean Basins, Geol. Soc. of Am., Boulder, Co., 1974.

Schmidt, P.W., The non-uniqueness of the Australian Mesozoic palaeomagnetic pole position, Geophys. J.R. astr. Soc., 47, 285-300, 1976.

Sclater, J.G., and R.L. Fisher, Evolution of the East Central Indian Ocean, Geol. Soc. of America Bulletin, 85, 683-702, 1974.

Smith, A.G., and A. Hallam, The fit of the southern continents, Nature, 225, 139-144, 1970.

Smith, A.G., J.C. Briden, and G.R. Drewry, Phanerozoic world maps, special papers in Palaeontology, No. 12, 1-42, 1973.

Takeuchi, H., and S. Sakata, Convection in a mantle with variable viscosity, J. Geophys. Res., 75, 921-927, 1970.

Thompson, R., South American palaeozoic palaeomagnetic results and the welding of Pangaea, Earth Planet. Sci. Letters, 18, 266-278, 1973.

Torreson, O.W., T. Murphy, and J.W. Graham, Magnetic polarization of sedimentary rocks and the Earth's magnetic history, J. Geophys. Res., 54, 111-129, 1949.

Van der Voo, R., R.B. French, and D.W. Williams, Palaeomagnetism of the Wilberns Formations (Texas) and the Late Cambrian palaeomagnetic field for North America, J. Geophys. Res., 81, 5633-5638, 1976.

Vine, F.J., and D.H. Matthews, Magnetic anomalies over oceanic ridges, Nature, 199, 947-949, 1963.

Wegener, A.L., Die Entstehung der Kontinente und Ozeane (The origin of continents and oceans), Cover Publications, New York, N.Y., 1929 (also 1966).

Wensink, H., The structural history of the India-Pakistan subcontinent during the Phanerozoic, in Progress in Geodynamics, Ed. by G.J. Borradale, A.R. Ritsema, H.E. Rondeel and O.J. Simon, North-Holland Publishing Co., Amsterdam, Netherlands, 190-207, 1975.

Wilson, J.T., A new class of faults and their bearing on continental drift, Nature, 207, 343-347, 1965.

Wilson, J.T., Did the Atlantic close and then re-open? Nature, 211, 676-681, 1966.

ON THE MECHANISM OF THE GRAVITATIONAL DIFFERENTIATION IN THE INNER EARTH

Vitaly P. Keondjan

P.P. Institute of Oceanology of USSR
Academy of Science
Krasikiva 23, 117218 Moscow, USSR

Abstract. The assumption that the Earth has been initially a homogeneous body and data on its present inhomogeneity leads to the conclusion that there is an effective mixing with in the Earth affected by its gravity field. This process (which is called gravitational differentiation) produces the core's growth. It transforms the planet into successive states with increasing thermodynamic stability. The process is accompanied by release of significant energy ($\sim 1.6 \cdot 10^{38}$ erg). This source of energy plays an essential role in the tectonic activity and evolution of the planet. Qualitative consideration shows the possibility of gravitational instability due to separation of matter within the layers of the core-mantle boundary. The result of such an unstable process is convective motion in the body of the mantle. This process is connected with density inversion of nonthermal origin. Construction of a mathematical model of this phenomenon based on multicomponent fluid dynamics is a necessity.

Equations of a two-component Newtonian fluid for the case of an axisymmetrical spherical layer with nonhomogeneity of rheological characteristics in the radial direction and with flux of one the components through the boundary have been formulated. The penetrable boundary can simulate the absorption of a heavy component by the growing core. Experiments have shown that large scale structure of the currents produced by instability have nonstationary dynamics. The order of convection and its energetics vary periodically. The scale of the cells might be compared with a horizontal scale of plates at the Earth's surface. The largest velocities of mantle flow reach several tens of cm/year. This highly developed process causes significant nonhomogeneity in the mantle because diffusive mixing appears to be noneffective. The energetics and kinematical characteristics of the convection oscillate with periods which correspond with geological cycles ($\sim 10^8$ years). The results of this study have shown that chemical convection may be considered as one of the main processes of the Earth's evolution.

Introduction

The evolution of the Earth is above all a change in its inner structure. There are two arguments that support the idea of an efficient mixing mechanism in the planet's interior. The first relates to the existence of the Earth's core and mantle. The second to the concept of a homogeneous, or quasi-homogeneous initial state of the Earth beginning around 4.5 billion years ago (Schmidt, 1948; Hoyle, 1948, 1960; Safronov, 1969).

The evolution of layers (we call it the gravitational differentiation) at subsequent stages of the Earth's history provided a large amount of potential energy. This source of energy seems to play a fundamental role in the thermal history of the planet and in the dynamics of the lithosphere. This source's power is governed first of all by the velocity of the core's accretion. Present ideas of fast accretion are derived from the data on the age of the most ancient magnetized rocks (3.7 billion years ago). However I believe that similarity in the magnetic properties of ancient and contemporary rocks does not prove equal dimensions of the ancient and contemporary core. It only follows from this evidence that the core already existed 3.7 billion years ago and that there was a mechanism to generate the magnetic field similar to the present-day mechanism and not much dependent on the core's dimensions. The core might still be growing slowly.

It was Urey (1952) using geochemical arguments for differentiation, who first pointed out this possibility. Later, this concept was qualitatively developed by Runcorn (1962) who indicated the possibility of chemical convection in the mantle, which could generate continental drift at the Earth's surface. Artyushkov (1968, 1970, 1979) and Sorokhtin (1972, 1974) studied gravitational differentiation as a process supporting the tectonic activity of the Earth. In 1975 a gross quantitative model of this process for the terrestrial planets was constructed (Keondjan, Monin, 1975). The evolution of a two-component spherically symmetrical planet was considered in that study. The heterogeneous reaction in a thin

Ψ 33 my.

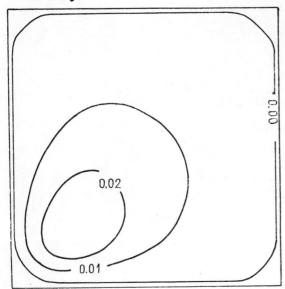

Ψ 100 my.

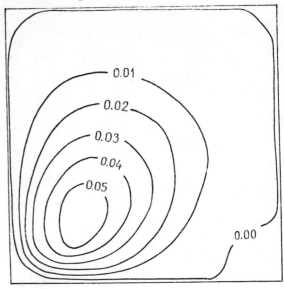

ρ 33 my.

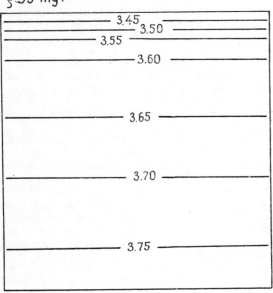

Fig. 1a.

ρ 100 my.

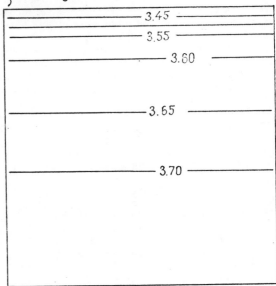

Fig. 1b.

Fig. 1 (a-g). Stream function ψ (cm^2/sec) and density field ρ (g/cm^3) in a square area with 3000-km side. Initial instability is placed in the left lower corner. Value at the top of each plot is the time from the beginning of the process.

layer between the core and the mantle was used to parameterize the movement of the heavy fraction towards the core. This approach enabled one to explicitly obtain as a first approximation the time dependence of energetic and structural parameters of planets and to estimate a relative velocity of evolution of the terrestrial planets (Keondjan and Monin, 1975, 1976, 1977). However, our study was of a provisional nature since we did not present a mechanism for the separation of components, nor the associated dynamic processes in the mantle. Further development of this concept requires solution of the chemical convection problem. Essential geophysical evidence to support this model is discussed next.

We assume that the mantle consists of a mixture of heavy and light components. Seismic data reveal that a relatively thin layer above the core-

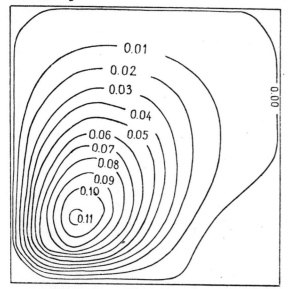

Ψ 133 my.

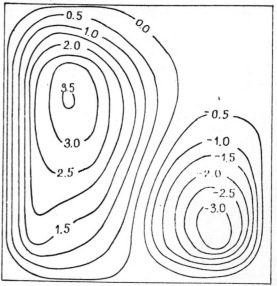

Ψ 333 my.

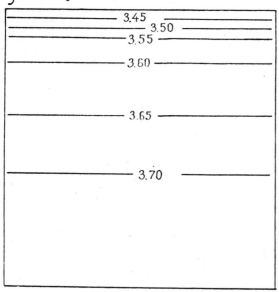

ρ 133 my.

Fig. 1c.

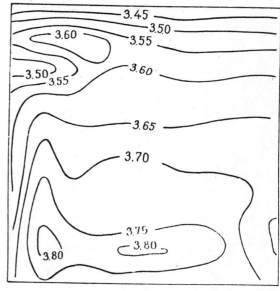

ρ 333 my.

Fig. 1d.

mantle interface has anomalous properties. Seismic velocities in this 100-200 km thick layer D" practically do not increase with depth. Phinney and Alexander (1969) pointed to the absence of the vertical gradient in V_p at this depth under the Atlantic, as well as to the horizontal inhomogeneity D". This layer is also associated with the abnormally strong damping of the Earth's free oscillations. In this aspect layer D" seems to be similar to the asthenospheric layer in the upper mantle. The above properties of layer D" may be produced by the partial melting of rocks in the lower mantle and by the decrease in its

effective viscosity of some orders of magnitude. This situation makes possible active separation of components of the mantle material. The sedimentation of a heavy component from the mixture is an example. Porous downward filtration of a heavy component towards the surface of the nonviscous core is also possible. In this case the flow of a dense fraction into the core should provide gravitational instability in the transitional layer. Diffusion in the mantle may be neglected. Thus the convective mixing must take place even at a slight inversion in density (Artyushkov, 1968; Sorokhtin, 1972; Keondjan and Monin, 1975).

Ψ 467 my.

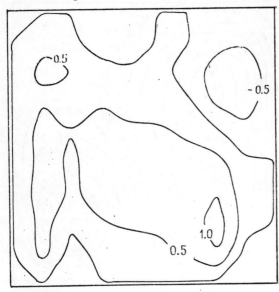

Ψ 733 my.

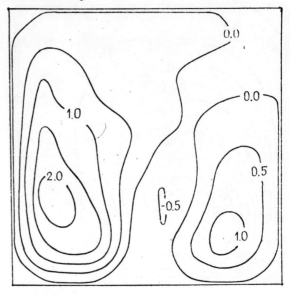

ρ 467 my.

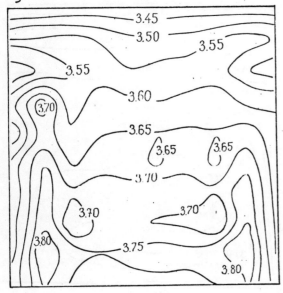

Fig. 1e.

ρ 733 my.

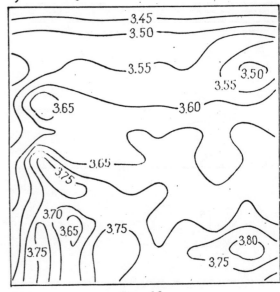

Fig. 1f.

The above discussion shows that the model of convection we require have the following properties:

1. Relative motion of the mantle components at least in the transitional layer should be described.

2. Heterogeneity in rheology of the mantle at least in the radial direction must be accounted for.

3. The absorption of a dense component by the accreting core after that component is separated in the transitional layer must be described.

The basic geophysical requirements to be answered by the model have been stated. We next turn to the construction of the model itself.

The Model.

Consider the two-component Newtonian fluid in a spherical axially symmetrical layer. Mass concentration of a dense component is C (r, θ, t); of the light component, $C_2 = 1-C$. Relative component volumes are respectively S (r, θ, t) and $1-S$. Partial densities are ρ_1 and ρ_2, t = time, r =

ψ 967 my.

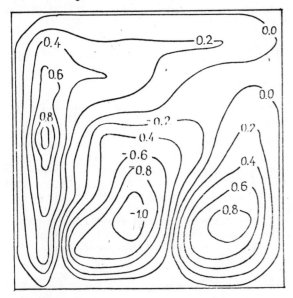

ρ 967 my.

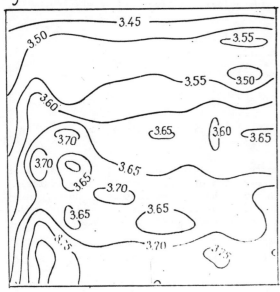

Fig. 1g.

radial coordinate, and θ = colatitude. We neglect the chemical interaction of components. The density of the component mixture is estimated from the equations of Birch and Magnitsky (Birch, 1952; Magnitsky, 1952).

$$\rho = \left\{ \frac{C}{\rho_1} + \frac{I - C}{\rho_2} \right\}^{-1} \qquad (1)$$

in which ρ_1 and ρ_2 are defined by Equations of State

$$\rho_1 = \rho_1(P,T)$$
$$\qquad (2)$$
$$\rho_2 = \rho_2(P,T)$$

There are also formulas

$$S\rho_1 = C\rho, \quad (I - S)\,\rho_2 = (I - C)\rho \qquad (3)$$
$$\rho = \rho_1 S + \rho_2 (I - S) \qquad (4)$$

for the partial characteristics.

Assume that each component treated separately is defined by its velocity field $\vec{V}(1,2) = \{V_r(1,2);\ V_\theta(1,2)\}$. Interaction of components in a unit volume is characterized by force, $\vec{F} = \{F_r\ ;\ F_\theta\}$. Write down the equation of conservation of impulse for each component (Barenblatt, 1953). For a dense component we have

$$\frac{\partial}{\partial t}\,\rho_1 S\,V_r^{(1)} + \frac{1}{r^2}\,\frac{\partial}{\partial r}\,r^2\,\rho_1 S V_r^{(1)} V_r^{(1)} +$$
$$+ \frac{1}{r\sin\theta}\,\frac{\partial}{\partial\theta}\,\sin\theta\,\rho_1 S V_r^{(1)} V_\theta^{(1)} - \frac{V_\theta^{(1)} V_\theta^{(1)}}{r}\rho_1 S$$
$$= -\frac{\partial\chi P}{\partial r} - \rho_1 g S + f_r^{(1)} + F_r \qquad (5)$$

$$\frac{\partial}{\partial t}\,\rho_1 S V_\theta^{(1)} + \frac{1}{r^3}\,\frac{\partial}{\partial r}\,r^3\,\rho_1 S V_\theta^{(1)} V_r^{(1)}$$
$$+ \frac{1}{r\sin\theta}\,\frac{\partial}{\partial\theta}\,\sin\theta\,\rho_1 S V_\theta^{(1)} V_\theta^{(1)} = -\frac{1}{r}\,\frac{\partial\chi P}{\partial\theta}$$
$$+ f_\theta^{(1)} + F_\theta \qquad (6)$$

For a light one similarly

$$\frac{\partial}{\partial t}\,\rho_2(1-s)V_r^{(2)} + \frac{1}{r^2}\,\frac{\partial}{\partial r}\,r^2\rho_2(1-s)V_r^{(2)}V_r^{(2)}$$
$$+ \frac{1}{r\sin\theta}\,\frac{\partial}{\partial\theta}\,\sin\theta\rho_2(1-s)V_r^{(2)}V_\theta^{(2)} - \frac{1}{r}\,\rho_2(1-s)$$
$$V_\theta^{(2)}V_\theta^{(2)} = -\frac{\partial}{\partial r}(1-\chi)P - \rho_2(1-s)g$$
$$+ f_r^{(2)} - F_r \qquad (7)$$

$$\frac{\partial}{\partial t}\,\rho_2(1-s)V_\theta^{(2)} + \frac{1}{r^3}\,\frac{\partial}{\partial r}\,r^3\,\rho_2(1-s)V_\theta^{(2)}V_r^{(2)}$$
$$+ \frac{1}{r\sin\theta}\,\frac{\partial}{\partial\theta}\,\sin\theta\rho_2(1-s)V_\theta^{(2)}V_\theta^{(2)}$$
$$= -\frac{1}{r}\,\frac{\partial}{\partial\theta}(1-\chi)P + f_\theta^{(2)} - F_\theta \qquad (8)$$

In these relations, XP is part of the pressure due to a dense component, g is gravity, and $f_r^{(1,2)}$ and $f_\theta^{(1,2)}$ are forces of viscous friction in each fluid.

Introduce $\vec{V} = \{V_r; V_\theta\}$, velocity of the center of masses of volume using the relation

$$\vec{V} = \frac{\vec{V}^{(1)} S\rho_1 + \vec{V}^{(2)}(1-S)\rho_2}{\rho} \qquad (9)$$

or

$$\vec{V} = C\vec{V}^{(1)} + (1-C)\vec{V}^{(2)} \qquad (10)$$

Summing equations (5) and (7), (6) and (8), we obtain equations of motion for the mixture of components

$$\frac{\partial}{\partial t}\rho V_r + \frac{1}{r^2}\frac{\partial}{\partial r} r^2 \{\rho_1 S V_r^{(1)} V_r^{(1)} + \rho_2(1-S)V_r^{(2)} V_r^{(2)}\} + \frac{1}{r\sin\theta}\frac{\partial}{\partial\theta}\sin\theta \times \{\rho_1 S V_r^{(1)} V_\theta^{(1)} + \rho_2(1-S)V_r^{(2)} V_\theta^{(2)}\} - \frac{1}{r}\{\rho_1 S V_\theta^{(1)} V_\theta^{(1)} + \rho_2(1-S)V_\theta^{(2)} V_\theta^{(2)}\} = -\frac{\partial P}{\partial r} - g + f_r \qquad (11)$$

$$\frac{\partial}{\partial t}\rho V + \frac{1}{r^3}\frac{\partial}{\partial r} r^3 \{\rho_1 S V_\theta^{(1)} V_r^{(1)} + (1-S)\rho_2 V_\theta^{(2)} + \frac{1}{r\sin\theta}\frac{\partial}{\partial\theta}\sin\theta \{\rho_1 S V_\theta^{(1)} V_\theta^{(1)} + \rho_2(1-S)V_\theta^{(2)} V_\theta^{(2)}\} = -\frac{1}{r}\frac{\partial P}{\partial\theta} + f_\theta \qquad (12)$$

where $f_r = f_r^{(1)} + f_r^{(2)}$ and $f_\theta = f_\theta^{(1)} + f_\theta^{(2)}$ are viscous forces in the mixture.

In the same way equations

$$\frac{\partial}{\partial t}\rho_1 S + \text{div } \rho_1 S\vec{V}^{(1)} = 0 \qquad (13)$$

$$\frac{\partial}{\partial t}\rho_2(1-S) + \text{div } \rho_2(1-S)\vec{V}^{(2)} = 0 \qquad (14)$$

express the law of conservation of mass for each component. Combination of these relations yields the equation of conservation of mass for the mixture.

$$\frac{\partial\rho}{\partial t} + \text{div } \rho\vec{V} = 0 \qquad (15)$$

and the equation of continuity

$$\text{div } \{S\vec{V}^{(1)} + (1-S)\vec{V}^{(2)}\} + \frac{S}{\rho_1}\frac{\partial\rho_1}{\partial t} + \vec{V}^{(1)}\text{grad } \rho_1\} + \frac{1-S}{\rho_2}\{\frac{\partial\rho_2}{\partial t} + \vec{V}^{(2)}\text{grad } \rho_2\} = 0 \qquad (16)$$

Together with the equation for conservation of energy, the system of equations (1), (2), (9), (11), (12), (15), and (16) should describe the dynamics of the two-component fluid. But this system is too difficult to be analyzed. Therefore, we will introduce additional simplifications. Assume that individual accelerations of particles in the fluid are much less than the acceleration of gravity and that the concentration of a dense component in the mixture is not strong (C << 1). In this case the angular components of velocities of fluids are equal while the radial components differ by some amount a.

$$V_i^{(1)} = V_i^{(2)} - a\delta_{ir}, \qquad i = r,\theta \qquad (19)$$

Now kinematical relations can be written:

$$V_i = V_i^{(1)} + \frac{\rho_2(1-S)}{\rho} a\delta_{ir} = V_i^{(2)} - \frac{\rho_1 S}{\rho} a\delta_{ir} \qquad (20)$$

or

$$V_i = V_i^{(1)} + (1-C)a\delta_{ir} = V_i^{(2)} - aC\delta_{ir} \qquad (21)$$

Parameter a plays the role of the velocity of the separation of components by some physical mechanism. Thus in the case of standard Stokes' motion of a particle with radius = 1 in a medium with kinematic viscosity ν, we have:

$$a = \frac{2}{9}(1 - \frac{\rho_1}{\rho}) g \frac{1^2}{\nu}. \qquad (22)$$

Having this in mind we derive instead of (11), (12), (15), or (16), the next equations for the mixture of components:

$$\frac{\partial V_r}{\partial t} + V_r\frac{\partial V_r}{\partial r} + \frac{V_\theta}{r}\frac{\partial V_r}{\partial\theta} - \frac{V_\theta^2}{r}$$

$$= -\frac{1}{\rho}\frac{\partial P}{\partial r} - \frac{1}{r^2}\frac{\partial}{\partial r} r^2 a^2 C(1-C) - g + \frac{fr}{\rho}$$

$$\frac{\partial V_\theta}{\partial t} + V_r\frac{\partial V_\theta}{\partial r} + \frac{V_\theta}{r}\frac{\partial V}{\partial\theta} + \frac{V_r V_\theta}{r} = -\frac{1}{\rho r}\frac{\partial P}{\partial\theta} + \frac{f_\theta}{\rho} \qquad (24)$$

$$\frac{1}{r^2} \frac{\partial}{\partial r} r^2 V_r + \frac{1}{r Sin\theta} \frac{\partial}{\partial \theta} Sin\theta\, V_\theta = -\frac{1}{r^2} \frac{\partial}{\partial r} r^2 aC$$

$$(1 - \frac{\rho}{\rho_1}) \tag{25}$$

$$\frac{\partial C}{\partial t} + V_r \frac{\partial C}{\partial r} + \frac{V_\theta}{r} \frac{\partial C}{\partial \theta} = -\frac{1}{\rho r^2} \frac{\partial}{\partial r} r^2 \rho aC(1-C)$$

$$+ \frac{1}{\rho r^2} \frac{\partial}{\partial r} r^2 \rho D \frac{\partial C}{\partial r} + \frac{1}{\rho r^2 Sin\theta} \frac{\partial}{\partial \theta} \rho D\, Sin\theta \frac{\partial C}{\partial \theta}$$

$$\tag{26}$$

We neglect the compressibility of components in the equation of continuity (25), but take into account the dependence of the density of the mixture on concentration. This latter effect seems important. Practically we obtain the Boussinesq approximation for each component separately.

Expression (26) is the equation of diffusion for a dense component. D is the coefficient of diffusion in a binary fluid.

Dissipative forces in the equation of motion are written as usual for a Newtonian fluid

$$f_r = \frac{1}{r^3} \frac{\partial}{\partial r} r^3 \sigma_{rr} + \frac{1}{r Sin\theta} \frac{\partial}{\partial \theta} Sin\theta \sigma_{r\theta} \tag{27}$$

$$f_\theta = \frac{1}{r} \frac{1}{r^3} \frac{\partial}{\partial r} r^3 \sigma_{r\theta} + \frac{1}{r Sin\theta} \frac{\partial}{\partial \theta} Sin\theta \sigma_{\theta\theta}$$

$$+ ctg\theta \frac{\sigma_{rr} + \sigma_{\theta\theta}}{r} \tag{28}$$

The components of the stress tensor σ_{ij} are expressed through the velocity of center of masses

$$\sigma_{rr} = 2\eta \frac{\partial v}{\partial r}; \quad \sigma_{\theta\theta} = 2\eta (\frac{V_r}{r} + \frac{1}{r} \frac{\partial V_\theta}{\partial \theta})$$

$$\sigma_{r\theta} = \eta (\frac{1}{r} \frac{\partial V_r}{\partial \theta} + \frac{\partial V_\theta}{\partial r} - \frac{V_\theta}{r}) \tag{29}$$

where coefficient of dynamic viscosity η allows for Einstein's correction

$$\eta = \eta^{(2)}(r)(1 + \frac{5}{2} \frac{C\rho}{\rho_1}) \tag{30}$$

It is well known that the Prandtl number ($P_r = \frac{\nu}{k}$, k is a coefficient of thermal conductivity of planetary interiors) are generally large i.e., of the order of 10^{24} and above. Therefore all known studies on thermal mantle convection justify the infinite Prandtl number approximation. In this case the Schmidt number ($Sh = \frac{\nu}{D}$) must be used

rather than Prandtl. The inequalities k >> D; Sh >> P_r >> 1 hold. These considerations provide the grounds to neglect inertial terms in the equations of motion.

In this type of convection the changes in the density field are supported primarily by the advection of concentration of denser matter. Our aim is to study this major process. Therefore we will neglect the thermal effects from further consideration and take into account the dependence of partial densities only on pressure in a simple form:

$$\rho_{1,2} = \alpha_{1,2}\, \tau(P) \tag{31}$$

For axisymmetric cases, stream function V_i can be defined as:

$$V_\theta = -\frac{1}{r} \frac{\partial}{\partial r}(\psi r);$$

$$V_r = \frac{1}{r Sin\theta} \frac{\partial}{\partial \theta}(\psi Sin\theta) - A \tag{32}$$

where

$$A = aC (1 - \frac{\rho}{\rho_1})$$

so that equation (25) is identically satisfied. Following the general practice in fluid mechanics, a new variable called vorticity is introduced. Vorticity is defined as:

$$\vec{\varepsilon} = curl\ \vec{V} \tag{33}$$

In the axisymmetric case it has only one non-vanishing component. Using Eq. (32) we obtain:

$$\varepsilon = \frac{\psi}{r^2 Sin^2\theta} + \frac{1}{r} \frac{\partial A}{\partial \theta} - \Delta\psi \tag{34}$$

Take the curl to exclude the pressure from the equations of motion (23), (24). Then the following equation for vorticity is obtained:

$$\nu \{\Delta\varepsilon - \frac{1}{r} \frac{\partial \varepsilon}{\partial r} - \frac{\varepsilon}{r^2}(1 + \frac{1}{Sin^2\theta})\}$$

$$+ \frac{\partial \nu}{\partial r} \{2 \frac{\partial \varepsilon}{\partial r} + \frac{3}{r} \varepsilon\} - \frac{\partial^2 \nu}{\partial r^2} \varepsilon$$

$$= \nu\{\frac{\psi}{r^4 Sin^2\theta} + \frac{2}{r^2 Sin^2\theta} \frac{\partial^2 \psi}{\partial r^2} + \frac{2}{r^3} \frac{\partial \psi}{\partial r}\}$$

$$- \frac{\partial \nu}{\partial r} \{\frac{4}{r}(\frac{\partial^2 \psi}{\partial r^2} + \frac{\partial}{\partial r} \frac{\psi}{r})\} + 2 \frac{\partial^2 \nu}{\partial r^2} \{\frac{\partial^2 \psi}{\partial r^2}$$

$$+ \frac{\partial}{\partial r} \frac{\psi}{r} \} - R - Q \tag{35}$$

where

$$R = \nu \left\{ \frac{\sigma}{r^3} \frac{\partial A}{\partial \theta} + \frac{4}{r^2} \frac{\partial^2 A}{\partial r \partial \theta} + \frac{8 ctg\theta}{r^3} A - \frac{2 ctg\theta}{r^2} \frac{\partial A}{\partial r} \right.$$

$$\left. - \frac{2 ctg}{r} \frac{\partial^2 A}{\partial r^2} - \frac{2}{r^3 Sin\theta} \frac{\partial \nu}{\partial r} \frac{\partial}{\partial \theta} Sin\theta \frac{\partial}{\partial r} r^2 A \right.$$

$$Q = \frac{1}{r^3} \frac{\partial^2}{\partial r \partial \theta} r^2 a^2 C(1-C) + \frac{g}{r\rho_0} \frac{\partial \rho}{\partial \theta}$$

Equations (1), (26), (32), (34), (35) represent the final governing equations to be solved for the six variables ρ, C, ψ, ε, V_r, V_θ.

This system requires the specification of boundary conditions. Since only the derivatives of stream functions are important in the problem, ν can thus be specified to within a constant. Consequently, they are chosen, for convenience, to be zero at all boundaries. At the axis of symmetry $\theta = 0, \pi$:

$$V_\theta = \frac{\partial V_\theta}{\partial} = \frac{\partial V_r}{\partial \theta} = \frac{\partial C}{\partial \theta} = 0$$

Because of symmetry. At the outer surface and the core interface $r = r_2$, r_1:

$$U_r = \frac{\partial V_r}{\partial \theta} = \frac{\partial}{\partial r} = \frac{V_\theta}{r} = 0 \text{ (shear stress free sur-}$$

face conditions). We require also lack of flux of a dense component through the upper boundary and its complete absorption ($C=0$) at the core interface.

Imitation of core growth in the first approximation may be accomplished by the upward transfer of the lower boundary. The velocity of the transfer is proportional to the velocity of the sinking of a dense component into the core.

We chose as the initial conditions a uniform stream function of $\psi = 0$ and small perturbations of uniform concentration of dense component C. The value of C is normally taken as 0.1.

In most numerical experiments, the mantle (a spherical shell $3500 \le r \le 6500$ km) was subdivided into 10 layers each 300 km thick into 36 5-degree latitudinal zones. A time step was taken as 10^{23} sec $\sim 1/3$ My. We also made numerical experiments with different initial and boundary conditions, geometry, space and time grids, profiles $\nu(r)$ and values of a. In particular, calculations with the space grid twice more fine and with the time step reduced according to KFL criterion show very small differences in the flow pattern, less than 10 percent.

The alternating-direction explicit finite-difference scheme was used to solve Eq. (35) numerically. Central space differences in both r and Q directions were used in the difference approximation of Eq. (34). Equation (26) was solved by the method of splitting (at each intermediate step the following parameters are sequentially taken into account: sedimentation, radial and latitudinal transfer and diffusion). The Lax-Wendroff scheme with a second-order approximation in space and times is used.

Numerical experiments and geophysical conclusions.

We now turn to the description of the numerical experiments. Variable model parameters are sedimentation rate a (C, ν), dimensions of the area and the viscosity profile ν (r). The kinematic viscosity of the main body of the mantle was taken as 10^{23} cm/sec in all experiments. This value was found from the analyses of polar wandering (Keondjan and Monin, 1977 a,b). The lower boundary layer 150 km thick was given a viscosity lower by 4-5 orders. We thus imitated layer D" and stimulated the development of instability.

Figure 1 shows the progress of instability in the square with a side of 3000 km. Specifically, the patterns of stream function and the initial density concentration was assumed at 1.0 everywhere except at the base. Motion was absent. The initial disturbance of concentration was of the order of 0.01 percent and was given at the left lower corner. The time step was assumed to be .33 million years and space step, 150 km.

The beginning of the process is characterized by a weak one-cell circulation which slowly intensifies with time. The density field is practically not disturbed for the initial 200 million years. When the process gets more active, a zone of positive rotation is gradually transferred to the left and a cell with negative rotation is growing at the right. Anomalies in the field of density arise due to the upward motion of light matter occur along the left boundary. In 400 million years the intensity drops abruptly and the process returns to the single cell regime. This stage takes about 200 million years. Then the convection again takes the form of a double-cell structure and the intensity increases. In the interval of 700-800 million years, stream lines gradually transform to the triple-cell structures. Average concentration at this stage drops to five percent.

This simple experiment demonstrates the basic features of the type of convection under study, these features include the instability of process and the tendency to oscillations in the pattern and intensity of convection.

Figure 2 shows an experiment in a spherical shell 6000 km thick ($3500 \le r \le 6500$ km). There is no motion initially. The upper 150 km thick layer does not contain a dense component. Everywhere outside this layer initial concentration is 10 percent. Initial disturbance is at the equator. Double-cell convection develops for the first 50 million years and turns to the four-cell structure with intense upwelling in the equatorial zone by 150 million years. The uprise has the pattern of a relatively narrow asthenolite which flows along the upper boundary. The situation is somewhat similar to the channel studied by Artyushkov (1979). Long lenses of dense matter are formed in the zones of sinking. From 300 to 400 million years the convection is again in the form of two cells. At 450 million years the symmetric pattern

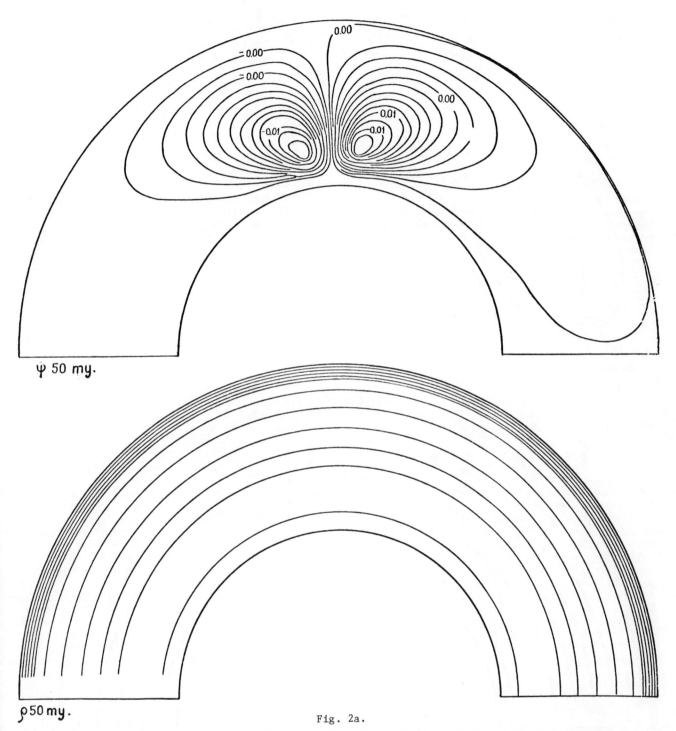

ψ 50 my.

ρ 50 my.

Fig. 2a.

Fig. 2 (a-f). Stream function ψ and density field ρ in a spherical shell. Initial instability is placed at the equator ($\theta = 90°$) at the base of the shell. Other notations the same as those of Figure 1.

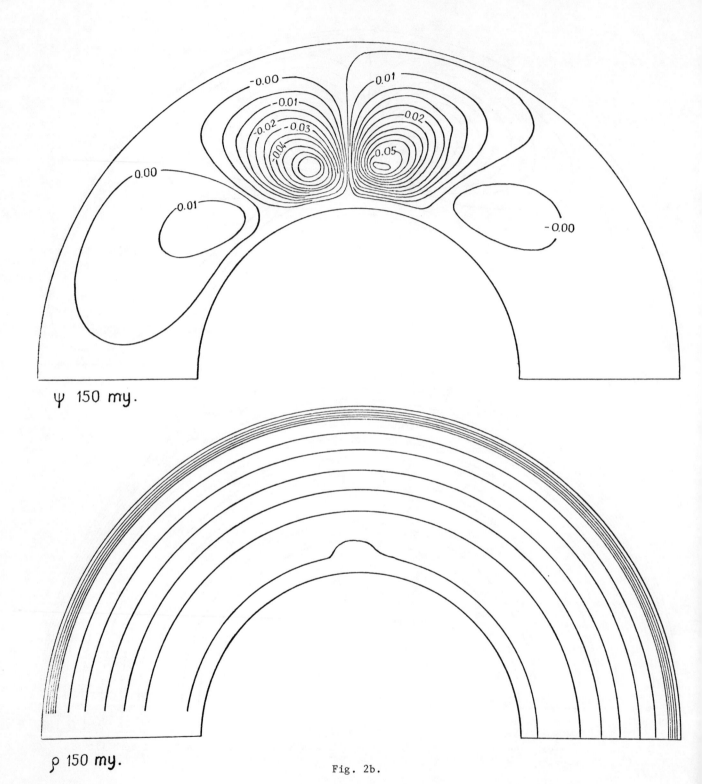

ψ 150 my.

ρ 150 my.

Fig. 2b.

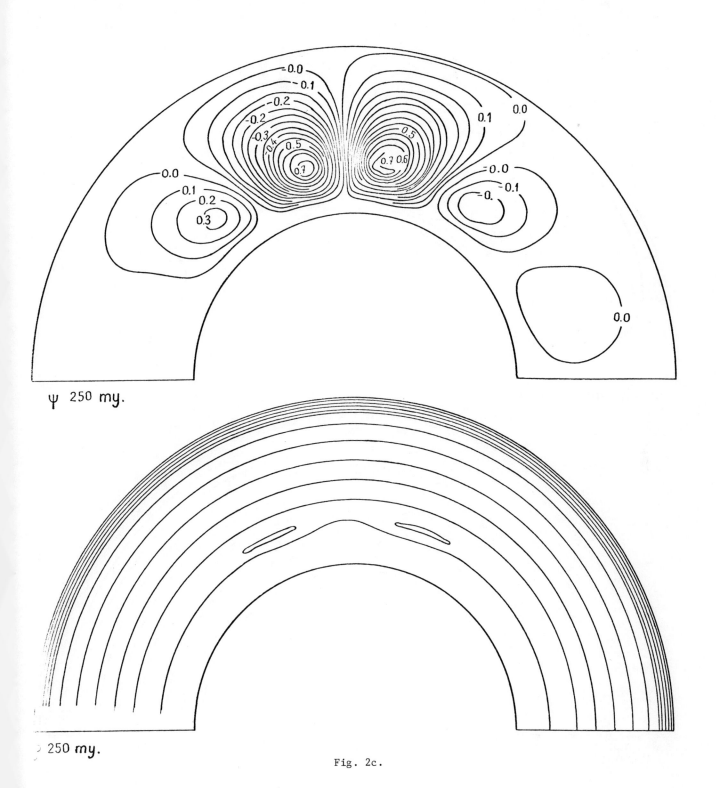

ψ 250 my.

250 my.

Fig. 2c.

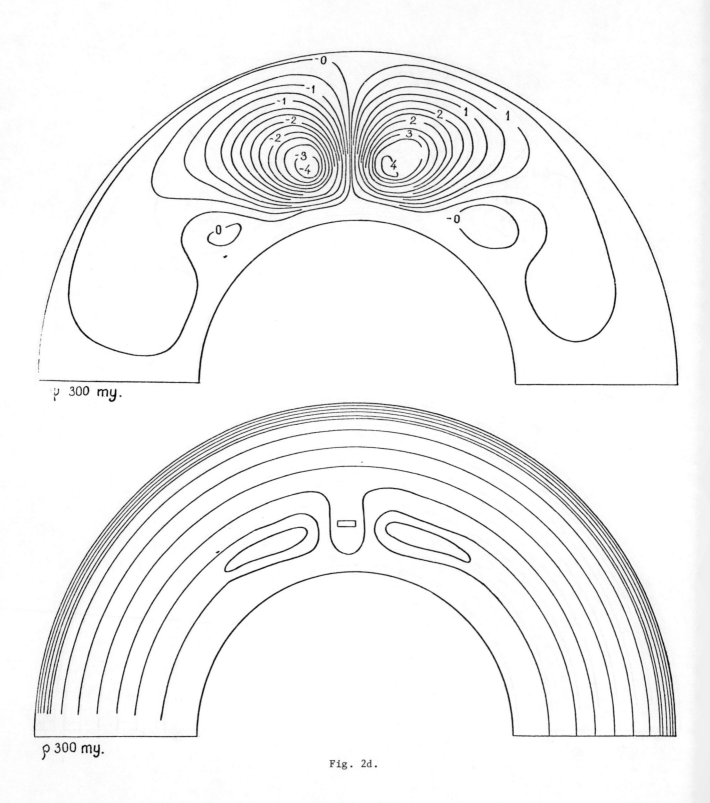

φ 300 my.

ρ 300 my.

Fig. 2d.

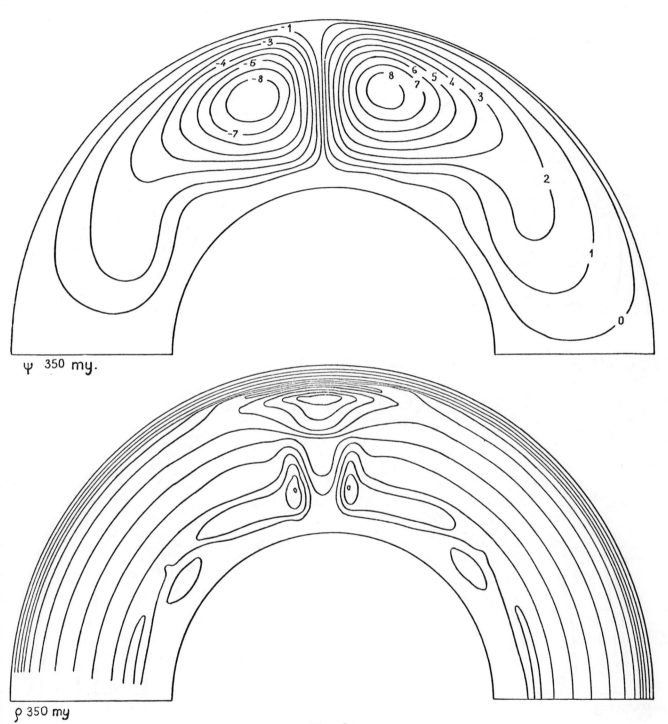

ψ 350 my.

ρ 350 my

Fig. 2e.

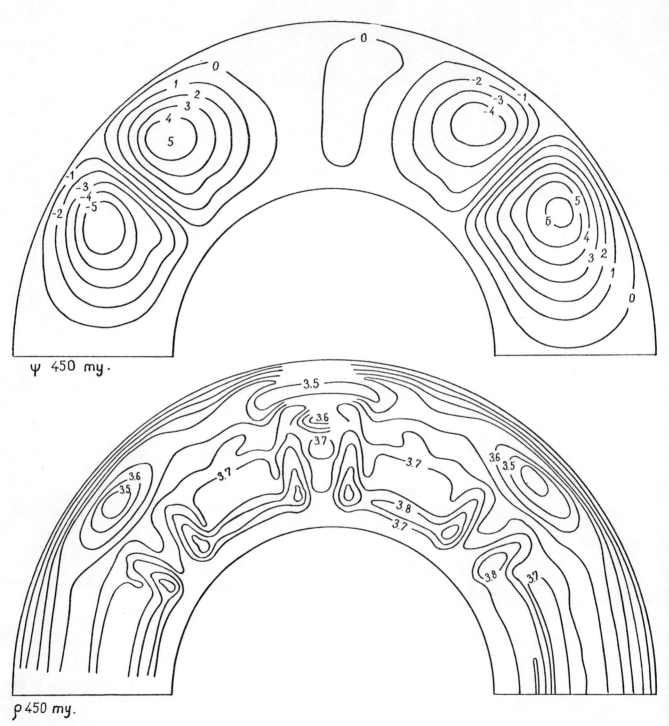

ψ 450 my.

ρ 450 my.

Fig. 2f.

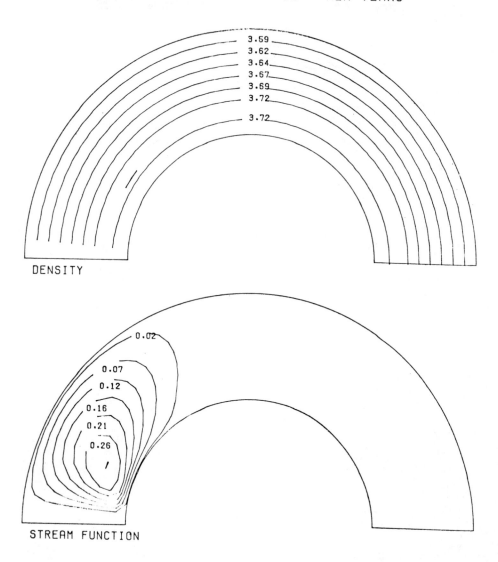

STEP NUMBER   10   30   MLN YEARS

DENSITY

STREAM FUNCTION

VARIANT 35   NOVEMBER 1979

VISCOSITY DISTRIBUTION ALONG Z-AXIS:

1E19 2E22 1E23 1E23 1E23 1E23 1E23 1E23 1E23 2E22 1E19

MEAN ENERGY (LN)   -3.53

MEAN CONCENTRATION   0.0897

NEW CORE   6   %

Fig. 3a.

Fig. 3 (a–m). Stream function ψ and density field ρ in a spherical shell. Initial instability is placed at the pole ($\theta = 0^{o}$) at the base of the shell. Viscosity profile, dimensionless mean kinetic energy (LOG), mean concentration of a dense component, and a fraction of a dense component lost by the mantle are also shown.

STEP NUMBER   20   6'0   MLN YEARS

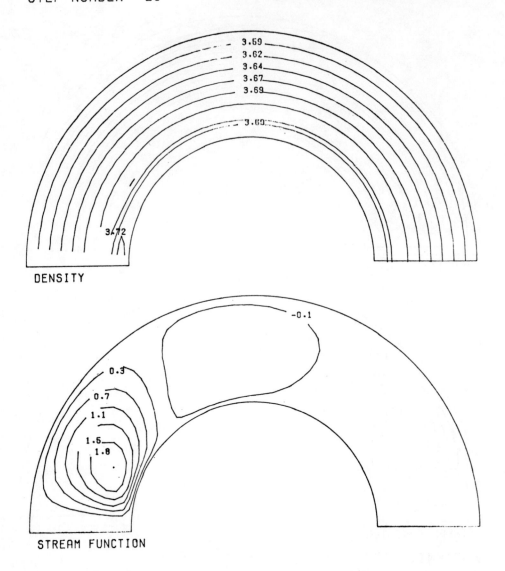

DENSITY

STREAM FUNCTION

VARIANT 35   NOVEMBER 1979

VISCOSITY DISTRIBUTION ALONG Z-AXIS:

1E19 2E22 1E23 1E23 1E23 1E23 1E23 1E23 1E23 2E22 1E19

MEAN ENERGY (LN)   -1.86

MEAN CONCENTRATION   0.0868

NEW CORE   9   %

Fig. 3b

STEP NUMBER 30   *9*0 MLN YEARS

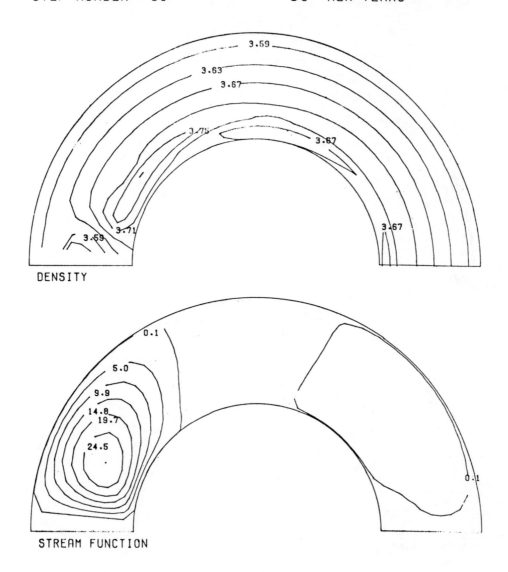

DENSITY

STREAM FUNCTION

VARIANT 35   NOVEMBER 1979

VISCOSITY DISTRIBUTION ALONG Z-AXIS:

1E19 2E22 1E23 1E23 1E23 1E23 1E23 1E23 1E23 2E22 1E19

MEAN ENERGY (LN)   1.12

MEAN CONCENTRATION   0.0848

NEW CORE   11   %

Fig. 3c.

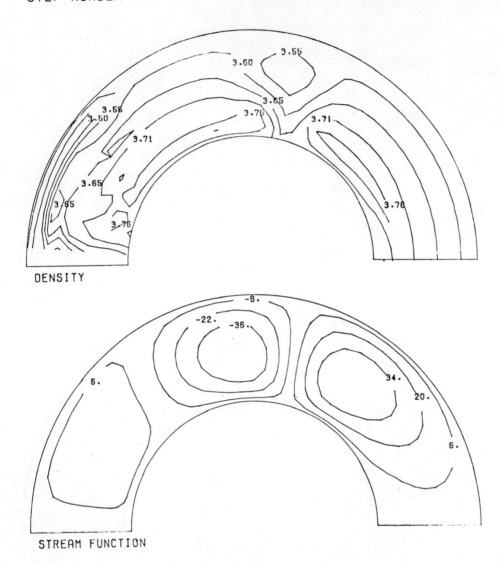

STEP NUMBER 40   120 MLN YEARS

DENSITY

STREAM FUNCTION

VARIANT 35   NOVEMBER 1979

VISCOSITY DISTRIBUTION ALONG Z-AXIS:

1E19 2E22 1E23 1E23 1E23 1E23 1E23 1E23 1E23 2E22 1E19

MEAN ENERGY (LN)   1.21

MEAN CONCENTRATION   0.0839

NEW CORE   12   %

Fig. 3d.

STEP NUMBER 50　　　　150 MLN YEARS

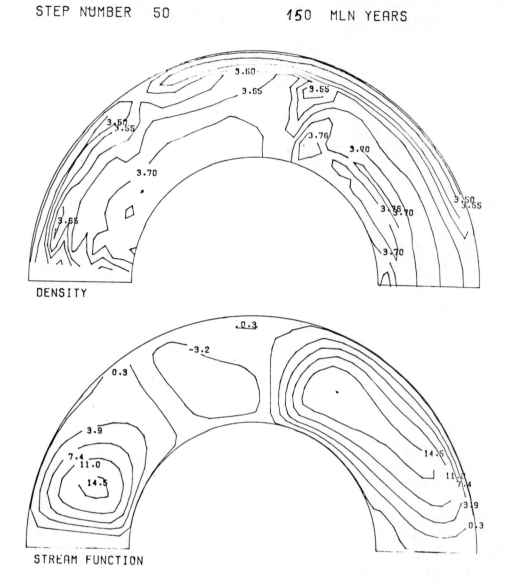

DENSITY

STREAM FUNCTION

VARIANT 35　　　NOVEMBER 1979

VISCOSITY DISTRIBUTION ALONG Z-AXIS:

1E19 2E22 1E23 1E23 1E23 1E23 1E23 1E23 1E23 2E22 1E19

MEAN ENERGY (LN)　　1.50

MEAN CONCENTRATION　　0.0808

NEW CORE　　16　　%

Fig. 3e.

STEP NUMBER 60   180 MLN YEARS

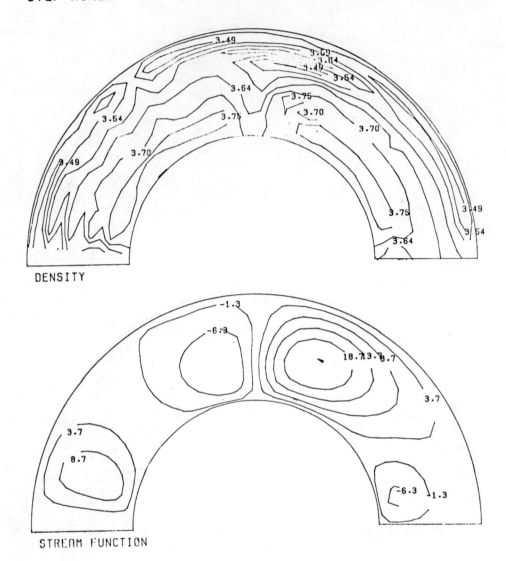

DENSITY

STREAM FUNCTION

VARIANT 35   NOVEMBER 1979

VISCOSITY DISTRIBUTION ALONG Z-AXIS:

1E19 2E22 1E23 1E23 1E23 1E23 1E23 1E23 1E23 2E22 1E19

MEAN ENERGY (LN)   1.08

MEAN CONCENTRATION   0.0772

NEW CORE   20   %

Fig. 3f.

STEP NUMBER 70   210 MLN YEARS

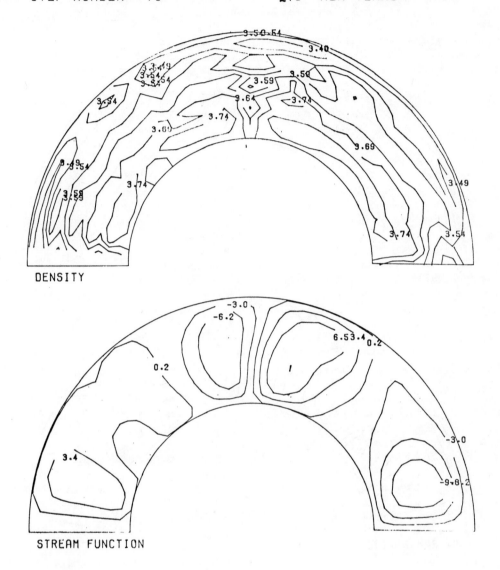

DENSITY

STREAM FUNCTION

VARIANT 35   NOVEMBER 1979

VISCOSITY DISTRIBUTION ALONG Z-AXIS:

1E19 2E22 1E23 1E23 1E23 1E23 1E23 1E23 1E23 2E22 1E19

MEAN ENERGY (LN)   1.08

MEAN CONCENTRATION   0.0730

NEW CORE   24   %

Fig. 3g.

STEP NUMBER   80   240 MLN YEARS

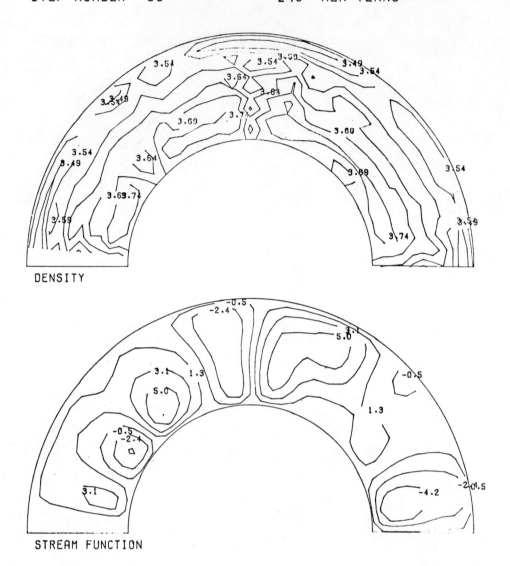

DENSITY

STREAM FUNCTION

VARIANT 35   NOVEMBER 1979

VISCOSITY DISTRIBUTION ALONG Z-AXIS:

1E19 2E22 1E23 1E23 1E23 1E23 1E23 1E23 1E23 2E22 1E19

MEAN ENERGY (LN)   -0.46

MEAN CONCENTRATION   0.0700

NEW CORE   28   %

Fig. 3h.

STEP NUMBER   90   270 MLN YEARS

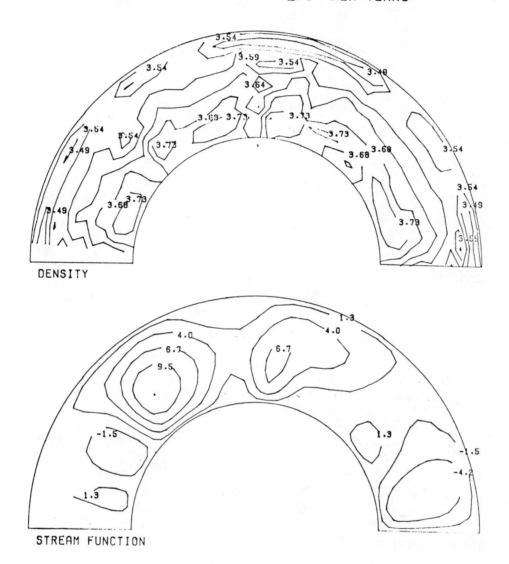

DENSITY

STREAM FUNCTION

VARIANT 35   NOVEMBER 1979

VISCOSITY DISTRIBUTION ALONG Z-AXIS:

1E19 2E22 1E23 1E23 1E23 1E23 1E23 1E23 1E23 2E22 1E19

MEAN ENERGY (LN)   0.22

MEAN CONCENTRATION   0.0666

NEW CORE   32   %

Fig. 3i.

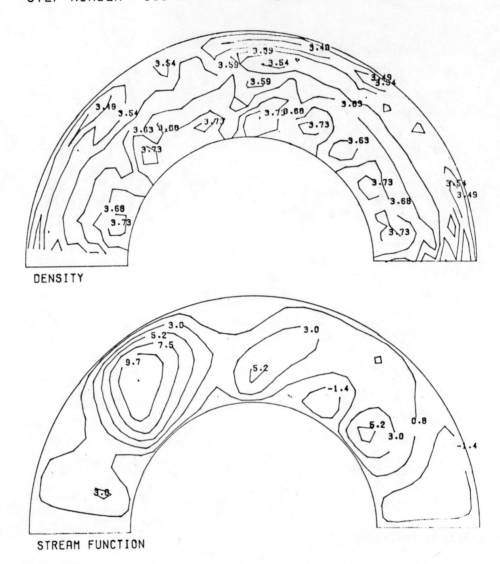

STEP NUMBER 100   300 MLN YEARS

DENSITY

STREAM FUNCTION

VARIANT 35   NOVEMBER 1979

VISCOSITY DISTRIBUTION ALONG Z-AXIS:

1E19 2E22 1E23 1E23 1E23 1E23 1E23 1E23 1E23 2E22 1E19

MEAN ENERGY (LN)   0.62

MEAN CONCENTRATION   0.0635

NEW CORE   35   %

Fig. 3j.

STEP NÜMBER   110   330   MLN YEARS

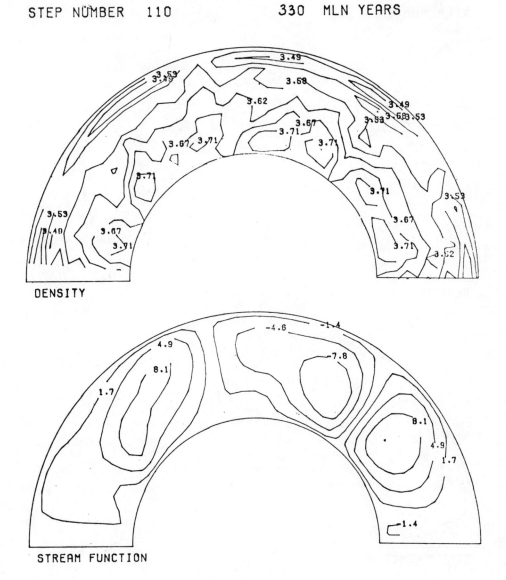

DENSITY

STREAM FUNCTION

VARIANT 35   NOVEMBER 1979

VISCOSITY DISTRIBUTION ALONG Z-AXIS:

1E19 2E22 1E23 1E23 1E23 1E23 1E23 1E23 1E23 2E22 1E19

MEAN ENERGY (LN)   0.14

MEAN CONCENTRATION   0.0610

NEW CORE   38   %

Fig. 3k.

STEP NUMBER   160   480   MLN YEARS

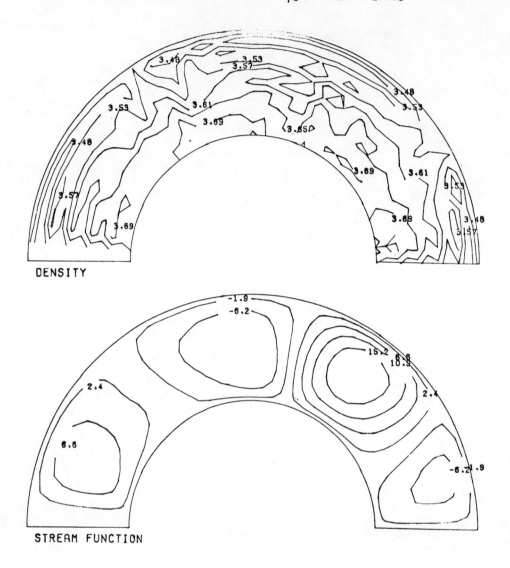

DENSITY

STREAM FUNCTION

VARIANT 35   NOVEMBER 1979

VISCOSITY DISTRIBUTION ALONG Z-AXIS:

1E19 2E22 1E23 1E23 1E23 1E23 1E23 1E23 1E23 2E22 1E19

MEAN ENERGY (LN)   0.25

MEAN CONCENTRATION   0.0529

NEW CORE   47   %

Fig. 31.

STEP NUMBER   260   780 MLN YEARS

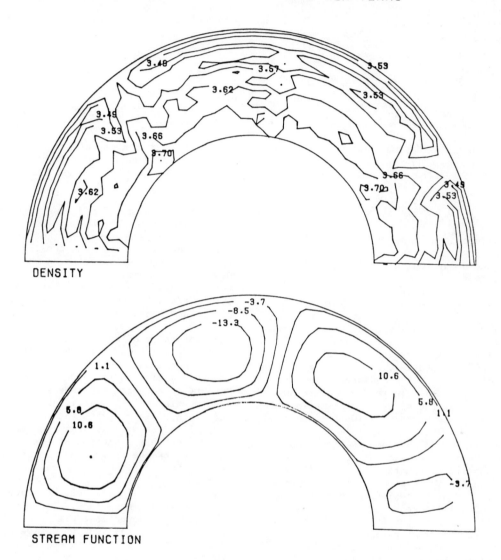

DENSITY

STREAM FUNCTION

VARIANT 35   NOVEMBER 1979

VISCOSITY DISTRIBUTION ALONG Z-AXIS:

1E19 2E22 1E23 1E23 1E23 1E23 1E23 1E23 1E23 2E22 1E19

MEAN ENERGY (LN)   0.12

MEAN CONCENTRATION   0.0466

Fig. 3m.

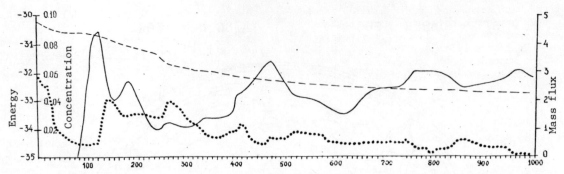

Fig. 4. A change with time in the following parameters: (1) mean density of logarithm of kinetic energy (solid line), (2) mass flux through the core/mantle interface (in percents of the initial mass for 10 My (dotted line), (3) mean concentration of a dense component (dashed line).

is destroyed and five cells of various intensity arise.

Figure 3 illustrates convection in the same spherical shell with the initial disturbance at the base in the polar region $\theta = 0$. In this experiment the initial concentration is uniform everywhere except the core/mantle interface. A low viscosity layer ($\nu = 10^{20}$ cm/sec) is placed at the upper boundary. In this case the convection turns rather fast to the quasi-stable regime. Further evolution of flows is governed by the concurrent struggle of three to four cells. At some moments such as 300 million years the process is effectively double-cell. Redistribution of masses if fulfilled again in the rather narrow upwelling zones. At the sites of sinking we see again the zonal dense structures in the form of lenses.

The last plot (Figure 4) shows integral characteristics of the process as a function of time. The average density of kinetic energy oscillates quasi-periodically within the limits of one order of magnitude. A typical period to this oscillation is 200-300 million years. There are also distinct oscillations in the flow of mass to the core roughly with the same period. Amplitude of the oscillations decreases as the layer is gradually depleted of the dense component.

We have shown only some numerical tests out of a large number of completed tests. The experience of dealing with the model described above enabled us to infer the typical features of the convection considered. These are:

1. Chemical convection is essentially non-stationary as its source of energy, that is the potential energy, decreases with time quasi-monotonously.

2. The specific pattern of velocity and mass fields at any given moment is controlled by three factors superimposed:
   a) Initial conditions,
   b) The velocity of decrease of a dense component in the volume.
   c) The specific characteristics of the area (geometry, a profile of viscosity).

3. At any choice of model parameters the process has the following stages.
   a) Monotonous increase in the intensity of motion and of non-zonality of the density field. Development of a number of cells on the planetary scale. Development of density anomalies as plumes in the zone of uplift and as lenses in zones of the sinking. This stage ends with the maximum intensity of streams appearing as a regular pattern of large scale cells.
   b) Intensity of streams diminish. Centers of cells migrate towards the upwelling zones and the cells become more narrow. Light inhomogeneities in density flow out along the upper boundary. Density inhomogeneities decrease in scale. Lenses are interlayered. Finally large cells are split into rather small, transient curls.
   c) Flows are again reorganized as quasi-stable cells. Further evolution of the process is characterized by the redistribution of energy among cells. One or two cells prevail at certain moments.

4. The kinetic energy of the process varies within one order of magnitude. A period of energy oscillations is compatible with time of geological cycles (10^8 years). During active epochs the velocity of convection amounts to some tens of cm per year.

5. If the asthenosphere is taken into account, a typical number of cells within the layer decreases.

Finally we briefly discuss the geophysical importance of the model suggested. The intensity of flows and their structure, which allow motions of one sign for hundreds of million of years on a scale of thousands of kilometres, enable explanation of the kinematics of lithospheric plates. The oscillating regime of continental drift can be explained (such events as interchange of large structure type, for example Laurasia and Gondwana with smaller scale structures). Geological cycles can be explained with the quasi-periodicity of convection (Monin and Sorokhtin, 1977; Monin, 1977; Keondjan and Monin, 1977 a).

A reasonable velocity of plate motions is pro-

vided by the high intensity of circulation in the
mantle, which attains some tens of cm per year.
The viscosity of the matter is only insignificantly
decreased (the viscosity is $4 \cdot 10^3$ poises). The
flow of light material along the base of the lithosphere as inferred from convection can be treated
as a separate problem.

We have shown only preliminary results of modeling the chemical convection. We shall further
develop the concept and make more numerical experiments. As we proceed into further investigation of this problem.

Acknowledgements

The foresight of Prof. A.S. Monin made it possible for me to accomplish the study reported in
this paper. The assistance of Dan Seidov in
computations is gratefully acknowledged.

References

Artyushkov, E.V., Gravitatsionnaya konvektsiya v
nedrakh Zemli, Izv. AN SSSR, Fizika Zemli, 9,
1968.
Artyushkov, E.V., Differentsiatsiya po plotnosti
veschestva Zemli i svyazanniye s nej yavleniya,
Isv. AN SSSR, Fizika Zemli, 5, 1970.
Artyushkov, E.V., Geodinamika, M. "Nauka", 1979.
Barenblatt, G.I., O dvigenii vzveshennih chastits
v turbulentnom potokye, Ac. Nauk SSSR, Priklad-
naya matematika i mehanika, v. XVI. vip. 3, 1953.
Birch, F., Elasticity and constitution of the
Earth's interior, J. Geoph. Res., 57, 2, 1952.
Hoyle, F., On the condensation of the planets,
Mon. Notic. Roy. Astron. Soc., 106, 5, 1946.
Keondjan, V.P. and Monin, A.S., Model gravitatsion-
noy differentsiatsii nedr planet, Dokl. AN SSSR,
220, 4, 1975.
Keondjan, V.P. and Monin, A.S., Reschyot evolutsii
nedr planet, Izv. AN SSSR, Fizika Zemli, 4, 1976.
Keondjan, V.P. and Monin, A.S., Calculations on
the evolution of the planetary interiors,
Tectonophysics, 41, 1977.
Keondjan, V.P. and Monin, A.S., Bluzhdaniye polusa
Zemli vsledstvii kontinentalnogo dreyfa, Dokl.
AN SSSR, 233, 2, 1977a.
Keondjan, V. P. and Monin, A.S., Dreyf continentov
i krupnomesshtabniye smescheniya polusa Zemli,
Izv. AN SSSR, Fizika Zemli, 11, 1977b.
Magnitsky, V.A., K voprosu o plotnosti i szhimae-
mosti obolochki Zemli, Voprosi Kosmogonii, v. 1,
M., AN SSSR, 1952.
Monin, A.S. and Sorokhtin, O.G., O tectonicheskoy
periodizatsii istorii Zemli, Dokl. AN SSSR, 234,
2, 1977.
Monin, A.S., Istoriya Zemli, M. "Nauka", 1977.
Phinney, R.A. and Alexander, S.S., The effect of
velocity gradient at the base of the mantle on
diffracted P-waves in the shadow, J. Geophys.
Res., 74, 20, 1969.
Runcorn, S.K., Convection currents in the Earth's
mantle, Nature, 195, 4848, 1962.
Safronov, V.S., Evolyutsija doplonetnogo oblaka
i obrazovaniye Zemli i planet, M. "Nauka", 1969.
Schmidt, O.Yu., Chetire lektsii o teorii proisho-
zhdenija Zemli, M. "Nauka", 1948.
Sorokhtin, O.G., Differentsiatsiya veschestva Z
Zemli i razvitiye tectonicheskih protsessov,
Izv. AN SSSR, Fizika Zemli, 7, 1972.
Sorokhtin, O.G., Globalnaya evolutsiya Zemli,
M. "Nauka", 1974.
Urey, H.C., The origin of the Earth and the planets,
London, Oxford Univ. Press, 1952.

THE ROLE OF OXIDATION-REDUCTION REACTIONS IN THE EARTH'S EARLY HISTORY

O.L. Kuskov

Vernadsky Institute of Geochemistry and Analytical Chemistry

USSR Academy of Sciences, Moscow, USSR

Abstract. Thermodynamical analysis of oxidation-reduction reactions for the undifferentiated
Earth in the early stage of its evolution has
been carried out. The possibility of reduction
of ferromagnesian silicates, magnesiowustite,
magnetite, and stishovite by carbon in the closed
and in the open systems at very high pressures
has been considered. It is shown that the
process of reduction in the closed system are
thermodynamically forbidden. Calculations of
thermal effects of oxidation-reduction reactions
in the open systems have been carried out. The
results of calculations show that a change in
sign of the thermal effect occurs in the open
system at a certain critical pressure on the
solid phases. Calculations of ΔH for several
reactions which are important for the thermal
history of the Earth are presented at pressures
up to 2 mbar and the change in temperature for
the undifferentiated Earth are estimated. The
data obtained show that the central part of the
Earth and the deep interior could become warm
due to reduction processes. The reducing conditions in the early history of the Earth during
the core formation imply that a large fraction
of chemical reaction heat was released in the
central part of the Earth. The existence of a
source of differentiation in the deep interior of
undifferentiated Earth is assumed.

Introduction

In a study of the Earth's thermal history, all
primary sources of energy should be evaluated.
An assessment has been made of these essential
contributions to the Earth's overall energy
balance. Included are: gravitational energy,
produced by its accretion, radioactive decay
energy, adiabatic compression energy, tidal
friction energy and core formation gravitational
energy [Lubimova, 1968, 1977; Lyustikh, 1948;
Levin et. al., 1972; Safronov, 1969, 1975; Beck,
1961; Urey, 1952; Birch, 1965].

However, until recently, no energy balance
estimates of the Earth have been made on the

energy of chemical reactions, an important factor
of thermal evolution.

Analysis of all the data obtained so far (but
only partially represented in the references)
prompts the conclusion that a realistic model of
the Earth (or any other planet) must be based on
the common foundation of cosmochemistry, geochemistry and geophysics in order that the
planet's development might be traced from its
origin to the present. Naturally we can not
cover the entire spectrum of this evolution or
offer a hypothesis complying with all requirements.

This paper puts special emphasis on the need to
define the main physico-chemical processes in the
Earth's interior. This understanding along with
existing data in geochemistry, geophysics and
tectonics, would progress toward constructing a
non-contradictory model of the Earth. This paper
presents a first attempt at a physico-chemical
approach.

I. A Description of the Processes Under Study

The physico-chemical processes in the Earth's
interior may occur in both closed and open
systems. It would be of great importance to
define the area of applicability of thermodynamic
systems governing the physico-chemical aspects
to the Earth's evolution. Closed systems determine the upper limits of the mantle mineral
stability and allow P-T parameters and divariant
mineral stability fields to be specified--The
thermodynamics of these systems characterize the
static life of the Earth. Open systems, the
processes of which have been studied by Kuskov
and Khitarov [1978a], are characterized by the
inflow and outflow of volatile substances. Their
thermodynamics being closely related to the most
important aspects of planetary evolution are
displayed in large-scale gravitational differentiation and in degassing.

The strict application of thermodynamic closed
and open system techniques to resolve uncertainties in the Earth's inner structure and chemical

196

evolution, are likely to involve two basic difficulties. These are:

(1) The degree of not knowing the equations of state for substances with the P-T parameters of the Earth's interior.

(2) Many details of the Earth's early geodynamic development are obscure. This is a major problem wherein the physico-chemical aspects may be specified only through a thermodynamic analysis employing the latest developments in experimental and theoretical geophysics and geochemistry.

The Schmidt hypothesis [1958] concerning cold quasi-homogenous protoplanet formation assumes that the differentiation of the Earth has caused the heavy core material to sink to the center and light material to surface and volatiles to be released, thus causing the formation of the primary atmosphere and hydrosphere. It should be readily apparent that systematic studies to clarify the volatile pattern and define primary and secondary buffer equilibria in the Earth's lower mantle are necessary for a comprehensive understanding of many problems relating to the planetary evolution of earth-type bodies.

To this end, calculations have been made of gas fugacities and the free energies of oxidation-reduction reactions with the participation of silicates, oxides, metallic iron and carbon that might occur during mass redistribution in the planet interior during the differentiation of its substance with reagent masses characterized by $n \cdot 10^{26} - 10^{27}$g. In other words, consideration is given to basic processes defining the Earth's inner structure, core and mantle compositions, primary atmosphere/hydrosphere composition that give the greatest possible chemical energy contribution to the planet's overall energy balance.

Gas fugacities and thermal effects have been calculated for the following reducing reactions of ferromagnesian solid solutions of silicates, magnesiowustite, magnetite and stishovite by carbon:

$$2FeSiO_{3(ss)} + C = 2Fe_{(\ell)} + 2SiO_2 + CO_2 \qquad (1)$$

$$2FeSi_{0,5}O_{2(ss)} + C = 2Fe_{(\ell)} + SiO_2 + CO_2 \qquad (2)$$

$$FeSiO_{3(ss)} + C = FeO_{(ss)} + Si_{(\ell)} + CO_2 \qquad (3)$$

$$2FeSi_{0,5}O_{2(ss)} + C = 2FeO_{(ss)} + Si_{(\ell)} + CO_2 \quad (4)$$

$$2FeSi_{0,5}O_{2(ss)} + 2MgSi_{0,5}O_{2(ss)} + C = 2MgSiO_{3(ss)}$$
$$+ 2Fe_{(\ell)} + CO_2 \qquad (5)$$

$$FeSiO_{3(ss)} + C = FeO_{(ss)} + [Si] + CO_2 \qquad (6)$$

$$2FeSi_{0,5}O_{2(ss)} + C = 2FeO_{(ss)} + [Si] + CO_2 \quad (7)$$

$$2FeO_{(ss)} + C = 2Fe_{(\ell)} + CO_2 \qquad (8)$$

$$1/2\ Fe_3O_4 + C = 3/2\ Fe_{(\ell)} + CO_2 \qquad (9)$$

$$SiO_2 + C = Si_{(\ell)} + CO_2 \qquad (10)$$

$$SiO_2 + C = [Si] + CO_2 \qquad (11)$$

where ss denotes a solid solution of pyroxene, γ – spinel and magnesiowustite with a ferruginous molar concentration equal to 0.1; ℓ – liquid [Si] – a silicon liquid solution in liquid iron with $X_{Si} = 0\ 2$; SiO_2 – stishovite, C – diamond.

II. The Thermodynamics and Energetics of Oxidation Reduction Processes in the Undifferentiated Earth

Equations of state

At present there is no experimental method for analyzing complex multicomponent chemical reactions with the participation of volatiles at extreme temperatures and pressures.

However, the advance of experimental techniques permits a study of substance properties to be carried out by investigative methods that for simple systems provide all the thermodynamic information necessary for calculating far more complex systems at the widest range of temperatures and pressures. These include experimental techniques for both low (calorimetry, the solubility method, e.m.f. method, etc.) and high (static and shock-wave methods) pressures. In recent years, these studies have produced a considerable amount of experimental data on the equilibrium constants of many reactions at high temperatures and compressibilities of different phases in the megabar range.

When studying the thermodynamic processes occurring in the Earth's interior, it appears more advantageous to attempt (along with a continued experimental study of simple systems) a generalization of all known experimental data on simple systems by means of chemical thermodynamics techniques and to carry out a thermodynamic analysis of natural processes on this basis. In this lies the advantages of comparing the results of the thermodynamic method with experimental techniques for studying a wide variety of complex polycomponent chemical equilibria at superhigh pressures and demonstrates the essence of the methodological approach used as the basis for studying the physico-chemical evolution of the Earth.

The vast amount of experimental data obtained over the last ten years by means of static and shock-wave compression techniques is essential to physico-chemical investigations of the Earth's interior. Processing of this data obtained directly from experiments has made it possible to assess thermodynamic characteristics, along with errors in their determination, and to use them for **analyzing** chemical reaction features as they change with pressure. This is the principal way of calculating chemical equilibria in the high pressure range.

Experimental data on ultrasonic measurements and static and shock compressions have been used to set up the equations of state in the form of $\int V dP$ as required for calculating in free energy and enthalpy of chemical reactions at superhigh pressures [Kuskov, 1979]. Equations of state have been developed for the following substances: Fe, Ni, Al, Si, C, Ti, Fe-Ni, Fe-Si, FeO, Fe_2O_3, Fe_3O_4, MgO, CaO, SiO_2, Al_2O_3, TiO_2, H_2O, $MgSiO_3$, $FeSiO_3$, $\alpha-F_2SiO_4$, $\gamma-Fe_2SiO_4$, $\alpha-Mg_2SiO_4$, $\gamma-Mg_2SiO_4$, $CaCO_3$, $MgCO_3$ and others.

In order to develop the equations, compressibility data have been taken directly from experimental work. Here, apart from a critical analysis of them, consideration should be given to the advisability of introducing corrections for temperature, the possibility of phase transitions, and and the consistency of static and shock compression data. The CO_2 equation of state has been experimentally derived by Shmulovich and Shmonov [1975] up to 10 kbar and analytically deduced (method of potentials) by Ostrovsky and Rizhenko [1978] up to 1 mbar.

The first calculations of chemical equilibria at superhigh pressures (hundreds and thousands kilobars) using valid (within the error margins of the test) equations of state have been made by us during the study of the core and mantle chemical composition [Kuskov, 1974], and large-scale physico-chemical processes of the Earth's gravitational differentiation [Kuskov, Khitarov, 1977; 1978 a,b; 1979].

The Thermodynamics of the Reduction Processes in the Closed System.

The results of the free energy calculations for several reactions at $P_S = P_{CO_2}$ (in a closed system) are given in Table 1. They show explicitly that the free energies of the reduction of solid solutions of spinel, pyroxene and magnesiowustite, as well as magnetite and stishovite, by carbon are always positive and appreciably above zero at all temperatures and pressures typical of the Earth's mantle and terrestrial planets. This implies that no reduction processes being prohibited by thermodynamics are possible in the Earth's mantle.

This very interesting result should prompt the conclusion that the primary matter of the homogeneous Earth in the early stages of its evolution has been oxidized, while all volatiles are in the associated form of condensed compounds. This conclusion appears however, to contain serious ambiguities, since all reduction processes should be prohibited despite the large-scale differentiation accompanied by the movement of enormous masses of mantle matter and inner heating of the planet. Yet this is impossible because no explanations of the Earth's metallic core, its impurities and powerful degassing during the Earth's early evolution leading to the formation

TABLE 1. The Free Energy Estimates of the Reduction of Silicates and Oxides in the Closed System

$P_S = P_{co}$ kbar	G_T^P, Kcal/mol		
	2000	3000	4000°K
$2FeSi_{0,5}O_{2(ss)} + C = 2Fe_{(\ell)} + SiO_2 + CO_2$			
0,001	5	-14	-34
200	86	101	117
400	111	133	154
$2FeSi_{0,5}O_{2(ss)} + C = 2FeO_{(ss)} + Si_{(\ell)} + CO_2$			
0,001	34	-14	-62
200	151	137	125
400	197	190	1 3
$2FeSi_{0,5}O_{2(ss)} + C = 2FeO_{(ss)} + [Si] + CO_2$			
0,001	13	-35	-83
200	114	100	88
400	149	142	135
$SiO_2 + C = [Si] + CO_2$			
0,001	-8	-65	-121
200	102	77	57
400	140	124	109
600	165	171	155
800	193	201	195
1000	214	249	240

of the primary atmosphere and hydrosphere, may be offered in this case.

The Thermodynamics of the Reduction Processes in the Open System

The geochemical and geophysical applications of thermodynamics involve analysis not only of the parameters of the processes within the system but also for all the possible energy interactions as well as the exchange of the system substance with the environment. These additional requirements on the analysis of outer processes imply that the most general data and ideas about the environment as a material object beyond the system under study are totally inadequate and must be further developed in greater detail.

During the construction of P-T diagrams for closed systems, the total system composition is assumed to be fixed. However, it is impossible to be sure of this for systems containing such volatile components as H_2O, H_2, CO_2, etc. In the large-scale physico-chemical processes occurring during the planet's differentiation, these components appear and disappear with ease. Thermodynamically, this means that the petrological system must be treated as an open one for volatile substances. Hence, the stability field of a mineral containing a volatile component, must be dependent not only on the temperature and pressure, but also the component's chemical potential or fugacity measured in its immediate environment [Thompson, 1955; Korzhnnsky, 1957].

A detailed analysis of the oxidation-reduction processes taking place in the Earth's lower mantle in the early stages of its evolution under the conditions of the open (for volatile substances) system has been carried out by Kuskov and Khitarov [1978a, 1979].

The estimates of carbon dioxide fugacity during the reduction of ferromagnesian solid solutions of silicates are given in Figures 1 and 2 for pressures of 200 kbar (conventional transition zone of the undifferentiatdd Earth - silicate composition) and of 600 kbar (conventional lower mantle of the undifferentiated Earth - oxide composition). In the transition zone[1] the carbon dioxide fugacity has been controlled by the spinel pyroxene-iron-carbon buffer equilibria. The stishovite and silicate reduction in an iron-silicon melt in the presence of metallic iron might be accomplished by the following reactions:

$$FeSiO_{3(ss)} + C = FeO_{(ss)} + [Si] + CO_2$$

$$2FeSi_{0,5}O_{2(ss)} + C = 2FeO_{(ss)} + [Si] + CO_2$$

$$SiO_2 + C = [Si] + CO_2$$

At 200 kbar, the equilibrium temperatures of silicate solid solution reduction are within 2700-3000 K (See Fig. 1). The carbon dioxide fugacity in the lower mantle has been controlled by the magnesiowustite-iron-carbon buffer equilibrium. The equilibrium temperatures of the stishovite-to-iron/silicon melt reduction at 600 kbar are about 3650 K (See Fig. 2). Thus, thermodynamic analysis shows that, if a system is open for volatiles, a silicate/stishovite-to-iron/silicon melt reduction is possible.

The Energetics of Oxidation-Reduction Reactions

Calculations of the thermal effects of mantle oxidation-reduction reactions makes it possible to review the Earth's thermal history from the reduction of considerable silicate and oxide masses approach and to assess the thermal contri-

---

1) Petrological studies show that the contemporary upper mantle is made up of olivine with the ratio of $MgO/MgO + FeO = 0.89 \pm 0.02$. This surprisingly stable composition is indicative of a powerful buffer equilibrium established in the mantle.

bution made by these processes to the overall energy balance of the Earth. Gas fugacities estimated for different reactions exchanging volatiles with the environment allow enthalpies to be calculated within the necessary temperature and pressure range.

The calculation methods and equations connecting gas fugacity with the enthalpy of a reaction in open systems at any P-T-X have been thoroughly discussed by Kuskov and Khitarov [1978a, 1979].

Figure 3 illustrates the relationship between the fugacity of carbon dioxide and and pressure for the following reaction:

$$SiO_2 + C = [Si] + CO_2$$

As may be seen from the Figure, fCO_2 increases with rise in temperature and a 400-450 kbar pressure build up. Higher pressures show a reverse dependence of fCO_2 decreasing with the increase in temperature.

Consequently, gas fugacity stops being a temperature function at a certain solid pressure (referred to as critical P_c). The independence of f_i from temperature at a specific pressure testifies that the oxidation-reduction thermal effect (this also applies, to hydration-dehydration, carbonization-decarbonization and other reactions in open systems) becomes equal to zero. The decrease of f_i with increasing temperature at $P_s > P_c$ indicates that reduction reactions (dehydration, decarbonization) are accompanied by heat release, while oxidation reactions (hydration, carbonization) are accompanied by heat absorption.

The thermal effect estimates made for some oxidation-reduction reactions important for the Earth's thermal history are given in Table 2--the thermal effect of reduction (oxidation) reactions changes its sign at the pressure rises and the process becomes exothermal (endothermal) The greater the pressure applied the more heat is released in reduction (or absorbed in oxidation). Thus, an increasing solid pressure in open (for volatiles) systems causes the thermal effect to vanish.

The data obtained show that, in order to understand the Earth's chemical evolution and thermal history, it is important to define the thermal balance of chemical reactions (including phase transformations), and thus assess the energetics of the chemical process. The data makes it possible to estimate temperature variations along the radius of the undifferentiated planet caused by its inner chemical processes.

To this end, the heat absorbed or released during the stishovite-to-silicon (dissolved in liquid iron) reduction has been calculated according to the core formation model now under development [Kuskov, Khitarov, 1977; 1978a, b; 1979].

Heat estimations of

$$Q = \int \rho(r) \Delta H(r) 4\pi r^2 dr$$

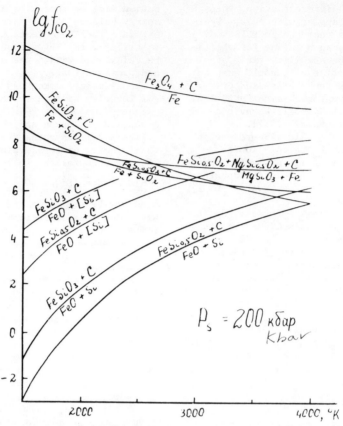

Fig. 1. The CO_2 fugacity in the "transition" zone of the undifferentiated Earth: P_s = 200 kbar.

have been made for an undifferentiated Earth's model with a Birch [1965] density distribution. All the data necessary for the Q estimations have been provided by Kuskov and Khitarov [1978a].

The estimation results are given in Table 3. The first and second columns characterize the thickness of the mantle layer under study and the pressure range. The third column gives estimates of the undifferentiated Earth's layer masses. The fourth column provides mass estimates for stishovite reduced to silicon dissolved in liquid iron. The estimations have been made in the following way. On the basis of earlier studies and geophysical data, it is assumed that the silicon mole fraction in the Earth's outer core is 0.2 or 11.2% Si. Now let us suppose that the entire silicon mass within the outer core has been formed by stishovite reduction. The reduced stishovite mass within the Earth's volume under study will then be $4,278 \cdot 10^{26}$g. Hence, knowing the layer i mass and total mass within the range of 0-5900 km (stishovite is metastable at lower pressures, the mass of the undifferentiated Earth amounts to $50,63 \cdot 10^{26}$g) the reduced SiO_2 mass can easily be estimated within the layer on the assumption that the reduction process is uniform.

The amount of heat absorbed or released by the reaction

$$SiO_2 + C = [Si] + CO_2$$

will be equal to (the fifth column):

$$Q_i = \Delta H_i (cal/g\ SiO_2) \cdot M_i (gSiO_2)$$

The sixth column gives estimates of the temperature variations in layer i (heating or cooling) with stishovite reduced by carbon:

$$\Delta T_i = \frac{Q_i}{M_i C_p}$$

where M_i is the layer mass of the undifferentiated Earth, calculated according to the Birch [1965] ρ-P relationship; C_p is the specific heat (C_p = 0.3 cal/g·grad = $1.25 \cdot 10^7$ erg/g·grad). The estimate reveals that the same reaction might cause some parts of the Earth's mantle to heat up and others to cool down.

It should be noted that stishovite reduction by carbon is only one of the possible reactions that bring about the formation of Fe-Si melt. Other versions are also possible. For example:

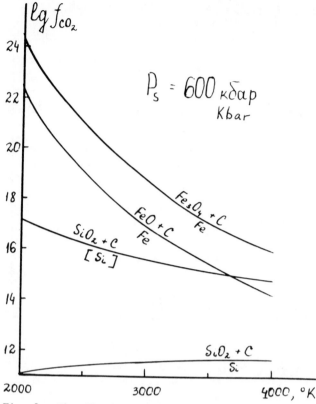

Fig. 2. The CO_2 fugacity in the "lower" mantle of the undifferentiated Earth: $P_s = 600$ kbar.

$$SiO_2 = [Si] + O_2$$

$$SiO_2 + 2H_2(CO) = [Si] + 2H_2O(CO_2)$$

Heat estimates for these reactions and temperature variations are also given in Table 3.

The estimation results show that the choice of the reducing agent, essentially the choice of a model, is of great importance for the total chemical reaction energy balance.

This comes from comparing the ΔT estimates obtained by three different reactions within the same layer. The choice of the reducing agent is not, however, so fundamentally important as the conclusion about the thermal effect sign reversal caused by the increased pressure in the Earth's interior. Now let us estimate the total energy contribution made by the chemical reactions within the range of 200-2140 kbar or within a radius of 0-5900 km.

For the SiO + C = [Si] + CO reaction,

$$\Delta T = \frac{Q_i}{C_p \; \Sigma \; M_i} = -\frac{3.364 \cdot 10^{37}}{1.25 \cdot 50.63 \cdot 1033} = -5300^2$$

---

2) The minus sign indicates an increase in the temperature of the layer, and a plus sign, a decrease in the temperature of the layer.

For the $SiO_2 = [Si] + O_2$ reaction,

$$\Sigma Q_i = + 1.137 \cdot 10^{37} \text{ erg and } \Delta T = +180^\circ$$

For the $SiO_2 + 2H_2 = [Si] + 2H_2O$ reaction,

$$\Sigma Q_i = - 2.466 \cdot 10^{37} \text{ erg and } \Delta T = - 390^\circ$$

This clearly shows that thermal contributions made by the reactions are different: in two cases we have additional heating and in one case, cooling.

The chemical energy effect is essentially reversible: the same reaction may be endothermal under some conditions and exothermal under others. This is one of our study's basic conclusions. The same process might cause some parts of the mantle to heat up and other parts to cool down. The Earth's interior and central part might be heated by reduction processes, while the mantle surface layers have witnessed heat absorption caused by the same processes. The ΔT absolute values characteri-

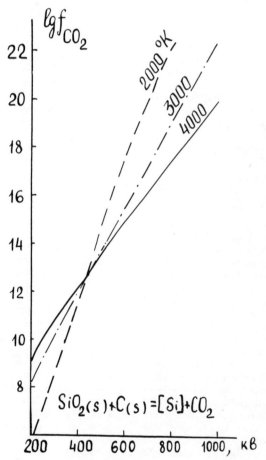

Fig. 3. Dependence of lgf_{CO_2} on temperature and pressure for the reaction $SiO_2 + C = [Si] + CO_2$.

TABLE 2.   The Thermal Effects of the Reduction Reactions
for a Spinel Solid Solution and Stishovite.

P_s, Kbar	H_{2000}^P, Kcal/mol			
	$SiO_2 + C =$ $[Si] + CO_2$	$2 -FeSi_{0.5}O_2(ss)+C=$ $SiO_2+2Fe(1)+CO_2$	$2 -FeSi_{0.5}O_2(ss)$ $+C = 2FeO(ss)+$ $Si(1)+CO_2$	$2 -FeSi_{0.5}O_2(ss)+$ $C = 2FeO(ss)+[Si]$ $+CO_2$
0,001	105.0	44.0	130.0	110.0
200	58.0	−29.0	92.0	55.0
400	6.0	−97.0	48.0	−2.0
600	−44.0			
800	−92.0			
1000	−138.0			
1400	−228.0			
2000	−358.0			

Note: SiO_2- stishovite, C-diamond, ss-solid solution with a ferruginous molar fraction of 0.1, 1-liquid,   [Si]-Fe-Si melt with X_{si} = 0.2.

zing additional heating or cooling (Table 3) disclose a very substantial energy contribution made by the chemical reactions to the planet's overall thermal balance and point to the need for it to be used in calculations of the thermal history of the Earth and other planets. Indeed, chemical energy estimates show that the heating of the Earth's central layer with a radius of 4000 km up to 1000-2000° is due solely to the reaction leading to the formation of the iron-silicon melt, as governed by the reduction of stishovite by carbon.

The stability of carbonates in the Earth's

mantle. The reduction of the Fe-Mg-silicates and stishovite by carbon and the formation of the iron-silicon core will cause the release of carbon dioxide, thus raising the question of whether it will interact with the mantle's silicates and oxides to form carbonates and whether the latter are stable at the Earth's upper and lower mantle pressures. Of great interest is thermodynamic analysis of stability of the crystalline carbonates at the Earth's upper and lower mantle pressures.

The important part of the carbon dioxide in the mantle melting processes has been repeatedly

TABLE 3.   Heat Estimates for Some Reactions in the Undifferentiated Earth.

R-interval Within the Layer, km	P-interval Within the Layer, km	Layer Mass, $M.10^{26}$, g	Reduced Styshovite Mass Within the Layer $M.10^{26}$, g	$SiO_2 + C =$ $[Si] + CO_2$		$SiO_2 = [Si] + O_2$		$SiO_2 + 2H_2 =$ $[Si] + 2H_2O$	
				$Q.10^{37}$ erg	ΔT	$Q.10^{37}$ erg	ΔT	$Q.10^{37}$ erg	ΔT
0–1400	2140–2000	0.81	0.068	−0.176	−1740	−0.070	−690	−0.127	−1250
1400–2560	2000–1700	4.06	0.343	−0.778	−1530	−0.273	−540	−0.566	−1120
2560–3360	1700–1400	5.88	0.497	−0.902	−1240	−0.226	−310	−0.644	− 880
3360–3800	1400–1200	4.53	0.383	−0.548	− 970	−0.064	−110	−0.395	− 700
3800–4250	1200–1000	5.70	0.482	−0.538	− 750	+0.032	+ 40	−0.383	− 540
4250–4650	1000– 800	5.96	0.504	−0.400	− 540	+0.152	+200	−0.276	− 370
4650–5060	800– 600	6.98	0.590	−0.276	− 320	+0.309	+350	−0.173	− 200
5060–5480	600– 400	7.99	0.675	+0.089	+ 90	+0.522	+520	−0.042	− 40
5480–5900	400– 200	8.72	0.737	+0.165	+ 150	−0.755	+690	+0.140	+ 130

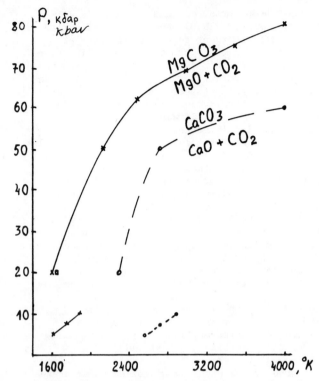

Fig. 4. The monovariant curves representing the upper stability limits of carbonates:

$MgCO_3 = MgO + CO_2$ - solid line
$CaCO_3 = CaO + CO_2$ - dotted line
      - experimental point for the magnesite dissociation reaction obtained from the Irving and Wyllie (1973) data;
X - P-T equilibrium parameters for reaction (13);
O - P-T equilibrium parameters for reaction (14).

discussed [Khitarov and Kadik, 1973; Kadik, 1975; Eggler, 1978; Wyllie, 1978; Brey and Green, 1976]. Recent years have seen new experimental work on crystalling carbonate stability at pressures up to 30 kbar (Irving and Wyllie, 1973; Huang and Wyllie, 1976; Newton and Sharp, 1975; Eggler, et al., 1976).

Let us consider a typical reaction of the carbon dioxide interaction with the mantle's silicate, leading to the formation of carbonate:

$$Mg_2SiO_4(c) + CO_2 = MgSiO_3(c) + MgCO_3(c) \quad (12)$$

where Mg_2SiO_4 - olivine at low pressures and γ - spinel at high pressures. The reactions that determine the upper carbonate stability limit in the mantle are as follows:

$$MgCO_3 = MgO + CO_2 \quad (13)$$

$$CaCO_3 = CaO + CO_2 \quad (14)$$

Free energy estimates have been made for the closed system at pressures $P_s = P_{CO_2} = P_{total}$ up to 1000 kbar and temperatures up to 4000 K.

The estimation results are given in the form of the P-T diagrams in Figures 4 and 5 [Kuskov, 1978]. Free energy estimates of reactions (12-14) within 5-10 kbar have been made according to the CO_2 equation of state [Shmulovich, and Shmonov, 1975] and within 20-1000 kbar according to the CO_2 equation of state [Ostrovsky and Rizhenko, 1978].

Figure 4 shows the upper limits of carbonate stability in the Earth's mantle (carbonate melting is not considered here because the melting temperature-pressure relationship is not known for a high P). The Irving and Wyllie [1973] experimental point for reaction (13) is in good agreement with the theory at 20 kbar, but does not correlate well at pressures up to 10 kbar, where the monovariant curves of reactions (13) and (14) should be displaced toward higher temperatures.

The P-T diagram reveals the thermodynamic stability of pure carbonates at pressures above 20-50 kbar; at higher pressures, the carbonate stability fields are considerably extended in terms of temperature. As the calculations show, $\Delta G_T^P \gg 0$ at pressures greater than 80 kbar for reaction (13) and greater than 60 kbar for reaction (14) for all temperatures and pressures representative of the Earth's mantle. Thus, pure carbonates $CaCO_3$, $MgCO_3$ (and their solid solutions) are thermodynamically stable in the mantle.

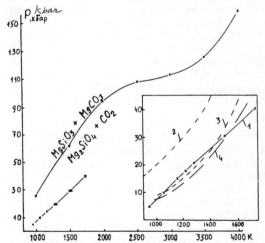

Fig. 5. The monovariant curve for the reaction $Mg_2SiO_4 + CO_2 = MgSiO_3 + MgCO_3$.

Symbols:

X - P-T equilibrium parameters (calculated);
O - Newton and Sharp (1975) experimental data;
      - Eggler, et al. (1976) experimental data.
Insertion: 1-reaction (12); geotherms: 2-continental: 3-oceanic (low temperature): 4-oceanic (high temperature).

Figure 5 Shows the monovariant curve of reaction (12). As seen from the figure, the carbonate stability fields become narrower in terms of temperature.

The experimental data obtained by Newton and Sharp [1975] and Eggler et al. [1976] check well with the P-T curve calculations made by extrapolating the CO_2 equation of state [Shmulovich and Shmonov, 1975], to 20-30 kbar. According to this data, the monovariant curve (12) should also have a temperature shift to the right at high pressures, calculated according to the CO_2 equation of state [Ostrovsky and Rizhenko, 1978].

Figure 5 also shows the oceanic and continental geotherms calculated according to the data obtained by Mercier and Carter, [1975]. The geotherms determine the stability of the divariant fields of the monovariant curve (12) defining the upper limit (in pressure) for the existence of free CO_2 in the mantle. As can be seen, the peridotite-carbonate association is stable in the subcontinental area throughout the P-T range. In the suboceanic area, the mantle's carbonates may be unstable at $P \lesssim 30$ kbar and dissociate to release free CO_2.

At higher pressures, the free energy of reaction (12) is appreciably below zero throughout the P-T range of the mantle, i.e. carbon dioxide interacts with the peridotites of the mantle to form stable carbonates.

The carbonate stability in the Earth's mantle is of great interest not only for geochemistry, but for geophysics as well, so carbonate thermodynamic characteristics at high pressures should be considered when building modern models of the Earth. It is the presence of carbonates that may be responsible for the inconsistency observed between the wave velocities in the mantle and those expected for the mixture of MgO, FeO, SiO_2 of the olivine or pyroxene composition [Davies and Dziewonski, 1975].

III. The Earth's Core

Studies of the physical properties of the Earth's core by Birch [1952] indicated that its density was 10%-20% lower than that of an iron/nickel alloy, while its seismic velocity was higher than that of the alloy under comparable P-T conditions. These conclusions have been supported by investigations of the density and seismic velocity in metals at extreme pressures in shock-wave experiments. There are two basic hypotheses to explain the core composition: an iron/nickel core with light element impurities and a core consisting of metallized silicates or oxides.

The Thermodynamics of oxide metallization under pressure and the metallized core hypothesis

The iron/nickel core hypothesis as a whole calls for different total compositions of earth-type planets. To get around this difficulty, Ramsey [1948] developed an alternative hypothesis according to which silicates at high pressures transform into a metallic state to form the Earth's core. Recently, the hypothesis has been further supported by superhigh pressure experiments carried out under both static and dynamic conditions [Vereshchagin, et.al., 1974, 1977; Simakov, et.al., 1973; Kawai, et.al., 1974a, b].

The works of Vereshchagin, et al. [1974, 1977] and Kawai, et al. [1974a, b] investigated oxide behavior at high static pressures to find a sharp decrease in resistance as the load increases. The resistance jump is considered to be a result of oxide metallization. The conduction state is observed only under pressure. With the pressure removed, the resistance does not return to its initial value, it drops even lower. This may be explained by oxide decomposition (reduction) irreversibility under pressure.

The theoretical method suggests that the resistance jump observed in high pressure tests is caused by the reduction of the oxide to metal and oxygen gas rather than its metallization [Kuskov and Khitarov, 1978b, c]. Let us consider the metallization test conditions in thermodynamic terms.

In the system under study (oxide under pressure), the pressure applied to solid phases and as fugacity (pressure) are independent thermodynamic parameters, so $P_S \neq P_{O_2}$. What's more $P_S > P_{O_2}$ because the gas is under its own pressure. The system is considered open with respect to oxygen--i.e., the latter may exchange with the oxygen of the environment. The reduction of oxide to metal and oxygen gas (and the corresponding resistance jump observed in metallization tests) takes place at a constant temperature and P_S under which the oxygen fugacity exceeds the partial pressure of the oxygen in the environment. There is only one case in which the critical pressure (P_c) of the reduction (decomposition) is not temperature dependent (See Fig. 6). At the intersection of the curves, the enthalpy of the reaction is equal to zero when $P_S = P_c$ (the sign of the thermal effect is reversed).

Consequently, zero thermal effect is indicative of a critical pressure at which the reduction of oxide is due to the fact that the oxygen fugacity in the system is greater than the oxygen partial pressure in the air at $P_S = P_c$.

To demonstrate the dielectric-metal transition in metallization tests [Vereshchagin, et al., 1974], SiO_2 (under pressure and in conduction state) was heated up. It was noted that the resistance increased from 10^2 ohm to the initial value of $10^8 - 10^9$ ohm. The authors believe that the return of the oxide's resistance to its initial value indicates either the inability of SiO_2 to decompose under pressure or a negligable decomposition.

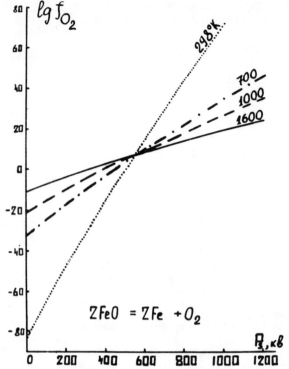

Fig. 6. Dependence of $\lg f_{O_2}$ on temperature and pressure for the reaction $2Fe_{(c)} + O_{2(g)}$.

There is another possible explanation. At pressures $P_s > P_c$, the SiO_2 (and other oxides) decomposition becomes exothermal, and according to the Le-Shatelier principle and our estimate, the reduction is facilitated by a temperature decrease rather than increase. The temperature increase takes us from the silicon (metal) stability field into the oxide (dielectric) stability field and, consequently, the resistance should, indeed, rise. Figure 7 illustrates the explanation with the estimate of $\lg f_{O_2}$ for the reduction of stishovite: $SiO_{2(s)} = Si_{(s)} + O_{2(g)}$ for the two isobars: $P_s < P_c$ and $P_s > P_c$ with the SiO_2 and Si stability fields indicated.

Thus, the experimentally observed fact (an increase of the resistance of the oxide during heating) cannot be used to support the idea of oxide metallization at lower temperatures. Moreover, the data confirm the possibility of decomposition of the oxides under pressure and of the reversal of the enthalpy sign of the decomposition process as a result of the transition ($Si_{(metal)} \rightarrow Si_{(dielectric)}$) at rising temperatures.

Analysis of the possibility of oxide reduction to metals has been made based on experimental data on compressibility for both open and closed systems. As was shown, the reduction of oxides in the closed system with a total pressure of 2.0 mbar is thermodynamically impossible.

The estimates obtained for the open system show the following sequence of oxide reduction under pressure:

$$FeO < TiO_2 < SiO_2 < Al_2O_3 < MgO$$

The reduction pressures are as follows: 550, 1450, 1700, 2300, and 2600 kbar.

This sequence may be compared with that obtained for oxides converting into a conducting state under pressure: $Al_2O_3 < SiO_2 < MgO$ [Vereshchagin, et al., 1974, 1977] and $FeO < TiO_2 < SiO_2 < MgO$ [Kawai, et al., 1974a, b].

The comparison prompts the conclusion that the succession noted for the oxide reduction pressures is practically the same as that for the dielectric-metal transition pressures. All of the above suggests that the resistance jump in high pressure tests does not characterize the oxide metallization, but is likely to relate to the reduction of oxides to metal and oxygen gas [Kuskov and Khitarov, 1978c].

It is obvious from the thermodynamic analysis that metallization experiments cannot support the assumption that the Earth has a metallized core. This has been recently confirmed by the Cupta and Ruoff [1979] experimental data on the Al_2O_3 compression. Their measurements of electrical resistance show that Al_2O_3 continues to be dielectric up to a pressure of 1.2 Mbar.

Silicon in the composition of the Earth's core. The hypothesis that the Earth's outer core is made up of an iron/nickel alloy with appreciable amounts of light elements has now gained wide support. Si, S, C, and O may be considered as possible candidates. The silicon entry into the core composition will here receive a rigorous thermodynamic treatment. The assumption regarding the Fe-Si core was made in the works

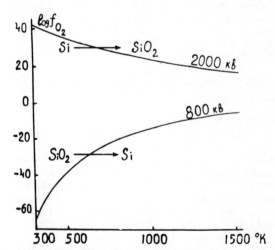

Fig. 7. The O fugacity for the reaction, $SiO_{2(c)} = Si_{(c)} + O_{2(g)}$: 1-isobar $P_s = 800$ kbar; 2-isobar $P_s = 2000$ kbar. The arrow shows stability fields, SiO_2 - dielectric, Si - metal.

of Macdonald and Knopoff, [1958]; Ringwood, [1959].

The free energy of the dissolution of silicon in liquid iron

$$Si_{(\ell)} = [Si] \qquad (15)$$

where [Si] denotes the liquid solution of silicon in liquid iron with X = 0.2 (11.2% Si by weight), can be estimated according to

$$\Delta \overline{G}_T^P = \Delta \overline{G}_T^O + \int_1^P \Delta \overline{V} dP$$

where $\qquad \Delta \overline{V} = \overline{V}_{[Si]} - V_{Si}$

The silicate activity in the Fe-Si alloy and the value of $\Delta G_T^O = \overline{G}_i - G_i^O = RT \ln a_i$ were estimated according to the equations of the quasi-chemical theory of solutions. Estimates of the free energy of the dissolution of silicon at high pressures were made considering the data on the compressibility of metals and Fe-Si alloys [Kuskov, 1974; Kuskov and Khitarov, 1978a]. The estimates of $\Delta G_T \rho$ for reaction (15) are given in Table 4.

The estimates show that the values of $\Delta G_T \rho$ are negative for the dissolution of silicon in liquid iron throughout the P-T range. This attests to the stability of the Fe-Si alloy under the P-T conditions of the Earth's outer core and of the cores of terrestrial planets [Kuskov and Khitarov, 1978b].

The Fe-Si alloy stability in the P-T range of the Earth's core is a necessary, but not a sufficient condition, because the process by which the silicon entered the core's composition is unknown, It is assumed (as shown above), that the silicon might have entered the composition of the core as a result of the reduction of ferromagnesian silicates and stishovite during the Earth's differentiation when an exchange of volatiles between the system and surrounding medium is apparently possible.

A physico-chemical model of the formation of the Earth's core.

So far, the formation of the core of the Earth and terrestrial planets has been conside-red in terms of geophysical and cosmogenic con-cepts, with none of the existing models or ap-proaches providing a satisfactory core formation theory. One reason is that no real considera-tion has been given to the chemical aspects and role and distribution of volatiles in the planet shells. This approach leads to contra-dictions that cannot be resolved when tackling the problem of the core composition and for-mation mechanism. Heterogenous accretion hypotheses call for a fundamental revision of the existing theories of planet formation and evolution and do not always consider geochemical

TABLE 4. Free Energy for the $Si_{(\ell)}$ = [Si] Reaction at High Temperatures and Pressures

P, Kbar	G_T^P, Kcal/mol		
	2000	3000	4000°K
0,001	−21.0	−25.3	−25.6
300	−43.3	−45.6	−47.9
500	−51.4	−53.7	−56.0
1000	−68.0	−70.3	−72.6
1400	−85.8	−86.1	−88.4
1700	−97.5	−99.8	−102.1
2000	−113.0	−115.3	−117.6
2200	−125.6	−126.0	−128.2
2500	−139.9	−142.2	−144.5
3000	−168.9	−171.2	−173.5

or geological proof of the general planetary differentiation taking place in the Earth's postaccumulation development period. All the models, as pointed out in the review made by Vityazev, et al. [1977], lack a specific definition of the physico-chemical and dynamic aspects of the differentiation. The hypotheses concerning the core formation mechanism and composition are not based on any quantitative estimate, but confined to the most general logical speculations. This approach does not appear promising, not only because it does not draw on the available data but also because many patterns observed at moderate or even high pressures change somewhat at superhigh pressures. The concept of the Earth's core formation presented in this paper may be considered as a synthetic one. First, it has the advantages of the models developed from the geophysical standpoint. Second, it considers physico-chemical aspects of the problem for the first time, these having been partially covered in the preceding sections and described in more detail by Kuskov and Khitarov [1977; 1978a, b; 1979].

To avoid duplication, let us formulate the radically new ideas that have been used as a basis for the model. These are:

1. The possibility of the dissolution of silicon (and carbon) in liquid iron under the P-T conditions of planet shells has been strong-ly suggested by thermodynamics. This condition is necessary, however, but not sufficient, even though experimental data on the shock compres-sion of iron/nickel alloys have revealed that the silicon share in the core composition (up to about 20% by weight) complies with geophysical data on the properties of the Earth's outer core.

2. Apart from proving thermodynamic stability there has been a need to assess the possibility of forming an iron-silicon alloy. This has been done by investigations of oxidation-reduction conditions in the Earth's mantle.

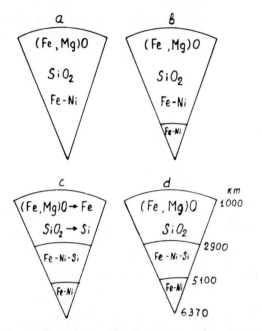

Fig. 8. Schematic model of the Earth's core formation: a - undifferentiated planet; b - inner core formation; c - outer core formation; d - differentiated planet.
Periods: a - pregratitational; b,c - gravitational, d - postgravitational.

3. A thermodynamic study of the heterogenous chemical reactions at superhigh pressures in open systems disclosed the exothermal nature of the reduction processes, which suggests the existence of primary chemical energy sources in the Earth's interior.

4. These physico-chemical conditions bring about the most important dynamic aspect of core formation. The Elsasser [1963] and Safronov [1969, 1975] concepts assumed that iron melting originated in the upper mantle and went down through the extremely viscous low mantle. The modernization of this concept, being extremely substantial and radical, states that the same melting taking place on the lower horizons of the undifferentiated Earth as a result of the exothermicity of the reduction processes and enormous heat release. Thus, the "drops" of the melt were formed throughout the mantle, the density decrease owing to a local melting with core formation thus being greatly facilitated and accelerated.

This brings us to the very interesting conclusion that a differentiation source in the lower mantle might have existed along with that in the upper mantle (at depths of about 500 km). Then, within the homogeneous accumulation model, core formation may be considered as a three-stage process (See Fig. 8). The first stage saw the formation of the undifferentiated planet caused by the condensation and accumu-

lation of the protoplanetary cloud; the planet temperature averaged $1500^{\circ}K$ in the pregravitational period. The next stage in the planet's evolution brought about the gravitational differentiation of matter. Fe-Ni alloy melting occurs in the upper mantle as the planet heats[3] up. This is the result of radioactive decay.

Sinking to the lower mantle's center, as assumed by the Safronov [1969, 1975] and Elsasser [1963] concepts would hardly, however, have been possible because of the high viscosity of the low mantle (drops and melting blocks were to cover a distance of about 6000 km).

From the assumption made concerning the second differentiation source, it is obvious that the heat-producing reduction reactions (like the reduction of stishovite by carbon and other reactions) result in the nickel/iron melting in the central zone that was responsible for the formation of the Earth's inner core being later solidified (See Fig. 8b). The inner core is commonly accepted to be an iron/nickel alloy, but this is unlikely to be true for such a multicomponent system as the Earth. The alloy is more likely to contain light element impurities (Si, C, etc.) but in small concentrations.

The next stage in the planet's evolution may be defined in the following way. The upper mantle differentiation, nickel/iron separation and inner core growth, as well as further development of the reduction processes, brought about a temperature increase throughout the mantle as a result of gravitational, radiogenic and chemical energy release and mantle viscosity decrease. The Fe-Ni-Si melt, unstable in the planet gravitational field, sinks to the center to form the Earth's outer core (See Fig. 8c). This process, spontaneous and accelerating, probably took place all over the Earth. It may be visualized in the following way. The colder and denser rocks of the upper mantle sank down into the high geothermal gradient zone, where reduction reactions accompanied by heat release began. The mantle's density decreased to push the matter upward. Thus, gravitational convection was accelerated by thermal convection. The overall thermogravitational convection probably involved the entire volume of the Earth and was responsible for the Earth's differentiation.

If the formation of the Earth's core took place during the first or second billion years of its evolution, then the core was formed by nickel/iron already in a metallic state and by an Fe-Ni-Si alloy produced by the reduction processes

---

3) Massive body collisions result in thermal heterogeneties during the Earth's accretion. The collision areas are hundreds of degrees hotter than the surrounding medium (Safronov, 1975), i.e. at this moment, thermally "weak" zones from in which iron melting could originate.

during that period. The core composition was represented by certain other ingredients: carbon oxygen, and sulphur (possibly), but in lesser quantities This process completes the global gravitational differentiation[4] to mark the next stage in the planet's development--the postgravitational period (See Fig. 8d).

Conclusions

On the basis of the data obtained, we attempt to present a concept of the Earth's early history. Many details of the Earth's inner development in the early stages of its evolution remain obscure so these speculations constitute only a hypothesis calling for further investigation. The possibility has been considered of reducing silicates and oxides by carbon under conditions of closed and open systems with the P-T parameters of the Earth's mantle. It has been demonstrated that reduction processes are thermodynamically impossible in the closed system. The reduction of silicates and stishovite to an iron silicon melt with the simultaneous release of carbon dioxide, takes place in the open system. The reduction of stishovite is accompanied by a heat release to increase temperature in the central zones which might lead to a softening or even partial melting (considering radioactive decay and gravitational differentiation energies) of this region.

It is assumed that, during the Earth's differention, the petrogenic system was likely to be open for volatiles. The core formation was accompanied by large-scale hydrodynamic movements and chemical reactions, with the simultaneous release of gravitational and chemical energies. The oxidation-reduction processes under study was displayed by the reduction of silicates and oxides, silicon entering the core composition with the simultaneous release of carbon dioxide, and its subsequent partial association with the mantle substance to form carbonates.

References

Beck, A.E. Energy requirements of an expanding Earth, J. Geophys. Res., 66, 1485-1490, 1961.

Birch, F. Elasticity and constitution of the Earth's interior, J. Geophys. Res., 57, 227-286, 1952.

Birch, F. Energetics of core formation, J. Geophys. Res., 70, 6217-6221, 1965.

Brey, G.P. and D.H. Green. Solubility of CO_2 in olivine melilitite at high pressure and role of CO_2 in the Earth's upper mantle, Contr. Mineral. Petrol., 55, 217-230, 1976.

---

4) The possibility cannot be ruled out completely that the lower mantle is still an undifferentiated substance, meaning that the core growth is not yet complete.

Davies, G.F. and A.M. Dziewonski. Homogeneity and constitution of the Earth's lower mantle and outer core, Phys. Earth Planet. Interiors, 10, 336-343, 1975.

Eggler, D.H., I. Kushiro, and J.R. Holloway. Stability of carbonate minerals in a hydrous mantle, Carnegie Inst. Washington Yearbook, 75, 631-636, 1976.

Eggler, D.H. The effect of CO_2 upon partial melting of peridotite in the system Na_2O-CaO-Al_2O_3-MgO-SiO_2-CO_2 to 35 kb with an analysis of melting in a peridotite-H_2O-CO_2-system, Am. J. Sci., 278, 305-343, 1978.

Elsasser, W.M. Early history of the Earth, in Earth Science and Meteorites, Amsterdam, 1963.

Gupta, M.C. and A.L. Ruoff. Static compression of Al_2O_3 to 1,2 Mbars, J. Appl. Phys., 50, 827-828, 1979.

Huang, W.L. and P.J. Wyllie. Melting relationships in the system CaO-CO_2 and MgO-CO_2 to 33 kilobars, Geochim. Cosmochim. Acta, 40, 129-132, 1976.

Irving, A.J. and P.J. Wyllie. Melting relationships in CaO-CO_2 and MgO-CO_2 to 36 kilobars with comments on CO_2 in the mantle, Earth, Planet. Sci. Lett., 20, 220-225, 1973.

Kadik, A.A. The influence of degassing of basic magmas on the regime of water and carbon dioxide in the crust and upper mantle in Geodynamic Investigations, No. 3, 67-86, Nauka, Moscow, 1975 (in Russian).

Kawai, N. and A. Nishiyama. Conductive SiO_2 under high pressure, Proc. Jap. Acad., 50, 72-75, 1974.

Kawai, N. and A. Nishiyama. Conductive MgO under high pressure, Proc. Jap. Acad., 50, 634-635, 1974.

Khitarov, N.I. and A.A. Kadik. Water and carbon dioxide in magmatic melts peculiarities of the melting process, Contrib. Mineral, Petrol., 41, 205-215, 1973.

Korzhinsky, D.S. Physicochemical Basis of the Analysis of the Paragenesis of Minerals. Moscow, 1957 (in Russian).

Kuskov, O.L. Chemical composition of the cores of the terrestrial planets, 1. Thermodynamics of the processes of nickel and silicon dissolution in liquid iron. Geochemistry, No. 12, 1809-1824, 1974 (in Russian).

Kuskov, O.L. Stability of carbonates in the Earth's mantle. Geochemistry, No. 12, 1813-1820, 1978 (in Russian).

Kuskov, O.L. Equations of state for some substances under high pressures, Geochemistry, No. 7, 963-983, 1979 (in Russian).

Kuskov, O.L. and N.I. Khitarov. The chemical composition of the core of the terrestrial planets and the moon. Presented at the Soviet-American Conference on Cosmochemistry of the Moon and Planets. NASA, Washington D.C., 1977, 231-242.

Kuskov, O.L. and N.I. Khitarov. Oxidation-reduction conditions and thermal effects of chemical reactions in the undifferentiated

Earth. Geochemistry, No. 4, 467-494, 1978a (in Russian).

Kuskov, O.L. and N.I. Khitarov. Thermodynamic description of metallization of oxides under pressure. JETP Lett., 27, 269-273, 1978c.

Kuskov, O.L. and N.I. Khitarov. Physiochemical model of the Earth's core, in high-pressure science and technology. Sixth AIRAPT Conf. v.2, 245-254, 1978b.

Kuskov, O.L. and N.I. Khitarov. Thermal effects of chemical reactions in the undifferentiated Earth. Phys. Earth Planet. Inter., 18, 20-26, 1979

Levin, B.Y., S.V. Majeva, and V.S. Safronov. Thermal history of the Earth and terrestrial planets, in Energetics of Geological and Geophysical Processes. Nauka, Moscow, 1972, 38-51 (in Russian).

Lubimova, E.A. Thermals of the Earth and Moon. Nauka, Moscow, 1968 (in Russian).

Lubimova, E.A. The development of geothermal models. Izv. Acad. Nauk U.S.S.R., Fizika Zemli, No. 1, 40-52, 1977 (in Russian).

Lyustikh, E.N. On the use of O.Y. Schmidt's Theory in geotectonics, Doklady Acad. Nauk U.S.S.R., 59, 1417-1419, 1948 (in Russian).

MacDonald, G.J.F. and L. Knopoff. The chemical composition of the outer core, Geophys. J., 1, 284-297, 1958.

Mercier, J.C. and N.L. Carter. Pyroxene geotherms, J. Geophys. Res., 80, 3349-3362, 1975.

Newton, R.C. and W.E. Sharp. Stability of forsterite $+CO_2$ in the mantle, Earth Planet. Sci. Lett., 26, 239-244, 1975.

Ostrovsky, I.A. and B.N. Rizhenko. Fugacities of gases and some mineraligical reactions at very high pressures and temperatures, Contrib. Mineral. Petrol., 2, 297-303, 1978.

Ramsey, W.H. On the constitution of the terrestrial planets, Mon. Not. Roy. Astron. Soc., 108, 406-413, 1948.

Ringwood, A.E. On the chemical evolution and densities of the planets, Geochim. Cosmochim. Acta, 15, 257-282, 1959.

Safronov, V.S. Evolution of the Protoplanetary Cloud and Formation of the Earth and Planets. Nauka, Moscow, 1969 (in Russian).

Safronov, V.S. Time scale for the formation of the Earth and Planets and its role in their geochemical evolution. Presented at the Soviet-American Conf. on Cosmochem. of the Moon and Planets, NASA, Wash. D.C., 1977, 797-803.

Schmidt, O.Y. A Theory of the Origin of the Earth: Four Lectures. Moscow: Foreign Languages Publ. House, 1958, London: Lawrence and Wishart, 1959.

Schmulovich, K.I. and V.M. Shmonov. Fugacity coefficients (fugacities) of CO_2 from 1.032 to 10000 bars and from 450° to 1300°K, Geochemistry, No. 4, 551-555, 1975 (in Russian).

Simakov, G.V., M.A. Poduretz, and R.F. Trunin. New data on compressibility of oxides and fluorides and a hypothesis about homogeneous composition of the Earth, Doklady Acad. Nauk U.S.S.R., 211, 1330-1332, 1973 (in Russian).

Thompson, J.B. The Thermodynamic basis for the mineral facies concept, Am. J. Sci., 253, (2), 1955.

Urey, H.C. The Planets. New Haven, Conn; Yale Univ. Press, 1952, 245.

Vereshagin, L.F, E.N. Yakovlev, B.V. Vinogradov, V.P. Sakun, and G.N. Stepanov. Transition of SiO_2 into conduction state. Pis'ma Zh. Eksp. Teor. Fiz., 20, 472-474, 1974 (in Russian).

Vereshagin, L.F., E.N. Yakovlev, Y.A. Timofeev, and B.V. Vinogradov. Dielectric-metal transitions in megabar range, in Metal-dielectric phase transition, Moscow-L'vov, 3-5, 1977 (in Russian).

Vityazev, A.V., E.N. Lyustikh, and V.V. Nikolaichik. The problem of the formation of the Earth's core and mantle. Izv. Acad. Nauk U.S.S.R., Fizika Zemli, 3-14, 1977 (in Russian).

Wyllie, P.J. Mantle fluid compositions buffered in peridotite-CO_2-H_2O by carbonates, amphibole and phlogopite. J. Geol., 86, 687-713, 1978.

A TWO-LAYER CONVECTIVE MANTLE WITH AN INTERNAL BOUNDARY LAYER

R. St J. Lambert

Department of Geology, University of Alberta, Edmonton, Alberta, Canada T6G 2E3

Abstract. A broad array of assumptions about the history of the Earth and some speculative reasoning about causes of its present state are combined in a model which calls for an Earth with a two-layer mantle. It is argued that the following combination of circumstances has occurred: heterogeneous accretion with contemporaneous core formation; development of a chemically-layered mantle segregated at 700 km depth into upper and lower mantles; a heat flow maximum around 3.0 to 2.8 Ga ago accompanied by rapid continental growth; convection in the lower mantle beginning around 2 Ga producing major disturbances of the upper mantle and continents; and continued fractionation of the lower mantle leading to a chemically heterogeneous upper mantle as material from the lower mantle gradually rises into the upper mantle. It is proposed that the region between the two parts of the mantle has some of the properties of a lithosphere. It is speculated that there may be circumstances appropriate to the production of a lower mantle asthenosphere, but the necessary combination of thermal gradients is unlikely.

Introduction

The problem of the present structure of the Earth is normally tackled by geophysical observation and modelling. Yet another class of geophysical models considers the whole history of the Earth, usually from analysis of probable thermal histories. A different approach has been used by some geochemists or petrologists, who attempt to account for present-day element distribution and patterns of Sr, Pb and Nd isotopes by invoking various present and/or past mantle models, with or without input from lunar analogy. In previous papers (1980, 1981) I have attempted development of an Earth model which--after reviewing various alternatives--merges a two-layer convecting mantle with a relatively cool initial Earth state and a thermal (heat-flow) maximum to give an episodic and unidirectional evolution of the mantle, lithosphere and crust. This model is meant to be comprehensive, but as much of the argument has been derived from consideration of our imperfect knowledge of the fragmentary geo-

logical record, it is necessarily speculative and very difficult to test. In this paper, the nature of the present Earth is examined in the light of the model, which was originally developed for the Archean, and some tests mentioned.

Review of the Model

Three principal lines of argument were used to develop the model:
(a) that the Sr, Pb and Nd isotopic evidence requires progressive but episodic continental growth, concentrated largely into the Archean;
(b) that the same evidence, plus variation of element abundances in basalts and related rocks requires that the mantle is heterogeneous, and
(c) that high surface heat-flow is an expression of the proximity of partially melted rock (magma) close to the surface and that the converse has always held true.

These arguments were rehearsed in some detail in the previous papers. It was argued that a hot spot model for the Archean, with continental growth accelerating towards the close of that era, provided suitable mechanisms for all known Archean geological phenomena (1981, Fig. 1). The Archean Earth had a negligibly thin lithosphere, but had a 50 km (or more) basalt composition crust, largely metamorphosed under conditions of moderate to high geothermal gradient. This crust was formed and continuously re-worked by several hundred plumes, which formed hot spots. Fragments of this crust are now preserved as greenstone belts. As continents grew and K, U and Th were transferred out of the upper mantle, a lithosphere began to develop by conductive cooling of the primitive continents. As this happened, it became impossible for mantle diapirs (at hot spots) to penetrate the continental lithosphere (at least 45 km thick), because no significant partial melting will occur when a diapir's ascent is physically stopped at >45 km depth (Fig. 1). There is therefore a critical cooling level at which a continent stabilizes, and it was first reached at some points on the Earth's surface at about 3000 Ma ago, perhaps 3400 Ma ago in the Pilbara area. The whole Earth could have been warming at that stage (1980, Fig. 1; Fig. 2 of

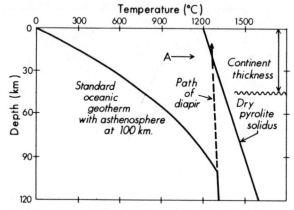

Fig. 1. Highly simplified evolutionary path for a mantle diapir. Partial melting will only begin at shallow depth: if the diapir is stopped below a barrier, for example, the continental crust at 45 km, no melting will occur.

this paper): all that is required is that enough sial had accumulated in one region, which was sufficiently large to maintain its identity and cool as one unit.

It was argued that a mean surface heat flux much in excess of 125 mW/m² would result in extraordinarily rapid continental growth, a feature of the Earth unknown before about 3200 Ma ago, unless recycling of the sialic component can occur. Such a phenomenon is unknown today and is theoretically extremely unlikely in a regime of higher geothermal gradients. The surface heat flux pattern must perforce have, at the most, modest values at any time. The only models published to date which fit the requirement are the cool-initial-state models of McKenzie and Weiss (1975) for a two-layer convective mantle. In such a mantle, convection is constrained by a barrier at the conventional depth of 700 km; convection begins in the upper mantle first and later in the lower mantle.

At first, it seemed most likely to me that the thermal maxima in such a model would both be in the Archean, but the global revolution (evidenced by wide-spread re-setting of argon clocks) known as the Hudsonian orogeny (Stockwell, 1964) does not have a cause in such a model. Solely for this reason (as it is impossible to review the evidence here in detail), an alternative thermal history is sketched in Fig. 2. The model is constrained by the present-day heat flow of 80 mW/m² (Williams and von Herzen, 1975) and thermal maxima of about 160 mW/m² at 3000 Ma and 140 mW/m² at 1800 Ma ago. The brief initial period of high heat flow of Fig. 2 reflects the initial temperature of the outer layers in pre-upper mantle convection time and is irrelevant to continental growth. The relationship of this pattern to theories of accretion of the Earth was discussed by me earlier; it is evident that this model best

relates to cool, heterogeneous accretion with core formation during growth (Smith, 1979).

Isotopes

At the risk of over-simplifying a very complex story, a brief review of the evidence from Sr, Pb and Nd isotopes is presented, because the model can be tested by further studies of these elements. Sr isotopes have been used to investigate mantle heterogeneity and apparent age, and have been held to indicate that most of the contintental crust evolved in the late Archean. Brooks, James and Hart (1976) showed that Sr isotope systematics from volcanics of a wide age range contained [87]Sr above that permitted by their Rb contents and age; they suggested that the anomalies reflected the age of the subcontinental lithosphere. Contamination by continental crustal material cannot be ruled out in the cases they discuss. However, oceanic islands (a.g. Galapagos; Hedge, 1978) and the Mid-Atlantic Ridge and Iceland show either further examples of "mantle isochrons" or Sr-isotope variations which cannot be explained except by mantle heterogeneity (Hofmann and Hart, 1978). Pb isotopes show similar anomalous patterns most easily interpreted by mixing (see review by Tatsumoto, 1978). The equally complex problem of crustal evolution has been tackled by Moorbath (1975, 1976, 1977) who concludes that the Archean continental crust grew continuously and has not had a long history at the surface of the Earth in anything like its present form. O'Nions and Pankhurst (1978) reached a comparable conclusion and McCulloch and Wasserburg (1976) used Nd isotopes to conclude that the period 2.5 to 2.7 Ga ago was a period of major continental crustal growth. Nd isotopes also show that the transfer of material from the mantle to the continental crust is a continuous process and that re-working is a

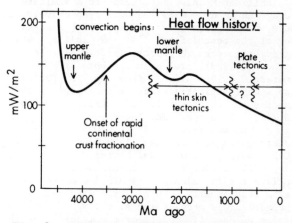

Fig. 2. Semi-quantitative heat flow history for the Earth. Based on McKenzie and Weiss (1975) model modified to initial state = 1500°C, heat flow = 80 mW/m² today and peaks adjusted to the two major Precambrian tectonic events.

restricted process, although it does occur (De Paolo and Wasserburg, 1976; Hart, 1979). Overall, these isotopic studies concur in indicating a heterogeneous mantle which has evolved over a considerable period, with peak fractionation in the late Archean.

The discussion above shows that there is extremely strong evidence in favour of a heterogeneous mantle which has produced continents in episodes, the major one being in the late Archean, with the remainder distributed in some unknown manner over subsequent time. The simplest relationship between these facts and my model is that the upper mantle provided the Archean continents and the lower mantle the remainder, perhaps with plenty of potential sial left behind in the lower mantle for future production. In heterogeneous model terms (Smith, 1979) this means that Rb/Sr will be higher in the upper mantle than in the lower, while U/Pb will show the reverse pattern. Sm/Nd will probably differ little between the two reservoirs, producing Nd isotopic compositions which cannot be distinguished by present-day analytical techniques.

Figs. 3 and 4 show what the Rb/Sr and U/Pb evolutionary diagrams for a two-layer model Earth should look like: in each case the two-layer (two-reservoir) model is simplified into two time stages only, whereas each reservoir will evolve continuously during any period of continent formation if recycling is absent. The Rb/Sr curves are constrained by achondritic meteorite values (~0.699) at t = 4.5 Ga ago and evidence from ocean-floor basalts (Hofmann and Hart, 1978; Hedge, 1978) and estimates of average mantle Rb/Sr (O'Nions and Pankhurst, 1978; O'Nions and others, 1979). The Sm/Nd argument leads to a bulk earth $^{87}Sr/^{86}Sr$ of 0.7047 today and Rb/Sr of 0.032, within the range shown on Fig. 3. Taking the simplest possible interpretation (one-stage model) for the Samoan and Tahiti or Reunion data leads to a limiting average Rb/Sr of 0.0375 and 0.03 respectively, a range which yields growth curves which bracket many of the best defined $^{87}Sr/^{86}Sr$ initial ratios for Archean rocks (Moorbath, 1976; O'Nions and Pankhurst, 1978). To reach present-day MORB figures (0.7025 - 0.7029) from the East Pacific Rise and Mid-Atlantic Ridge, a two-stage model requires the upper mantle Rb/Sr falls to 0.008 to 0.012 at 2.7 Ga ago. The lower mantle, starting with an assumed lower Rb/Sr of 0.025, and fractionating slightly at, say, 1.95 Ga to Rb/Sr = 0.022, would yield material having 0.7033 today, which will be almost indistinguishable from upper mantle material which has remained comparatively little fractionated. Such mantle material (lower or upper) could provide the slightly ^{87}Sr enriched basalts of Hawaii, Iceland or the majority of the Pacific islands. Note that the mantle evolution curves shown do not give us much hope of being able to prove or disprove this hypothesis from studies of mid to late Proterozoic rocks, as in the age range 1.8 to 0.6 Ga there is little difference between the two reservoirs as modelled,

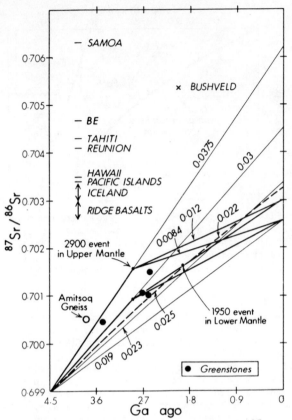

Fig. 3. Some one-stage and two-stage ^{87}Sr evolution curves for the mantle, leading to the range of ocean floor and ocean island basalts $^{87}Sr/^{86}Sr$ found today. BE = bulk earth (O'Nions and others, 1979), Rb/Sr ratios given on evolution curves.

while in the Archean there is only one reservoir. The U-Pb system offers more hope, however. Fig. 4 shows the generalized main trend line for Pb from modern oceanic islands and crust, deviations from this trend mostly being a little above this line (see Fig. 11, Tatsumoto, 1978, for data base). As shown, this trend lies to the high 206/204 side of the zero isochron for single-stage Pb evolution curves. The only way to develop Pb lying in this position today is for the U/Pb system to undergo at least one fractionation during the course of Earth history to a higher U/Pb (or $\mu = {}^{238}U/{}^{204}Pb$) value. For simplicity, Fig. 4 shows two two-stage growth curves which originate at the Canyon Diablo troilite Pb (Tatsumoto et al., 1973), age 4.57 Ga, and evolve in two separate reservoirs which undergo increases of μ, one at 2.9 Ga and the other at 1.95 Ga. These curves produce Pb having isotopic compositions at the two ends of the trend line, suitable for mixing and production of the observed line. Pb from MORB lies at the low 206/204 end of the trend, corresponding to Pb from an upper mantle reservoir which has always had a lower μ than the lower mantle.

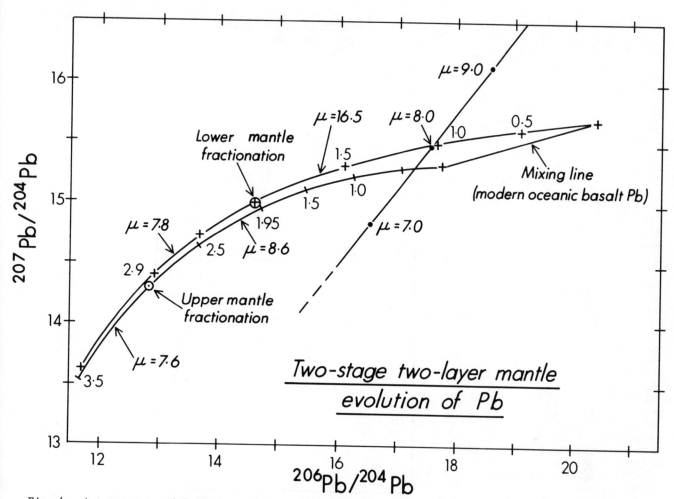

Fig. 4. A two-stage model of Pb isotope evolution, involving two source regions with different U/Pb and different history, designed to produce most of the Pb isotopic compositions found in ocean floor and ocean island basalts today.

Those other modern oceanic basalt Pb isotopic compositions which are not shown on Fig. 4 could be produced from parts of a two-layer mantle having only slightly different μ values and fractionation dates. It is interesting to note that mantle fractionation causing μ to fall at any time does not appear to be likely, indeed its absence is almost certain in the case of the lower mantle. The geochemistry of U and Pb seems to require that Pb loss to the core in sulphide during fractionation is the dominant cause of changes in μ.

The growth curves of Fig. 4 are not unique, but were constructed by constraining the fractionation dates to geologically pertinent ages, thus defining μ_1 and μ_2 if modern ocean basalt Pb is to be produced. The values shown are reasonable, but not by any means the only ones: in general, of course, fractionation will not occur at one instant, but will be spread over a period during which there will be a continuous evolution of μ. The particular values of μ and fractionation

date chosen yield mantle evolution curves which differ significantly only after 1.95 Ga. The model could, in theory, be tested on any igneous complex having an age <1.95 Ga and which was not contaminated by continental crust.

A Mantle with an Internal 'Lithosphere'

The two-layer mantle model with a boundary zone at the conventional depth of 700 km provides us with reasonable answers to the geological and isotopic evidence discussed above (cf Schubert, 1979). The thermal structure of such a mantle is not well constrained by physical facts, but a version suitable for discussion is presented in Fig. 5. The thermal gradient in the upper lithosphere is shown as reaching 1400°C at 150 km, approaching the wet garnet-peridotite solidus (0.1% H_2O) at a depth of perhaps 100-120 km to provide an asthenosphere. I consider it to be likely that close to the wet solidus, minerals

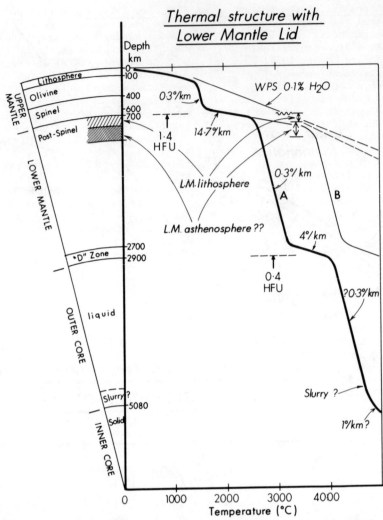

Fig. 5. An average geotherm for the mantle with suggestions for a lower mantle lithosphere and in an extreme case, a lower mantle asthenosphere.

will be hydrolytically weakened and actual melting need not occur to produce an asthenosphere of low strength. For those who prefer a partially-melted asthenosphere, the thermal gradient in the lithosphere would have to be a little higher than shown on Fig. 5, producing a curve just slightly to the right of the one shown. The difference is negligible. The upper mantle is shown as convective between 150 and 650 km with an average thermal gradient at an adiabat of 0.3°/km (Ringwood, 1975). The average temperature thus reaches 1600°C at 650 km allowing for thin regions of higher gradient in the upper and lower regions of the upper mantle convective system. The lower, slower moving region of the cell begins where spinel and other phases begin to invert to oxide phases and/or where the transition to a different chemistry begins (Hazen and Finger, 1978; Liu, 1979). If Schubert et al. (1975) are correct, convection through this layer is retarded by neg-

ative dP/dT of the inversion curves. Material arriving at this layer from above either plates-out or is recirculated as it is warmed by the heat flux from below (shown as 16 mW/m² corresponding to a model with 60 mW/m² at the surface: if higher heat fluxes are assumed, the same geotherm can of course be maintained by making the 700 km thermal boundary layer thinner – not a critical problem at this stage of development of the model). Material approaching this boundary from below can only break through this layer if it has exceptionally high velocity or low density, or both, because of the great density contrast.

Turning next to the core-mantle boundary, for which estimates of thermal properties vary widely (Boschi, 1975; Jones, 1977; Loper, 1978; Jeanloz and Richter, 1979) a figure of 4000°C has been chosen for curve A of Fig. 5. This is slightly above that assumed by Jones (1977), well above that of Jeanloz and Richter, and at the bottom end of the range considered for a NiFe core by Boschi. The core may be producing heat from latent heat of crystallization at a considerable rate ($3.6.10^{11}$ W according to Loper, 1978) to which may be added a cooling component. Jones (1977) estimated $8.7.10^{12}$ W with a corresponding CM boundary heat flow of 55 mW/m^2 for this cooling component. Loper argued against retention of a cooling component over the history of the Earth. Jeanloz and Richter assumed a closely similar CM boundary heat flux and estimated that the D^{11} zone possesses a thermal gradient of 3°C/km. Taking an arbitrary intermediate value of core heat production of $2.5.10^{12}$ W and thermal conductivity in D^{11} of 0.04 j/cm.sec.$^{\circ}$C then the gradient is 4°C/km and the heat flux 16 mW/m^2. The thickness of D^{11} is assumed to be 200 km (Bullen, 1950).

If an adiabat of 0.3°C/km is also assumed for a convecting lower mantle (Jeanloz and Richter, Fig. 2 op cit, use 0.2°C/km giving a lower temperature change in the lower mantle) then we are faced with a temperature gap of around 1100°C between the estimates of temperature of the top and bottom of the 700 km discontinuity. Faced with this same problem, although their version was numerically smaller, Jeanloz and Richter discussed several alternatives including the one proposed here, namely a significant thermal boundary layer at the junction between the upper and lower mantle convective systems. Taking a mantle K U Th composition as in Lambert (1979), core heat production of $2.5.10^{12}$ W, conductivity in the 700 km boundary layer as the lowest possible reasonable value (0.025 j/cm.sec.$^{\circ}$C) we find a gradient of 6.7°C/km. The layer would then be 165 km thick for a 1100°C temperature drop. This boundary layer or lid to the lower mantle has some of the properties of a lithosphere, in that its thickness will depend on the local convective behavior of the mantle above and below, and in that it will be a mechanically distinct unit.

Discussion: Consequences of a Lower Mantle Boundary Layer

From the numerous lines of approach previously reviewed it appears that the mantle may well be physically divided into two systems. The lower mantle accreted first, warmed up to convecting level second, began to differentiate second and is still evolving today. It is the source of one component of high-μ igneous rocks and ^{87}Sr enriched ocean-floor and island basalts today. If we regard the upper surface of the lower mantle as a thermally-controlled lithosphere, its thickness will vary according to the quantity of heat being supplied to its base. Above rising limbs of lower mantle convection systems or plumes the LM lithosphere will be thin, and it will be thick over lower heat flow regions. Assuming curve A of Fig. 5 or the lower geotherm of Jeanloz and Richter

(Fig. 6c, 1979) there will be no partial melting of the lower mantle. However, lower mantle diapirs rising from, say, 1200 km and penetrating the 'thin' regions of the LM lithosphere (analogous to oceanic ridges at the top of the upper mantle) will be at such high temperature that they may well begin to melt at 500 km or so. This will give abundant opportunity for that melt to mix with upper mantle partial melt, producing such geochemical results as those we see along the Reykjanes Ridge away from Iceland. The return flow to the lower mantle would be by a gradual, widespread body-flow process involving the denser fractions of the upper mantle, plating-out on the upper surface of the LM lithosphere. However, the ascent of a lower mantle diapir into the upper mantle is going to be a mechanically restricted and perhaps rare process, density differences being what they are.

If, however, the CM boundary is at a very much higher temperature (5000°C), towards the upper limit of recent suggestions, then an average geotherm such as curve B (Fig. 5) may apply. Such a geotherm must approach the solidus (even allowing for phase changes below 300 km depth) and partial melting might occur in diapirs or in ascending limbs of lower mantle convection cells, and cause the necessary reduction of density for such material to pass through the LM lithosphere into the upper mantle. There might even be a LM asthenosphere. Seismic evidence for discontinuities, phase changes and layering in the mantle in the appropriate depth range has been discussed and referenced by Liu (1979) and Jeanloz and Richter (1979). If it is correct to associate geochemical anomalies with two separate mantle reservoirs as discussed above, then it is intuitively attractive to have mantle temperatures as high as possible to produce an environment conducive to upward transport of the enriched lower mantle.

In the preceding discussion, a highly simplified view of the isotopic evidence was presented. A much more complex view of the mantle and its numerous sub-regions can be derived from in-depth studies of the isotopic evidence; but it is difficult to relate these details to physical reservoirs. The multi-reservoir concept does not conflict, however, with the model presented above.

References

Boschi, E., The melting relations of iron and temperatures in the Earth's core. Riv. Nuovo Cimento, 5, 501-531, 1975.

Brooks, C., D.E. James, and S.R. Hart, Ancient lithosphere: its role in young continental volcanism, Science, 193, 1086-1094, 1976.

Bullen, K.E., Compressibility-pressure hypothesis and the earth's interior, Mon. Notic. Roy. Astron. Soc. Geophys. Suppl., 5,355-368, 1949.

DePaolo, D.J., and G.J. Wasserburg, Nd isotopic variations and petrogenetic models, Geophys. Res. Letters, 3, 249-252, 1976.

Hart, S.R., Archean crust-mantle chemical evolution: an overview, Geol. Assoc. Can. Program with Abstracts, 4, 56, 1979.

Hazen, R.M., and L.W. Finger, Crystal chemistry of silicon-oxygen bonds at high pressure: implications for the Earth's mantle mineralogy, Science, 201, 1122-1123, 1978.

Hedge, C.E., Strontium isotopes in basalts from the Pacific Ocean basin, Earth Planet. Sci. Lett., 38, 88-94, 1978.

Hofmann, A.W., and S.R. Hart, An assessment of local and regional isotopic equilibrium in the mantle, Earth Planet. Sci. Lett., 38, 44-62, 1978.

Jeanloz, R. and F.M. Richter, Convection. composition and the thermal state of the lower mantle, J. Geophys. Res., 84, 5497-5504, 1979.

Jones, G.M., Thermal interaction of the core and the mantle and long-term behavior of the geomagnetic field, J. Geophys. Res., 82, 1703-1709, 1977.

Lambert, R. St J., The thermal history of the Earth in the Archean, Precambrian Res., 11, 199-213, 1980.

Lambert, R. St J., Earth tectonics and thermal history: review and a hot-spot model for the Archean. In Precambrian Plate Tectonics, ED. A. Kröner, Elsevier, 1981.

Liu, L.-G., On the 650-km seismic discontinuity, Earth Planet. Sci. Letters, 42, 202-208, 1979.

Loper, D.E., Some thermal consequences of a gravitationally powered dynamo, J. Geophys. Res., 83, 5961-5970, 1978.

McCulloch, M.T., and G.J. Wasserburg, Sm-Nd and Rb-Sr chronology of continental crust formation, Science, 200, 1003-1011, 1978.

McKenzie, D.P. and N.O. Weiss, Speculations on the thermal and tectonic history of the Earth, Geophys. J. Roy. Astr. Soc., 42, 131-174, 1975.

Moorbath, S., The geological significance of early Precambrian rocks, Proc. Geol. Assoc. Lond., 86, 259-279, 1975.

Moorbath, S., Age and isotope constraints for the evolution of Archaean crust, pp. 351-360 in The Early History of the Earth, ed. B.F. Windley, John Wiley N.Y., 1976.

Moorbath, S., Ages, isotopes and evolution of Precambrian continental crust, Chem. Geol., 20, 151-187, 1977.

O'Nions, R.K., S.R. Carter, N.M. Evensen, and P.J. Hamilton, Geochemical and cosmochemical applications of Nd isotope analysis, Ann. Rev. Earth Sci., 7, 11-38, 1979.

O'Nions, R.K. and R.J. Pankhurst, Early Archean rocks and geochemical evolution of the Earth's crust, Earth Planet. Sci. Lett., 38, 211-236, 1978.

Ringwood, A.E., Composition and petrology of the earth's mantle, McGraw-Hill, N.Y. p.618, 1975.

Schubert, G., Subsolidus convection in the mantles of terrestrial planets, Ann. Rev. Earth Sci., 11, 289-342, 1979.

Schubert, G., D.A. Yuen, and D.L. Turcotte, Role of phase transitions in a dynamic mantle, Geophys. J. Roy. Astr. Soc., 42, 705-735, 1975.

Smith, J.V., Mineralogy of the planets: a voyage in space and time, Mineral. Mag., 43, 1-89, 1979.

Stockwell, C.H., Fourth report on structural provinces, orogenies, and time-classification of rocks of the Canadian Precambrian Shield: in Age determinations and geological studies, pt. II, Geol. Surv. Canada, Paper 64-17 (II), 1-21, 1964.

Tatsumoto, M., Isotopic composition of lead in oceanic basalt and its implication to mantle evolution, Earth Planet. Sci. Lett., 38, 63-87, 1978.

Tatsumoto, M., R.J. Knight, and C.J. Allègre, Time differences in the formation of meteorites as determined from the ratio of lead-207 to lead-206, Science, 180, 1279-1283, 1973.

Williams, D.L. and R.P. von Herzen, 1975, Heat loss from the Earth; new estimate, Geology, 2, 327-328, 1975.

TERRESTRIAL HEAT FLOW HISTORY AND TEMPERATURE PROFILES

E. A. Lubimova and O. Parphenuk

Institute of Physics of the Earth, Moscow, U.S.S.R.

Abstract. Terrestrial heat flow values are still the main constraints for any evolutionary thermal model. The accepted value of mean global heat flow increased from 64 to 79 mW/m^2 during the last decade. This corresponds to a global heat flow of about 4×10^{13} watts as a boundary condition for the Earth's thermal history balance. An empirical negative relationship between heat flow and the age of the latest geotectonic event is indicated for oceanic and for continental lithosphere. Apart from the anomalous central parts of mid-oceanic ridges, the equation $q \propto t^{-1/2}$ works well for oceanic lithosphere that is less than 120 m.y. old and predicts the relationship between topography and tectonic age. For continents the problem is more difficult. There may be several types and intensities of thermotectonic, metamorphic, and magmatic events. The time variation of continental heat flow is associated with cooling processes as well as with erosion and removal of heat sources. The superposition of the processes has no simple time scale representation.

Unfortunately the deviations of observable heat flow values from those predicted by plate models are usually analyzed without any consideration of deviations in the age of the latest tectonomagmatic event, which can be expected to have a wide range of influence. The nature of this event is still not quantitatively determined for land.

Interest in the Earth's thermal history has revived, with three particular aspects being:
1. Emphasis on a possible large energetic pulse from core-mantle segregation or density differentiation processes;
2. Large body impacts on the primitive surface of the primordial Earth in its growing stages. As a result of this the initial temperatures may have reached partial melting in the upper mantle of the undifferentiated Earth;
3. A parameterized convection model for heat transfer beneath a mainly conductive lithosphere.

Numerical reconsideration of the Earth's evolutionary thermal model is given here on the basis of an accretion model and convective heat transfer inside the melting layer followed by density differentiation. It is shown that the
mainly conductive lithosphere strongly controls the global heat loss into space.

Two approaches are being used for estimation of geotherms within the Earth's interior: a solid-state or seismological one, and construction of an evolutionary thermal model, based on heat flow data and heat generation-heat transfer models. A discrepancy between geotherms remains significant for both approaches: the first approach gives concave curves in the C-layer, while the second leads, as a rule, to convex curves and positive temperature gradients.

1.Introduction

The theory of the Earth's thermal history has been a long story. Significant contributions to it were made by Kelvin (1984), Holmes (1915), Jeffreys (1929), Slichter (1941), Tichonov (1937), Urey (1952), MacDonald (1959), Lubimova (1958, 1967, 1977), Birch (1965), Fricker et al. (1967), Tozer (1967), McKenzie et al. (1974), Sharpe and Peltier (1979) and many others.

The mean heat flow value, which is evaluated by Chapman and Pollack (1979) as 75 mW/m^2 observed on the surface, was usually the main constraint for thermal history theory and for global energy balance calculations. The terrestrial heat flux as a global boundary condition $Q = 3.8 \times 10^{13}$ W (or $Q = 8.91 \times 10^{12}$ cal/sec) constrains the temperature distribution inside planetary interiors.

A decade ago a conductive thermal evolution model with variable physical parameters and heat sources seemed to have been constructed successfully on the basis of a cosmogonical concept of planetary accumulation from dust and solid bodies (Lubimova, 1958, 1967; Jacobs and Allan, 1954; MacDonald, 1959; Levin and Majeva, 1970; Majeva, 1977; and Lee, 1967).

In current thermal evolution models a very large heat pulse is expected from iron core segregation (Birch, 1965; Keondjan and Monin, 1975, 1977; Vitjazev, 1973; Kalinin and Sergeeva, 1977; Runcorn, 1967; Iriajma, 1978). It consists of about 1.6 to 2.3×10^{38} ergs. It is equal to or even a little more than the planetary radioactive heat generation, and it amounts to

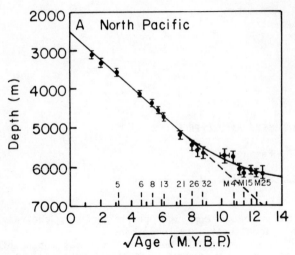

Fig. 1. Oceanic topography versus age for the North Pacific according to the plate models of McKenzie and Sclater (1969), and Sclater and Tapscott (1979).

200 erg/cm^{3}y in the upper mantle and 400 erg/cm^{3}y near the core-mantle boundary. There is evidence that the process of gravitational differentiation has been one of the main mechanisms of geodynamics. In this case convective heat transfer should be postulated as a mechanism for the release of heat from the Earth's interior.

Geophysical observations directly and indirectly imply that solid-state convection is presently occurring and has occurred in the early history of planetary mantles. The motion of the plates constitutes clear evidence for large-scale mantle convection. There are several additional observations, specifically terrestrial oceanic heat flow and bathymetry, that strongly suggest operative mantle convection.

We will study here the "blanketing" effect of the conductive lithosphere for the energy balance of the underlying convective mantle in different models with core segregation at an early stage as well as long-term global differentiation. We will concentrate attention on the fine structure of heat transfer coefficients and heat balance near the surface of the planet, and the history of surface heat flow.

2. Evolutionary heat flow relationships

For land areas it is usually assumed that simple conductive models work well as predictors of temperature profiles inside the Earth's crust and the solid lithospheric region, where tectonic extension is not rapid. An empirical negative relationship between q, topography and τ, (q is terrestrial heat flow; τ is the age of the crust) showing an increase in heat flow and elevation as the age of the latest geotectonic event decreases is widely known (Figure 1). However for a long time the difference in time-scales of this relationship between that for oceanic and that for continental lithosphere has not been properly accounted for. To understand this time-scale discrepancy Lachenbruch (1978) and Vitorrello and Pollack (1979) stressed that it was important to remove the background due to heat generated by radioactivity in the crust. Vitorrello and Pollack (1979) have interpreted the decrease of continental heat flow in a model with three components.

The next step was the analysis of the heat flow beneath the ocean in terms of $t^{-1/2}$ (Lister, 1975; Sclater et al. 1976). The latest relation between heat flow and age was found to be (Sclater and Tapscott, 1979):

$$q = 11.3 \ t^{-1/2} \qquad (1)$$

(q is the heat flow in 10^{-6} cal/cm^{2}s; t is the age in m.y.).

Apart from the anomalous central-ridge average, this relation works well for oceanic lithosphere that is less than 120 million years old. Sclater and Tapscott (1979) have shown that their plate model predicts

$$q = 0.9 + 1.6 \ \exp(-t/62.8) \qquad (2)$$

which works well enough for any time.

These thermomechanical models for processes near the axis of the mid-oceanic ridge provide the initial conditions necessary to describe the time-evolution of oceanic heat flow as the conductive decay of a thermal transient (McKenzie, 1967; Turcotte and Oxburgh, 1967; Sclater and Francheteau, 1970; Forsyth and Press, 1971). For the continents the problem is more difficult. There may be several types and intensities of thermotectonic and metamorphic events and there is no satisfactory general thermo-mechanical model of any of them.

The secular variation of continental heat flow is associated with the cooling process as well as with the erosion and removal of heat sources. The superposition of the two processes has no simple time scale representation (Vitorrello and Pollack, 1979). A correction is also necessary for radioactivity of the continental crust.

A plot of mantle heat flow versus $t^{-1/2}$ is given in Figure 2. It is in terms of $t^{-1/2}$ that the continental and oceanic heat flow begin to be in accord (Hamza, 1978,1979). This defines the reduced heat flow as the flux from below the upper continental crust. The radiogenic heat generation of lower continental crust is estimated to be in the range 0.2 µW/m^3 to 0.6 µW/m^3. Because the surface heat flow varies with age between about 90 and 45 mW/m^2, this radiogenic component also varies with age between about 36 and 18 mW/m^2, a net decrease of some 18 mW/m^2. The absolute decrease with time of this component is achieved by erosional removal of radioisotopes from the surface. The erosion,

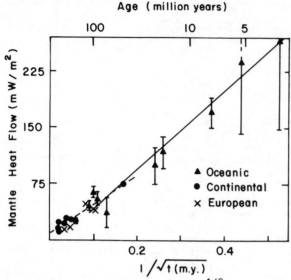

Fig. 2. Mantle heat flow versus $t^{-1/2}$ for oceanic (triangles) and continental (points) lithosphere due to Hamza (1978) using heat flow data for the east European plate (crosses) due to Kutas et al.(1978, 1979) and Lubimova (1979).

exponentially decreasing with age, has a time-constant of some 300 – 400 m.y. (Vitorello and Pollack, 1979).

In contrast to continents the radiogenic heat production in oceanic crust is very small in relation to surface heat flow. The basaltic oceanic crust of 5 km thick produces negligible heat flow and oceanic heat flow values can be considered as representing the residual heat of cooling of the oceanic lithosphere. Using a conductive model, Vitorello and Pollack (1979) concluded that heat flow from the continental asthenosphere, q_A, is approximately 0.4 of the mean terrestrial heat flow. As surface flux diminishes with time, so also must q_A.

The linear relationship between the terrestrial heat flow and heat production by long-lived radioactive isotopes in the surficial rocks:

$$q = q_0 + Ab \qquad (3)$$

implies conductive heat transfer in the continental lithosphere (Birch et al., 1967; Decker et al., 1973; Lachenbruch, 1972). It was estimated that the thickness of the enriched zone of the upper crust, b, is about 7 km on average.

It has been shown that the continental heat flow history is closely connected with a sedimentary basin history (Sleep, 1976; Turcotte et al., 1976), climatic variations (Jessop and Cermak, 1971) and age of the latest metamorphic and tectonic events (Polyak and Smirnov, 1969; Hamza, 1979), which could be considered in terms of conductive heat transfer and these are in agreement with a conductive lithosphere model.

The continents do not participate in mantle convection, therefore the radioactive heat production within the continental crust must reach the surface by conduction (Turcotte and Burke, 1978). The continental component of surface heat flow associated with tectogenesis yields approximately 30% of the heat flow in Cenozoic tectonic zones.

For the regions where deep seismic profiles and petrological models are available a heat production model and thermal conductivity model are usually constructed as a basis for calculations of geotherms using heat flow values for the region. After that the mantle heat flow and Moho temperatures can be determined. In this way the heat flow from the upper mantle and temperature at the base of the continent were obtained for the East European plate (Lubimova, 1979). The Moho-temperature was reported to be equal to $400^{\circ}C$ for the Precambrian part and $800^{\circ}C$ to $1000^{\circ}C$ for the actively forming belt in the Carpathians and Caucasus (Kutas et al., 1978, 1979). Some points on Figure 3 give us the relationship between heat flow and age of the latest tectonomechanic event for the above area (Lubimova, 1979). The analysis of all these data indicates that the East European mantle is much colder than the West European mantle.

Locally the lithosphere can be penetrated by volatiles, magma intrusions, or hydrogeothermal jets, and it is faulted in its upper part. Heat flow anomalies can very often be explained by all these effects. Faulting, inelastic deformation, and movements of magma and water generally result in relative vertical motion of mass in tectonically active parts of the lithosphere. Lithosphere extension can result in local convective transport of large amounts of heat. Lachenbruch (1978, 1979) shows that under certain circumstances lithosphere extension of regions 1000 km wide at rates on the order of only 1 cm/yr could double the conductive heat loss at the surface. Estimates of upper mantle temperatures could be reduced by hundreds of degrees and estimates of lithosphere thickness could be increased by tens of kilometers compared to predictions from simple heat conduction models. Therefore it seems likely, for example, that the West European lithosphere is much warmer than its eastern part because it is more faulted

The two-dimensional theory for the prediction of ridge crest heat flow and topography was developed by Lubimova and Nikitina (1975). This was a more detailed version of the plate model of oceanic lithosphere cooling than previous models of McKenzie (1967); McKenzie and Sclater (1969, 1971); and Sclater et al. (1971), which gave infinite heat flow of the ridge crest itself. The solution for the ridge heat flow values was constructed by Lubimova and Nikitina (1974, 1975) from two parts: a normal one dimensional model and a two dimensional model with active vertical intrusion at the spreading center. Fluid of variable temperature, transported along the

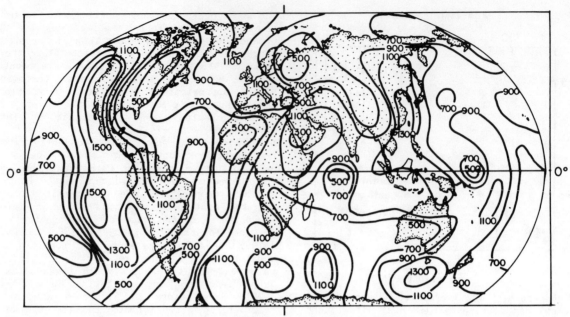

Fig. 3. The global distribution of temperature at the depth of 50 km, based on the spherical harmonic analysis of the terrestrial heat flow data by Suetnova (1979) and effective thermal conductivity of the extensive lithosphere by Lubimova: 0.05 cal/cm Cs for oceanic ridges and rifts, 0.065 for continental orogenic regions.

intrusive channel, were postulated. The three principal physical parameters are:

$$m = q_m/q^o \; ; \quad h = L-L/m \; ; \quad \delta \; ; \; \kappa \; ; \quad (4)$$

(q_m is the maximum heat flow value over the crest; q^o is heat flow far from the ridge; L is the thickness of the lithosphere; h is the depth of the melt; δ is the halfwidth of the heat flow anomaly. McKenzie's (1967) model was obtained as a special case of the more general solution by Lubimova and Nikitina (1975). The total heat loss at the ridge crest was found to be

$$\Delta q = LdqS(0,m) \qquad (5)$$

where d is the total length of the world ridge system.

For typical parameters in the North Pacific: m = 2.8; L = 100 km; L – h = 36 km; and $T - T_0 = 1200^oC$. One can find the heat loss over the world ridge system as q = 2.2 x 10^{12} cal/sec or 30% of the total terrestrial heat flow.

This evaluation is very close to the value 2.54 x 10^{12} cal/s given by Turcotte and Burke (1978) on the basis of boundary layer theory. Anderson et al. (1977) indicated the limiting ages where crustal hydrothermal circulation no longer needs to be considered. Hobard et al. (1978) evaluated the a width of convective hydrothermal transfer zone over the ridge axis as 4 km.

Summarizing the evolutionary relation between heat flow and age for the oceanic lithosphere one can observe the discrepancy between the boundary layer and plate models for ages greater than 120 m.y. A comparison of the continental mantle heat flow (CMHF) and oceanic mantle heat flow (CMHF) shows that CMHF corrected for radiogenic heat produced within the lithosphere plotted against $t^{-1/2}$ gives slope and intercept values of 489 ± 32 and -8 ± 7 respectively by a least squares fit (Hamza, 1978; 1979). It seems to be in good agreement with value of 500 predicted by the boundary layer model of Lister (1975).

It is desirable to stress the importance of variation of the thickness of the lithosphere. It seems likely that the variability of the thickness of the thermal lithosphere between oceans and continents is both a partial product of, and an influence on, the pattern of mantle convection. The very thick continental root beneath cratons, according to Jordan (1975), would seem likely to modify mantle convective flow. The continental roots are colder and more viscous than upper mantle at the same depths and probably retard the motion of the plates.

3. Model of lithosphere thermal properties

The relationships between heat flow and heat generation in crustal rocks and part of lithosphere were mentioned above, as well as the tectonic history and age of geological structures. These factors affect mainly the geothermal gradient. Below we shall dwell on the second component of heat flow associated with thermal properties of the strata where heat transport occurs. These properties are determined not only by the thermal conductivity but also by the amount of lithosphere penetration and extension, which are difficult to determine.

The thermal conductive zone of the Earth has nonuniform structure. Heat transfer inside this zone is determined not only by the thermal conductivity and radiative transport coefficients of the terrestrial rocks, but also by degree of permeability of the lithosphere. Experimental data for the crustal rocks and rock-forming minerals are given in Figure 4 to illustrate the wide range of variation in the thermal conductivity coefficients versus temperature for lithospheric conditions. Temperature is the predominant factor in the lithosphere compared to the relatively smaller pressure rise. High temperature data for granites, olivines, dunites, forsterites, gabbro-diabases, diabases and others were given by Birch and Clark (1941), Kawada (1964), Kanamori (1967), Shatz and Simmons (1972), Simmons and Horai (1971). For olivine spinel these coefficients were given by Jurchak (1979); for ceramic MgO and Mg_2SiO_4 by Kingery et al. ; and for pure coesite-minerals by Beck (1979). Two lithospheric conductvity models (EI, EII) for the thermal and radiative components are shown as solid lines in Figure 4.

These lines show an approximate upper conductivity for the lithospheric layer near 0.02 $cal/cms\,^oC$ and a lower limit near 0.004 $cal/cms\,^oC$.

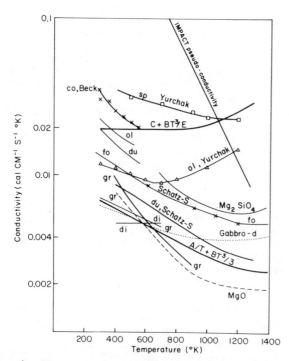

Fig. 4. The thermal conductivity of rocks, main rock-forming minerals and models for the lithosphere thermal properties. Notation: du, gr, di, ol, co, fo, sp correspond to dunite, granite, diorite, olivine, coesite, forsterite, spinel.

Parameters of heat-mass-transfer have been changing during the evolution of the Earth's temperatures. For comparison with thermal conductivity transfer coefficients for real material two models for the "pseudo" conductivity of the upper layer of the primitive Earth are given on the Figure 4; these take into account the effect of impacts on the early thermal evolution of the Earth (Kaula, 1979; Safronov, 1979). Based on the mechanism of mixing by impacts of bodies during planetary accretion these are important for calculation of the initial temperature distribution $T_0(r)$.

4. Initial Heating of the Earth

The heating of the Earth during its accretion from solid bodies was evaluated by Safronov (1969, 1978) and Kaula (1978, 1979). The accretionary heating occurred due to impacts of falling bodies on a growing planet. A cooling effect followed, which was caused by impact mixing. This can be characterized by a pseudo-diffusivity coeffficient, which should decrease from 2.7 km^2/sec at the surface to 0.01 cm^2/sec at the bottom of layer of mixing. The result depends substantially on the sizes of impacting bodies and the stage of the Earth's growth. Toward the end of the Earth's accumulation the melting state of the outer part of the primitive Earth may be reached in the outer 60 km (Safronov, 1978). The evaluation of both the initial temperature of the Earth and planets and a pseudo-diffusivity coefficient may vary within wide ranges depending on mass. As Greenberg et al .(1978) noted neither the initial nor final stages of the accumulation processes has yet been studied. Thus the evaluation of initial temperature due to impacts of protoplanetary bodies are very preliminary and the question of a "hot" or "cold" origin of the Earth is still open. A feature of all current initial temperature curves $T_0(r)$ determined by the accretion process is melting of the outer 50 to 400 km of the planet with a decrease in temperature toward the center of the core. How long could this state have persisted? The answer depends on the possibility and extent of convective heat transfer inside the primitive asthenosphere. Contributions of chemical reactions, core segregation and adiabatic compression should also be included in the initial temperature $T_0(r)$. On Figure 5 different determinations of $T_0(r)$ are given: the effect of heating from impacts according to Kaula (1978) and Safronov (1978) are included for comparison. One can see from Figure 5 that a melting region is present in the outer 400-600 km. Table 1 gives the comparative contributions of different heat sources.

5. Convective heat transfer conditions

A lot of geophysical observations can be explained by a dynamic theory based on thermal

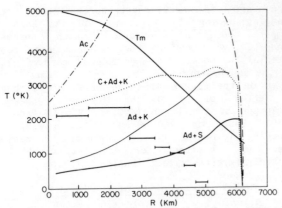

Fig. 5. Different models for initial temperature of the Earth: Ac is core segregation and accretion model due to Sharpe and Peltier (1979); Ad + K is the net effect of impact heating from Kaula (1979) plus adiabatic compression due to Lubimova (1979); C + Ad + K is Ad + K plus chemical heating due to Kuskov (1979). Steps signify chemical heating due to iron oxidation-reduction after Kuskov (1979). Ad + S are impact heating due to Safronov (1978) and adiabatic compression due to Lubimova (1979).

convection. They include: (1) the value of the mean terrestrial heat flow through the ocean bottom, approximately equal to 58 mW/m^2; (2) velocity of the relative plate movements between 1 and 10 cm/yr; (3) large scale gravity anomalies, and topographic and heat flow ocean floor anomalies.

In studies of thermal convection two main approaches have been taken, one using nonlinear theory and the other a linear one (McKenzie et al. (1974)). The former is used to study the structure of developed convection and the geometry of convection cells at various values of characteristic parameters and boundary conditions. The latter supposes that convective perturbations are small relative to the equilibrium state. Linear theory for a two layered model is reviewed by Phillips and Ivins (1979). They studied the effect of inhomogeneous viscosity on the convective cell size, for convection extending throughout the entire mantle (d ~ 2900 km) with the uppermost 300 km having a viscosity of 10^{21} cm^2/s and the lower mantle having 10^{24} cm^2/s. Even though the viscosity jump is as large as 10^3, convection should extend through the whole mantle with the lateral size of the cell being of the same order as the mantle thickness.

Schubert and Turcotte (1976) have studied the stability of the upper mantle including a phase change of olivine to olivine spinel, taking into account the latent heat and density difference at the phase boundary. They have shown the destabilizing effect of the phase change on convective stability and that this effect

facilitates the occurrence of convection for mantle viscosity in the range of 10^{24} to 10^{26} cm^2/s. This result was confirmed by Schubert and Young (1976).

Richter and Johnson (1974) studied a two layer model of convection and considered the effect of chemical composition on convective instability. The difference in density is supposed to be at 650 km and to be related to an increase of iron content in the lower mantle. They have shown that chemical inhomogeneity of the mantle appears to be a barrier to large scale convection.

Any increase in viscosity does not increase the width of the convective cell more than twice in comparison with the uniform viscosity case. Both the critical value of Rayleigh number and the wave number are determined mainly by lower viscosity. The model also shows the convective flow to be concentrated in the layer of low viscosity.

In terms of spherical models consideration of both convective instability of the Earth's mantle and large-scaled convection currents indicates that the thickness of the convective zone is comparable with the radius of the planet. The current study shows that different inhomogeneities, inner heat sources and phase transitions are not an obstacle to the occurrence of largescale convection covering the whole mantle and having a lateral size of the cell of the order 3000 - 4000 km.

Nevertheless when considering thermal convection as a driving mechanism for the lithospheric plate movement many authors assume the convection to occur only in the upper 700 km of the mantle. Since the sizes of the largest lithosphere plates are of the order of 10,000 km, in order for the convective cell to be beneath the plate we have to expect the lateral size of the convective cell to be approximately ten times its vertical size.

TABLE 1. Earth Energetics

	10^{38}erg
Energy loss by conductive heat flow into space (Lubimova, 1968; Sharpe and Peltier, 1978)	1.2
Accretion energy: impacts, adiabatic compression, (Safnonov, 1978)	0.7
core segregation (Sharpe and Peltier, 1978)	12.0
Heat production by radioactive isotopes	1.2
Gravitational differentiation (Kalinin and Sergeeva, 1977)	1.6
Tidal friction (Lubimova, 1968)	0.4
Heat capacity at T_{melt} (Lubimova, 1968)	2.8
Thermal effect of chemical reactions (Kuskov, 1978)	0.5

The Rayleigh number must be calculated to establish actual instability. The Rayleigh number may be computed from

$$Ra = \frac{\rho \alpha g L^3 \Delta T_s}{\kappa \cdot \eta}$$

where ΔT_s is the superadiabatic temperature difference; α is the thermal expansion coefficient; L is the thickness of convective layer; η is viscosity; and is the thermal diffusivity. Following Tozer (1966) one may use the Rayleigh number for a region containing heat sources

$$Ra = \frac{\rho \alpha g L^5 \cdot Hk}{\kappa \cdot \eta}$$

where H is the heat generation rate.

Once convective motion occurs, the following relations relate the Nusselt number, Nu, (dimensionless heat transport) and the Rayleigh number, Ra, (Sharpe and Peltier, 1979):

$$Nu = (Ra/Ra_c)^{1/3}$$

$$\text{if } Ra \geq 10^5 ,$$

where Ra_c is the critical value for the onset of convection. This relation was obtained for steady-state conditions. For $Ra < Ra_c$, Nu = 1 and only conductive heat transport occurs. According to our analysis mantle convection is characterized by a Rayleigh number of 10^6 to 10^7 suggesting that convection is vigorous.

Trubitzin et al.(1979) considered unstable convection in the mantle and showed that the thermal time constant for mantle convection is not small and is comparable to the age of the earth. Many current convective theories give the characteristic time for mantle convection as a few hundred million years.

6. Heat transfer evolutionary model

Assuming radial symmetry the energy equation becomes (Sharpe and Peltier, 1979):

$$\rho(r,T) \cdot C(r,T) \frac{\partial T}{\partial T} = \frac{1}{r^2} \frac{\partial}{\partial r} \left[r^2 (Nu \cdot K) \frac{\partial T}{\partial r} \right] + H(r,T)$$

with boundary conditions

$$T(R,t) = T_0 = \text{constant}; \quad 0 < t < t_0$$

$$T(r,0) = T_0(r); \quad \left. \frac{\partial T}{\partial r} \right|_{r=0} = 0$$

The Nusselt number is used to calculate the heat transfer. The total diffusivity coefficient K contains three parts:

$$K = Nu \cdot K_c + K_i + K_r$$

TABLE 2. Heat Transfer Parameters

	cgs units
Surface thermal conductivity	$4 - 8 \; 10^5$
Opacity coeffecient in K_r	$20 - 100 \; cm^{-1}$
Specific heat C_p	$1.3 \; 10^7$
Impact pseudo-diffusivity k	0.5
	0.02
Critical Rayleigh number Ra	1700
Range of Nusselt numbers Nu	5 - 19
Heat of fusion,silicates,erg/g C	$4 \; 10^9$
Density differences between iron and silicate phases)	3.5

where
- K_c -- lattice conductive diffusivity
- K_i -- impact pseudo-diffusivity
- K_r -- radiative diffusivity

The lattice conductivity and radiative transfer contribution are (Lubimova, 1967)

$$k_c + k_r = Af/T + \sigma n T^3 / \varepsilon$$

where A is nearly constant, ε is the coefficient of opacity, f(p) is the correction for pressure which is close to 1 for the lithosphere.

The heat capacity is a function of temperature and it is increased by an amount equal to $L/\Delta T$ in regions where melting occurs (L is the latent heat of melting and ΔT is the range of temperature where melting occurs). Heat transfer parameters are given in table 2.

The heat generation rate also has three parts:

$$H(r,t) = H_{HG} + H_i + H_{GD}$$

where

H_{HG}: radioactive heat generation (for the abundances of Wasserburg et al.,1964): 5×10^{-8} W/m^3;

H_i: heat generation by impacts (Safronov, 1978; Kaula, 1979) during early evolution only;

H_{GD}: heat generation by gravitational differentiation process (Keonjan and Monin, 1977; Vitjazev and Majeva, 1977).

The heat flow equation can be solved efficiently using, for example, the Samarsky algorithm (Samarsky and Nikolaev, 1977). Precipitous temperature drops due to extremely high Rayleigh numbers can be avoided by an iteration routine that reduces the time step of the calculation when necessary. Discontinuous changes in the heat capacity, the Nusselt number at the boundaries of the convecting regions are

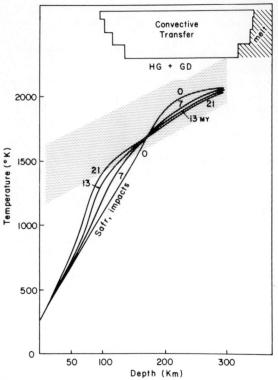

Fig. 6. Geotherms according to numerical calculation of the thermal history of the Earth on the base of the initial temperature curves $T_0(r)$, taking into account melting, convection and differentiation of the Earth's upper mantle; dotted area is subsolidus zone.

smoothed and hence do not present any difficulties.

7. Numerical results for the thermal evolutionary model

The remarkable feature of the initial heating is the melting by the end of accretion that results in a maximum of $T_0(r)$ in the upper mantle, and a minimum of the initial temperature inside the Earth's core even when core segregation occurs (Hanks and Anderson, 1969; Iriyama, 1970; Sharpe and Peltier, 1979). Therefore subsequent thermal evolution should result in heating of the deep interior.

The curves of temperature during the first stage of heating of the undifferentiated Earth (just after accumulation of its whole mass) are given in Figures 6 and 7. Safronov's (1979) and Kaula's (1979) initial temperature values $T_0(r)$ from impacts, adiabatic compression and chemical reactions are included. The potentially unstable conditions for convection are controlled by the super-adiabatic gradient. A convective heat transfer region occurred at a depth of 120 km during the first 10^6 years and it moved up to 30 – 60 km during the period from 3.8 my to 7.9 my.

The thermal conductivity-radiative transfer model was EII (see above) with opacity $\varepsilon = 40\ cm^{-1}$ and the surface value of conductivity was twice the normal lattice conductivity of rocks in order to model the permeability due to faulting and metamorphic reactions in the upper shell. The convective parameters inside the convective zone vary in the range (see table 3): viscosity in melted region: from 10^{21} to 6×10^{20} poise; Rayleigh number Ra: from 8×10^5 to 2×10^6; Nusselt number Nu: 7.5 – 19.8. The critical Rayleigh number Ra_c was assumed to be 2000; the width of the convective transfer zone is shown in Figures 6 and 7.

Discussion

Empirical relationships between heat flow, q, the age of the last tectonomagmatic event in a given province, and heat generation, H, offer a basis for understanding the current thermal state of the Earth's crust and lithosphere. The mean value of the parameter defining the thickness of the enriched upper crustal zone appears to be about 7 km (more precisely, from 5 to 14 km) for the continental crust. Heat production falls off rapidly with depth. A notion about different thermal states of the upper and lower crust becomes visible in outline.

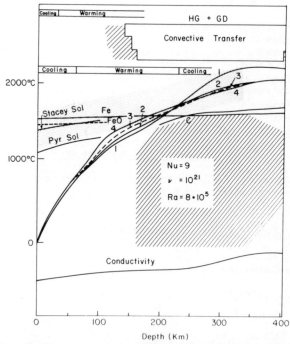

Fig. 7. Upper mantle geotherms according to early thermal history of the earth on the base of $T_0(r)$ = C + Ad + K from Fig. 5. Notation: dotted is subsolidus area; dashed is area where Nu > 1; number 1 corresponds to $T_0(r)$; numbers 2, 3, 4 to geotherms 3.5, 5.2, 7.9 m.y. after accretion.

In the lower part of the crust in localities with weakened zones or zones with developing dehydration, heat transfer should be more intensive due to penetrative pore fluid convection. These phenomena are most likely to be in zones of lithosphere extension.

A reason for a discrepancy between time scales relating heat flow and age of continental and oceanic lithospheres is more clear now. The inverse root mean square relationship $q \sim t^{-1/2}$ seems to be well satisfied for oceanic crust although the time scale for continents was longer. However, if the Earth's crust radioactive heat production background is removed and the mantle heat flow is considered to depend on the age of a deep tectonic block, then the time-scales in oceanic and continental relationships are closer. If continental heat flow data are considered as a separate set, the least squares fit gives a slope as 433 ± 31, whereas for the oceanic data set the slope value is 520 ± 35 (Hamza, 1978). These figures are near to the theoretical relationship $q \sim 500\ t^{-1/2}$ predicted by Lister (1975) on the basis of a boundary layer model. Such results testify that deep convective processes affect the history of heat flow emerging from the upper mantle and asthenosphere into the lithosphere. But extrapolation of the present relationships to the past is valid only for the first hundreds of million years. Numerous data indicate that the major internal process in the mantle and asthenosphere is solid-state convection.

Sharpe and Peltier (1978, 1979) tried to study the simplest parameterized Earth model containing solid lithosphere and a convecting upper mantle. This model disregarded any internal heat sources as well as differentiation processes. The Earth's cooling is simply followed from the initial pressure dependent melting temperature assuming convection within the mantle and thermal conduction through a solid lithosphere with a thickness of 64 to 100 km. This model can

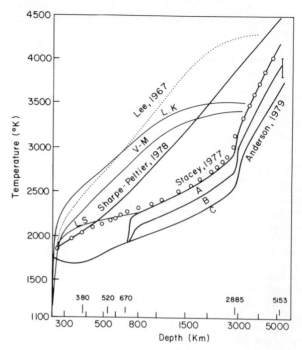

Fig. 8. Temperature profiles inside the Earth after Stacey (1977), Anderson (1979) based on the seismological models of Dziewonski (1979), and geotherms from different kinds of thermal models due to Lee (1967), Sharpe and Peltier (1979), Vitjazev and Majeva (1976) noted by V-M, and Lubimova (1979) for the early stage after accretion. L.S. corresponds to curve 21 from Figure 6; L.K. corresponds to curve 4 from Figure 7.

TABLE 3.
Parameters of the thermal evolutionary model

Surface temperature	273 K
The age of the Earth, after accretion stage	4.5 AE
Period of core segregation, model I	2 AE
Radioactive heat generation, W/m^3	
Lithosphere	$5\ 10^{-8}$
Mantle	$1\ 10^{-8}$
Heat sources in undifferentiated Earth (g/g)	
U	$2\ 10^{-8}$
Th	$8\ 10^{-8}$
K	$8\ 10^{-4}$
Mean mantle viscosity (poise)	$10^{20}-10^{22}$

scarcely be accepted as a satisfactory one to depict even the simplest heat transfer in a structure as intricate as the Earth, because the whole lithosphere is described by merely one step of the spatial grid. The model includes no detail concerning distribution of heat transfer coefficients, or the decrease of thermal conductivity of the lithosphere with increasing temperature. Nevertheless the important result by Sharpe and Peltier is that they show distinctly a close interrelation between the Earth's thermal evolution and variation in the mean viscosity of the Earth's interior.

Sharpe and Peltier (1978) determine the temperature at the core mantle boundary as $4500^{\circ}C$. However, their curves do not reflect the real evolution in time, ignoring any inner heat sources to a large extent. In fact, they do not include the complicated early evolution of the primordial Earth resulting from impacts and multiple remelting of the lithosphere. On the other hand, the results of Trubitsin et al. (1979), which consider unstable convection, indicate that convection over the whole mantle can develop for viscosities around 10^{23} poise, a value larger than that obtained by Sharpe and Peltier (1978).

Another approach to geotherm calculation for large depths is based on seismic models. The seismic model of Dziewonski et al. (1979) is used by Stacey (1977) and O. Anderson (1979) to deduce lower temperatures at the core-mantle boundary than thermal evolution models indicated (see Figure 8).

The evolutionary thermal models in Figures 6 and 7 are based on an algorithm that takes into account migration of radioactive elements, heat generation during gravitational differentiation, adiabatic compression and accretional energy from impacts. Most important is that the thermal model proposed includes different heat exchange processes in the solid lithosphere and convecting mantle, along with a detailed lithosphere structure at various levels, especially near the surface. The latter is of importance for a more fundamental study of heat flow. The lithosphere thermal conductivity is assumed to be affected by penetrative convection, degassing, and lithosphere extension. Such an effective thermal conductivity should be several times higher than the lattice thermal conductivity of terrestrial rocks. It was shown that, even with an increased effective thermal conductivity, the lithosphere

considerably retards the loss of heat to space from the Earth's surface at early evolution.

Two approaches have been developed for the estimation of geotherms within the Earth's interior: a solid-state physical or seismological one (Gilvarry, 1969; Artjushkov and Magnitsky, 1975; Stacey, 1977; Anderson, 1979) and the construction of the thermal evolution model based on heat flow data and heat generation-heat transfer models (MacDonald, 1979; Levin and Majeva, 1970; Vitjazev and Majeva, 1976; Lubimova, 1967, 1969; Lee, 1968; Sharpe and Peltier, 1978, 1979).

A discrepancy between calculated geotherms remains significant for both approaches. It can be noted, however, that the first approach gives concave curves in the C-layer and negative vertical gradients, while the second based on heat flow generation model, leads, as a rule, to convex curves and positive temperature gradients everywhere. According to Trubitzin et al. (1979) negative vertical gradient of temperature can appear only occasionally, and then only locally during unstable convective motion.

APPENDIX

Numerical model

For the computation of the surface heat flow history, the numerical integration of the heat equation (Samarsky, Nikolaeb, 1977; Parphenuk and Lubimova, 1979) was performed using an implicit numerical scheme, with a time step satisfying the stability criterion:

$$h^2/2 = (\min \; c)/\max K(T)$$

with a safety factor = 1. The minimum space step was 2 km near the Earth's surface, and then 5.9 km etc.

In previous works the surface interval above 25 km was not included in order to economize on computer time. It leads to the computational inaccuracies for the surface heat flow and heat loss balance.

Acknowledgements

We thank Dr. Frank Stacey, Dr. Lin-gun Liu and Professor W. Fyfe as well as Dr. Valeri Trubitsyn for critical remarks and valuable discussions, and Acad. A. A. Samazsky for his help with the computer program, particularly points of discontinuities.

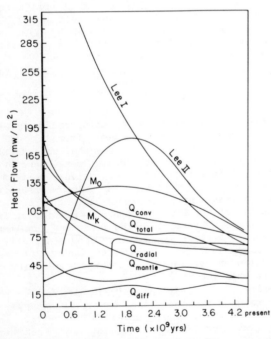

Fig. 9. Heat flow versus time from several thermal models: M_O, M_k are MacDonald (1959) for oceanic and continental crust; L is Lubimova's model (1967); Lee I and Lee II are models by Lee (1967) for undifferentiated and differentiated Earth; Q_D, Q_M, Q_R, Q_C and Q_t are diffusional, mantle, radiative, convective and total heat flux according to Sharpe and Peltier (1978).

References

Anderson, O. L., The temperature profiles in the Earth, this volume.

Anderson, O. L., The high-temperature acoustic Grueneisen parameter in the Earth's interior, Phys. Earth and Planet. Int., 18, 221-231, 1979.

Artyushkov, E. V. and V. A. Magnitzky, A general question of the Earth's dynamics, in: Tectonosphere, edited by V. V. Beloussov, Nauka, Moscow, 487-525, 1978.

Birch, F., Energetics of core formation, J. Geophys. Res., 70, 6217-6221, 1965.

Birch, F., R. F. Roy and E. R. Decker, Heat flow and thermal history in New England and New York, in: Studies of Appalachian Geology, Northern and Maritime, edited by E. Zen et al., Interscience, New York, 437-451, 1968.

Chapman, D. S. and H. N. Pollack, Heat flow and heat production in Zambia: evidence for lithospheric thinning in Central Africa, Tectonophysics, 41, 79-100, 1977.

Clark, S. P., Effect of radiative transfer on temperature in the earth, Bull. Geol. Soc. Am., 67, 1123-1124, 1956.

Green, D. H., Magmatic activity as the major process in the chemical evolution of the Earth's crust and mantle, Tectonophysics, 13, 47-71, 1972.

Hamza, V. M. and R. K. Verma, The relationship of heat flow with age of basement rocks, Bull. Volcanol., 33, 123-152, 1969.

Hamza, V. M., Variation of Continental Mantle Heat Flow with Age: Possibility of Discriminating between Thermal Models of the Lithopshere, Pageoph. 117, Nos. 1-2, 65-74, 1978-79.

Hanks, T. C. and D. L. Anderson, The early thermal history of the Earth, Phys. Earth Planet. Int., 2, 19-29, 1969.

Hayashi, C., K. Nakazawa and I. Adachi, Earth's melting due to the blanketing effect of the primordial dense atmosphere, Earth Planet. Sci. Lett. 73, 22-28, 1979.

Hurtig, E. and C. Oelsner, Heat flow, temperature distribution and geothermal models in Europe: some tectonic implications, Tectonophysics, 41, 147-156, 1977.

Iriyama, J., Thermal history of the Earth with consideration of the formation of its core, J. Phys. Earth, 18, 295-311, 1970.

Jordan, T. H., The continental tectosphere, Rev. Geophys. Space. Phys., 13, 1-12, 1975.

Jordan, T. H., Composition and development of the continental tectosphere, Nature, 274, 544-548, 1978.

Kalinin, V. A. and N. A. Sergeeva, Variations of inner structure of the Earth according to its evolution, Izvestija Acad. Nauk, USSR, Phys. Earth, 5, 3-16, 1977.

Kaula, W. M., Thermal evolution of Earth and Moon growing by planetesimal impacts, J. Geophys. Res., 83, 999-1008, 1978.

Kaula, W. M., The beginning of the Earth's thermal evolution, in The Continental Crust and its Mineral Deposits, edited by D. W. Strangway, Geological Association of Canada, Special Paper 20, Toronto, 1980.

Kennedy, G. C. and G. H. Higgins, Melting Temperatures in the Earth's Mantle, Tectonophysics, 13, 221-232, 1972.

Keonjan, V. P. and A. S. Monin, Model of gravitational differentiation of the interiors of planets, Doklady Academy Nauk USSR, 221, 1-6, 1975.

Keyes, R. W., High temperature thermal conductivity of insulating crystals: relationship to the melting point, Phys. Rev. 115, 3, 1959.

Kingery, J., The variation of thermal conductivity with temperature for some silicates, J. Am. Cer. Soc., 37, 2, 1954.

Kuskov, O. L. and N. I. Khitarov, Oxidation-reduction conditions and thermal effects of chemical reactions in the undifferentiated Earth, Geochemistry, 4, 467-494, 1978.

Lachenbruch, A. H., Crustal temperature and heat production: implications of the linear heat-flow relation, J. Geophys. Res., 75, 3291-3300, 1970.

Lachenbruch, A. H., Heat flow in the Basin and Range Province and thermal effects of tectonic extension, Pageoph, 117, Nos. 1-2, 34-50, 1978-79.

Levin, B. Y. and S. V. Majeva, Some calculations of the thermal history of the Earth, Doklady Akademy Nauk, USSR, 133, 44-47, 1970.

Liu, L. and W. A. Bassett, The melting of iron up to 200 kbar, J. Geophys. Res., 80, 3777-3782, 1975.

Lister, C. R. B., The heat flow consequences of the square-root law of ridge topography, Proc. XVI General Assembly IUGG IASPEI, Grenoble, 1975.

Lubimova, E. A., Thermal history of the Earth with consideration of the variable thermal conductivity of its mantle, Geophys. J. R. astr. Soc., 1, 115-134, 1958.

Lubimova, E. A., Theory of thermal state of the earth's mantle, in: The Earth's Mantle edited by T. F. Gaskell, NY, Academic Press, 231-323, 1967.

Lubimova, E. A., Thermics of the Earth and Moon, Nauka, Moscow, 1977.

Lubimova, E. A., Heat flow map for the European plate, Geothermics, 3, 1975.

Lubimova, E. A., Developing of geothermal models, Physics of the Solid Earth, 13, 25-32, Moscow, 1977.

MacDonald, G. J. F., Calculations on the thermal history of the Earth, J. Geophys. Res., 64, 1967-2000, 1959.

McKenzie, D. P., J. M. Roberts, N. O. Weiss, Convection in the Earth's mantle: towards a numerical simulation, J. Fluid Mech., 62, 465-538, 1974.

O'Connell, R. J., On the scale of mantle convection, Tectonophysics, 38, 119-136, 1977.

Parsons, B. and J. G. Sclater, An analysis of the variation of the ocean floor bathymetry and heat flow with age, J. Geophys. Res., 82, 804-827, 1977.

Phillips, R. J. and E. R. Ivins, Geophysical observations pertaining to solid-state convection in the terrestrial planets, _Phys. Earth. Planet. Int.,_ 19, 107-148, 1979.

Pollack, H. N. and D. S. Chapman, On the regional variation of heat flow, geotherms, and the thickness of the lithosphere, _Tectonophysics,_ 38, 279-296, 1977.

Polyak, B. G. and Y. B. Smirnov, Relationship between terrestrial heat flow and the tectonics of continents, _Geotectonics,_ 4, 205-213, 1968.

Rao, R. U. M., G. V. Rao and H. Narain, Radioactive heat generation and heat flow in the Indian Shield, _Earth Planet. Sci. Lett.,_ 30, 57-64, 1976.

Richter, F. M. and B. Parsons, On the interaction of two scales of convection in the mantle, _J. Geophys. Res.,_ 80, 2529-2541, 1975.

Richter, F. M. and C. E. Johnson, Stability of a chemically layered mantle, _J. Geophys. Res.,_ 79, 1635-1639, 1974.

Roy, R. F., D. D. Blackwell and F. Birch, Heat generation of plutonic rocks and continental heat flow provinces, _Earth Planet. Sci. Lett.,_ 5, 1-12, 1968.

Safronov, V. S., _Evolution of the protoplanetary cloud and formation of the earth and the planets,_ Isr. Program Sci. Trans., NASA Technical Translation, 206 pp., 1971.

Safronov, V. S., The heating of the Earth during its formation, _Icarus,_ 33, 3-12, 1978.

Sammis, C. G., J. C. Smith, G. Schubert and D. A. Yuen, Viscosity-depth profile of the earth's mantle: effects of polymorphic phase transitions, _J. Geophys. Res.,_ 82, 3747-3761, 1977.

Schatz, J. F. and G. Simmons, Thermal conductivity of earth materials at high temperatures, _J. Geophys. Res.,_ 77, 6966-6983, 1972.

Sclater, J. G., L. A. Lawver and B. Parsons, Comparisons of long- wavelength residual elevation and free air gravity anomalies in the North Atlantic and possible implications for the thickness of the lithospheric plate, _J. Geophys. Res.,_ 80, 1031-1052, 1975.

Sclater, J. G. and J. Francheteau, The implications of terrestrial heat flow observations on current tectonic and geochemical models of the crust and upper mantle of the Earth, _Geophys. J. R. astr. Soc.,_ 20, 509-542, 1970.

Sclater, J. G., J. Crowe and R. N. Anderson, On the reliability of oceanic heat flow averages, _J. Geophys. Res.,_ 81, 2997-3006, 1976.

Sclater, J. G. and C. Tapscott, The history of the Atlantic, _Sci. Am.,_ 240, 156-174, 1979.

Stacey, F. D., A thermal model of the Earth, _Phys. Earth Planet. Int.,_ 15, 351-358, 1977.

Sharpe, H. N. and W. R. Peltier, Parameterized mantle convection and the earth's thermal history, _Geophys. Res. Lett.,_ 5, 737-740, 1978.

Sharpe, H. N. and W. R. Peltier, A thermal history model for the Earth with parameterized convection, _Geophys. J. R. astr. Soc.,_ 59, 171-203, 1979.

Smith, O. Y., _Four lectures on the theory of the Earth's origin,_ Academy of Sciences USSR, Moscow, 1948.

Tozer, D. C., Towards a theory of thermal convection in the mantle, in: _The Earth's Mantle,_ edited by T. F. Gaskell, NY, Academic Press, 325-351, 1967.

Trubitzin, A. A., P. P. Vasiliev, A. A. Karasev, A remark on viscosity and convection in the mantle, this volume.

Turcott, D. L. and K. Burke, Global sea-level changes and the thermal structure of the Earth, _Earth and Planet. Sci. Lett.,_ 41, 341-346, 1978.

Uffen, R. J., A method of estimating the melting-point gradient in the Earth's mantle, _Trans. Amer. Geophys. Un.,_ 33, 893-896, 1952. K.PAGE

Vityazev, A. V. and S. V. Maeva, The model of the early evolution of the Earth, _Izvestiya Acad. Nauk, USSR, Physics of the Earth,_ 2, 3-12, 1976.

Vitorello, I., and H. N. Pollack, On the variation of continental heat flow and the thermal evolution of continents, _J. Geophys. Res.,_ 85, 983-995, 1980.

Wasserburg, G. J., G. J. F. MacDonald, F. Hoyle, W. A. Fowler, Relative contributions of uranium, thorium and potassium to heat production in the Earth, _Science,_ 143, 465-467, 1964.

Williams, D. L., R. P. von Herzen, J. G. Sclater and R. N. Anderson, The Galapagos spreading center: lithospheric cooling and hydrothermal circulations, _Geophys. J. R. astr. Soc.,_ 38, 587-608, 1974.

SURFACE PLATES AND THERMAL PLUMES:
SEPARATE SCALES OF THE MANTLE CONVECTIVE CIRCULATION

W.R. Peltier[1]

Department of Physics, University of Toronto, Toronto, Ontario   M5S 1A7 Canada

Abstract The whole mantle convection hypothesis follows naturally from the inference using post-glacial rebound data that there is no dramatic increase of viscosity with depth in the mantle. If convection fills the entire mantle then those properties of the circulation which are accessible to direct measurement are easily explicable in terms of the same scaling laws which govern high Rayleigh and Prandtl number convection in the laboratory. The viscosity contrasts which exist across the thermal boundary layers appear to exert no active influence on the flow, apart from stabilizing the top boundary layer (lithosphere) against secondary convective instability and destabilizing the bottom boundary layer adjacent to the core-mantle interface. Destabilization of the bottom boundary layer, which is observed using seismic methods as the D" region, may lead to the formation of thermal plumes which rise so rapidly to the base of the lithosphere that their ascent is almost adiabatic. Such events provide a natural explanation for volcanism in plate interiors since intense partial melting is expected where they impact the base of the lithosphere. Temperature dependent viscosity therefore leads to the observed style of mantle convection which is dominated by two widely separated spatial scales associated respectively with surface plates and thermal plumes.

Introduction

  Although the ideas of continental drift and sea-floor spreading have become almost universally accepted during the past decade, and the paradigm of plate tectonics now directs most current geophysical thinking, there remains at least one major impediment which has yet to be overcome if we are to be in any position to claim to understand <u>why</u> the surface geological processes we see have their observed characteristics. This impediment has to do with difficulties attending the construction of a dynamical model which is able to predict "observables" and which must supplant the purely descriptive kinematic model of plate tectonics. The specific problems involved in constructing such a model have been discussed in detail in Peltier (1980b) and the purpose of the present paper is to further elaborate some of the arguments developed there.

  In spite of the existing general consensus to the effect that it is thermal convection which causes the creation and destruction of plates and drives them in their relative motion, this idea does not in itself lead to the construction of a dynamical model which is in any sense unique. Important ambiguities remain. We do not know, for example, whether the dominant energy source responsible for maintaining the motion against dissipation is distributed radioactivity in the mantle and core or whether the circulation is driven by the secular cooling of the planet (as recently suggested by Sharpe and Peltier,1978,1979). A second and equally important question, concerns the vertical extent of the region in which the material motion is significant. The prevailing view seems currently to be that the dominant circulation is confined above the seismic discontinuity at 670 km. depth (McKenzie et.al., 1974;Richter, 1978). De Paolo and Wasserburg (1976) and O'Nions et.al. (1979) have recently given geochemical arguments which appear to favour this view.

Any theory of mantle convection is obliged to satisfy several observational constraints if it is to be considered successful and some of these are listed in Table I, for convenience, with the first five set apart to indicate that they should perhaps carry greater weight than the remainder. Of these, the first three are obtained from more or less direct measurements whereas the last two constraints on mean plate thickness and mean viscosity in the underlying mantle are inferences based upon mechanical models of plate bending (Walcott,1971) and postglacial rebound (Peltier, 1974, 1976; Peltier and Andrews, 1976, Peltier et.al., 1978). The other constraints have been variously employed by different authors but of these that numbered 6 in Table I, when coupled with deep earthquake focal mechanism data which demonstrate 12 (Isacks and Molnar, 1971), has played the most important

_____
1.   Alfred P. Sloan Foundation Fellow

TABLE 1

Constraints on models of the mantle convective

circulation

---

1. mean horizontal plate scale $1 \sim 4000$ km
2. mean plate speed $u \sim 4$ cm yr^{-1}
3. mean surface heat flow $q \sim 75$ mWm^{-1}
4. mean plate thickness $\delta \sim 100$ km
5. mean mantle viscosity $\nu \sim 10^{22}-10^{23}$ Poise

6. seismicity ceases below the 670 km seismic discontinuity
7. heat flow and bathymetry $\alpha \sqrt{\overline{AGE}}$ of the ocean floor
8. volcanism observed in plate interiors - hot "spots"
9. bathymetry deviates from $(AGE)^{-1/2}$ for ocean floor AGE $> 7.10^7$ yrs
10. viscosity increases from 10^{22} P to $\sim 10^{23}$P through transition region
11. boundary layer nature of D" region above core mantle interface
12. deep seismic focal mechanism compressive
13. seismic low velocity zone and high attenuation zone exist only beneath young oceanic lithosphere (low viscosity?)
14. Rb/Sr and Nd/Sm constraints on the degree of mantle mixing

---

role recently. These data have been interpreted as implying (McKenzie and Richter, 1976) that the relatively cold material in the down-thrust lithospheric slab does not penetrate beneath the seismic discontinuity at 670 km depth, and therefore that there is no exchange of material between the upper and lower mantles. It has been argued that this could be a consequence of a very large increase of viscosity across the 670 km discontinuity (McKenzie and Weiss, 1975) which in turn has been supported by arguments in favour of a large increase of creep activation energy there assuming that the discontinuity is a phase boundary. This hypothesis is consistent neither with direct inference based upon recent modelling results from postglacial rebound (Peltier, 1980,a,b; Wu and Peltier, 1981, a,b) which provide the basis for entries 4 and 5 in Table I, nor with direct calculation based upon oxygen ion systematics assuming once more that the boundary marks a phase change (Sammis et.al.; 1977). Since knowledge of the viscosity profile plays such an important role in determining the expected depth extent of the circulation, the recent modelling efforts which are contributing to our knowledge of $\nu(r)$ will be reviewed in Section 2.

In Section 3, we will consider the extent to which knowledge gained from the study of convection in the laboratory may be employed to understand the main observations 1 - 5 which are

listed in Table I. One often reads statements in the literature on mantle convection to the effect that it is completely different from convection in the laboratory. As we will show, this is manifestly incorrect since the observations 1-5 are in fact related just as they would be if they were made on the surface of a heated below convection cell in a laboratory apparatus. As we shall show in reviewing and extending arguments which have been published elsewhere (Sharpe and Peltier, 1978; 1979), mantle convection does scale like a laboratory system if the depth of the circulation is taken equal to the mantle thickness and if the scaling law which one employs is that appropriate for a heated below model. The implications of this agreement are considered at some length since they clearly reflect upon the second major question mentioned above concerning the extent to which the circulation in the mantle is driven by heating from within i.e. by radioactive decay processes. This is perhaps the outstanding issue in mantle dynamics and I shall argue that the simple observations 7 and 9 may allow us to place strong bounds on the fraction μ of the total surface heat loss which is contributed by radioactive decay processes in the mantle.

If one is able to accept the arguments presented here in support of the existence of a mantle wide convective circulation which is sustantially driven by heating from below then there must of course exist a sharp thermal boundary layer at the base of the mantle in consequence of the heat loss from the core. Because of the thermally activated nature of the creep process the viscosity in this region may be sufficiently low to render the boundary layer violently unstable against secondary convective instability. In Section 4 I will discuss such instability from two points of view, firstly using linear stability analysis, and secondly results from a finite difference numerical model to investigate the nonlinear evolution of these fast growing small scale disturbances. These analyses demonstrate that the temperature dependence of viscosity could support the existence of instabilities of the lower boundary layer with growth times which are orders of magnitude shorter than the time required for a single large scale convective overturn and that they mature into fast rising plume like structures. This process appears to provide a natural explanation for the observation of volcanism in plate interiors (entry 8 in Table I). According to boundary layer arguments developed in Yuen and Peltier (1980 a,b) such plumes should rise so quickly that the process will be essentially adiabatic and thus they should be expected to produce intense partial melting at the base of the lithosphere.

In the view of the mantle convective circulation expressed here and in Peltier (1980) the oceanic lithosphere is the cold surface thermal boundary layer of the large scale flow which effects plate creation, destruction, and relative motion. The motion of the lithosphere and of the mantle beneath it are strongly coupled and there is no shear between them. In Section 4 we show that the ability of this upper boundary layer to remain intact, in spite of the large temperature drop across it, is again due to the temperature dependence of viscosity. The high viscosity of the cold lithosphere inhibits the development of buoyantly driven instabilities in the boundary layer itself and the region beneath is stable because the temperature field is adiabatic. It follows from this analysis that the existence of any significantly energetic "second scale" of convection in the near surface region is highly unlikely, in particular that described by Richter and Parsons (1975). The second scale of convection which is required to explain the existence of hot spots in plate interiors is derivative of secondary instability of the bottom boundary layer as discussed above. We will show in this section following Jarvis and Peltier (1980,1981 a,b) that observation 9 of Table I can be understood as a consequence of partial internal heating of the convective circulation and can be used to infer the concentration of radioactivity itself. This inference turns out to be in accord with a mantle convective circulation driven to a considerable extent by heating from below.

To the extent that the convective overturn time for the mantle is shorter than the timescale over which significant changes in the mean temperature of the planetary interior occur, one may investigate the impact of convection upon the thermal history of the Earth using the parameterization scheme introduced in Sharpe and Peltier (1978,1979). Application of such schemes to studies of the cooling history show that a sustained flow of heat across the core mantle boundary which is comparable to that at the surface is not at all inconceivable in that, for a reasonable initial state, it does not lead to solidification of the core within a time equal to the Earth's age. The thermal histories predicted by such models are strongly governed by negative feedback due to the temperature dependence of viscosity and are strongly constrained by the mantle viscosity required to fit postglacial rebound data. These ideas are discussed in Section 5 and they raise anew the dilemma of the thermal history as it was first encountered in the work of Lord Kelvin (e.g. see Burchfield, 1975, for a review). The view of the thermal history developed here, in which the present day convective circulation is forced significantly by the secular cooling of the planet, has the added desireable characteristic that it provides an immediate explanation of the source of energy which sustains the motions in the outer core which are required to maintain the geodynamo. Growth of the inner core as the planet cools drives a compositional convection which is very efficient in sustaining the dynamo process (Loper and Roberts, 1981).

2. Mantle Viscosity and convection

The importance of determinations of the variation of mantle viscosity with depth as a means of distinguishing between competing convection models can best be illustrated using a simple scaling argument in terms of a specific example. The "degree" of convective instability of a layer of fluid is determined by the Rayleigh number which, for a heated below geometry, takes the form $Ra = g\alpha\Delta TL^3/\kappa\nu$. In this expression, g is the gravitational acceleration, α the coefficient of thermal expansion, ΔT the temperature drop across the layer, L the layer depth, κ the thermal diffusivity, and ν the kinematic viscosity. Consider the competing hypotheses of upper mantle and whole mantle convection and suppose that the upper mantle is characterized by kinematic viscosity ν_1 and length scale d and the lower mantle by kinematic viscosity ν_2 and length scale (D-d). We expect convection limited to the upper mantle if the Rayleigh number for this region alone exceeds that for the mantle as a whole. Call these Rayleigh numbers R_1 and R_2. If we assume that g, α, and κ are independent of depth then
$R_2/R_1 = (\Delta T_2/\Delta T_1) (D/d)^3 (\nu_1/[(D-d)\nu_1/D+\nu_1 d/D])$.
With $d/D \sim 0.2$ as the ratio of the thickness of the upper mantle to the thickness of the whole mantle and $(\Delta T_2/\Delta T_1) = 1$, which is certainly an underestimate, it follows that $R_2/R_1 > 1$ unless $\nu_1/\nu_2 < 10^{-2}$. The viscosity of the lower mantle must therefore exceed the viscosity of the upper mantle by more than two orders of magnitude if convection is to be confined to the upper region at all. This estimate is confirmed as an underestimate of the contrast required by linear stability analysis (Peltier, 1972). If we can measure the viscosity contrast across the transition zone, it follows that we should be able to distinguish which of the two convection hypotheses is most likely. The above argument of course relies upon the assumption that the mantle is chemically homogeneous and therefore upon the assumption that the 670 km discontinuity is a phase change. If the boundary were not a phase change, but were rather a chemical boundary, then mixing might be prevented even if there were no large increase of viscosity across the boundary.

The only geophysical method which is capable in principal of providing information concerning the viscosity of the mantle from depths in excess of 670 km is that which uses postglacial rebound data. Such data are obtained from analyses of the deformation of the planet forced by the last large scale deglaciation event which began ca. 18,000 years BP. Triggered by a major change in global climate, this unloading event constitutes a natural stress relaxation experiment, the results from which have been conveniently recorded in the geological record. The data set consists mainly of relative sea level histories obtained by the ^{14}C dating of raised or submerged beach material. When the height above present sea level ($\pm$) of each of a sequence of such beaches at a given geographic location is plotted as a function of its ^{14}C age (corrected to give proper sidereal age), one obtains a single relaxation curve for that site which measures the time dependent separation between the geoid and the surface of the solid Earth. Massive quantities of such data are now available from a global distribution of sites and these data may be inverted using a physical model of the process to recover a viscosity profile.

It is not my purpose here to review the structure of the model since recent and complete discussions are available elsewhere (Peltier et.al.,1978; Wu and Peltier,1981,a,b;Peltier 1981) but it is important to restate the major inferences which have been made in terms of it. The calculations completed to date show quite clearly that it is possible to fit the relative sea level data from both the Canadian and Fennoscandian centres of rebound using a model in which the stress strain relation is Newtonian (viscosity independent of stress) and in which the effective viscosity is independent of time (steady state rheology). Although this does not prove that Newtonian steady state creep does in fact govern the observed stress relaxation, it does demonstrate that non-Newtonian and/or transient effects are not discernible in the data themselves. A series of comparisons between observed relative sea level data and predictions made using a particular combination of melting history and viscosity profile is shown in Figure 1. The observed data are from a sequence of sites including locations within the Laurentide depression centred upon Hudson's Bay and from points outside the depression extending along the eastern seaboard of North America. The specific locations are noted in the Figure caption. Both regions of submergence and emergence are well fit by the model as are the data at sites at which the sea level history is non-monotonic (Bay of Fundy). Non-monotonic sea level histories are observed at sites near the edge of the original ice sheet and are produced by the inward migration of the forebulge following deglaciation (Peltier, 1974). The viscosity profile for this model which fits the sea level data has infinite viscosity in a layer which is 120 km thick at the surface (the lithosphere) and a viscosity of 10^{22} Poise (c.g.s. units) from the base of the lithosphere to the core mantle boundary. The ice sheet history is a modification of that tabulated in Peltier and Andrews (1976), effected by reducing the Laurentide ice thickness by approximately 30% and delaying melting by approximately 2000 years to remedy the discrepancies between theory and ob-

servation found by Peltier et.al. (1978). A detailed discussion of this process will be found in Wu and Peltier (1981 b) where it is furthermore shown that the sea level data themselves do not strongly constrain the viscosity of the mantle at great depth beneath the transition region. For example, an increase of viscosity of two orders of magnitude at a depth of 1000 km does not lead to significant misfits between theory and observation. Such a large contrast at a depth of 650 km would, however, be reflected by RSL observations. The viscosity of the upper mantle is very well constrained by the sea level data to the value of about 10^{22} Poise which has been obtained from most previous analyses of strandline data since Haskell (1937).

In order to strongly constrain the viscosity beneath the transition region, we are obliged to consider data in addition to that from RSL. The most useful additional data set is that concerning the magnitude of the free air gravity anomaly over the once ice covered area. This anomaly provides a direct measure of the extent of current isostatic disequilibrium and as such is a signal which is dominated by the longest relaxation time modes excited by the deglaciation event. Such modes (Peltier, 1976; Wu and Peltier, 1981 a) are most sensitive to the viscosity distribution at depth in the mantle. In Figure 2, I compare the observed free air gravity anomaly over Hudson's Bay (Walcott,1970) with that predicted using the same model employed to fit the RSL data. Although the general form of the anomaly mimics the observed (coincidence of the zero anomaly contours) the magnitude of the anomaly in the central depression (-22 mgal) is somewhat smaller than that observed (-33 mgal).

Calculations discussed at length in Wu and Peltier (1981b) show that this misfit between the observed and predicted free air anomalies can be removed by an increase of viscosity which is approximately one order of magnitude across the 670 km seismic discontinuity but definitely less than two orders of magnitude. An increase of two orders of magnitude is too large since it both destroys the fit to the RSL data and predicts a present day free air gravity anomaly which is excessive. On the basis of the postglacial rebound data it therefore seems that the increase of viscosity across the transition region is much to small to confine convection to the upper mantle although it could well be large enough to explain the compressive nature of the deep seismic focal mechanisms (entry 12 in Table I). Although further increases of viscosity with depth beneath the transition region cannot be ruled out by the data if they occur at sufficiently great depth, there is some evidence that this is unlikely (Sammis et.al., 1977). Recent analyses of the polar motion forced by the deglaciation event have furthermore shown that one can fit the polar wander evident in the ILS pole path with the same

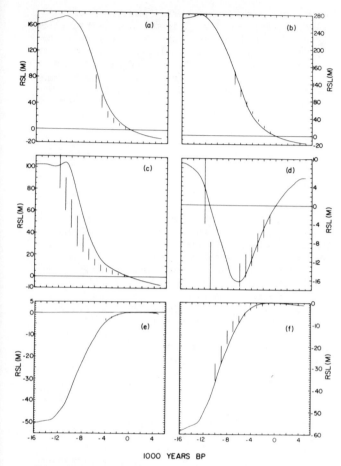

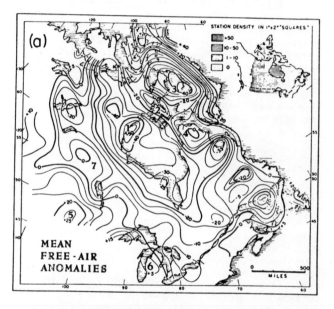

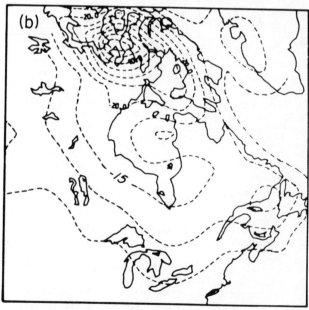

boundary layer at the interface between them in which the temperature would rise rapidly as a function of depth. Since the preferred geochemical model has the upper mantle essentially depleted of radioactive heat sources the temperature increase through this boundary layer would be expected to be on the same order as that across the lithosphere itself i.e. about 10^3 °C. Since the viscosity depends exponential

Fig. 1. Comparison of predicted and observed relative sea level curves for a series of sites along the eastern seaboard of North America, ranging from the region of continuous emergence to the region of continuous submergence. Note between these extremes RSL is not monotonic. The locations of the sites are (a) Ipik Bay, 69°N 74°W; (b) Churchill, 58°N 96°W; (c) N.W. Newfoundland, 51°N 56°W; (d) Bay of Fundy, 45°N 65°W; (e) Southport, N.C., 34°N 78°W; (f) Bermuda 32°N 65°W.

viscosity profile required by the uplift data (Sabadini and Peltier, 1981).

The fact that mantle viscosity appears to in-crease somewhat across the 670 km boundary argues rather strongly against the simple 2-box models which geochemists have recently been employing to interpret the isotopic differences between mid-ocean ridge basalt (MORB) and oceanic island basalts (OIB), although these models are in accord with the picture of the mantle convective circulation preferred by some authors (McKenzie et.al., 1974; Richter, 1978). If the upper and lower mantles were convecting separately then there would be a sharp thermal

Fig. 2. Observed free air gravity anomaly over eastern Canada (top) and the predicted gravity anomaly map using the model with uniform mantle viscosity which fits the RSL data as shown in Figure 1.

upon temperature the boundary should therefore be marked by a sharp viscosity <u>decrease</u>, even allowing for a substantial increase of the creep activation energy in the more dense phase. This is ruled out by the rebound data. Although one might invoke an increase of activation energy which is just sufficient to keep the viscosity constant across the phase boundary this might be considered a somewhat artificial explanation.

3. Convection in the laboratory and in the Earth

In the laboratory, convective flows are described quite accurately in terms of the Boussinesq equations in which the transport coefficients κ and ν are taken as constants and variations of density are neglected everywhere except in the buoyancy term of the momentum balance equation. In this approximation, the non-dimensional field equations are (Peltier, 1972)

$$\frac{Ra}{Pr} \frac{du_i}{dt} = \frac{1}{\alpha'} (-\partial_i p + \rho \lambda_i) + \nabla^2 u_i \qquad (1a)$$

$$\partial_j u_j = 0 \qquad (1b)$$

$$\frac{dT}{dt} = \frac{1}{Ra} (\nabla^2 T + h) \qquad (1c)$$

$$\rho = [1 - \alpha(T-T_0)] \qquad (1d)$$

respectively for momentum, continuity, internal energy, and state. The variable h is the non-dimensional internal heating rate and $\alpha' = at$ where t is the temperature scale. If the layer is heated entirely from below then h = 0 and $Ra = g\alpha\Delta T L^3/\kappa\nu$ as before, but if $h \neq 0$ and the heat flow through the bottom boundary vanishes, then $Ra = g\alpha L^5 (H/\rho c_p)/\kappa^2\nu$ so that the Rayleigh number goes as the fifth power of the thickness L rather than as the cube. Regardless of the way in which the layer is heated the Prandtl number $Pr = \nu/\kappa$. For the Earth's mantle this number is on the order of $10^{23} - 10^{24}$ (using $\nu = 3.10^{21} cm^2 s^{-1}$ from postglacial rebound, and $\kappa = 10^{-2} cm^2 s^{-1}$ from direct laboratory measurement.) It follows from the large value of the Prandtl number that the inertial force in (1a) is completely negligible for mantle flows.

Equations (1) are relatively easy to solve, in spite of the fact that they are strongly non-linear. Two dimensional steady state solutions are the simplest to construct and such solutions exist (with $\underline{u} \neq 0$) only for Ra > Rc where Rc is the critical Rayleigh number deduced on the basis of linear stability theory. Such solutions may be obtained using a variety of numerical methods, the most common of which is finite differencing. Here we will illustrate the form of such solutions using some which were constructed (Peltier, 1980b) using the Galerkin formalism which has been exploited with such success by Busse (1978, for a review) and his

co-workers. In Figure 3 are shown two such solutions, one for a flow driven entirely by heating from below and one driven entirely be heating from within. Both calculations are for $Ra \sim 10Rc$ and each is illustrated by a conventional stream function, isotherm, and mean temperature field sequence. The main difference between the two heating configurations is seen in the structure of the mean temperature field (horizontal average as a function of height). The heated below circulation has thermal boundary layers adjacent to both horizontal boundaries whereas, in the heated within case, the lower boundary layer is absent. A second and related difference can be seen in the two streamfields where an asymmetry is evident in the internally heated case. This asymmetry consists of the migration of the stagnation point in the cell core away from the centre and towards the corner from which the cold thermal boundary layer descends into the interior. An experimental observation of this effect at high Rayleigh number, where it is much more pronounced, is described in De la Cruz(1976). In the laboratory, the horizontal scale of the realized flows is such that the aspect ratio of the circulation is near 1 for either heating configuration. Cells with either much smaller or larger aspect ratio are unstable (Busse, 1978) and the instabilities return the circulation to one with aspect ratio near 1. This non-linear scale selectivity is an important property of convection and one difficulty which hounds the upper mantle convection hypothesis is that it requires aspect ratios in excess of 10 to explain the observed plate scales. No fully satisfactory explanation for the realization of such large aspect ratio cells has ever been provided. It will be noted that the assumption of whole mantle convection, however, immediately explains the observed mean plate scale (entry 1 of Table I).

At high Rayleigh number the heat transport across the layer is completely dominated by the vertical thermal plumes which mark the edge(s) of each cell. Even at Ra=10Rc (see Figure 3) these plumes are very well developed and at higher Rayleigh number the lateral thickness of these regions decreases as does the vertical thickness of the thermal boundary layers. In Figure 4 these effects are illustrated by superpositions of isotherm and horizontal and vertical average temperature profiles for heated below configurations in which $Ra = 10^3 Rc$ and $Ra = 10^4 Rc$. Plate (b) is the parameter range most relevant to the Earth's mantle. This calculation was done using a modification of the finite difference model described by Jarvis and McKenzie (1980) and further high resolution results from its application are described elsewhere (Jarvis and Peltier, 1980, 1981a,b). Here, I wish only to point out that at such high Rayleigh numbers the mean temperature field develops a region of reversed curvature (positive stability) adjacent to the thermal boundary

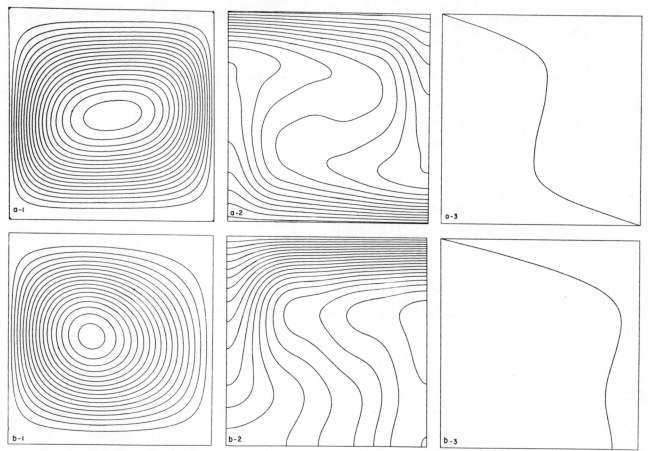

Fig. 3. Results from the Galerkin calculation for both heated below and heated internally flows at Ra $\sim$ 10 Rc. Sequence (a) is for the heated below case and (b) for heated internally. In each sequence plate (1) is the streamline pattern, (2) the field of isotherms, and (3) the horizontally averaged temperature profile as a function of height. Comparing a3 and b3 it is clear that no thermal boundary layer forms adjacent to the lower adiabatic boundary in the internally heated case.

layers themselves. Given the hypothesis stated in the introduction, that the surface lithosphere is synonymous with the thermal boundary layer of the large scale flow, it would be interesting to see if one might discover geophysical evidence for the existence of such a region beneath the lithosphere. Because of the temperature dependence of viscosity, this region in which the temperature overshoots the mean value characteristic of the core of the circulation would correspond to a low viscosity zone. Furthermore, this low viscosity zone actually fades in definition (becomes less pronounced) with distance away from the hot plume (oceanic ridge). This could provide a very nice explanation of the fact that the seismic low velocity zone is found only under relatively young oceanic lithosphere whereas it is conspicuously absent beneath continental lithosphere (entry 13 in Table I). It will be noted that this explanation depends upon the association of the oceanic

lithosphere with the thermal boundary layer of the large scale flow. Furthermore, this circulation would have to be driven substantially by heating from below because it is only in this case that any substantial hot rising plume exists. I shall return to this point below.

For values of the Rayleigh number which are as large as those we expect to be characteristic of the mantle $(0(10^{7}))$ it is rather expensive computationally to construct hydrodynamic solutions numerically. In this limit, however, we may invoke boundary layer theory to obtain estimates for the characteristic properties of the circulation. These ideas, recently reviewed in Peltier (1980b) were first introduced in the context of discussions of mantle convection by Turcotte and Oxburgh (1967). Jarvis and Peltier (1981) have recently provided a systematic investigation of the validity of the assumptions upon which boundary layer theories are based and point-out several important quantitative deficiencies.

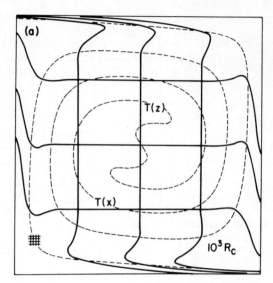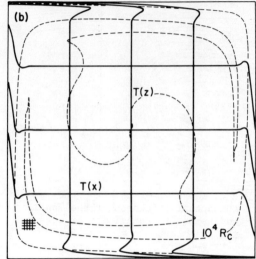

Fig. 4. Vertical and horizontal temperature profiles through heated from below convective circulations at Ra = 10^3 Rc (a) and Ra = 10^4 Rc (b). The dashed lines are reference isotherms for the two flows and demonstrate that at these high Rayleigh numbers the cell cores are very nearly isothermal.

Boundary layer theory, for high Rayleigh and Prandtl number heated below convection in a layer between stress free boundaries, leads to explicit relations between the Rayleigh number and the circulation properties. These relations may be written in the form

$$\delta = a_1(\Delta) \ L \ (Rc/Ra)^{1/3} \qquad (2a)$$

$$u = a_2(\Delta)Ra^{2/3} \ \kappa/L \qquad (2b)$$

$$w = a_3(\Delta) \ Ra^{2/3} \ \kappa/L \qquad (2c)$$

$$q = a_4(\Delta) \ (Ra/Rc)^{1/3} \ k\Delta T/L \qquad (2d)$$

where δ is the boundary layer thickness in the mean temperature field, w and u are characteristic vertical and horizontal velocities and q is the total heat flow across the layer. The constants $a_i(\Delta)$ are functions of the aspect ratio of convection Δ which can be determined by a detailed matching of the boundary layer solutions for the vertical plumes and horizontal boundary layer to the isothermal Stokes flow which obtains in the core of the cell (e.g. Roberts, 1979). Olson and Corcos (1980) have recently given a reanalysis of the boundary layer theory in which the qualitative form of the mean temperature profile shown in Figure 4 is recovered (i.e. the stably stratified regions outside the boundary layers themselves). For an aspect ratio Δ_m slightly larger than that at marginal stability the coefficients are $a_1(\Delta_m) = .5$, $a_2(\Delta_m) = .143$, $a_3(\Delta_m) = .251$, and $a_4(\Delta_m) = 1$. From (2d) we obtain an expression for the Nusselt number Nu which is Nu $= a_4(\Delta)(Ra/Rc)^{1/3}$ and this fits the

Nusselt number versus Rayleigh number curve from numerical calculations (Moore and Weiss, 1973: Jarvis and Peltier, 1981a) quite nicely.

Equations (2) can be employed to determine the extent to which the properties of convection in the laboratory are similar to the properties of thermal convection in the Earth's mantle. To do this comparison we note that the observed lithospheric thickness provides a good estimate of the thermal boundary layer thickness δ, the mean plate speed provides a good estimate of the characteristic horizontal velocity u, and the surface heat flow q is measured directly. The left hand sides of (2a,b,d) are therefore geophysical observables. The simplest way of investigating the compatibility of the geophysical data with the convection hypothesis is to combine relations (2) together in such a way as to eliminate the unknown temperature drop ΔT across the system which appears in the Rayleigh number Ra = $g\alpha\Delta T L^3/\kappa\nu$. This can be done by combining (2b) and (2d) and solving for the depth of convection L as

$$L_1 = \frac{a_4}{a_2}^{1/2} \left[\frac{c_p \rho \nu}{Rc^{1/3} q\alpha g} \right]^{1/2} \qquad (3a)$$

or by combining (2a) and (2b) to obtain

$$L_2 = \frac{(\delta^2/\kappa)}{a_2 a_1^2 Rc^{2/3}} \cdot u \qquad (3b)$$

Clearly equations (3a) and (3b) provide two independent estimates of the depth of convection L in terms of geophysical observables. If in (3a) we substitute u $= 4 \text{cm yr}^{-1}$ as the mean plate speed, q = .75 erg cm s as the surface heat flow, $\rho\nu = 10^{22}$ Poise for the mantle viscosity from postglacial rebound, $g = 10^3$ cm s^{-2}, $\alpha = 3 \times 10^{-5}\rho$K^{-1}, $c_p = 1.2 \times 10^7$ erg gm^{-1} °K^{-1}, Rc $\sim 10^3$, then we obtain $L_1 \sim 8 \times 10^3$ km. If we use a boundary layer thickness $\delta = 100$ km (postglacial rebound) and a thermal diffusivity $\kappa = 10^{-2}$cm^2 s^{-1} in (3b) with the other parameters as previously stated we obtain the second estimate $L_2 \sim 3.6 \times 10^3$ km. That both these estimates give values for the depth of convection which are near the thickness of the mantle may be taken as evidence of the compatability of the whole mantle convection hypothesis with the observations when an aspect ratio near one is assumed. Such disagreements as do exist between these estimates and the actual thickness of the mantle may be understood as due to sphericity or weak variations of the transport properties. For example, the heat flow we use in (3a) for the plane layer scaling should be increased by a factor of about 2 over the observed surface value to give the mean flow through the shell since the flow itself decreases geometrically from inside to outside. This will decrease the estimate L_1. The fact that the boundary layer scaling gives two independent estimates of the depth and that these are in agreement with the whole mantle hypothesis was pointed out in Peltier (1980b) after an analysis in Sharpe and Peltier (1978) in which the boundary layer scaling was first employed to demonstrate compatability of observations with this hypothesis. Insofar as the mean properties of the observed circulation are concerned they are related precisely as they would be if they were made on the surface of a heated below convection apparatus in the laboratory (except for effects due to non stress free boundary conditions in the lab.).

Having established $L_1 \sim L_2$ we may restate (3a,b) by equating them and solving for the viscosity of the fluid in terms of the other observables. This leads to the result

$$\rho\nu = \delta^4 q \, \frac{(\alpha \, g)}{(c_p \kappa^2)} \cdot \frac{1}{a_1^4 \, a_4 \, Rc} \qquad (3c)$$

It should come as no surprise that when we substitute previously stated values for quantities on the right hand side we get a molecular viscosity near 10^{22} Poise (closer to 10^{23} Poise with necessary parametric corrections) which is the value which we observed from analyses of postglacial rebound. Indeed, we may now state that if the viscosity were much different from the effective viscosity obtained in such studies then the convection hypothesis would not be nearly as attractive as it is. Postglacial rebound must be seeing essentially the steady state viscosity and not merely a transient creep value.

From the boundary layer arguments given above it is quite clear that the first 5 constraints on models of the mantle convective circulation which are listed in Table I are met extremely well by the simple heated below whole mantle convection model. From the structure of (3a) and (3b) the importance of constraints 4 and 5 in the Table is clear. Without these two extra constraints, which are both obtained from postglacial rebound studies, (3a) and (3b) could not be employed to make any significant statement concerning the convective circulation. With them, however, the scaling laws become important tools of inference. Application of these scaling laws verifies quite conclusively that thermal convection driven mainly by radial superadiabaticity accounts remarkably well for the observed mean properties of the convective circulation. To the extent that these scaling laws are valid for the mantle one might argue that whole mantle convection is preferred.

Direct application of the boundary layer equations (2) to the problem of mantle convection, however, involves several implicit assumptions and before the validity of these assumptions is understood we will be in no position to proceed with the development of our conceptual model. The first and perhaps most important is that the flow behaves as if the viscosity within it were a constant equal to that which obtains in the adiabatic core of a cell, in spite of the fact that there exists a large viscosity contrast through the thermal boundary layer (lithosphere). The success of the constant viscosity scaling laws may be taken to imply that this is in fact the case - although precisely how it could be the case remains unclear. I have suggested elsewhere (Peltier, 1980b) that it is necessary to invoke melting to understand how hot mantle derived material rising beneath the ridge crest is able to reach the surface. Without melting, the surface would remain unbroken and plates could not form. Once a system of plates is established, however, the large viscosity contrasts across them appear to exert no active influence on the flow. These plates, which are hypothesized to exist in very roughly a 1:1 relation with deep mantle convection cells, are to be associated with the thermal boundary layers of these cells. I have previously referred to this as the "passive" viscosity hypothesis in connection with discussions of the relation between the surface lithosphere and the surface thermal boundary layer (Peltier, 1980b).

The second major assumption which is implicit in the direct application of equations (2) to the mantle is that the circulation is driven entirely by heating from below. In the context of the whole mantle convection model this would imply that the rate of heat loss from the core of the Earth were equal to the rate of heat loss from the planet as a whole if the system were in a steady state. Such a steady state could exist only if the heat loss from the core were precisely compensated by insitu heat generation (e.g. radioactive decay or latent heat release due to the freezing of the inner core) and this is impossible on a sufficiently long timescale since the intensity of such processes decreases with time. Heat loss due to secular cooling (Sharpe and Peltier, 1979) must appear at the surface of the planet where it contributes to the observed surface heat flow and this effect is not included in the steady state scaling laws embodied in equations (2). To the extent that this secular cooling or heating occurs on a timescale significantly longer than that characteristic of the convective circulation in the mantle, however, it can be demonstrated that this effect will not substantially modify the heated from below assumption. If the mantle itself contains a significant complement of radioactivity, though, the heated from below model of the convective circulation would be singularly inappropriate.

The manner in which the Earth's radioactivity is distributed throughout its volume and the magnitude of its net complement of such elements are characteristics of its chemical structure which are imperfectly understood. The following positions are nevertheless rather widely accepted as true: (1) the amount of radioactivity in the Earth's core is small and may be effectively zero, (2) the total radioactive heat production throughout the entire volume of the planet is much less than would be required to explain the observed surface heat follow, (3) about 1/3 of this radioactivity is now found in the continental crust which has been stabilized over the past 3.5 billion years from material derivative of continuous chemical differentiation of the mantle. Quantitative estimates of the fraction of the total heat flow which could be explained by radioactive heating have been published by O'nions et.al. (1979) who obtain 1/2 and Ringwood (1979) who prefers 1/3. Since these are estimates of total radioactivity and since much of this is now found in continental crust (a chemical boundary layer) where it cannot contribute to driving the mantle convective circulation, it seems clear that the geochemically preferred convection model is one which is forced predominantly by heating from below. A priori geochemical argument therefore suggests the appropriateness of the scaling laws (2).

That heated from below forcing must be important can be argued from a still more direct point of view if one accepts the previously stated hypothesis that there is a more or less 1:1 relation between surface plates and deep mantle convection cells and that the oceanic plates are in fact the thermal boundary layer of the large scale flow. We must then argue on the basis of entry 7 in Table I that heating from below is important. Only if there is a substantial hot plume rising beneath the ridge crest will the characteristic square root of the age variation of heat flow and topography be observed. If the flow were driven entirely by heating from within, for example, there would be no variation of heat flow and topography with age except near the descending cold plume. This is because no

hot rising plume exists for a heated within flow and this in turn is due to the fact that the heated within flow has no bottom boundary layer (see Figure 2). The observed variation of heat flow and topography with the square root of ocean floor age can therefore be used directly to infer that heating from below is important.

In fact these data can be employed to obtain a quantitative estimate of the ratio (μ) of internal heat production to surface heat loss by requiring a numerical convection model to fit observation 9 in Table I. Jarvis and Peltier (1980, 1981a,b) have recently shown that convection cells which are partially heated from within have associated surface topography which flattens away from the $\sqrt{AGE}$ variation which is characteristic of a purely heated from below model. When the effect of constant surface plate speed (which is itself a consequence of the temperature dependence of viscosity) is taken into account, then the heat flow has been shown to follow the $\sqrt{AGE}$ behaviour across the entire cell. This is also characteristic of the observations made on the ocean floor: bathymetry flattens but heat flow does not. By fitting the observed percentile flattening in the bathymetry data we obtained an estimate of the ratio μ of ≈ 0.2. This estimate is in accord with the independent estimate discussed above which was obtained on the basis of a geochemical argument. The ocean floor data therefore provide strong support for the present day validity of the heated from below scaling embodied in equations (2).

Additional restrictions on the use of (2) must also be mentioned besides those discussed above. Of these, the most important concerns the fact that the relations obtain only for Cartesian geometry and we might expect that the effect of sphericity could be important. Such effects have been discussed in Peltier (1980b) where it is argued that so long as one employs in (3a) the heat flow q which is an average for the shell then the plane layer scaling should not be terribly inaccurate. This correction was in fact already discussed above in connection with the application of (3a). An equally important effect of sphericity is that which occurs in the structure of the thermal boundary layers and this can be demonstrated only by direct numerical calculation of the fully non-linear steady state flows (e.g. Young, 1974). Because the surface area of the core mantle boundary is smaller than the surface area of the Earth by approximately a factor of 4 it is quite clear that in the steady state the heat flow across the lower boundary must be 4 times the heat flow across the Earth's surface for a heated below configuration. If the thermal conductivities were constant in the two regions then either the temperature drops across the two layers are the same and the bottom boundary layer is one quarter the thickness of the upper, or the boundary layer thicknesses are the same and the temperature drop across the bottom boundary layer is four times that across the upper. Numerical calculations (Young, 1974) show that the boundary layer structure actually adjusts in the second way. I should point out here that the opposite assumption was made in Sharpe and Peltier (1979) although it does not effect the results. This effect of sphericity may be important for the arguments which we will present below in support of the idea that the bottom boundary layer is highly unstable. We should also note, however, that temperature dependence of viscosity would lead to a marked thinning of the lower boundary layer so that the net result of these two competing effects is unclear.

The last qualification to the use of (2) which I shall discuss here concerns affects due to the compressibility of mantle material. The importance of such effects is measured by the dissipation number $\tau = g\alpha d/c_p$ which was introduced in Peltier (1972). This non-dimensional group determines the importance of both compression work and viscous dissipation (shear heating) on the energy balance of the flow such that the larger τ the greater the importance. In Peltier (1972) the effect of τ was discussed from the point of view of linear theory in which limit the effect of the adiabatic gradient first appears. Recent finite difference solutions to the fully non-linear anelastic system by Jarvis and McKenzie (1980) have shown that this in fact the only important effect for $0.5 < \tau < 1$ such as would obtain for whole mantle convection. It therefore follows that if one simply replaces ΔT in the definition of Ra by $\Delta T - \Delta T_s$, where ΔT_s is the adiabatic temperature drop across the layer, then the boundary layer results (2) continue to hold. In deriving (3a,b) from (2), however, the unknown ΔT was explicitly eliminated so that the results of applying (3a,b) are unaffected by the finite compressibility of the mantle.

Having demonstrated here that mantle convection does appear to scale like a laboratory system, if heated below whole mantle convection is assumed, in the next section we will proceed to argue that any small scale motions which do co-exist with the large scale flow must be derivative of local thermal instabilities within the boundary layers, and that only the lower boundary layer (which must exist to the extent that the flow is heated from below) is susceptible to such instability. Such spontaneous disturbances, we shall argue, might evolve in a highly non-linear fashion into the sharp thermal plumes which produce the phenomenon of volcanism in plate interiors (e.g. Hawaii).

4. Boundary layer stability: surface plates and mantle plumes

The basis of the boundary layer approximation for heated from below flows is that at high Rayleigh number the flow near the surface is such that vertical diffusion of heat is perfectly balanced by horizontal advection. In this approx-

imation the steady state dimensional form of the energy equation (1c) is

$$u \frac{\partial T}{\partial x} = \kappa \frac{\partial^2 T}{\partial z^2} \qquad (4)$$

where $\partial^2/\partial x^2 \ll \partial^2/\partial z^2$ has been assumed. Under the similarity transformation $y' = z/x^{1/2}$ (4) reduces to an o.d.e. in y' which may be non-dimensionalized by writing $y' = y.2(\kappa/u)^{1/2}$ to give

$$\frac{d^2T}{dy^2} + 2y \frac{dT}{dy} = 0 \qquad (5)$$

so that subject to boundary conditions $T = T_s$ on $x = 0$ and $T \to T_m$ as $z \to \infty$ the dimensional solution is

$$T = T_s + (T_m - T_s) \ \mathrm{erf} \ [\frac{z}{2} (\frac{u}{\kappa x})^{1/2}] \qquad (6)$$

and this is the boundary layer solution for the temperature field in the boundary which cools horizontally as it moves away from the ridge crest. T_s is the temperature at the ridge crest $x = 0$ and T_m is the temperature of the material in the core of the convection cell beneath the boundary layer. The surface heat flow associated with the temperature distribution (6) is

$$q_s = K \frac{\partial T}{\partial z} \Big|_{z=0}$$

$$= K(T_m - T_s) [\frac{u}{\pi \kappa x}]^{1/2} \qquad (7)$$

Since the age of the lithosphere is just x/u, (7) predicts that the heat flow will decrease inversely as the square root of the age (item 7 in Table I). From (b) we may also show, assuming isostatic compensation, that the bathymetry of the sea floor also increases as the square root of the age. A detailed analysis of the errors incurred in these approximations is given in Jarvis and Peltier (1981a) who show that they may be extreme although they are somewhat mitigated by the rigid behaviour of surface plates.

Here what we wish to do is to use the boundary layer temperature profile (6) in conjunction with the passive viscosity assumption to construct a basic state boundary layer temperature and viscosity field and then to investigate the stability of this state. It is important to note that the simple error function profile (6) is monotonic and therefore does not include the stably stratified region seen in the high Rayleigh number numerical profile of Figure 3. However, as previously stated, this feature is found only at young age and for our present purposes the monotonic profiles will suffice. Given the temperature field (6) through the boundary layer we may determine the corresponding viscosity profile by invoking the fact that the creep mechanism is thermally activated to write

$$\nu = \nu^{*} \exp \left\{ \frac{Q^{*}}{RT} \right\} \qquad (8)$$

In order to force (8) to give the observed viscosity ν_{m} in the core of the cell (beneath the thermal boundary layer) we may rewrite (8) in the form

$$\nu = \nu_{m} \exp \frac{Q^{*}}{R} (\frac{1}{T} - \frac{1}{T_{m}}) \qquad (9)$$

where T is the absolute temperature, ν_{m} is the "asymptotic" viscosity of the mantle beneath the boundary layer as measured by postglacial rebound. and T_{m} is the temperature in the cell core. $Q^{*} = E^{*}_{m} + pV^{*}$ is the activation enthalpy where E^{*} is the activation energy (125 kcal/mole for Olivine, Kohlstedt and Goetze, 1974, Kohlstedt et al., 1976) and V^{*} is the activation volume (10.6 cm^{3}/mole < V^{*} < 15.4 cm^{3}/mole; Ross et al., 1979).

Seen from a frame of reference in which the mean flow is at rest, the boundary layer temperature profile is a slowly varying function of time and we may assume it to be stationary in time for present purposes, an assumption which can be verified a-posteriori. As in Yuen and Peltier (1980) and Yuen, Peltier and Schubert (1981) we may investigate the stability of the boundary layer (6) in which the viscosity is as in (9) by solving the appropriate two point boundary value problem. When the secondary motion is described through a stream function Ψ and a temperature perturbation Θ , each of which varies as exp $[\mathrm{i}1 x + \Omega t]$, where the growth rate Ω and the horizontal wave number ℓ are both real, then Ψ and Θ are determined by the following coupled ordinary differential equations.

$$D^{4}\Psi + 2 \; D\mathrm{ln}\nu \; D^{3}\Psi + (\nu^{-1}D^{2}\nu - 21^{2})D^{2}\Psi - 21^{2}D\mathrm{ln}\nu D\Psi$$

$$+ (1^{4} + 1^{2}\nu^{-1}D^{2}\nu)\Psi = (-1Ra/\nu)\Theta \qquad (10a)$$

$$D^{2}\Theta + (\Omega - 1^{2})\Theta = 1DT\Psi \qquad (10b)$$

where $D = d/d\xi$ in which ξ is the vertical co-ordinate made non-dimensional with respect to the depth δ at which the temperature has reached 97.5% of its' asymptotic value. The boundary conditions subject to which equations (10) are to be solved are $\Theta = \Psi = D^{2}\Psi = 0$ on $\xi = 0$ (the Earth's surface or the core mantle boundary) and $\Theta = \Psi = D\Psi = 0$ so $\xi \to \infty$. In (10) ℓ is also non-dimensionalized with respect to the length scale δ. For given Ra = $(\Delta T/2)\delta^{3}\alpha g\rho/\kappa\nu_{m}$ which is the local Rayleigh number, in which $\Delta T/2$ is the temperature drop across the boundary layer, the non-dimensional growth rate (characteristic time δ^{2}/κ) is the ℓ dependent eigenvalue of the differential system (10). The dimensional horizontal wavelength of the perturbation is $\lambda = 2\Pi\delta/\ell$.

We can use the model embodied in (10) to study the stability of the thermal boundary layers

associated with the global scale flows discussed in the last section and the same formulation can be used to investigate the cold viscous upper boundary layer as is employed with the relatively hot inviscid lower boundary layer and will look at these two cases separately. In figure 5, typical results from a stability analysis of the lower boundary layer are shown for a case in which the viscosity of the lower mantle is 5 × 10^{23} Poise, the temperature drop across the boundary layer is 1500°C and the boundary layer thickness is taken to be 100 km (approximately equal to the observed thickness of D''), and the activation energy is 200 kcal/mole (an extrapolation to depth using the Olivine data, Sammis et al. 1977). This Figure shows the basic state temperature and viscosity profiles and the eddy heat flow which is obtained from the correlation between the vertical velocity and temperature perturbation eigenfunctions $< \Theta \; v \; >$. The peak of the correlation marks the region of the basic state in which the instability resides and the correlation itself has been calculated for the fastest growing mode which has a horizontal wavelength $\lambda = 102.\mathrm{km}$ (on the order of the boundary layer thickness) and a growth time Ω^{-1} of approximately 2.10^{6} years. Clearly this instability is focussed strongly into the relatively inviscid region of the boundary layer nearest the core mantle boundary. Furthermore the growth time of the mode is sufficiently short and its horizontal wavelength sufficiently small that our initial

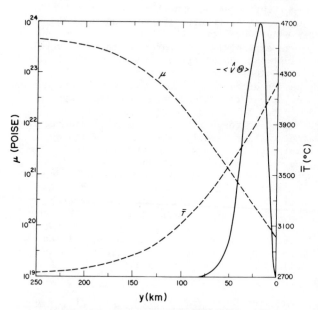

Fig. 5. Temperature ($\bar{T}$) and viscosity (μ) profiles through the thermal boundary layer adjacent to the core mantle boundary. y is the height above the interface. The correlation $<v\theta>$ is the eddy heat flow for the fastest growing mode of instability of the boundary layer.

assumptions that the basic state is stationary in time and parallel in space are justified a-posteriori. In Figure 6 the way in which the growth rate of these violent instabilities changes as a function of the asymptotic viscosity in the mantle overlying the boundary layer is illustrated. The higher the asymptotic viscosity the slower the growth rate for given activation enthalpy. These calculations are discussed in greater detail in Yuen and Peltier (1980b). If partial melting were to occur within the lower boundary layer adjacent to the CMB then the instability would be violent indeed.

The important question is whether the above described calculations, which suggest strong instability in the lower boundary layer, are based upon realistic estimates for conditions at the base of the mantle. I would suggest that the parameters are realistic indeed. The thickness of the lower boundary layer is constrained by the observed thickness of about 100 km for D''. The lower mantle viscosity is a best upper bound based upon the observed rise of viscosity across the 670 km discontinuity extrapolated downward using the method of Sammis et.al. (1977). The choice $\Delta T = 1500°C$ is conservative since it would imply that only about one quarter of the heat leaving the surface of the planet actually crosses the core mantle boundary. The lower thermal boundary layer should be strongly destabilized by the temperature dependence of viscosity and further enchanced if partial melting were to occur.

The question as to how these instabilities might be expected to evolve in time is also an important one. In order to understand this evolution in detail we are obliged to include the full effects of non-linearity upon them and to

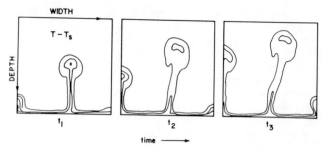

Fig. 7. Isolines of constant potential temperature in a sequence of thermal plumes formed by thermal instability of a thermal boundary layer at the base of the computational domain. Three time slices (t_1, t_2, t_3) through the temporal evolution are illustrated.

follow their transient development. Although such solutions are not yet available for the temperature dependent viscosity case, we are able to answer the question if this effect is suppressed. In Figure 7 is shown a time sequence of snapshots in the evolution of a constant viscosity boundary layer instability in which the disturbance is depicted through isolines of constant potential temperature (Jarvis, unpublished). The calculation was done with the same finite difference numerical model used to produce Figure 4 and the instability was initiated in response to the sudden imposition of a large heat flow from below. After application of the step in heat flow, a thermal wave propagates slowly into the interior of the fluid and this wave has a boundary layer form like (6) with the boundary layer thickness increasing with time. Eventually the critical Rayleigh number for the boundary layer is exceeded and if the initial step in heat flow was large enough a violent instability ensues which develops into thermal plumes as shown in Figure 7. These thermal plumes are remarkably similar to those obtained in the laboratory models of diapiric structures constructed by Whitehead and Luther (1975). Our suggestion here is that these plumes might originate in instabilities of the bottom boundary layer of the large scale mantle convective circulation and that surface hot spots and intraplate volcanism are the consequence of their rise to the surface. Instability of the bottom boundary layer is made possible by the strong temperature dependence of mantle viscosity.

The opposite is the case with the **top** boundary layer since in it the viscosity is very much higher than that which obtains in the core of the circulation and this is a stabilizing rather than a destabilizing effect. That this is the case is illustrated in Figure 8 which is analagous to Figure 5 for the bottom boundary layer. This figure illustrates the basic state viscosity and temperature profiles through the upper boundary layer for a boundary layer thickness of 100

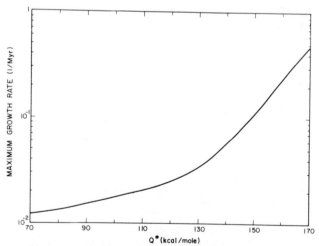

Fig. 6. Growth rate of the fastest growing mode of instability of the thermal boundary layer at the core mantle interface for models with an asymptotic lower mantle viscosity of 5×10^{23} Poise.

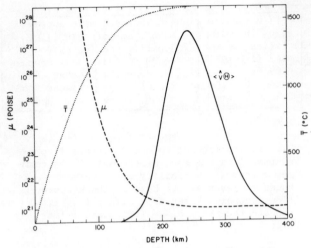

Fig. 8. Temperature ($\overline{T}$) and viscosity (μ) profiles through the thermal boundary layer (lithosphere) adjacent to the Earth's surface. The correlation $\langle v\theta \rangle$ is the eddy heat flux for the fastest growing mode of instability with a sub-lithospheric mantle viscosity of 10^{21} Poise. Note that the temperature dependence of viscosity forces the instability into the deep upper mantle where no available potential energy exists.

km and a temperature drop $\Delta T = 1600^{0}C$ (an over-estimate) in which the activation enthalpy Q^* in (9) has been taken equal to a constant 140 kcal/mole. The asymptotic viscosity in the mantle for this basic state is taken equal to 10^{21} Poise which is an order of magnitude less than the value obtained in postglacial rebound studies, and the boundary layer age is 10^8 years which is greater than the age of 7×10^7 years which Richter and McKenzie and others have found to mark the age beyond which deviations from the $t^{1/2}$ variation of the sea floor topography occur. Although this boundary layer is unstable, the instability is so weak that its existence is a mere mathematical curiosity. This is seen in the perturbation heat flow $\langle v\theta \rangle$ for the fastest growing mode in Figure 8 which has $\Omega^{-1} \sim 10^8$ years and $\lambda \sim 600$ km. The peak in the correlation function is pushed to great depth out of the thermal boundary layer itself and into the region in which the temperature gradient vanishes. There is therefore no available potential energy upon which the instability can feed and this accounts for the small growth rate of the mode. That even this weak instability exists, however, is an artifact of the choice of the low asymptotic viscosity of 10^{21}P for the deep mantle beneath the boundary layer. In Figure 9 I show the growth rate of the fastest growing mode of linear theory as a function of this asymptotic viscosity. If the viscosity of the deep mantle is 10^{22}Poise as rebound analyses suggest then <u>no</u> instability is possible. It seems to me therefore that the suggestion of

the existence of a second scale of convection confined to the upper mantle and presumably driven by boundary layer instability (Richter and Parsons, 1975; Parsons and McKenzie, 1978) is completely untenable. This idea was originally introduced into the mantle convection literature in order to explain entry 9 in Table I, the argument being that the extra heat flow obtained from the second scale would reconcile the observed deviations from the $t^{1/2}$ dependence. A more likely explanation of this observation is that recently advanced by Jarvis and Peltier (1980, 1981a,b) and discussed previously. The observed flattening of bathymetry is a natural consequence of the fact that the mantle convective circulation is partly forced by heating from within. The observed percentile flattening can be employed to measure the concentration of radioactivity in the mantle.

5. Mantle climate and convection

Because thermal convection provides such an efficient means of transporting heat to the surface of the planet from its deep interior, it

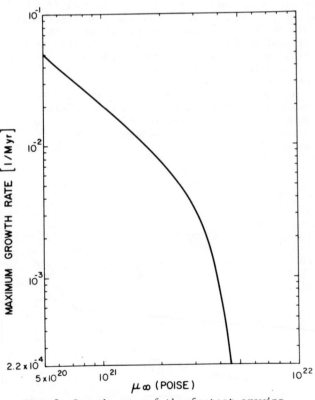

Fig. 9. Growth rate of the fastest growing mode of instability of the upper thermal boundary layer (lithosphere) as a function of the asymptotic viscosity of the underlying mantle. Note that no instability is possible if the mantle viscosity equals the postglacial rebound value of 10^{22} Poise.

completely governs the way in which internal temperatures evolve in time. Having shown in previous sections that the whole mantle convection hypothesis fits most of the observational constraints insofar as present day mean conditions are concerned, in this section I will briefly review recent attempts to demonstrate that the same process probably governed the circulation in the most remote geological past (excepting perhaps for the first 10^9 years or so of Earth history). What one could do in order to make this argument is simply (!) to integrate the full Navier-Stokes equations forward in time from a given set of initial conditions and so produce a full thermal and dynamical history of the planet since accretion !! Even disregarding the enormous technical difficulties which such a program would involve, the detailed dynamical and thermal histories predicted would be completely useless since they depend strongly upon the precise initial conditions which are unknown. If we are willing to focus our attention on the long timescale evolution of the "thermal climate" of the mantle, however, then we may in fact assess the effect of convective heat transport without actually having to solve the dynamical equations. To do this we may employ methods which are commonplace in the modelling of atmospheric climate (e.g. North, 1975) in which one "parameterizes" (Sharpe and Peltier 1978, 1979) the heat transport by convection in terms of the instantaneous temperature field itself. The validity of the parameterization scheme which we have constructed is determined mostly by the existence of a sufficiently sharp separation between the dynamic timescale for convective overturning and the longer timescale on which significant secular changes of mean temperature occur. In the last section we showed that disturbances of the mantle circulation with very much shorter timescales than that for a single convective overturn might exist in consequence of the thermal destabilization of the bottom boundary layer of the mantle wide flow. These fast growing disturbances evolve into thermal plumes which, although they transport a negligible fraction of the total heat to the surface, produce important observational effects. Here we are discussing the very long timescale impact of the radial heat transport by the main flow upon the mean temperature of the mantle and core.

The most general form of the parameterized convection models which we have so far employed is obtained by appropriate averaging of the non-dimensional form of the fully compressible energy equation derived in Peltier (1972) which is

$$\frac{dT}{dt} - \frac{\tau T}{\rho} \frac{dp}{dt} = \frac{1}{\rho Ra} [\nabla^2 T + h] + \frac{2\tau}{\rho} [e_{ij} e_{ij} - \frac{1}{3} \Delta^2]$$ (11)

The terms which scale like the dissipation number $\tau = g\alpha L/c_p$ are respectively the compression

work term (on the left) and the viscous dissipation term (on the right). The anelastic approximation, in which one takes $\rho = \bar{\rho}(r)$ in (11), simplifies the energy equation considerably when coupled with the very accurate additional approximation that $dp/dt \underset{\sim}{} -w\bar{\rho}$ where w is the radial component of flow velocity. Full non-linear time dependent solutions for this system have recently been computed by Jarvis and McKenzie (1980) for base heated thermal convection in liquids. For $.5 < \tau < 1$, as is the case for whole mantle convection, their results show that the dissipation term in (11) has no effect upon the depth dependence of the horizontally averaged temperature profile just as was assumed in the construction of the parameterization scheme in Sharpe and Peltier (1978). There, and in Sharpe and Peltier (1979), it is shown that the energy equation (11), when averaged over spherical shells concentric with the origin, and integrated in radius from the base of the mantle to the Earth's surface can be written dimensionally as

$$\rho \ c_{p,m} \frac{\partial}{\partial \tau} \int_{r_c}^{r_p} r^2 <T_m> dr = r_p^2 q_s - r_c^2 q_c$$
$$+ \frac{1}{3} (r_p^3 - r_c^3) \ Q_m$$ (12)

when ρ_m is the mean density of the mantle, $c_{p,m}$ is the specific heat capacity, $<T_m>$ the height dependent horizontal mean temperature in the mantle, q_c and q_p the heat flow across the core mantle boundary and the Earth's surface respectively, and Q_m the (assumed constant) rate of heating per unit volume in the mantle. We may of course write a similar expression for the core, since (12) is just a simple statement of conservation of energy, and this (with Q_c the rate of internal heating in the core) is

$$\rho_c c_{p,c} \frac{\partial}{\partial \tau} \int_0^{r_c} r^2 <T_c> dr = r_c^2 q_c + \frac{1}{3} r_c^3 Q_c \cdot$$ (13)

The important point in deriving (12) and (13) from (11) concerns the interpretation of the heat flows q_c and q_p. If the whole mantle is convecting and if the convection is driven by heating from below ($Q_m = 0$ or all mantle heat sources concentrated in a thin layer near the surface) then q_c is given by (Sharpe and Peltier 1978,1979)

$$q_c = K_m \frac{(T_c - T_s)}{(r_p - r_c)} \ Nu \ \frac{((r_c + r_p)^2)}{(4r_c^2)}$$ (14)

where K_m is the effective thermal conductivity of the mantle, T_c and T_s are the temperatures at the core mantle boundary and planetary surface respectively, r_c and r_p are the radii of the same surfaces, and Nu is the Nusselt number for convection in the mantle which is given by boundary layer theory (2) in terms of the mantle Rayleigh number by

$$Nu = \frac{(Ra)^{1/3}}{(Rc)} \cdot a_4(\Delta) \qquad (15)$$

with the Rayleigh number in turn defined by

$$Ra = \frac{g\alpha(r_p - r_c)^3(\Delta T - \Delta T_s)}{\kappa \nu_m} \qquad (16)$$

in which the effective mantle viscosity ν_m is a strong function of mantle temperature which may be determined from the homologous temperature form of the creep relation (8) which is

$$\nu_m = \nu^* \exp(gT_0/T_m) \qquad (17)$$

where T_0 is the mantle melting temperature at some reference pressure (depth) and T_m is the actual temperature on the geotherm at that depth. If Q_m is in fact zero we may obtain the surface heat flow from (12) as

$$q_s = \frac{r_c^2}{r_p^2} q_c + \rho_m c_{p,m} \frac{\partial}{\partial \tau} \int_{rc}^{r_p} r^2 <T_m> dr \qquad (18)$$

which shows that the surface heat flow contains a contribution due to the secular cooling of the planet on the long time scale τ, and the use of (15) for Nu assumes that this secular cooling does not alter the empirical heat transfer relation. In (13) we may assume the $<T_c>$ profile to be spatially self-similar at all times to write $<T_c> = C(t) \cdot T^i_c(r)$ where $T^i_c(r)$ is the initial shape of the $<T_c>$ profile (adiabatic) and where $C(t)$ describes its time variation. With an appropriate adjustment of heat capacity we may in fact use $T^i_c(r) = T^i_c(r_c)$ (isothermal core), then (13) reduces to

$$\rho_c c_{pc} T_c(r_c) \frac{\partial C}{\partial \tau} = \frac{3q_c}{r_c} + Q_c , \qquad (19)$$

which describes the cooling of the core by mantle convection. Mantle geotherms $<T_m>$ may be constructed by matching an adiabatic temperature profile in the mantle through conduction boundary layers to the surfaces $r = r_c$ and $r = r_p$. This involves solution of a transcendental equation as in Sharpe and Peltier (1979) which should however, be constrained by the assumption of equal boundary layer thickness rather than equal boundary layer temperature drop if the mantle properties (like thermal conductivity and viscosity) were constant. Such thermal history models, which are governed by the negative feedback due to temperature dependent viscosity, may be modified to account for non-zero Q_m which includes the latent heat of freezing associated with the growth of the inner core. It is clear from the structure of the model that if the viscosity is initially low, Ra and thus Nu will be large and there will therefore be a large heat flow across

the core mantle boundary. This, however, will lead to a rapid secular cooling of the system so that ν will increase exponentially as in (17). This will rapidly inhibit the cooling. The system therefore has a strong negative feedback.

A sample of such a thermal history from Sharpe and Peltier (1979) is shown in Figure 10a in which hot initial conditions are assumed (in accord with the predictions of modern accretion models such as that described by Wetherill (1976) and Q_c has been calculated on the basis of the assumption that there is .2% ^{40}K in the core. As discussed in Sharpe and Peltier (1979), this heat source concentration is so high that mantle convection is not sufficiently vigorous at first to keep pace with the heating that core radioactivity produces and consequently the core heats up initially and the inner core does not begin to form until about 2 billion years ago. The hatched area of the Figure represents the region of partial melt in p - T space. The boundary layer structure of the mantle geotherm is clearly evident in the Figure, with well defined thermal boundary layers adjacent both to the Earth's surface (the lithosphere) and the core-mantle boundary (the D" region). The curvature of the adiabatic profile in the mantle itself is a consequence of the fact that $\tau = g\alpha L/c_p$ has been assumed constant so that the adiabatic gradient $g\alpha T/c_p$ increases with depth. Since α/c_p is expected to decrease somewhat with depth, a better assumption would probably be a constant gradient with $dT_s/dr = .3°K/km$. As pointed out in the last section, the boundary layer at the base of the mantle may be expected to act as the source of small scale fast rising thermal anomalies which are able to ascend so quickly that the process is essentially adiabatic.

In Figure (10b) I show what might be the adiabatic trajectory of a typical plume and it is clear from the Figure that the trajectory can be expected to emerge from the region of partial melting above the liquidus, follow a mostly subsolidus path through the mantle and then re-enter the partially molten state near the base of the lithosphere. As discussed in Sharpe and Peltier (1979) such thermal histories fit the present day surface heat flow constraint (entry 3 in Table I), the observed mean mantle viscosity (entry 5) obtained from postglacial rebound analyses, and the present day boundary layer thickness (entry 4). The high concentration of radioactivity in the core, which is completely unrealistic according to current geochemical thinking (e.g. Taylor et al., 1981), was required in order to inhibit the too rapid cooling of the interior which otherwise occurs when the mantle is assumed to be devoid of internal heat sources. As suggested in Peltier (1980), however, the same delayed cooling is obtained for models with internal radioactivity in the mantle and the results from thermal history calculations which include this effect will be described elsewhere.

Inspection of Figure (10a) illustrates an ex-

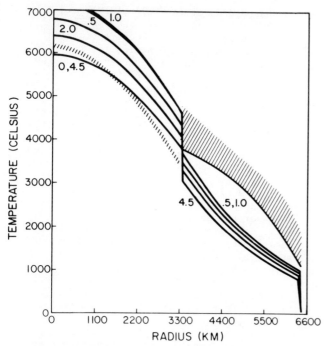

Fig. 10a. Thermal history obtained using a parameterized convection model (Sharpe and Peltier, 1978, 1979). The hatched region in p-T space is the regime of existence of partial melt.

tremely important point arising from the relation between the mean geotherm and the solidus, which is predicted by models in which the form of the geotherm is governed by whole mantle convection. One expects the mean geotherm to most closely approach the solidus near the base of the litho- sphere and adjacent to the core mantle boundary, and furthermore that the extent of partial melt- ing in these regions would have been greater in the past when the mean mantle temperature was higher than it is at present. It will be noted that the schematic geotherms drawn for this sim- ple thermal history do not contain the mean temp- erature field overshoot above (below) the inter- ior temperature which is predicted by exact sol- utions to the Navier-Stokes equations such as those shown previously in Figure 4. This raises the interesting possibility that early in the thermal history there might have existed a sub- stantial layer of partial melt beneath the prim- itive lithosphere which was global in extent. If this were the case, as seems rather likely, we expect that surface tectonic processes then must have been quite different from the large horizontal scale plate dynamics which we observe today, with the first billion years or so of the thermal history being characterized by a highly unstable surface with widespread volcanism. One consequence of this high degree of dynamical in- stability near the surface would be a marked inability to stabilize continental crustal mat-

erial formed by chemical differentiation. This would be rapidly reingested into the interior and remixed with the more refractory phases. This convection scenario accords well with cur- rent geochemical ideas (Taylor et.al., 1981) and provides an immediate explanation as to why sur- face rocks older than about 3.8 billion years are absent (or at least are hard to find!). The timescale of about 1 billion years over which continental crust could not be stabilized is predicted by the parameterized convection models of the thermal history first introduced by Sharpe and Peltier (1978,1979). It is the length of time which such models take to cool suffici- ently that they enter the "regulated" state in which the rate of secular cooling of the interior has become extremely slow due to the high value of the viscosity $(O(10^{22}\text{Poise}))$ which obtains in the presently "cold" state.

It is equally important to recognize the conse- quences of partial melting in the thermal bound- ary layer adjacent to the core mantle boundary. The partial melting which is expected here will markedly increase the violence of the boundary layer instabilities which we invoked previously as the mechanism by which thermal plumes are gen- erated. When these plumes reach the Earth's sur- face, after their adiabatic ascent from depth, they produce oceanic island hot spots character- ized by basaltic volcanism and presumably also

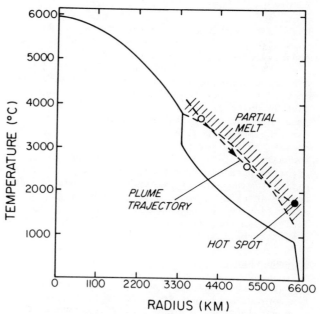

Fig. 10b. Adiabatic trajectory of a thermal plume formed by instability of the thermal boundary layer at the core mantle interface. Note that the trajectory intersects the solidus near the base of the upper thermal boundary layer (lithosphere) and may origi- nate from above the solidus near the core mantle boundary.

the surface hot spots (characterized by alkalic basalts) found in continental regions. It should be clear that these plumes produced by partial melting in the lower thermal boundary layer will be strongly enriched in the less refractory elements in accord with the observed geochemical characteristics of ocean island basalts (OIB) (O'nions et.al., 1979).

It does not seem, however, that this process would be sufficient by itself to explain the history of the process of mantle differentation which has been revealed by the important recent work on the isotopic systems Sm-Nd and Rb-Sr (DePaolo and Wasserburg, 1976; O'Nions et.al., 1979; Allegre et.al., 1981). These observations have been interpreted as requiring a physical separation between two distinct chemical reservoirs, and mass balance arguments have been employed (Wasserburg and DePaolo, 1979) to argue that these two reservoirs are to be associated with the upper and lower mantles (separated by the 670 km seismic discontinuity). The data do not require that the two reservoirs have remained separated throughout the age of the Earth, however, and there is in fact some evidence that the mantle must have been well mixed for at least the first 7×10^8 yrs., that is prior to the first stabilization of continental crust. Since no serious attempt has been made to quantify the uniqueness of the constraint on the degree of mantle mixing allowed by the isotopic data (entry 14 of Table 1) it would therefore seem premature to accept uncritically the simple two box model which these data seem to imply. At present they are the only data which may be reasonably invoked to support the idea of a layered convective circulation in the mantle. As I have discussed above, all of the geophysical evidence appears to favour the whole mantle model and in fact is quite strongly suggestive that the layered model is unacceptable. The main argument concerns the lack of evidence in the mantle viscosity profile for the existence of a thermal boundary layer at 670 km depth. Although it might be possible to explain the absence of significant viscosity variation through this boundary by invoking a very large increase of creep activation energy to compensate for the sharp increase in temperature, this possibility seems somewhat contrived.

6. Conclusions

In the preceeding review of the problem of mantle convection I have focussed upon several recent development from the theory of convection itself, from studies of mantle viscosity using postglacial rebound data, and from geochemistry. In my view this recent work has demonstrated that most of the observations which can be employed to constrain models of the convective circulation are explained extremely well by the simple whole mantle model which is forced to an important degree by heating from below, as was first pointed

out in Sharpe and Peltier (1978,1979), and Peltier (1980). This conceptual model has been strongly reinforced by the recent geochemical concensus that the Earth contains far too little radioactivity to explain the present day observed surface heat flow. Estimates of the fraction of the heat flow which could be in equilibrium with the present rate of heat production vary from 0.5 (O'nions et.al., 1978), to ∿0.4 (Taylor et. al., 1981), to 0.33 (Ringwood, 1979). The difference between the rate of radioactive heat supply and the surface heat loss should be made up primarily by the secular cooling of the planet. As mentioned previously, this scenario has the further attractive feature that it immediately provides the energy to the core which is necessary to power the geodynamo. As the planet cools, the inner core grows and in so doing drives a chemical convective circulation which Loper and Roberts (1981) have most recently argued to be an extremely efficient means of generating magnetic field. Although there is clearly a great deal of work which needs to be done to further quantify our understanding of the fundamental dynamical and chemical processes which have governed the evolution of the planet through time, the hypothesis of whole mantle convection appears to reconcile most of the existing observations. The extent to which this interpretation is uniquely required by the observations, however, will likely be cause for much enjoyable disputation in the future.

Acknowledgements. I have very much enjoyed conversations with Gary Jarvis, Howard Sharpe, Patrick Wu, and David Yuen concerning the ideas discussed in this paper.

References

Allegre, C.J., B. Dupre, B. Lambret, and P. Richard, The subcontinental versus suboceanic debate, I. Lead-neodynium-strontium isotopes in primary alkali basalts from a shield area: the Ahaggar volcanic suite, Earth Planet. Sci. Lett., 52, 85-92, 1981.

Burchfield, J.D., Lord Kelvin and the Age of the Earth, Science History Publications, New York, 260 pp., 1975.

Busse, F.H., Non-linear properties of thermal convection, Rep. Prog. Phys., 41, 1929-1967, 1978.

De la Cruz-Reyna, S., The thermal boundary layer and seismic focal mechanisms in mantle convection, Tectonophysics, 35, 149-160, 1976.

De Paolo, D.J. and G.J. Wasserburg, Nd isotopic variations and petrogenesis models, Geophys. Res. Lett., 3, 249-252, 1976.

Haskell, N.A., The viscosity of the asthenosphere, Am. J. Sci., 33, 22-28, 1937.

Isacks, B., and P. Molnar, Distribution of stresses in the descending lithosphere from a global survey of focal mechanism solutions of mantle earthquakes, Rev. Geophys. Space Phys., 9, 103-165, 1971.

Jarvis, G.T., and D.P. McKenzie, Convection in a compressible fluid with infinite Prandtl number, J. Fluid Mech., 96, 515-583, 1980.

Jarvis, G.T. and W.R. Peltier, Oceanic bathymetry profiles flattened by radiogenic heating in a convecting mantle, Nature, 285, 649-651, 1980.

Jarvis, G.T. and W.R. Peltier, Mantle convection as a boundary layer phenomenon, Geophys. J. R. astr. Soc., to appear, 1981.

Jarvis, G.T. and W.R. Peltier, Effects of lithosphere rigidity on ocean floor bathymetry and heat flow, Geophys. Res. Lett., submitted, 1981.

Kohlstedt, D.L., and C. Goetze, Low stress and high temperature creep in Olivine single crystals, J. Geophys. Res., 79, 2045-2051, 1974.

Kohlstedt, D.L., C. Goetze, and W.B. Durham, Experimental deformation of single crystal olivine with application to flow in the mantle. London: Wiley, in The Physics and Chemistry of Minerals and Rocks, ed. by R.G. Strens, J. Wiley and Sons, Inc. New York, 35-50, 1976.

Loper, D.E., and Paul H. Roberts, Compositional convection and the gravitationally powered dynamo, Geophys. Astrophys. Fluid Dyn., to appear, 1981.

McKenzie, D.P., J.M. Roberts, and N.O. Weiss, Convection in the Earth's mantle: Towards a numerical solution, J. Fluid Mech., 62, 465-538, 1974.

McKenzie, D.P., and F.M. Richter, Convection currents in the Earth's mantle, Sci. Am. 235, 72-89, 1976.

McKenzie, D.P., and N.O. Weiss, Speculation on the thermal and tectonic history of the Earth, Geophys. J.R. astr. Soc., 42, 131-174, 1975.

Moore, D.R., and N.O. Weiss, Two-dimensional Rayleigh - Benard convection, J. Fluid Mech., 58, 289-312, 1973.

Olson, P. and G.M. Corcos, A boundary layer model for mantle convection with surface plates, Geophys. J.R. astr. Soc., 62, 195-219, 1980.

O'Nions, R.K., N.M. Evenson, P.J. Hamilton and S.R. Carter, Melting of the mantle past and present: isotope and trace element evidence, Phil Trans. R.S. London A, 258, 547-559, 1978.

O'Nions, R.K., N.M. Evenson, and P.J. Hamilton, Geochemical modelling of mantle differentiation and crustal growth, Jour. Geophys. Res., 84, 6091-6101, 1979.

Parmentier, E.M. and D.L. Trucotte, Two-dimensional mantle flow beneath a rigid accreting lithosphere, Phys. Earth Planet Int., 17, 281-289, 1978.

Parsons, B., and D.P. McKenzie, Mantle convection and the thermal structure of the plates, J. Geophys. Res., 83, 4485-4496, 1978.

Peltier, W.R. Penetrative convection in the planetary mantle, Geophys. Fluid Dyn., 5, 47-88, 1972.

Peltier, W.R. The impulse response of a Maxwell Earth, Rev. Geophys. Space Phys., 12, 649-669, 1974.

Peltier, W.R. Glacial-isostatic adjustment - II. The inverse problem, Geophys. J.R. astr. Soc., 46, 669-706, 1976.

Peltier, W.R. Models of glacial isostasy and relative sea level, Dynamics of Plate Interiors, R.I. Walcott, ed., A.G.U. Publications, 1980a.

Peltier, W.R. Mantle convection and viscosity, in "Physics of the Earth's Interior", Proceedings of the Enrico Fermi International School of Physics (Course LXXVIII), ed. by A. Dziewonski and E. Boschi, North Holland, New York, 1980b.

Peltier, W.R. Ice age geodynamics, Ann. Rev. Earth Planet. Sci., 9, 199-225, 1981.

Peltier, W.R. and J.T. Andrews, Glacial-isostatic adjustment - I: The forward problem, Geophys. J.R. astr. Soc., 46, 605-646, 1976.

Peltier, W.R., W.E. Farrell, and J.A. Clark, Glacial isostasy and relative sea level: a global finite element model, Tectonophysics, 50, 81-110, 1978.

Richter, F.M., Mantle convection models, Ann. Rev. Earth Planet Sci., 6, 9-19, 1978.

Richter, F.M. and B. Parsons, On the interaction of two scales of convection in the mantle J. Geophys. Res., 80, 2529-2541, 1975.

Ringwood, A.E., Origin of the Earth and Moon, Springer-Verlag, New York, pp. 295, 1979.

Roberts, G.O., Fast Viscous Benard Convection, Geophys. Astrophys. Fluid Dyn., 12, 235-272, 1979.

Ross, J.V., H.G. Ave Lallemant, and N.L. Carter, Activation volume for creep in the upper mantle, Science, 203, 261-263, 1979.

Sabadini, R., and W.R. Peltier, Pleistocene deglaciation and the Earth's rotation: implications for mantle viscosity, Geophys. J.R. astr. Soc., 66, to appear, 1981.

Sammis, C.G., J.C. Smith, G. Schubert, and D.A. Yuen, Viscosity depth profile of the Earth's mantle: effects of polymorphic phase transitions, J. Geophys. Res., 82, 3747-3761, 1977.

Sharpe, H.N. and W.R. Peltier, Parameterized mantle convection and the Earth's thermal history, Geophys. Res. Lett. 5, 737-774, 1978.

Sharpe, H.N., and W.R. Peltier, A thermal history model for the Earth with parameterized convection, Geophys.J.R.astr.Soc., 59, 171-203, 1979.

Taylor, S.R., and S.M. McLennan, The composition and evolution of the continental crust: Rare earth element evidence from sedimentary rocks, Phil. Trans. Roy. Soc., to appear, 1981.

Turcotte, D.L., and E.R. Oxburgh, Finite amplitude convective cells and continental drift, J. Fluid Mech., 28, 29-42, 1967.

Walcott, R.I., Isostatic response to loading of the crust in Canada, Can. J. Earth Sci., 7, 716-727, 1970.

Walcott, R.I., Flexural rigidity, thickness, and viscosity of the lithosphere, J. Geophys. Res., 75, 3941-3954, 1971.

Wasserburg, G.J., and D.J. DePaolo, Models of earth structure inferred from neodynium and strontium isotopic abundances, Proc. Nat. Acad.

Sci. USA, 76, 3594-3598, 1979.

Wetherill, G.W., The role of large bodies in the formation of the Earth and Moon, Proc. Lunar Sci. Conf. 7th, 3245-3257, 1976.

Whitehead, J.A., Jr., and D.S. Luther, Dynamics of laboratory diapir and plume models, J. Geophys. Res. 80, 705-717, 1975.

Wu, P., and W.R. Peltier, Viscous gravitational relaxation, Geophys. J.R. astr. Soc., submitted, 1980.

Wu, P., and W.R. Peltier, Glacial isostatic adjustment and the free air gravity anomaly as a constraint on deep mantle viscosity, Geophys. J.R. astr. Soc., submitted, 1980.

Young, R.E., Finite amplitude thermal convection in a spherical shell, J. Fluid Mech., 63, 695-721, 1974.

Yuen, D.A. and G. Schubert, Mantle plumes: a boundary layer approach for Newtonian and non-Newtonian temperature dependent rheologies, J. Geophys. Res., 81, 2499-2510, 1976.

Yuen, D.A., and W.R. Peltier, Temperature dependent viscosity and local instabilities in mantle convection, in "Physics of the Earth's Interior", Proceedings of the Enrico Fermi International Shcool of Physics (Course LXXVIII, ed. by A. Dziewonski and E. Boschi, North Holland, New York, 1980a.

Yuen, D.A. and W.R. Peltier, Mantle plumes and thermal instability of the D" layer, submitted to Geophysical Research Lett., 1980b.

Yuen, D.A., W.R. Peltier and G. Schubert, On the existence of a second scale of convection in the mantle, Geophys. J.R. astr. Soc., submitted, 1981.

INITIAL STATE OF THE EARTH AND ITS EARLY EVOLUTION

V. S. Safronov

O.J. Schmidt Institute of the Physics of the Earth, Academy of Sciences USSR, Moscow

Abstract. Dynamical investigation of planetary accumulation shows that the Earth probably took 10^8 yr to grow. Hence the inference of a high initial temperature of the Earth based on the assumption of a much shorter time scale of the Earth formation ($\sim 10^5$ yr) is groundless. Another source of initial heating of outer parts of the Earth is the impacts of large bodies, which does not depend on the rate of accumulation. With the fall of a body larger than ~ 1 km into a planet, the size of the crater and the impact stirring of material increase with the size of the body slower than the depth of the release of energy. This effect leads to heating more than proportionate to the size of the body. The largest accumulated bodies might have reached a lunar size, but their total mass was very small. Because of the tidal disruption effect, it is reasonable to assume an effective radius of the largest body in the inverse power law mass distribution of infalling bodies of about 100 km. From the equation of conductivity for the outer layer of the growing Earth heated and stirred by impacts, it is found that the central part of the Earth was heated up to 1000°K and the outer region containing about a half of the total mass was heated up to the temperature of the beginning of melting, corresponding to the kinematic viscosity $\sim 10^{16}$ cm^2/sec. Further heating was prevented by convection. Gravitational differentiation in the layer probably began at the end of accumulation. Thermal inhomogeneities of the mantle more than a thousand kilometers across formed by impacts of the largest bodies probably triggered the beginning of core differentiation.

Introduction

The early evolution of the Earth falls between planetary physics (cosmogony) and the geosciences. The main features of the evolution were determined first of all by the initial state of the Earth resulting from its formation. According to the concept formulated most explicitly by O. J. Schmidt in the 1950's and generally accepted, terrestrial planets were accumulated from solid particles and bodies of different sizes. Quantitative treatment of planetary accumulation shows that the time scale of the process and the sizes of bodies were the most important parameters which have determined an initial temperature of the Earth and inhomogeneities of its mantle. Only a small fraction of gravitational energy of the Earth was retained in its interior, most of it being radiated into space. The higher the rate of liberated gravitational energy, the higher should be the temperature of the radiating surface to remove all this energy. It seemed to some authors that it would be possible to obtain a high initial temperature of the Earth assuming a short time scale of its formation. Several years ago such opinion was widespread among geochemists and geophysicists. Ringwood [1966] conjectured a hot initial state of the Earth from geochemical considerations. Hanks and Anderson [1969] saw in the idea of rapid accretion the possibility of accelerating formation of the Earth's core. Turekian and Clark [1969] made use of it to suggest a hypothesis of inhomogeneous accretion of the Earth. A different version of the hypothesis was proposed by Anderson and Hanks [1972]. They assumed the accumulation time of the Earth to be 50,000 years, i.e. less than one thousandth of the time scale found by us from examination of the dynamics of planetary formation [Safronov, 1969].

The short scale of accumulation was then obtained in some cosmogonical models. However, the analysis of these models has shown [Safronov, 1977] that such result was related to some arbitrary assumptions about the parameters of the model. When these parameters are taken from the dynamical consideration of evolution of the system of protoplanetary bodies, a much longer time scale is obtained. The calculations by Weidenschilling [1976] have confirmed once more that the time scale of accumulation for the Earth was of the order of 10^8 years. Therefore the attempts to obtain a high initial temperature of the Earth at the cost of the assumption of its short formation time (less than 10^5 years) are groundless.

However, there was another source of an effective heating of the growing Earth--the impacts of large bodies [Safronov, 1964, 1969] which have also created considerable initial inhomogeneities of the Earth's mantle. Quantitative study of the inhomogeneous accumulation is very complicated. More recent estimates [Safronov, 1978; Kaula, 1979] have led to a higher initial temperature of the Earth up to melting of a wide outer layer. But the results depend on the values of the parameters involved and are still not quite certain. We shall consider below some points most important for the problem.

Mass Distribution and Random Velocities of Preplanetary Bodies

Due to some similarity of the processes of planetary accumulation, coagulation of colloids, and growth of rain droplets, the method of coagulation theory have been adopted to study the mass distribution function of preplanetary bodies. Because of the specificity of the problem only general relations could be used. Especially difficult is consideration of the later stage of accumulation when the velocities of bodies are high and fragmentations at collisions should be taken into account. In this more general case, the integral-differential equations of coagulation for the distribution function of masses $n(m,t)$ can be written in the form [Safronov, 1969].

$$\frac{\partial n(m,t)}{\partial t} = \int_0^{m/2} w(m',m-m')A(m',m-m')n(m',t)n(m-m',t)$$

$$dm' - n(m,t)\int_0^\infty A(m,m')n(m',t)dm' + \int_m^\infty n_1(m,m'')$$

$$\int_0^m [1-w(m',m''-m')]A(m',m''-m')n(m',t)n(m''-m',t)$$

$$dm'dm'' \tag{1}$$

where $A(m,m')$ is the coagulation coefficient which characterizes the probability of collision of two bodies with masses m and m', and $w(m,m')$ is the probability of coalescence of these bodies at the collision; hence $1-2(m,m')$ is the probability of their fragmentation. Finally $n_1(m,m'')$ is the mass distribution of fragments which form after collisional disintegration of two bodies with the sum of their masses m''.

The coagulation coefficient $A(m,m')$ depends on random velocities of bodies. The velocities in turn depend on the distribution of masses. The degree of fragmentation also depends on the velocities and affects the distribution of masses. Strictly speaking these characteristics should be determined simultaneously. However, an analytical solution of such a coupled problem seems to be impossible. So the problem was divided in two parts: for a given mass distribution the relative velocities of bodies were estimated, and for given velocities the mass spectrum was determined.

A rigorous analytic solution of the coagulation equation without fragmentation for $A(m,m') = C(m+m')$ [Safronov, 1969], a qualitative investigation of the equation for more general assump-

tions [Zvjagina and Safronov, 1971; Zvjagina, et al., 1973] and a numerical solution [Pechernikova, et al., 1976] have shown that the mass distribution can be approximated by the product of an inverse power function and an exponential function which cuts off the distribution in the large mass region. The inverse power law

$$n(m) = Cm^{-q} \tag{2}$$

is a good approximation for the region of smaller bodies. In the course of accumulation this region extends to larger and larger values of m. So the power function (2) is an asymptotic solution of the coagulation equation (1). For the system without fragmentation $q \approx 1.55 \pm 0.15$ and for a smooth variation of parameters describing fragmentation $q \approx 1.8$.

The power function (2) is not valid for the description of the distribution of largest bodies. An exponential factor also does not help much because the continuous function cannot represent satisfactorily the distribution of a few large bodies. The gravitational collisional cross-section of these bodies was considerably larger than geometrical, being proportional to the 4th power of their radii. Hence, the largest body grew more rapidly than the others, i.e., the ratio of its mass to the mass of the second largest body in its zone of feeding increased with time. This process of run-away of the largest bodies in mass led to formation of potential planet embryos moving on less eccentric orbits. Originally the masses of the embryos were small; their ring-shaped feeding zones were narrow. There were many embryos in the whole zone of the planet. They swept out the surrounding material and grew in mass. Their zones of feeding and gravitational influence widened and overlapped. Smaller embryos grew slower and departed from their regular orbits. Before falling into a larger body, a smaller embryo had several times higher probability of close encounters with the larger body, inside the Roche limit, so that tidal forces caused disintegration into smaller pieces-- some tens of kilometers in diameter. The anomalous obliquity of Uranus probably is connected with the more rare case of a fall of a neighbouring embryo which did not undergo such a close encounter.

In a system which is differentially rotating, gravitational interactions between bodies during encounters tend to increase their relative (random) velocities, while inelastic collisions (and coalescence, in particular) decrease velocities. As velocities increase, gravitational perturbations become less effective and relative importance of collisions increases, and vice versa. Hence the system tends to quasi-equilibrium velocity distribution which depends on the properties of the system, first of all on the distribution of masses. The average equilibrium velocity of bodies relative to circular Keplerian motion for a large mean free path can be written in the form [Safronov, 1969]

$$v = \sqrt{Gm/\theta r} = v_e/\sqrt{2\theta} \qquad (3)$$

where m and r are the mass and radius of the largest body in the zone considered and θ is a dimensionless parameter. In a system of bodies of equal mass which coalesce on collision, $\theta \approx 1$. For a power law distribution of masses (2) with q < 2 it was found $\theta \approx 3$ to 5. For q < 5/3 the larger bodies have somewhat smaller velocities. In the presence of gas slowing down the bodies, θ can be much larger than 10, dependent on body size. During accumulation, the velocities of bodies increase proportionally to the radius of the largest body and are about one-third of the velocity of escape v_e on its surface. If there is a runaway growth of the largest body, or if the size distribution is more steep and q > 2, then parameter θ is higher (or in other words, in Eq. (3) a mass and a radius of some effective body smaller than the largest one should be taken for m and r).

The parameter θ is an important factor because it determines the rate of growth of the planet: $dm/dt \propto (\theta +1)$. At $\theta = 3$ the Earth acquires 98% of its mass in $1 \cdot 10^8$ years, at $\theta = 5$, in $6 \cdot 10^7$ years.

Recently, this picture of a quasistationary evolution with approximately constant q and θ has been much debated. B. Levin [1978] has argued that the runaway growth of the largest body should have led to very large values of θ and accordingly to a very rapid accumulation of the planets. However his conjecture that in the whole wide zone of the planet there was only one rapidly growing planet embryo is physically unreliable. On the contrary G. Wetherill [1976, 1978] has found that at the latest stage of growth of the planet $\theta \approx 1$. An interesting numerical simulation of an intermediate stage of the accumulation process has been fulfilled by R. Greenberg, et al. [1978]. They have calculated a coupled evolution of size and velocity distributions of protoplanetary bodies. The initial swarm consisted of equal kilometer-sized planetesimals with velocities about the escape velocity at the surface of the body. In such a system a few 500-km bodies grow rapidly in $\sim 10^4$ yr but most of the mass during this time continues to reside in bodies of about original size and their relative velocities almost do not increase (q > 2). The initial mass distribution assumed by the authors is very different from a "steady-state" one of the type (2) and does not approach it up to the end of the simulation when the statistical method used in the model fails (a few 10^4 yr). Therefore, the character of evolution of the system depends on initial conditions and a question arises as to how long this dependence continues.

We shall not discuss the assumption of an initial equality of all masses here since it seems to us physically not reliable. More important is that, even in the case considered by Greenberg, et al., a substantial fraction of the mass of small bodies is acquired by the largest ones, as it can be shown, during some 10^5 years, i.e. in a time much shorter than that of accumulation of the planet. The distribution of masses then tends to approach (2) with decreasing q, the velocities of bodies increasing and θ in Eq. (3) decreasing.

So we should consider two types of bodies--a few big ones, which are the potential embryos of planets, and all other bodies which due to runaway growth of the former are much smaller. A gap in mass between these bodies is continuously filled by smaller embryos leaving their regular orbits and by their fragments. The total number N of embryos m in the zone of a planet with final mass m_p is

$$N \approx \frac{\Delta R_p}{\Delta R} \sim (\frac{m_p}{m})^{1/3}(\frac{\theta}{\theta_p})^{1/2} \qquad (4)$$

where ΔR_p and ΔR are the half widths of the whole zone of the planet and of one planet embryo respectively, θ_p is the value of θ at the end of accumulation (at $m \rightarrow m_p$). One can take $\theta/\theta_p \approx (m_p/m)^5$. The ratio of the mass Nm of the embryos to those of all other bodies in the zone of the planet is:

$$\frac{Nm}{m_p - Nm} \sim [(\frac{m_p}{m})^{2/3-5/2}-1]^{-1} \qquad (5)$$

Originally this ratio is small, but it increases with the increase of m. At $S \approx 0.4$ a two-fold decrease of N corresponds to about a four-fold increase of m. The mass of late embryos $\int mdN$ is several times less than the mass Nm of remaining embryos. Therefore we can see from (5) that when the largest embryo reaches the mass $m \sim 0.1 \, m_p$ it accretes an appreciable fraction of material in the form of comparatively large bodies (late embryos, to some extent disintegrated during close encounters). This result is important for the estimation of the initial temperature of the Earth.

Initial Temperature of the Earth

Most of the energy of impacts of a body accreted by the growing Earth is released inside a layer with the thickness of the order of its diameter. In the case of small bodies and particles, the layer is very thin, and almost all the energy is radiated into space at the surface temperature T_o about 300°K. Large bodies deposit their impact energy in much thicker layer, and some part of it is retained. A considerable temperature gradient arises in the layer and under it the temperature is much higher than at the surface. The larger the infalling bodies, the thicker the layer and the higher the temperature of the planet.

The heating of the growing Earth can be estimated using the equation of thermal conductivity which takes into account the increasing size of the globe. But there are some difficulties in evaluation of a depth dependence of energy deposited at impacts of different sizes and of a thermal conductivity due to impact stirring in the outer layer [Safronov, 1969]. We need to have a good physical theory of impacts and crater formation and to know the mass distribution of infalling bodies. One should also take into account the release of radioactive heat and an additional heating of material due to its compression during the increase of the Earth's mass. The latter effect was first evaluated by E. Lubimova [1968]. This heating is proportional to the temperature of the material which is compressed. In the center of the Earth (maximum compression) the temperature is doubled. Preliminary consideration of all these factors [Safronov, 1969] showed that to the end of formation of the Earth its central part was heated up to about $1000°K$ approaching the melting point.

Recently the problem has been considered anew. The conductivity equation for a plane parallel layer with moving boundaries heated by impacts was solved and a quasistationary solution for the temperature distribution in the outer parts of the growing Earth was obtained [Safronov, 1978]. This approximate model has helped to understand the relative importance of various factors and parameters. The heating increased with the sizes of bodies and practically did not depend on the time scale of accumulation. A similar calculation was fulfilled for the spherically symmetric layer [Safronov and Kozlovskaya, 1977]. Results for the two models differ less than 10 percent, even when falling bodies reach 300 km across. The quasistationary equation of conductivity for the outer spherical layer with the coordinate z counted from the moving surface is

$$\frac{d^2T}{dz^2} - a\frac{dT}{dz} + b = 0 \qquad (6)$$

where

$$a = \dot{R}/K + 2/R - dK/dz, \quad b = E/K, \quad E = \varepsilon/c\rho$$

K is the thermal diffusivity (mainly connected with impacts), E is the energy deposition rate per unit volume. At constant a and b for boundary conditions

$$T(0) = T_0, \quad dT(h)/dz = 0$$

the solution of Eq. (6) is

$$T(z) = T_0 + \frac{b}{a}\left[z - \frac{1}{a}e^{-ah}(e^{az}-1)\right]. \qquad (7)$$

For the thickness h of the layer smaller than a^{-1}, i.e. for radii of bodies $r \lesssim 100$ km, the temperature at the bottom of the layer is

$$T(h) \approx T_0 + \frac{Eh^2}{K}\left(1 - \frac{ah}{3}\right).$$

The heating by impacts of bodies with radii smaller than some critical value r_1 is only about $30°$. The

radius r_1 is connected with a gravity of the planet and decreases with the increase of its radius R as $R^{-5/3}$. For the present Earth $r_1 \approx 1$ km. The heating by impacts of larger bodies is roughly proportional to their radii. At the same average values $\bar{K}$ and $\bar{E}$ the heating is smaller when K and E decrease with the depth z. At linear decrease of both quantities to zero at z = h, temperature is two times lower than at constant values $K(z) = \bar{K}$ and $E(z) = \bar{E}$. Impacts of equal bodies with $r \approx 30$ km produce melting of the outer layer at depths $100 < z < 300$ km. If the bodies have the power law mass distribution (2) with q = 1.8, the functions K(z) and E(z) can be approximated by linear functions and the heating is the same as in the case of equal bodies for an effective mean radius about half a radius of the largest body r_M in the distribution (2). At $r_M = 100$ km the thickness of the melted layer exceeds 1000 km. In the absence of convection in the layer, the temperature at its bottom then would reach $2800°K$. However, due to the large negative temperature gradient, convection in the layer begins much earlier, and really T(z) cannot exceed appreciably the melting temperature $T_m(z)$.

W. Kaula [1979a] has applied the same method using new data on the energy partitioning at impact crater formation and assuming a somewhat different model of crater. Using the same mass distribution of bodies (2) with q = 1.8 he has taken much higher mass for the largest body: 0.002 of the mass of the growing Earth. By numerical computation he has found the temperature more than $3000°K$ for the upper layer of about 1500 km thick, though a convective energy transfer was included in the equation. In a later paper Kaula [1979b] imputes such high temperatures to computational imperfections and concludes that the temperature of the outer layer should be close to melting.

As it was pointed out above an inverse power law (2) fails in the region of largest bodies. Late embryos originally have a small total mass but to the end of accumulation it reaches 10 – 20 percent. They disintegrate at close encounters due to tidal attractions, into pieces of a size depending on their strength T. At $T = 10^5$ dyne/cm^2 the radii of fragments are about 20 km, while at $T = 10^7$ dyne/cm^2 they are about 100 km [Ziglina, 1978]. The contribution to the thermal evolution of the Earth of a few bodies which avoid such a disintegration is not important, though they could create considerable local inhomogeneities of the initial Earth. All other bodies are on the average smaller than these late embryos. Nothing more can be said about the mass distribution of largest bodies. It seems reasonable to assume as a first approximation an upper value of radii r_M about 100 km and a mass distribution in the form (2) with q < 2 (about 1.8). We have taken these values to estimate the heating of the growing Earth. The positions of the lower and upper boundaries of the melted zones are given in Table 1 for different r_M. The depth $z_{M1} = h(R_{m1}) + R_\oplus - R_{m1}$ characterizes the depth of the first melts (formed at $R = R_{m1}$) in the Earth completely formed ($R = R_\oplus$). The depth z_{m2} pertains to the end of formation ($R = R_\oplus$) and is found from the condition that in the melted zone $T \approx T_m$, all impact energy released in it should be transferred to the upper solid layer.

TABLE 1. Boundaries of the Melted Zones

$r_{M\oplus}$	60	80	100 km
z_{m1}	300	800	1300 km
z_{m2}	75	70	60

The melting temperature is taken according to the expression given by Kaula [1979a]. The calculations do not take into account radioactive and compressional heating, nor the convective transport of energy.

Convection takes place when the Rayleigh number

$$Ra = \frac{\alpha g}{\nu \kappa} L^4 \nabla T_{sa} \qquad (9)$$

exceeds the critical value $Ra_c \approx 1700$. Considerable uncertainty is connected with the evaluation of kinematic viscosity ν and thermal diffusivity κ. In principle κ should include all kinds of diffusivity, first of all that of impact stirring K. But in the lower part of the layer h impacts are rare (only from an infall of big bodies) and K is small. Kaula [1979b] takes the thickness L of convective layer larger than h and does not include K in κ. The viscosity ν is highly dependent on the temperature. The following relations can be used [Kaula, 1979a]: at $T < T_m$.

$$\nu = a \exp(b T_m / T) \qquad (10)$$

and at $T > T_m$

$$\nu = \nu_m \exp[-c(T - T_m)] + \nu_o \qquad (11)$$

where $\nu_m = a \exp b - \nu_o$, $a = 1.4 \cdot 10^9 cm^2/s$, $b = 25$, $c \approx 0.4 °K^{-1}$, $\nu_o \sim 10^2 cm^2/s$. For $T = T_m$ we have $\nu = \nu_m = 10^{20} cm^2/s$. Thus T_m is the temperature at the very beginning of melting. Assuming for the end of the Earth's formation $f_{M\oplus} = 100$ km one can find that the temperature curve $T(z)$ touches $T_m(z)$ at the depth z slightly lower than h when the Earth grows up to 0.8 $R_\oplus$, i.e., to a half of its present mass. For this zone $\nabla T \approx \nabla T_m$ and $\nabla T_{sa} \approx 0.6 \nabla T_m \approx 0.6°$ per km. From Eq. (9), we find that Ra reaches the critical value at $\nu \approx 5 \cdot 10^{17}$ in the layer with L = 100 km and at $\nu \approx 3 \cdot 10^{16}$ for L = 50 km. From (11) we find that such a viscosity corresponds to the temperature exceeding T_m only by 13° and 20° respectively.

Convection begins in the lower part of the layer h. Gradually its upper boundary lifts up. It should be emphasized that at a, kinematic viscosity $\nu \approx 10^{17} cm^2/s$, in spite of $T > T_m$, the proportion of a melted material, is still small-- probably not more than a few percent.

One can estimate what temperature is needed for steady state convection when all the energy brought into the layer by impacts is transferred outwards. The convective transport can be evaluated by introducing a convective diffusivity κ_v which is expressed by the Nusselt number [Kaula, 1979a]

$$\kappa_v = Nu \cdot \kappa \approx 2\kappa (Ra/Ra_c)^{1/3} \qquad (12)$$

At the balance of the heat added and the heat removed we have

$$\kappa_v \nabla T \approx E'L' \qquad (13)$$

where $E'c\rho$ is the average energy deposition rate per unit volume in the layer L' between z and h. For $T \sim T_m$ the superadiabatic gradient can be written in the form

$$\nabla T_{sa} = \nabla T - \nabla T_a \approx \nabla T_m (1 - \nabla T_a / \nabla T_m) = (1-\chi)\nabla T_m. \qquad (14)$$

Inserting κ_v, Ra and other quantities in (14) we find the expression for viscosity of steady state convection

$$<\nu^{-1} \kappa^{-1}> \approx Ra_c (E'L'/\kappa L)^3 / \{\chi(1 - \chi)\alpha g L \nabla T_m^4\} \qquad (15)$$

This expression gives only an average value of $(\kappa \nu)^{-1}$. Corresponding average values of κ and $E'L'$ for the whole layer L should be taken. In the upper part $\kappa \approx K$ and $E'L' \approx EL$. Then $\nu \approx 10^{16}$ for $L \approx 200$ km. In the lower part $E'L'$ decreases with the depth more rapidly than κ and ν is higher. Therefore the average value of ν should be higher than 10^{16}. The temperature curve in this case lays only about $20°$ above T_m.

Subsolidus convection, as treated by G. Schubert et al. [1979], creates some difficulties for a gravitational differentiation of the Earth's material. The most favorable conditions for its beginning would be a melting of one phase (e.g. of a heavy one), while the other remained solid. A small percent of melts means a very slow differentiation. It seems to us that more effective differentiation could begin only at a final stage of the Earth formation. It could be stimulated by some additional source of energy, for example, lunar tides.

C. Hayashi et al. [1979] have suggested another mechanism of the early heating of the Earth. The authors have developed a model of planetary formation in which an important role belongs to the surrounding gas. They believe that the Earth was formed during less than 10^{-7} yr when the gas have not yet dissipated from the solar system. The gas filled all Hill's sphere around the planet and when the Earth reached one tenth of its present mass its atmosphere became very dense and opaque due to water vapors and hydrogen molecules. The outer atmosphere was in radiative equilibrium as inner atmosphere was convecting. The retention of large energy of infalling bodies caused the growing Earth to become very hot. Toward the end of the Earth's formation, the temperature at the bottom of the atmosphere (and thus in the upper layer of the Earth) according to their calculations reached 4000°K and the total mass of the atmosphere -10^{26}g.

This result leads to serious geophysical consequences. But it raises also some serious doubts: (1) Near the boundary of the Hill's sphere, the thermal velocities of molecules considerably exceeded the velocity of escape. Hence this gas does not belong to the planet. Such an outer atmosphere only exerts pressure, and does not participate in the motion of the planet relative to the gas. Radiative equilibrium can be supported only in the inner atmosphere, which is retained by planetary gravitation. The temperature of the surrounding medium should be maintained therefore at this inner atmosphere boundary, not at the Hill's sphere. The real thickness of this atmosphere should be an order of magnitude less. (2) The gas probably was dissipated from the solar system before the Earth had reached its final mass. Instead of the values suggested by Hayashi et al. for time scales, 10^6-10^7 yr for the Earth's growth and 10^8 yr for the dissipation of gas by the T Tauri solar wind, one can expect rather the contrary correlation: 10^8 and 10^7 yr respectively. (3) Let us assume that these time scales are correct; then new doubts appear: a) how could the solar wind dissipate so massive an Earth's atmosphere as 10^{26}g? b) if it could, then why do we see no traces of this atmosphere: excess of heavy noble gases, etc.? Their absence is, in our opinion, strong argument against the model and against the conclusions of a very high initial temperature of the Earth.

Primary Inhomogeneities and Early Evolution of the Earth

The theory of planetary accumulation from solid bodies has given us a valuable inference of the initial state of the Earth: its temperature distribution and inhomogeneities of the interior. On the base of these data, its early evolution can be studied. An important role of primary inhomogeneities formed by an infall of large bodies has been shown [Safronov, 1964, 1972]. Thermal history of the Earth taking into account of the energy of differentiation released during $(1-2)\cdot 10^9$ yr and of convective energy transport has been calculated [Vitjazev and Majeva, 1977]. The possibilities of inhomogeneous accumulation has been discussed [Makalkin, 1979]. The main features of the early evolution of the Earth have been pointed out [Safronov et al., 1978]. In these papers the initial temperature was taken lower than the melting temperature and the material began to melt at a

depth 300-500 km a few 10^8 yr after the Earth's formation. The melting zone slowly widened downwards and a differentiation of substances began in the zone. When the thickness of the zone reached about 500 km the process was accelerated due to release of the energy of differentiation.

According to the new models described above, at the final stage of the Earth's formation the temperature in its outer part reached the melting point. Therefore the differentiation could also begin at this stage, probably to its end. A separation proceeds more easily when two components of different density are in different phase state: solid and liquid. In the presence of large inhomogeneities the process is determined by a hydrodynamic heat and mass transfer. In both cases a more dense component drops at first to the bottom of the melted layer. Then because of Rayleigh-Taylor instability, it fragments into separate drops which begin to sink down through the solid material. At this stage the mechanism resembles that of Elsasser [1963]. However, 1000-kilometer primary inhomogeneities, the excessively heated regions of impacts of the largest bodies, now play a role of initial perturbations. At first, differentiation proceeds actively only in the upper mantle. The core forms only later because silicates can be removed from the central region only after its appreciable heating, when the viscosity has become lower. Convection also at first takes place in the upper layer of superadiabatic temperature gradient. Gradually it extends into deeper regions. In general convection and differentiation are concurring processes which hinder each other. Convection slows down differentiation and differentiation creates a gradient of concentration of the heavy component which can stop convection until the end of separation. Afterwards convection can begin in each component separately. At the present time differentiation probably is going to the end and convective motions are more active.

References

Anderson, D.L., and T.C. Hanks. Formation of the earth core, Nature, 237, 387-388, 1972.

Elsasser, W.M.. Early history of the earth, in Earth Science and Meteorites, J. Geiss and E.P. Goldberg, eds., pp. 1-30, Amsterdam, 1963.

Greenberg, R., W.K. Hartmann, C.R. Chapman, and J.F. Wacker. The accretion of planets from planetisimals, in Protostars and Planets, T. Gehresl, ed., pp. 599-622, Univ. Arizona Press, 1978.

Hanks, T.C. and D.L. Anderson. The early thermal history of the earth, Phys. Earth Planet. Inter., 2, 19, 1969.

Hayashi, C., K. Nakazawa and H. Mizuno. Earth's melting due to the blanketing effect of the primordial dense atmoshpere, Earth Plan. Sci. Lett., 43, 22-28, 1979.

Kaula, W.M. Thermal evolution of earth and moon growing by planetisimal impacts, J. Geophys. Res., 84, 999-1008, 1979a.

Kaula, W.M. The beginning of the earth's thermal evolution, in *Proc. Wilson Conf. "The Continental Crust and its Mineral Deposits"*, Toronto, 1979b.

Levin, B.J. Relative velocities of planetesimals and the early accumulation of planets, *Moon and Planets, 19*, 289-296, 1978; Lett. to *Astron. Zhurn., 4*(2), 102-107, 1978.

Lubimova, E.A. *Thermics of the Earth and Moon*, Moscow, Nauka (Russ.), 279 pp., 1968.

Makalkin, A.B. Possibility of formation of the originally inhomogeneous earth, *Phys. Earth Planet. Inter., 22*, 302-312, 1980.

O'Keefe, J.D. and T.J. Ahrens. Impact-induced energy partitioning, melting and vaporization on terrestrial planets, *Proc. Lunar Sci. Conf. 8th*, 3357-3374, 1977.

O'Keefe, J.D. and T.J. Ahrens. Impact flows and crater scaling on the moon, *Phys. Earth Planet Inter., 16*, 341-351, 1978.

Pechernikova, G.V. and A.V. Vitjazev. *Lett. to Astron. Zhurn.* (Russ.), *5*(1), 54-59, 1979.

Pechernikova, G.V., V.S. Safronov, and E.V. Zvjagina. Mass distribution of protoplanetary bodies. II. Numerical solution of the generalized coagulation equation, *Astron. Zhurn., 53*, 612-619, 1976.

Ringwood, A.E. The chemical composition and origin of the earth. In *Advances in Earth Science*, P.M. Hurley, ed., MIT Press, Cambridge, 1966, 287-356.

Safronov, V.S. The primary inhomogeneities of the earth's mantle, *Tectonophysics, 1*, 217-221, 1964.

Safronov, V.S. *Evolution of the Protoplanetary Cloud and Formation of the Earth and Planets*, Moscow, Nauka (Russ.), NASA TT F-677, 206 pp., 1972.

Safronov, V.S. The initial state of the earth and certain features of its evolution, *Izv. Akad. Nauk SSSR, Fizika Zemli, No. 7*, 35-41, 1972.

Safronov, V.S. Time scale for the formation of the earth and planets and its role in their geochemical evolution, In *Proc. Soviet-American Conf. Cosmochin. Moon and Planets*, NASA, Washington, 797-803, 1978.

Safronov, V.S. The heating of the earth during its formation, *Icarus, 33*, 3-12, 1978.

Safronov, V.S. and S.V. Kozlovskaya. Heating of the earth by the impact of accreted bodies, *Izv. Akad. Nauk, Fizika Zemli, No. 10*, 677-684, 1977.

Safronov, V.S., A.V. Vitjazev, and S.V. Majeva. Problems of initial state and early evolution of earth, *Geochimia, No. 12*, 1763-1769, 1978.

Schubert, G., P. Cassen, and R.E. Young. Core cooling by subsolidus mantle convection, *Phys. Earth Planet. Inter. 20*, 194-208, 1979.

Turekian, K.K. and S.P. Clark. Inhomogeneous accumulation of the earth, *Earth Planet. Sci. Lett., 6*, 346-348, 1969.

Vitjazev, A.V. and S.V. Majeva. A model of the early evolution of the earth, *Tectonophysics, 41*, 217-225, 1977.

Wetherill, G.W. The role of large bodies in the formation of the earth and moon, *Proc. Lunar Sci. Conf. 7th*, 3245-3257, 1976.

Wetherill, G.W. Accumulation of the terrestrial planets by the sweeping of planetisimals, *Carnegie Inst. Wash. Yearbk., No. 77*, 421-428, 1978.

Weidenschilling, S.J. Accretion of the terrestrial planets II, *Icarus, 27*, 161-170, 1976.

Ziglina, J.N. Tidal destruction of bodies in the neighborhood of a planet, *Izv. Akad. Nauk SSSR, Fizika Zemli, No. 7*, 3-10, 1978.

Zvjagina, E.V. and V.S. Safronov. Mass distribution of protoplanetary bodies, *Astron. Zhurn., 48*(5), 1023-1032, 1971.

Zvjagina, E.V., G.V. Pechernikova, and V.S. Safronov. A qualitative solution of the coagulation equation with allowance for fragmentation, *Astron. Zhurn., 50*, 1262-1273, 1973.

ELECTRICAL CONDUCTION IN MANTLE MATERIALS

T. J. Shankland

Geophysics Group, Los Alamos National Laboratory, University of California,
Los Alamos, New Mexico 87545, USA

Abstract. It has been established for several years that electromagnetic methods readily provide information on lateral variations of physical properties in the earth's upper mantle, and that high electrical conductivities correlate well with other evidence of elevated mantle temperatures. This paper reviews the laboratory and theoretical evidence that connects conductivity with temperature, pressure, and the chemical activities of different mantle components.

Introduction

Electromagnetic methods share with seismology the fact that they geophysically probe the physical properties of the earth in the present instant of geological time. This paper discusses electrical conductivity in the materials thought to compose the earth's mantle. These are silicates and oxides of relatively high density, notably: olivine, pyroxene, plagioclase, spinel, garnet, their high-pressure modifications, the rocks that these minerals form, and their melt phases. Various aspects of low-frequency conductivity have been reviewed by Shankland [1975], Duba [1976], Shankland [1979], and Haak [1980], while Olhoeft [1976, 1979] has summarized high-frequency dielectric properties.

Hutton[1976] reviewed not only laboratory data but also field techniques and results. It should be mentioned that high electrical conductivity in the mantle correlates strongly with tectonic province [Garland, 1975; Filloux, 1979] with sea-floor age [Filloux, 1980; Oldenburg, 1980], with indicators of high temperatures such as high heat flow and low upper-mantle siesmic velocity [Gough, 1974], and with recent volcanism such as that in continental rifts or geothermal areas like Yellowstone [Leary and Phinney, 1974].

Electrical Conduction in Minerals

Temperature and Pressure Effects

With some rare exceptions, conduction in oxides is a thermally activated process and is well represented by an equation of the form

$$\sigma = \sigma_o \exp \left[-(E + P\Delta V)/kT \right] . \qquad (1)$$

Here E is the activation energy, ΔV an activation volume expressing pressure dependence, k the Boltzmann constant, and T the absolute temperature. The pre-exponential factor σ_o in general contains a relatively weak temperature dependence that is ordinarily ignored. The common conductivity unit in the geophysical literature is the siemens/meter (1 S/m = 1 ohm^{-1} m^{-1} = 10^{-2} ohm^{-1} cm^{-1}). [An expression of equation (1) in practical units is

$$\sigma = \sigma_o \exp \left[-(11,605\ E + 12.075\ P\Delta V)/T \right] \qquad (2)$$

where E is in electron volts, P in kbar-s, ΔV in cm^3 $mole^{-1}$ and T in kelvins.] The exponential temperature dependence in equation (1) makes electrical conductivity a sensitive indicator of high temperature and provides a physical basis for the association of high conductivity with other geophysical evidence for high mantle temperatures.

Possible charge carriers in minerals are electronic (electron or holes) or ionic (ions or lattice vacancies). For electronic mechanisms it is possible to rule out intrinsic conductivity where the number of electrons equals the number of holes. This is because of the wide energy gap E_g between valence and conduction bands (the oxygen p-like bonding states and cation s-like antibonding states); see Marfunin [1979] for a discussion of electronic energy levels in minerals. Figure 1 illustrates the large E_g in dense minerals, of the order of 8 ev. However, a variety of other electronic conduction mechanisms can exist as illustrated in Figure 2; these mechanisms principally involve transition metal cations such as Fe^{2+} and its defect state Fe^{3+}. Electronic conduction in minerals bears some similarity to that in disordered materials such as glasses [Mott and Davis, 1971]. An example of detailed optical and electrical interpretation is that for virtually pure forsterite by Morin et al. [1977, 1979] who conclude that its highly anisotropic conduction is electronic along the b-axis while it is ionic for the a- and c-directions. However, for olivine there is other, indirect evidence for

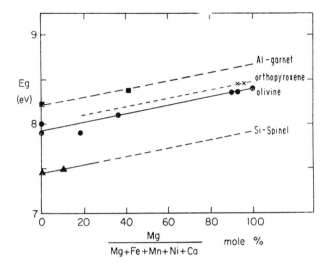

Fig. 1. Fundamental gap E_g between conduction and valence bands in dense minerals as a function of composition and crystal structure [Nitsan and Shankland, 1976].

ionic conduction at higher temperatures (above about 1100–1200°C) that comes from diffusion measurements. It is possible to approximately calculate electrical conductivity from ionic interdiffusion at the appropriate iron contents and oxygen fugacities (fO_2) using the Nernst-Einstein relation, $\sigma/D = \bar{n}q^2/kT$. This is shown in Figures 3 and 4 for periclase and olivine [Shankland and Ahrens, in preparation, 1982]. Here n is the number of carriers per unit volume having charge q. The result is only approximate because the effective binary diffusion coefficient D* [Brady, 1975] is used rather than the self-diffusion coefficient D. Bender [1976] measured ionic transport numbers for olivine at high temperatures and determined a dominantly ionic transport mechanism. These results and those of Schock et al. [1980] show anisotropic conduction highest along the c-axis. The anisotropy is in agreement with the vacancy mechanism inferred by Morin et al. [1979] for forsterite, although Schock et al. [1980] prefer an electronic mechanism in olivine.

Despite the fact that most energy levels in solids change by about ±10⁻⁶ eV/bar, there is little pressure effect seen in olivine and pyroxene [Schock et al., 1977; Dvořák, 1973] or in mantle rocks [Laštovičkova and Parkhomenko, 1976] at pressures in the range 30–50 kbar (Figure 5). However, there are more striking changes at still higher pressures that may be related to instability of Fe^{2+} [Mao and Bell, 1972] or to phase changes [Akimoto and Fujisawa, 1965].

Chemical Effects

There are several effects on conductivity that result from chemical activity of different ionic species. First, it is not surprising that con-

ductivity should tend to increase with increasing iron content; the changeable valence states of Fe^{2+} – Fe^{3+} can clearly alter the crystal electronic structure [Shankland, 1969]. Second, and related to the first, are effects of fO_2 on defect structure as observed in forsterite by Parkin [1972, cited in Figure 3 of Shankland, 1975] or in pyroxene by Duba et al. [1973], as shown in Figure 6. Nitsan [1974] demonstrated that olivine, and presumably other iron-bearing minerals, have definite T-fO_2 boundaries outside of which the crystal structure becomes unstable.

A third chemical activity is that of Si; Stocker and Smith [1978] schematically illustrated possible defect structures in olivine that would result from excess Si that might occur when olivine is in contact with enstatite (Figure 7) as it is in most upper mantle rocks. There are experimental observations on sintered forsterite [Pluschkell and Engell, 1968] and olivine [Cemič et al., 1978; Will et al., 1979; Cemič et al., 1980] that indicate excess Si increases conductivity markedly. Schock et al. [1980] also argued that different Mg/Si ratios cause different transport mechanisms to prevail in mantle olivine.

Figure 3 summarizes conductivities of some possible mantle components. Although comparisons

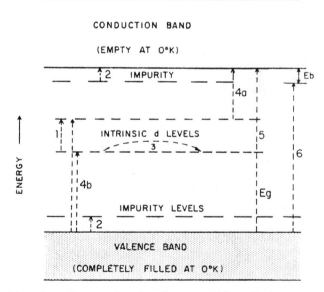

Fig. 2. Energy levels for possible electron or hole conduction mechanisms in silicates containing transition metals. Broken lines symbolize the following electronic transitions: 1. crystal field transitions; 2. transitions between bands and impurity or defect levels; 3. transitions between d states localized on different transition metal ions; 4. transitions between wide bands and intrinsic localized states; 5. valence-conduction band transitions; 6. interband exciton transition. E_b is the exciton binding energy [Nitsan and Shankland, 1976].

must be made with caution, the higher conductivities of the ultrabasic rocks peridotite and eclogite relative to that of monomineralic olivine suggests that chemical activities of minerals in contact with each other are important.

Volatiles, principally H_2O, CO_2, and SO_2 constitute a chemical effect that is only beginning to be experimentally investigated. Since water has a documented effect on conduction in crustal rocks [for example, Lebedev and Khitarov, 1964; Olhoeft, 1981] it is reasonable to expect similar enhancement in mantle materials. Morin et al. [1979] noted that forsterite conductivity increases by about an order of magnitude in an atmosphere of H_2 compared to its value in O_2; they attributed the increase to an increase of mobile interstitial ions induced by hydrogen. Duba and Heard [1980] reported increased conduction in Red Sea peridot when H_2 was used with CO_2 for buffering oxygen. However, some ambiguity remains because San Carlos olivine showed no such increase in the same atmosphere [Duba and Nicholls, 1973]. Ordinarily the effect of equilibration in a buffering atmosphere is diminished conductivity, presumably because of diminished oxygen fugacity; thus Duba and Nicholls [1973] measured far lower conductivity for olivine equilibrated in a H_2-CO_2 atmosphere than in the crystals during their initial heating. The drop above 600°C presumably comes from equilibration in the laboratory of

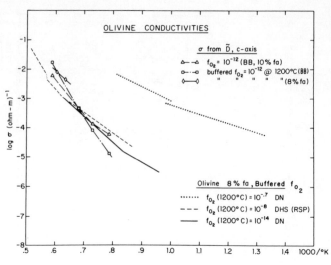

Fig. 4. Comparison of electrical conductivity in olivine as measured and as calculated from the Fe-Mg effective binary diffusion coefficient D* [Shankland and Ahrens, in preparation, 1982]. The curves have been adjusted to known fO_2 conditions. BB is Buening and Buseck [1973]; DN is Duba and Nicholls [1973]; DHS is Duba et al. [1974].

defects acquired in situ, during transport to the surface, or after surface emplacement. Olhoeft [1981] has argued for increased conductivity in the presence of sulfur; it may be further enhanced in the presence of water.

Electrical Conduction in Mantle Rocks

The mantle rocks peridotite and eclogite [Rai and Manghnani, 1978], illustrated in Figure 8, appear to behave in a more comprehensible manner than do more silicic compositions [Rai and Manghnani, 1977; Schloessin, 1977] just below the solidus. These conductivities may have been measured before thermal equilibrium in the samples was achieved. Nevertheless, they demonstrate that it is possible to use laboratory data to obtain reasonable, that is, subsolidus temperatures in the oceanic lithosphere from conductivity determinations in the range 0.001-0.01 S/m [Filloux, 1979, 1980]. Conductivities about 1-1/2 orders of magnitude above Red Sea peridot are at least plausible. This is illustrated in Figure 9.

High Conductivity Layers and Partial Melting

With presently available rock and mineral conductivites it is difficult to explain anomalously high-conductivity layers (HCL) of $\sigma > 0.1$ S/m--although it is conceivable that new data will change this conclusion. The large contrast between mineral and melt conductivities in Figure 8 makes partial melting an attractive hypothesis [Chan et al., 1973]. Shankland and Waff [1977], Chelidze [1978], and Honkura [1975] showed how

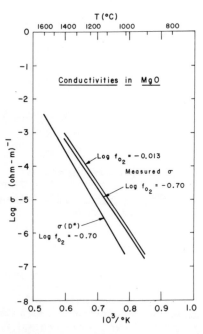

Fig. 3. Comparison of electrical conductivity in periclase as measured and as calculated from Fe-Mg effective binary diffusion coefficients D* [Shankland and Ahrens, in preparation, 1982]. The curves have been normalized to the same fO_2 conditions.

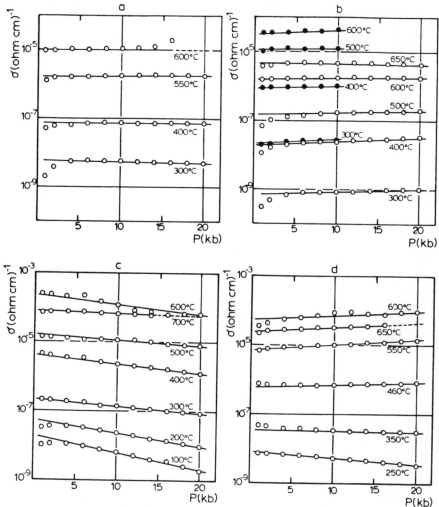

Fig. 5. Pressure effects on conductivity in olivine-rich rocks from Dvorak [1973]: (a) olivinite; (b) peridotite (open circles) with values taken during repeated heating on the same sample (full circles); (c) dunite; and (d) dunite.

theories of composites could be used to quantitatively explain such HCL's. Although a given conductivity in Figure 10 can be achieved at either high temperature or high melt fraction, the petrological partial melt curves permit only a single temperature and melt fraction to be associated with a given conductivity. The fact that melt conductivities vary little with chemical composition or fO_2 [Murase and McBirney, 1973; Waff and Weill, 1975; Rai and Manghnani, 1977] helps reduce variability of the model. A critical question is how to achieve interconnection of the melt fraction over considerable distances. In addition, if a minimum melt fraction is required for interconnection as suggested in the measurements of Manghnani and Rai [1979], then large melt percentages would be required, of the order of 20%. In this case, widespread partial melting would seem unlikely to persist for long geological

times from the point of view of mechanical stability against fluid loss unless a stabilizing method exists [Waff, 1980; Stolper et al., 1981].

Conclusions

With the most reliable laboratory data, conductivities obtained in the field, $10^{-3} - 10^{-2}$ S/m, adequately correspond to laboratory measurements on the major constituents of the upper mantle in the "lid" above the HCL at high but subsolidus temperatures. For the HCL it is hard to find a conductivity-enhancing mechanism with present laboratory data, apart from melting, that could take place in ultramafic rocks and that would give $\sigma \geq 0.1$ S/m. Time-dependent changes such as those measured in albite [Piwinskii and Duba, 1974] may exist, but the more abundant materials seem relatively well-behaved. The upper mantle is not

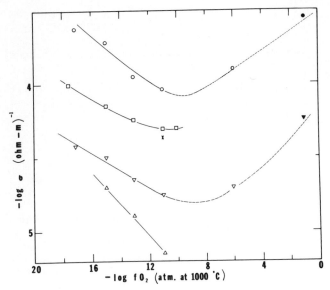

Fig. 6. Effects of oxygen fugacity, fO₂, on conductivity of pyroxene, from Duba et al. [1973], reprinted from Journal of Geology by permission of the University of Chicago press.

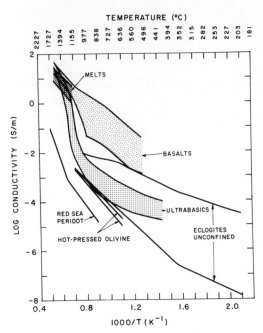

Fig. 8. Conductivities of possible upper mantle constituents [Shankland and Duba, 1978]. Ultrabasic rock values from Rai and Manghnani [1978].

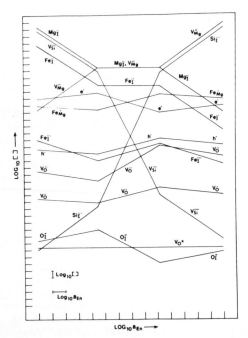

Fig. 7. Effect of enstatite (excess Si) activity on populations of different defects in olivine [Stocker and Smyth, 1978]. Superscript dots refer to a net positive charge, primes to a net negative charge with respect to the normal lattice.

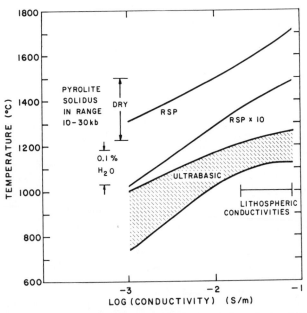

Fig. 9. Temperatures corresponding to given conductivities for single and polycrystal olivine (RSP & RSP x 10) and for ultrabasic rocks. Lithospheric conductivities (10^{-3} - 10^{-1} S/m) can be achieved at subsolidus temperatures with peridotite and eclogite but not with the best olivine measurements [Shankland and Duba, 1978].

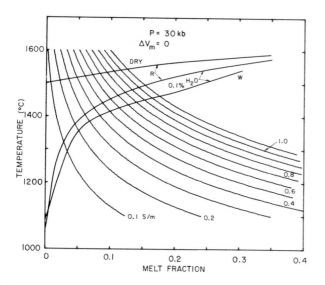

Fig. 10. Temperatures and melt fractions required to achieve a given high conductivity anomaly in the mantle [Shankland and Waff, 1977]. Addition of partial melt curves from experimental petrology [Ringwood, 1975] allows both temperature and melt fraction to be estimated.

likely to have enough albite to form a connected phase. Hence, present electrical measurements are consistent with this picture of a partially melted HCL at the depth of the LVZ and high but subsolidus temperatures in the lithosphere above.

Nevertheless, a "low-temperature" model has some strong supporting arguments, particularly from elastic properties. Minster and Anderson [1979] and Anderson and Minster [1979] presented purely solid-state mechanisms for achieving the high attenuation and low velocities characteristic of the seismic low-velocity zone (LVZ); O'Connell [1977] summarized laboratory data indicating that subsolidus creep can account for the mantle rheology implied by plate tectonics. Are there mechanisms acting in the real mantle and not yet observed in the laboratory that could raise electrical conductivity to values of 10^{-1} S/m or above at temperatures well below the peridotite solidus? For example, Tozer [1979] suggested that amphibole dehydration could cause both the HCL and the LVZ at temperatures near 600°C. With recent progress in understanding the nature of silicates as resistive oxides, conduction in mantle minerals has a far stronger experimental base than it did a decade ago. However, such questions indicate that much work is necessary (and possible).

Refinements include better definition of the chemical effects of one mineral in contact with another. Further, it is highly desirable to investigate the possible effects of volatiles; although water in particular can be expected to enhance conduction in minerals, rocks, and melts, we need quantitative confirmation of such a result. Crustal materials require far more study,

even though measurements presently exist [for example, Parkhomenko, 1967]. Silica-rich minerals have a variety of time-dependent effects that are influenced by slow diffusion rates in such minerals, and it would be reassuring to have evidence for attainment of equilibrium in laboratory measurements [Haak, 1980]. Another strong need, especially for extrapolation to high temperatures and pressures, is better models of conduction mechanisms. Present models, even for olivine, seem to vary widely from lightly iron-enriched forsterite to fayalite.

Acknowledgments

I thank A. G. Duba and R. N. Schock for detailed comments and for some literature references. This work was supported by the Division of Basic Energy Sciences of the Department of Energy through Contract No. W-7405-ENG-36 with the University of California and by supporting research funds of the Los Alamos National Laboratory.

References

Akimoto, S.-I. and H. Fujisawa, Demonstration of the electrical conductivity jump produced by the olivine-spinel transition, J. Geophys. Res., 70, 443-449, 1965.
Anderson, D.L., and J.B. Minster, The physical mechanism of subsolidus creep and its relation to seismic wave attenuation, EOS Trans. AGU, 83, 378, 1979.
Bender, N., Electrical conductivity of single crystals of $Mg_{1.8}Fe_{0.2}SiO_4$, Erlangen Research Abstracts in Materials Science, edited by B. Ilschner, Institute of Materials Science I, Department of Engineering, University of Erlangen-Nürnberg, Erlangen, Federal Republic of Germany, 1976.
Brady, J.B., Reference frames and diffusion coefficients, Amer. J. Sci., 275, 954-983, 1975.
Buening, O.K., and P.R. Buseck, Fe-Mg lattice diffusion in olivine, J. Geophys. Res., 78, 6852-6862, 1973.
Cemic, L., E. Hinze, and G. Will, Messungen der electrischen leitfahigkeit bei kontrollierten sauerstoffaktivitaten in Druckapparaturen mit festen Druck ubertragungsmedien, High Temperature, High Pressure, 10, 469-472, 1978.
Cemic, L., G. Will, and E. Hinze, Electrical conductivity measurements on olivines Mg_2SiO_4-Fe_2SiO_4 under defined thermodynamic conditions, Phys. Chem Minerals, 6, 95-107, 1980.
Chan, T., E. Nyland, and D.I. Gough, Partial melting and conductivity anomalies in the upper mantle, Nature Phys. Sci., 244, 89-90, 1973.
Chelidze, T.L., Structure-sensitive physical properties of partially melted rocks, Phys. Earth Planet. Interiors, 17, 41-46, 1978.
Duba, A., Are laboratory electrical conductivity data relevant to the Earth? Acta Geodaet., Geophys. et Montanist. Acad. Sci. Hung., Tomus 11, 485-495, 1976.

Duba, A., J.N. Boland, and A.E. Ringwood, Electrical conductivity of pyroxene, J. Geol., 81, 727-735, 1973.

Duba, A.G., and H.C. Heard, Effect of hydration on the electrical conductivity of olivine, EOS Trans. AGU, 61, 404, 1980.

Duba, A.G., and I.A. Nicholls, The influence of oxidation state on the electrical conductivity of olivine, Earth Planet. Sci. Lett., 18, 59-64, 1973.

Dvořák, Z., Electrical conductivity of several samples of olivinites, periodotites, and dunites as a function of pressure and temperature, Geophysics, 38, 14-24, 1973.

Filloux, J.H., Magnetotelluric and related electromagnetic investigations in geophysics, Rev. Geophys. Space Phys., 17, 282-294, 1979.

Filloux, J.H., Magnetotelluric soundings over the northeast Pacific may reveal spatial dependence of depth and conductance of the astnenosphere, Earth Planet. Sci. Letters, 46, 244-252, 1980.

Garland, G.D., Correlation between electrical conductivity and other geophysical parameters, Phys. Earth Planet. Interiors, 10, 220-230, 1975.

Gough, D.E., Electrical conductivity under western North America in relation to heat flow, seismology, and structure, J. Geomag. Geoelectr., 26, 105-123, 1974.

Haak, V., Relations between electrical conductivity and petrological parameters of the crust and upper mantle, Geophys. Surveys, 4, 57-69, 1980.

Honkura, Y., Partial melting and electrical conductivity anomalies beneath the Japan and Philippine Seas, Phys. Earth Planet. Interiors, 10, 128-134, 1975.

Hutton, V.R.W., The electrical conductivity of the Earth and planets, Rep. Prog. Phys., 39, 487-572, 1976.

Laštovičkova, M., and E.I. Parkhomenko, The electric properties of eclogites from the Bohemian massif under high temperatures and pressures, Pageoph., 114, 451-460, 1976.

Leary, P., and R.A. Phinney, A magnetotelluric traverse across the Yellowstone region, Geophys. Res. Letters, 1, 265-268, 1974.

Lebedev, E.B., and N.I. Khitarov, Dependence of the beginning of melting of granite and the electrical conductivity of its melt on high water vapor pressure, Geochem. Int., 1, 193-197, 1964.

Manghnani, M.H., and C.S. Rai, Electrical conductivity of a spinel lherzolite and a garnet peridotite to 1550°C: relevance to the effects of partial melting, Bull. Volcanologique, 41-4, 328-332, 1978.

Mao, H.K., and Bell, P.M., Electrical conductivity and the red shift of absorption in olivine and spinel at high pressure, Science, 176, 403-406, 1972.

Marfunin, A.S., Physics of Minerals and Inorganic Materials, an Introduction, translated by N.G. Egorova and A.G. Mischenko, Springer-Verlag, New York, 1978.

Minster, J.B., and D.L. Anderson, Diffusion-controlled dislocation damping in the mantle, EOS Trans., AGU, 60, 378, 1979.

Morin, F.J., J. R. Oliver, and R.M. Housley, Electrical properties of forsterite, Mg_2SiO_4, Phys. Rev. B., 16, 4434-4445, 1977.

Morin, F.J., J. R. Oliver, and R.M. Housley, Electrical properties of forsterite, Mg_2SiO_4. II, Phys. Rev. B., 19, 2886-2894, 1979.

Mott, N.F., and E.A. Davis, Electronic Processes in Non-crystalline Materials, Oxford, 1971.

Murase, T., and A.R. McBirney, Properties of some common igneous rocks and their melts at high temperatures, Geol. Soc. Amer. Bull. 84, 3563-3592, 1973.

Nitsan, U., Stability field of olivine with respect to oxidation and reduction, J. Geophys. Res., 79, 706-711, 1974.

Nitsan, U., and T.J. Shankland, Optical properties and electronic structure of mantle silicates, Geophys. Jour. R.A.S., 45, 59-87, 1976.

O'Connell, R.J., On the scale of mantle convection, Tectonophysics, 38, 119-136, 1977.

Oldenburg, D.W., Conductivity structure of oceanic upper mantle beneath the Pacific plate, Geophys. J. R. Astr. Soc., in press, 1981.

Olhoeft, G.R., Electrical properties of rocks, in The Physics and Chemistry of Minerals and Rocks, edited by R.G.J. Strens, Wiley, London, 261-268, 1976.

Olhoeft, G.R., Electrical properties, Initial Report of the Petrophysics Laboratory, U.S. Geol. Surv. Circ., 789, 1-25, 1979.

Olhoeft, G.R., Electrical properties of granite with implications for the lower crust, J. Geophys. Res., 86, 931-936, 1981.

Parkhomenko, E.I., Electrical Properties of Rocks, Plenum, New York, 1967.

Parkin, T., The electrical conductivity of synthetic forsterite and periclase, Thesis, School of Physics, The University of Newcastle Upon Tyne, 1972.

Piwinskii, A.J., and A. Duba, High temperature electrical conductivity of albite, Geophys. Res. Lett., 1, 209-211, 1974.

Pluschkell, Von.W., and H.J. Engell, Ionen- and elektronenleitung im magnesiumorthosilikat, Ber. Deutsch. Keram. Ges., 45, 388-394, 1968.

Rai, C.S., and M.H. Manghnani, Electrical conductivity of basalts to 1550°C, in Magma Genesis, edited by H.J.B. Dick, Bulletin 96 of Oregon, Dept. of Geology and Mineral Industries, Portland, 219-232, 1977.

Rai, C.S., and M.H. Manghnani, Electrical conductivity of ultramafic rocks to 1820 kelvin, Phys. Earth Planet. Interiors, 17, 6-13, 1978.

Ringwood, A.E., Composition and Petrology of the Earth's Mantle, McGraw-Hill, New York, 1975.

Schloessin, H.H., On the pressure dependence of solidus temperatures and electrical conductivity during melting of JOIDES Leg 37 samples, Can. J. Earth Sci., 14, 756-767, 1977.

Schock, R.N., A.G. Duba, H.C. Heard, and H.D. Stromberg, The electrical conductivity of polycrystalline olivine and pyroxene under pressure, in *High-pressure Research, Applications in Geophysics*, edited by M. Manghnani and S. Akimoto, Academic Press, New York, 39-51, 1977.

Schock, R.N., A.G. Duba, and R.L. Stocker, Defect production and electrical conductivity in olivine, *Lunar and Planetary Sci. XI*, Lunar and Planetary Institute, Houston, 987-989, 1980.

Shankland, T.J., Transport properties of olivine, in *The Application of Modern Physics to the Earth and Planetary Interiors*, edited by S.K. Runcorn, Wiley-Interscience, New York, 175-190, 1969.

Shankland, T.J., Elecrical conduction in rocks and minerals: parameters for interpretation, *Phys. Earth Planetary Int., 10*, 209-219, 1975.

Shankland, T.J., Physical properties of minerals and melts, *Revs. Geophys. Space Phys., 17*, 792-802, 1979.

Shankland, T.J., and A.G. Duba, Electrical conduction in upper mantle rocks and minerals, *EOS Trans. Agu, 59*, 269, 1978.

Shankland, T.J., and H.S. Waff, Partial melting and electrical conductivity anomalies in the upper mantle, *J. Geophys. Res., 82*, 5409-5417, 1977.

Stocker, R.L., and D.M. Smyth, Effect of enstatite activity and oxygen partial pressure on the point-defect chemistry of olivine, *Phys. Earth Planet. Inter., 16*, 145-156, 1978.

Stolper, E., D. Walker, B. H. Hager, and J.F. Hays, Melt segregation from partially molten source regions: the importance of melt density and source region size, *J. Geophys. Res., 86*, in press, 1981.

Tozer, D.C., The interpretation of upper-mantle electrical conductivities, *Tectonophys., 56*, 147-163, 1979.

Waff, H.S., Effects of the gravitational field on liquid distribution in partial melts within the upper mantle, *J. Geophys. Res., 85*, 1815-1825, 1980.

Waff, H.S., and D. F. Weill, Electrical conductivity of magmatic liquids; effects of temperature, oxygen fugacity and composition, *Earth Planet. Sci. Lett., 28*, 254-260, 1975.

Will, G., L. Cemic, E. Hinze, K.-F. Seifert, and R. Voight, Electrical conductivity measurements on olivines and pyroxenes under defined thermodynamic activities as a function of temperature and pressure, *Phys. Chem. Minerals, 4*, 189-197, 1979.

A THERMODYNAMIC APPROACH TO EQUATIONS OF STATE AND MELTING
AT MANTLE AND CORE PRESSURES

Frank D. Stacey

Department of Physics, University of Queensland, Brisbane 4067, Australia

Abstract. The precise relationship between the high temperature thermodynamic Grüneisen parameter γ and elastic properties of a solid has still to be derived. The essential reason for favouring the free volume formulation as the best available approximation for materials at high pressures is that the central atomic force assumption, upon which the free volume approach is based, is good for materials that are close-packed. Conversely the non-linearity of elastic mode interactions invalidates the lattice mode approach to γ at high temperatures. Given such a relationship, the value of γ is automatically specified by elastic constant data (as a function of pressure) or by an equation of state. Alternatively, thermodynamic evidence of the density dependence of γ can be used to derive an equation of state. However it is pointed out that the second approach depends upon the extrapolation to megabar pressures of thermo-dynamic information obtained only at much lower pressures and is necessarily less reliable than the first. The theory of melting is also simplified at high pressures by the fact that if both solid and liquid phases are approximately close-packed, there is little change in atomic coordination on melting. Then γ can be used to relate the volume change on melting to the latent heat, and therefore to the pressure dependence of melting point by the Clausius-Clapeyron relationship. The result is the differential form of Lindemann's melting law, seen to have a sound thermodynamic basis and provides justification for application of the dislocation theory of melting at very high pressures.

Introduction

The existence in principle of a relationship between the elastic and thermal properties of solids has been recognised for many years. The connection is represented in terms of the thermo-dynamic Grüneisen parameter γ,

$$\gamma = \alpha K_T / \rho C_V = \alpha K_S / \rho C_P \qquad (1)$$

(α is volume expansion coefficient, ρ is density, K_T, K_S are isothermal and adiabatic incompressib-ilities, and C_V, C_P are specific heats at constant volume or constant pressure). γ is a dimensionless parameter that must be related to the pressure derivatives of elastic constants, and although we still lack a completely satisfactory formulation there is now reasonable numerical agreement between various estimates of its values for the lower mantle and core, constituting most of the Earth's volume.

The geophysical significance of γ is that we have good data on the elastic moduli in the Earth's deep interior and can therefore infer the value of γ without any direct thermal observation. Many thermodynamic identities can be represented quite simply in terms of γ [Stacey, 1977a], in particular, equations representing adiabatic temperature gradient and the contribution of thermal energy to the pressure of a material at specified density, so that the values of γ calcu-lated from seismological data can be used directly to establish a thermal model of the Earth [Stacey, 1977b]. In this paper, the assertion that an understanding of γ is central to our understanding of the thermal problems of the Earth is re-emphasised in two applications – equations of state and melting at high compressions.

The use of a relationship between γ and elastic-ity to infer γ from seismological data can be inverted to obtain a theoretical equation of state if we have independent evidence of the density dependence of γ. Two new approaches to this dens-ity dependence have appeared quite recently. One is the direct observation of γ as a function of pressure by sudden partial adiabatic decompression of samples in high pressure systems [Boehler et al., 1977, 1979; Ramakrishnan et al., 1978]. As is pointed out in section 4, these observations re-quire more careful interpretation than has usually been recognised and for this reason they have not yet yielded useful data of direct geophysical interest, although we must expect that in due course they will do so. The second approach (con-sidered in section 3) is to appeal to observations of the temperature dependence of thermodynamic parameters at laboratory pressure, and to use thermodynamic identities to relate these to the density dependence of γ [Anderson, 1979b; Brennan and *Stacey*, 1979]. If we can assume both $\gamma(\rho)$ from thermodynamics and γ as a function of incompressibility and its pressure dependence, $K(P)$, then we have a theoretical equation of state, that is $K(P)$ or its integral $P(\rho)$ is specified without further assumption [*Brennan and Stacey*, 1979]. This approach is outlined in section 5 of the present paper.

Lindemann's [1910] melting law provides a convenient means of extrapolating melting points to the pressures of the Earth's deep interior in terms of γ. Its validity has received strong support in recent years. In particular *Stacey and Irvine* [1977a] pointed out that with certain assumptions it can be derived from the Mie-Grüneisen equation by a straightforward thermodynamic argument. The assumptions and implications in this derivation are discussed in section 6. Although Lindemann's law is not immediately applicable to complex materials, especially to mixtures of interacting phases, there is a feature that particularly commends it to the attention of geophysicists. The validity of the assumptions in its derivation improves with increasing pressure, so that difficulties in applying it to laboratory materials with very open crystal structures disappear at pressures of the Earth's deep interior, at which all materials must be close-packed.

Theory of the Grüneisen Parameter

The essential idea for the relationship between γ and elastic constants is contained in Fig.1, which represents the mutual potential energy of neighbouring atoms in a solid. The potential well is asymmetrical, that is energy, ϕ, increases more sharply on the compression side of the equilibrium spacing, a, than on the extension side. Two consequences of this asymmetry are (1) incompressibility K increases with pressure P, that is compression

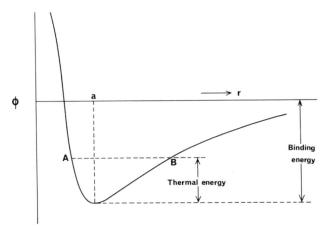

Fig.1. An anharmonic potential well. The asymmetry of the bond energy $\phi(r)$ with atomic spacing r is responsible for thermal expansion and for the volume changes on melting. AB represents the amplitude of thermal oscillation of a bond.

becomes increasingly more difficult as density is increased, and (2) thermal vibration causes thermal expansion. The necessity for thermal expansion is seen by considering the bond represented in Fig.1 to oscillate between the points A and B. The bond extension is greater than the bond compression and also the restoring force, that is the gradient of the energy curve, is less in extension, so that the bond spends more time in the extended state. Thus α and dK/dP are related consequences of the bond asymmetry. The relationship also involves thermal energy and incompressibility, so that when it is formally written out, we find that we are actually relating γ to dK/dP.

This idea has quite a long history. It appears in *Mott and Jones* [1935] whose analysis contains an error corrected by *Slater* [1939]. Slater's theory yielded the quantity, usually represented by γ_S, which he claimed to be appropriate to materials in which Poisson's ratio is independent of density

$$\gamma_S = \frac{1}{2}\frac{dK}{dP} - \frac{1}{6} \qquad (2)$$

Dugdale and MacDonald [1953] recognised a defect in the Slater formulation and partially corrected it. They pointed out that γ_S gave positive thermal expansion for a harmonic lattice, that is one in which the potential function is symmetrical or parabolic, and argued that it should be zero in such a case. They considered a chain of atoms in classical vibration and obtained the result

$$\gamma_{DM} = \left(\frac{1}{2}\frac{dK}{dP} - \frac{1}{2} + \frac{1}{9}\frac{P}{K}\right) \Big/ \left(1 - \frac{2}{3}\frac{P}{K}\right) \qquad (3)$$

It is of some interest to see why this result differs from γ_S. Slater's approach was to consider the vibrational modes of a lattice in terms of the elastic constants, that is the properties of the medium treated as a continuum. For this purpose the elastic modes of the lattice are all independent of one another. However, on the atomic scale we can see that this is not a valid approach, because the bond asymmetry (Fig.1) gives anharmonic (non-sinusoidal) atomic oscillations at the higher frequencies, or shorter lattice wavelengths, so that the high frequency modes of a lattice are not independent but interact. This interaction reduces the value of γ.

Vashchenko and Zubarev [1963] derived a third variant from a completely different approach, often referred to as the free volume formulation

$$\gamma_{VZ} = \left(\frac{1}{2}\frac{dK}{dP} - \frac{5}{6} + \frac{2}{9}\frac{P}{K}\right) \Big/ \left(1 - \frac{4}{3}\frac{P}{K}\right) \qquad (4)$$

and *Irvine and Stacey* [1975] later obtained the same result by a generalization of the Dugdale-MacDonald approach to three dimensions. Irvine and Stacey pointed out the relationship between γ_S, γ_{DM} and γ_{VZ} and showed that γ_{DM} was only a partial correction of γ_S and that γ_{VZ} was a more complete correction. Dugdale and MacDonald had effectively allowed for mutual interaction of the various compressional modes of a lattice but restriction of their analysis to a linear chain obscured the fact that the mutual interaction of longitudinal and transverse modes is also important. This can be seen in terms of Fig.2. Two neighbouring atoms, A and B, are displaced from their equilibrium positions, A_o and B_o, by thermal vibration in the y direction, which is perpendicular to the x direction of the bond for the undisplaced atoms. Unless the two y displacements of the atoms are equal, the bond between them is stretched by the relative displacement. Thus the y component of thermal vibration causes an average stretching of the bond that must be compensated by a contraction of the lattice. Thermal displacements in the x direction cause a thermal expansion in this direction as calculated by Dugdale and MacDonald, and interaction with the transverse vibrations introduces an additional negative contribution to the thermal expansion. The Irvine-Stacey derivation of equation (4) considered the three dimensional motion of atoms in a lattice and so included this negative contribution.

All of these presentations are subject to the objection by *Anderson* [1979a] that they relate thermal properties only to incompressibility and ignore the shear properties of a lattice. Anderson preferred to generalise the Slater approach by considering a composite formula that weighted the compressional and shear modes to obtain the "acoustic γ". In terms of K and Poisson's ratio, ν, this can be written [*Stacey* 1977a]:

$$\gamma_A = \frac{1}{2}\frac{dK}{dP} - \frac{1}{6} - \left[\frac{4 - 5\nu}{(1 + \nu)(1 - \nu)(1 - 2\nu)}\right]\frac{K}{3}\frac{d\nu}{dP} \quad (5)$$

This form of γ_A, obtained directly from the definition used by *Anderson* [1979a], emphasises that it is a generalization of γ_S to the case of $d\nu/dP \neq 0$. It is subject to the same criticism as γ_S with respect to neglect of the mode interactions.

The role of shear modes in γ is a function of crystal structure, being strong in open structures, such as silicates at ordinary pressures, but having only a minor effect in close-packed structures. The reason for this can be seen qualitatively by referring again to the transverse atomic displacement argument (Fig.2). In an open lattice, especially the diamond structure and equivalently

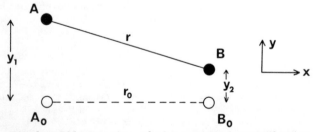

Fig.2. Illustration of the negative contribution to thermal expansion arising from motion transverse to atomic bonds. Motion of atoms A and B in the y direction stretches the x-oriented bond between them, causing contraction in the x direction.

α-quartz, vibrations transverse to a bond direction are relatively easy because the atoms can vibrate toward spaces in the lattice. This effect is particularly noticeable in germanium and silicon at low temperatures, at which such "soft" modes of the lattice are thermally excited but the "harder" compressional modes are frozen out, so that there is a temperature range over which α and γ are negative for these materials. The transverse interactions then predominate and the bond anharmonicity does not suffice to give positive expansion. In a close-packed structure, atoms are arranged in intersecting planes of hexagonal pattern and the lattice lacks the openness that would permit this effect. It follows that simple bond compressions and extensions or equivalently compressional modes are more important and the shear properties relatively less important in close-packed structures.

However, it remains true that, even in close-packed structures, shear properties should be allowed for in a complete formulation of γ, and that we still lack a satisfactory formulation that takes account of both shear properties and non-linearity (or mode interaction). This point is emphasised by the fact that Poisson's ratio is not the same for all atomic close-packed elements. A calculation in progress by A. Falzone and the author indicates that it may be possible to generalise the Irvine-Stacey central force approach for specific crystal structures by allowing for an intrinsic rigidity of bond angles and also (very importantly) what we call the "atomic Poisson's ratio effect" [*Falzone and Stacey*, 1980]. This is a recognition that the electron orbitals of an atom that bind it to its neighbours are not independent, but influence one another mutually, so that if one bond to an atom could be compressed without applying any force to the other bonds, they would nevertheless respond by extending. It is necessary to allow for this to explain the elasticity of close-packed crystals in which the intrinsic bond angle rigidity can have no relevance.

Until this problem is resolved we are faced with a choice of alternative formulations for γ. My own preference for the Vashchenko-Zubarev form (eq. 4) for the deep interior of the Earth is justified by the relative unimportance of shear properties in close-packed materials. We can also note that recent calculations of γ for the lower mantle and core from different formulations now agree within about 10% and it is improbable that a completely correct formulation would give values significantly outside this range. Thus, in spite of the remaining doubt about γ, we have a most important handle on the thermal properties of the Earth's deep interior from the well determined elastic properties.

Density Dependence of γ-Inference from Thermodynamic Relations

Brennan and Stacey [1979] have given a number of thermodynamic identities that relate the temperature and volume dependences of (αK_T) and γ in terms of more familiar or experimentally more accessible quantities. The most significant for the present purpose is the equation for $(\partial \ln \gamma / \partial \ln V)_T$ which is here written in terms of ρ instead of volume V:

$$\left(\frac{\partial \ln \gamma}{\partial \ln \rho}\right)_T = -1 - \left(\frac{\partial \ln C_V}{\partial \ln \rho}\right)_T + \frac{1}{\gamma \rho C_V}\left(\frac{\partial K_T}{\partial T}\right)_\rho \quad (6)$$

The two differentials on the right hand side of eq. (6) are experimentally accessible quantities. We expect the first to approach zero as long as T remains above the Debye temperature θ, although we note that θ increases with ρ. Without this assumption, the first of these differentials can be determined from the relationship

$$\left(\frac{\partial \ln C_V}{\partial \ln \rho}\right)_T = -\left(\frac{\partial \ln C_V}{\partial \ln(T/\theta)}\right)_\rho \frac{d \ln \theta}{d \ln \rho} \quad (7)$$

which is a differential identity if $(\partial C_V / \partial V)_{T/\theta} = 0$, as we must expect since compression at constant (T/θ) means at a fixed point on the characteristic Debye specific heat curve. Equation (6) is of interest following the analysis by *Anderson* [1980] of $K_T(T)$ data for close-packed oxides. He found that, for these materials, as the classical limit $(T > \theta)$ was approached with increasing T, K_T became virtually constant. Thus for these materials the last term in eq. (6) vanishes at high temperatures, so that

$$\left(\frac{\partial \ln \gamma}{\partial \ln \rho}\right)_T = -1 \quad (8)$$

Brennan and Stacey [1979] used this result as the basis of an equation of state, as outlined in section 5. As *Anderson* [1980] has noted, eq. (8) implies constancy of the product (αK_T) at high temperatures.

Density Dependence of γ-Interpretation
of Laboratory Observations

The definition of γ (eq.1) implies an identity that gives γ in terms of the temperature change during adiabatic compression or decompression:

$$\gamma = \left(\frac{\partial \ln T}{\partial \ln \rho}\right)_S \quad (9)$$

It follows that a simple and direct way of determining γ is to measure the temperature changes resulting from sudden small compressions or decompressions. Measurements of this kind have been reported by *Boehler et al.* [1977, 1979], and *Ramakrishnan et al.* [1978], whose range of pressures was such that they were able to observe the pressure dependence of γ. They represented their results as a power law dependence

$$\gamma = \gamma_0 (\rho_0 / \rho)^q \quad (10)$$

with the supposition that to a sufficient approximation over the observed pressure range, $q =$

$-(\partial \ln \gamma / \partial \ln \rho)_T$ is a constant. The observations referred to in the previous section suggest that, at least for several close-packed materials, $q = 1$ is a reasonable approximation. Some of the values of q reported by *Boehler et al.* [1979] appear at first sight to be surprisingly high, especially for fluorite for which they obtained $q = 4.6$, but it must be recognised that the measurements were all made below the Debye temperatures of the materials investigated and that in this circumstance the temperature dependence of γ has a strong influence on the observed value of q.

Since γ is a function of the vibration modes of a lattice that are excited at any particular temperature, its temperature dependence at any density is best represented as a function of normalised temperature, T/θ. To a good approximation we expect $\gamma(T/\theta)/\gamma_\infty$, at any fixed density, to be a unique function, independent of density, γ_∞ being the value of γ in the high temperature limit. Thus the density dependence that we wish to determine is not $(\partial \ln \gamma / \partial \ln \rho)_T = -q$ but $(\partial \ln \gamma / \partial \ln \rho)_{T/\theta} = -q*$. It is $q*$ that has the more fundamental significance and presumably gives the density dependence of γ at high temperatures $(T > \theta)$, at which γ becomes independent of T.

We may relate q and $q*$ by starting with a differential identity

$$\left(\frac{\partial \gamma}{\partial \rho}\right)_{T/\theta} = \left(\frac{\partial \gamma}{\partial \rho}\right)_T - \left(\frac{\partial \gamma}{\partial (T/\theta)}\right)_\rho \left(\frac{\partial (T/\theta)}{\partial \rho}\right)_T \quad (11)$$

Taking θ to be a function of ρ but not T and substituting for $(\partial \gamma / \partial T)_\rho$ by another identity

$$\left(\frac{\partial \gamma}{\partial T}\right)_\rho = \left(\frac{\partial \gamma}{\partial T}\right)_P + \left(\frac{\partial \gamma}{\partial \rho}\right)_T \cdot \alpha \rho \quad (12)$$

we obtain

$$q* = q\left[1 + \alpha T \frac{d \ln \theta}{d \ln \rho}\right] - \left(\frac{\partial \ln \gamma}{\partial \ln T}\right)_P \frac{d \ln \theta}{d \ln \rho} \quad (13)$$

The square bracketed factor differs from unity by too little to be significant and the difference between $q*$ and q is determined by the temperature dependence of γ. The factor $(d \ln \theta / d \ln \rho)$ is identified as the Debye γ, similar to but not identical to the acoustic γ; a value of about 2 is reasonable for most materials of interest. Reliable data on $(\partial \gamma / \partial T)_P$ are hard to pin down, but certainly exist for NaCℓ, for which q was reported by *Boehler et al.* [1977] to be 1.29 and for copper $(q = 1.33 - Ramakrishnan et al., 1978)$. From a data review by *Barron et al.* [1980], $(\partial \ln \gamma / \partial \ln T)_P \approx 0.26$ for NaCℓ at laboratory temperature $(T/\theta = 0.31)$ and for copper $(\partial \ln \gamma / \partial \ln T) = 0.07$. These give for NaCℓ, $q* \approx 0.77$ and for copper $q* = 1.17$. These values of $q*$ would certainly be improved if $(\partial \ln \gamma / \partial \ln T)_P$ were measured on the same specimens as q. However, the trend is clear. $q*$ is less than q, as measured at room temperature, and the difference is greatest for open crystal structures which have strong temperature dependence of γ. Thus derivation of fundamental conclusions from decompression measurements on geological materials requires particular attention to the effect of temperature dependence.

A Thermodynamic Equation of State

If $\gamma(\rho)$ is a known function (at temperatures high enough for γ to be independent of T), then by identifying it with eq. (4) (or any similar equation) we have, in differential form a $P(\rho)$ equation of state. Taking $\gamma(\rho)$ to be given by eq.(8), i.e.

$$\gamma = \gamma_o x^{-1} \tag{14}$$

where $x = \rho/\rho_o$ and γ_o, ρ_o are zero pressure values, the differential equation for $P(x)$ from (4) is

$$9x^3 P'' - (6x^2 + 18\gamma_o x)P' + (4x + 24\gamma_o)P = 0 \tag{15}$$

and the solution, following *Brennan and Stacey* [1979], with substitution of boundary conditions $P = 0$, $xP' = K_o$ at $x = 1$, is

$$P = \frac{K_o}{2\gamma_o} x^{4/3} \left[e^{2\gamma_o(1-x^{-1})} - 1 \right] \tag{16}$$

This equation is of interest, not merely because of possible applicability to the lower mantle, but because its derivation has a totally different basis from any other $P(\rho)$ equation of state. It depends upon the thermodynamic evidence of the variation of γ with ρ and, of course, the $\gamma(K,P)$ equation, for which (4) was used, but nothing else. Thus comparison of the equation with $P(\rho)$ data for the Earth, and also of its differential form for $K(\rho)$ with earth model tabulations, permits direct assessment of the thermodynamic input. *Brennan and Stacey* [1979] concluded from such a comparison that in the lower mantle γ depended upon ρ less strongly than ρ^{-1}. It will be a matter of considerable interest to pursue this further when an improvement of eq.(4) becomes available.

Melting at High Pressure

The pressure dependence of melting point is given by the thermodynamically rigorous Clausius-Clapeyron relationship

$$\frac{dT_M}{dP} = \frac{\Delta V}{\Delta S} = T_M \frac{\Delta V}{L} \tag{17}$$

where ΔV and ΔS are the volume and entropy increments on melting and L is the latent heat of melting. This equation is derived by asserting that since the solid and liquid states coexist in equilibrium at the melting point their free energies are equal at all pressures and therefore that the increments to their Gibbs free energies due to a pressure increment are equal. However, eq. (17) is not immediately suitable for extrapolation to lower mantle and core pressures

without constraints to the pressure dependence of ΔV and L that are not provided by this rigorous thermodynamic approach. As a result several more or less empirical relationships have been used in geophysics. One, due originally to *Lindemann* [1910], turns out to have a clear thermodynamic basis [*Stacey and Irvine* 1977a] that was not envisaged in its original "derivation". This thermodynamic basis is summarised here.

The thermodynamic identity

$$\left(\frac{\partial P}{\partial T}\right)_V = \gamma\rho C_V \tag{18}$$

can be integrated at constant volume over the temperature range T_1 to T_2:

$$P_2 - P_1 = \int_{T_1}^{T_2} \gamma\rho C_V \, dT \tag{19}$$

Now if both T_1 and T_2 are well above the Debye temperature, we may take γ to be constant over this range and then

$$P_2 - P_1 = \Delta P = \gamma\rho \int_{T_1}^{T_2} C_V \, dT = \gamma\rho\Delta E_V \tag{20}$$

where ΔE_V is the thermal energy increment per unit mass of material (maintained at constant volume). Equation (20) is a special case of the Mie-Grüneisen equation. In normal heating of a material the thermal energy is energy of atomic vibration and is almost exactly equipartitioned between kinetic energy of the motion and potential energies of the atoms in their mutual force fields. We can therefore rewrite eq. (20) in terms of the potential energy increment ΔE_P (per unit mass):

$$\Delta P = 2\gamma\rho\Delta E_P \tag{21}$$

We can now consider melting as a process in which a material acquires thermal energy without a temperature rise; the energy is therefore entirely potential energy, associated with rearrangement of atoms in their mutual force fields. Thus to a simple first approximation we may write the pressure increment due to melting at constant volume in terms of eq. (21) by identifying ΔE_P as the latent heat of melting, L. If melting occurred instead at constant pressure there would be a volume increment ΔV, which is related to the pressure increment due to melting at constant volume by the incompressibility K along the liquidus:

$$\Delta P = K\rho\Delta V \tag{22}$$

where ΔV refers to unit mass of material. Combining (21) and (22) and identifying the result with the Clausius-Clapeyron equation (17), we find

$$\frac{1}{T_M} \frac{dT_M}{dP} = \frac{\Delta V}{L} \approx \frac{\Delta V}{\Delta E_P} \approx \frac{2\gamma}{K} \tag{23}$$

Equation (23) has the form of the differential Lindemann melting law:

$$\frac{1}{T_M}\frac{dT_M}{dP} = \frac{2\left(\gamma - \frac{1}{3}\right)}{K} \qquad (24)$$

The approximations made in deriving (23), which we can improve, are neglect of the difference between the latent heats of melting at constant pressure and constant volume and the fact that during melting at constant volume, as melting proceeds so the pressure rises and so does the melting point. (There is also a basic assumption, examined later, that we can use γ in this way.) *Stacey and Irvine* [1977a] dealt with both problems simultaneously by relating the heat of melting at constant volume, Q, to the latent heat of melting at constant pressure, L, by the thermodynamic cycle of Fig.3, which involves melting at constant pressure P (limb AB), resolidification at constant volume, V + ΔV, (limb

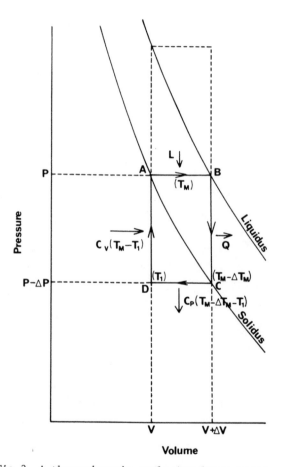

Fig.3 A thermodynamic cycle involving melting at constant pressure and resolidification at constant volume. The cycle is completed by a path in the solid phase, so that the energy of melting at constant volume may be related to solid properties and to the normal latent heat.

BC), cooling as solid at constant pressure to the original volume V (limb CD) and reheating at volume V to the original pressure P (limb DA). By this means the two melting processes, involving the heat exchanges L and Q, are related by the thermodynamic properties of the solid over the other two limbs of the cycle, in particular the expansion coefficient over CD which relates ΔV to the temperature difference, and the Grüneisen parameter over limb DA which relates ΔP to a temperature increment. The net heat exchange over one cycle is equal to the mechanical work done, ΔPΔV. The algebraic details of this calculation are not followed here. In the approximation of constant γ over the P-V range considered, the result is

$$\frac{L}{\Delta V} = \frac{K}{2\gamma}\left[1 + \frac{\Delta V}{V}(1+\gamma) + 2\gamma\alpha T_M\right] \qquad (25)$$

It is seen that eq. (25) has the form of (23) with an additional factor involving two correction terms that are both small compared with unity. (αT_M) is typically 0.06 for metals and less for materials such as silicates, and $\Delta V/V$ is similarly only a few percent for most materials. Thus we may as a good approximation invert the equation to obtain a melting equation of Lindemann form:

$$\frac{1}{T_M}\frac{dT_M}{dP} = \frac{\Delta V}{L} \approx \frac{2(\gamma - \eta)}{K} \qquad (26)$$

where $\eta = (\Delta V/V)(1 + \gamma) + 2\gamma\alpha T_M$.

This calculation is useful, not only for providing a clear basis for a melting law suitable for extrapolation of melting points to the highest pressures, but because it gives an insight into the significance of γ as a fundamental material property. Not all materials satisfy Lindemann's law, but the exceptions are those that undergo a substantial change in atomic coordination on melting and are close to a phase transition in the solid state. Thus if few atomic bonds are broken or formed in melting, but the energy of melting is consumed in stretching and compressing existing bonds in a process atomic rearrangement, then we have a situation similar to that arising from thermal vibration, which also causes stretching and compression of atomic bonds. Expansion on melting, like normal thermal expansion, is a consequence of bond asymmetry which makes the extensions easier than compressions. In either case Grüneisen's ratio relates the pressure required to prevent expansion to the thermal energy, the only difference being the factor 2 introduced in the melting case to account for the fact that the energy is all bond potential energy whereas in normal expansion half of the energy is kinetic energy of the atomic vibrations.

The nature of the atomic rearrangement implied by these conclusions was considered by *Stacey and Irvine* [1977b]. They calculated the ratio of volume increment to energy for introduction of a dislocation to a crystal and identified this ratio with the Clausius-Clapeyron equation (17) for the equilibrium phase boundary between a perfect crystal and a completely dislocated crystal, obtaining the Lindemann melting formula (24). This permits identification of melting as a free proliferation of dislocations, mobility of which accounts for fluidity of a liquid, and melting point as the temperature at which the Gibbs free energy of dislocations vanishes.

At the pressures of the lower mantle and core it is safe to assume all materials to adopt close-packed structures with no possibility of major coordination changes on melting. Thus the conditions required for validity of Lindemann's law are necessarily satisfied at the pressures of the core and deeper parts of the mantle. This is the basis of melting point tabulation in the author's thermal model of the Earth [*Stacey* 1977b].

Conclusions

The quantitative study of the thermal state of the Earth's deep interior depends crucially on an understanding of the thermal Grüneisen parameter γ at high pressures. The essential point is that γ is specified in principle by the pressure dependences of elastic constants, which are well determined by seismology. Although a completely satisfying formulation of the relationship is still lacking, alternative approaches have converged to near numerical agreement, so that thermodynamic studies of the Earth's interior using current estimates of γ are justified.

There is still some doubt about the density dependence of γ from laboratory data, the principal difficulty being that measurements made below the Debye temperature are confused by the temperature dependence of γ. Although the thermodynamic identity to "correct" $(\partial\gamma/\partial\rho)_T$ for the temperature effect is given (equation 13), available data are only marginally adequate to derive useful conclusions. More immediately useful is the inference of the density dependence of γ at high temperatures from other measurable quantities, especially $(\partial K_T/\partial T)_V$, using thermodynamic identities relating these quantities.

Given a constraint on the density dependence of γ from thermodynamic considerations, the relationship between γ and elasticity, especially K, implies directly a particular equation of state. Equation (16) was derived in this way by *Brennan and Stacey* [1979]. But we cannot reasonably hypothesise that the true equation of state for any material is really so simple a function. Therefore we cannot suppose that as simple a functional dependence of $\gamma(\rho)$ as equation (14) can be more than a convenient approximation. Thus it is dangerous to assume too confidently that such a relationship applies to the Earth's deep interior because it happens to look convincing in the laboratory, and we should not be surprised that *Brennan and Stacey* [1979] report an indifferent fit of equation (16) to terrestrial data. Rather we should reverse the argument and accept the values of γ inferred from seismological data and so base thermal models of the Earth directly on the seismological models without implied equation-of-state assumptions.

The validity of the Lindemann melting formula at high pressures is here reasserted and the implication for the nature of γ and its fundamental physical significance is explored in section 6. It perhaps needs to be emphasised that alternative empirical melting laws that have been used in geophysics (Simon-Glatzel and Kraut-Kennedy) have no sound fundamental basis. Of course Lindemann's law permits a wide range of melting curves according to the density dependence of γ. Thus any other 'law' can be accommodated by particular $\gamma(\rho)$, but extrapolation on this basis cannot be justified.

It is apparent that the next advance in our understanding of γ will lead to a corresponding advance in our understanding of the Earth's interior.

References

Anderson, O.L. The High Temperature Acoustic Grüneisen Parameter in the Earth's Interior. Phys. Earth Planet. Interiors 18, 221-231, 1979a.

Anderson, O.L. Evidence Supporting the Approximation $\gamma\rho$ = const for the Grüneisen Parameter of the Earth's Lower Mantle. J. Geophys. Res. 84, 3537-3542, 1979b.

Anderson, O.L. An Experimental High Temperature Equation of State Bypassing the Grüneisen Parameter, Conference on High Pressure Physics and Core-Mantle Dynamics, Potsdam, GDR, October 1978. Phys. Earth Planet. Interiors 22, 173-183, 1980.

Barron, T.H.K., J.G. Collins and G.K. White. Thermal Expansion of Solids at Low Temperatures, Advances in Physics 29, 609-730, 1980.

Boehler, R., I.C. Getting and G.C. Kennedy. Grüneisen Parameter of NaCl at High Compressions. J. Phys. Chem. Solids, 38, 233-236, 1977.

Boehler, R., A. Skoropanov, D. O'Mara and G.C. Kennedy. Grüneisen Parameter of Quartz, Quartzite and Fluorite at High Pressures J. Geophys. Res., 84, 3527-3531, 1979.

Brennan, B.J. and F.D. Stacey. A Thermodynamically Based Equation of State for the Lower Mantle. J. Geophys. Res. 84, 5535-5539, 1979.

Dugdale, J.S. and D.K.C. MacDonald. Thermal Expansion of Solids. Phys. Rev. 89, 832-834, 1953.

Falzone, A.J. and F.D. Stacey. Second Order Elasticity Theory: Explanation for the High Poisson's Ratio of the Inner Core. Phys. Earth Planet. Interiors 21, 371-377, 1980.

Irvine, R.D. and F.D. Stacey. Pressure Dependence of the Thermal Grüneisen Parameter. Phys. Earth Planet. Interiors 11, 157-165, 1975.

Lindemann, F.A. Über die Berechnung Molecular Eigenfrequenzen. Phys. Zeits. 11, 609-612, 1910.

Mott, N.F. and H. Jones. The Theory of the Properties of Metals and Alloys. London: Oxford University Press, 1935.

Ramakrishnan, J., R. Boehler, G.H. Higgins and G.C. Kennedy. Behavior of Grüneisen's Parameter of Some Metals at High Pressures. J. Geophys. Res. 83, 3535-3538, 1978.

Slater, J.C. Introduction to Chemical Physics. New York: McGraw-Hill, 1939.

Stacey, F.D. Applications of Thermodynamics to Fundamental Earth Physics. Geophys. Surv. 3, 175-204, 1977a.

Stacey, F.D. A Thermal Model of the Earth. Phys. Earth Planet. Interiors 15, 341-348, 1977b.

Stacey, F.D. and R.D. Irvine. Theory of Melting: Thermodynamic Basis of Lindemann's Law. Australian J. Phys. 30, 631-640, 1977a.

Stacey, F.D. and R.D. Irvine. A Simple Dislocation Theory of Melting. Australian J. Phys. 30, 641-646, 1977b.

Vashchenko, V.Ya and V.N. Zubarev. Concerning the Grüneisen Constant. Sov. Phys. Solid State 5, 653-655, 1963.

COOLING OF THE EARTH - A CONSTRAINT ON PALEOTECTONIC HYPOTHESES

Frank D. Stacey

Department of Physics, University of Queensland, Brisbane 4067, Australia

Abstract. The rate of cooling of the Earth can be represented quite simply in terms of the difference between heat generation by radioactivity (which is a decreasing function of time, t) and the convective heat loss, which decreases with increasing mantle stiffness as the mantle temperature, $\overline{T}$, falls. The result is a differential equation for T(t). The solution is remarkably independent of mantle properties and convective pattern and shows that the loss of residual heat by the Earth is a significant fraction of the total heat flux (37% in the preferred solution). However, it is shown that this rate of heat loss demands only a modest change in convection with time, i.e. less than a factor 1.5 in convective speed over the past 2×10^9 years.

Introduction - The Heat Budget Equation

The Earth derives internal heat from only two sources, radioactivity and, if it is shrinking or differentiating, from the energy of gravitational collapse. In analysing the heat budget we can treat the gravitational energy arising from thermal skrinking in terms of a simple adjustment to the heat capacity instead of regarding it as a heat source, that is we can refer to the energy release per degree of cooling and simply add it to the thermal heat that must be lost to cause that cooling. Continuing differentiation may be significant in the core, but since it is then also related to the cooling rate, we can treat it too in terms of adjustment to the heat capacity. Thus the only source that must be explicitly considered as a source is radioactivity. The power of this source $\dot{Q}_R$ is a decreasing function of time, "t"

$$\dot{Q}_R(t) \approx \dot{Q}_{R_o} e^{-\lambda t} \qquad (1)$$

where λ is the weighted average decay constant for the thermally important isotopes. We must admit that the heat conveyed to the surface by mantle convection may differ from $\dot{Q}_R$. It is here represented simply as $-\dot{Q}(\overline{T})$, the total rate of heat loss from the Earth, being controlled by the mean mantle temperature, $\overline{T}$. This control is via the viscosity or creep strength of the mantle, so that high temp-

erature means low viscosity, rapid convection and high heat flux. Representing the effective heat capacity of the Earth by mC we can write the heat balance equation as

$$mC \frac{d\overline{T}}{dt} = \dot{Q}_R(t) - \dot{Q}(\overline{T}) \qquad (2)$$

The essential point of the present paper can be seen qualitatively from this equation. Suppose that at the present time $\dot{Q}_R(t)$ and $\dot{Q}(\overline{T})$ are balanced, so that $\overline{T}$ is constant. This would be only the present day situation and since $\dot{Q}_R$ was greater in the past, it follows that $d\overline{T}/dt$ was positive and therefore that $\overline{T}$ was lower. But if $\overline{T}$ was lower, convection was slower and $\dot{Q}$ smaller, further increasing $d\overline{T}/dt$. Extending the argument further back in time, both effects, increasing $\dot{Q}_R$ and decreasing $\dot{Q}$, become stronger and we are left with no alternative to the conclusion that the Earth has never been as hot inside as it is now and that convection and geological processes were less effective in the remote past. How much so will be demonstrated by assigning numerical values to the terms in equation (2) in the following sections. Is this conclusion plausible? If not then we must discard the assumption of a balance between heat generation and the heat flux. Why has this assumption been so popular? It originates from an argument by Tozer [1972] that the very strong temperature dependence of mantle viscosity meant that the Earth's temperature profile was stabilized by the mantle convection. As it stands without elaboration, this is a correct and important idea but the apparently logical extension to assert that a decrease in radioactive heat could result in a correspondingly decreased heat flux by virtue of slower convection without a significant change in mantle temperature is erroneous. The resulting decrease in mantle temperature is significant. Following estimates give the present-day rate of loss of residual heat as 10% to 50% of the total heat, with 30% to 40% as the most plausible range, and a cooling rate of about 5.7 K per 10^8 years, sufficient to maintain the geomagnetic dynamo without any radioactivity in the core, as originally pointed out by Verhoogen [1961].

Although at least a large part of the Earth may

have accreted cold, compaction during the later-stages of accretion and gravitational settling out of the core, even without short-lived radioactivity, sufficed to bring the whole Earth to melting point within about 10^8 years of its origin, and it has been cooling ever since. Recognition that the cooling is still significant avoids some very serious difficulties that face the hypothesis of a steady state Earth.

Heat Capacity of the Earth

Taking the mass of the mantle-plus-crust to be 4.03×10^{24} kg of mean atomic weight 20.2, i.e. 2.0×10^{26} moles, its classical (high temperature) heat capacity (24.9 J K^{-1} mol^{-1}) is 4.97×10^{27} J K^{-1}. For the core (1.95×10^{24} kg of mean atomic weight 50) the heat capacity is 0.97×10^{27} J K^{-1}, making the total heat capacity of the Earth 5.9×10^{27} J K^{-1}.

The effective heat capacity is larger than this by virtue of the gravitational energy release accompanying the thermal contraction as the Earth cools. Values of the volume expansion coefficient throughout the Earth are tabulated by Stacey [1977a], the mean being about 12×10^{-6} K^{-1}, corresponding to a mean linear expansion coefficient of 4×10^{-6} K^{-1}. Thus the Earth contracts in radius by 4 parts in 10^6 for each degree of average temperature drop. The total gravitational potential energy of the Earth is

$$U = - f \frac{GM^2}{R} \qquad (3)$$

M being the mass, R is the radius, G is the gravitational constant and f = 0.66 is a numerical factor, slightly greater than 3/5, which is the value for a sphere of uniform density. The energy released by a small uniform contracted $\Delta R = 4 \times 10^{-6}$ R

$$\Delta U = \frac{dU}{dR} \Delta R = f \frac{GM^2}{R} \cdot \frac{\Delta R}{R} = 1.0 \times 10^{27} \text{ J} \qquad (4)$$

Since this is the energy release resulting from one degree of cooling, it must be added to the heat capacity in cooling calculations, so that the total effective heat capacity is

$$mC = 6.9 \times 10^{27} \text{ J } K^{-1} \qquad (5)$$

17% greater than the value obtained from thermal properties alone.

Temperature Dependence of the Convective Heat Flux

Convection is a remarkably self-stabilizing phenomenon. If convection occurs in quasi-steady state, its rate of heat transfer is hardly dependent upon the properties of the convecting medium, or geometry of boundary layers, because the temperature profile of the medium and the convective geometry, including the boundary layers, are self-adjusted to whatever is required by the con-

vection. This will be illustrated by assuming that the convective creep of the mantle obeys a creep law of a very general form with several arbitrary constants that can be varied quite widely without affecting the conclusion and similarly an arbitrary power law relationship between the speed of convection and the convective heat flux.

The steady state creep law now generally favoured for the mantle [e.g. Weertman, 1978] has the form

$$\dot{\varepsilon} = A\sigma^n \exp(- g\, T_M/T) \qquad (6)$$

where $\dot{\varepsilon}$ is the strain rate at stress σ, A, n and $g \approx 20$ are constants, T_M is melting point and T is absolute temperature. It will be shown that the cooling history of the Earth requires no knowledge of A and is hardly affected by major changes in the assumed values of n, g and T_M. The time dependence of T is the quantity that we wish to determine. We also need to make some assumption about the interdependence of $\dot{\varepsilon}$ and the convective heat flux $\dot{Q}$. Stacey [1980] showed that treatment of the lithosphere as a diffusive boundary layer to mantle convection led to the conclusion that $\dot{Q} \propto \dot{\varepsilon}^{\frac{1}{2}}$, or $\dot{\varepsilon} \propto \dot{Q}^2$. Here a more general dependence is allowed

$$\dot{\varepsilon} = C\dot{Q}^m \qquad (7)$$

where C and m are arbitrary constants, although the plausible range of m is limited. We also know that, averaged over the mantle

$$\dot{\varepsilon}\sigma = \eta\dot{Q}/V \qquad (8)$$

since $(\dot{\varepsilon}\sigma)$ is the rate of mechanical energy dissipation per unit volume of mantle material and $(\dot{Q}/V)$ is the convective heat transfer per unit volume. The thermodynamic efficiency η is a constant for any particular depth of mantle convection and depends only upon the thermodynamic properties of the mantle material and the depth of the convective zone [Stacey 1977b]. Combining (7) and (8) we have

$$\sigma = (\eta/CV)\, \dot{Q}^{1-m} \qquad (9)$$

Substituting for both $\dot{\varepsilon}$ and σ by (7) and (9) in (6), and making $\dot{Q}$ the subject of the resulting equation, we have

$$\dot{Q} = \left[\frac{A}{C}\left(\frac{\eta}{CV}\right)^n\right]^{1/(mn+m-n)} \cdot \exp\left\{- \frac{g}{(mn+m-n)} \cdot \frac{T_M}{T}\right\} \qquad (10)$$

Notice that the pre-exponential factor is simply a constant and that it is effectively calibrated out of the equation by substituting the present-day conditions, i.e. convection heat flux $\dot{Q}_o$ at mean mantle temperature T_o. Then

$$\dot{Q} = \dot{Q}_o \exp\left\{\frac{g T_M}{(mn+m-n)}\left(\frac{1}{T_o} - \frac{1}{T}\right)\right\} \qquad (11)$$

If we accept that $\dot{Q}_o$ is a reasonably well-observed quantity then equation (11) relates the

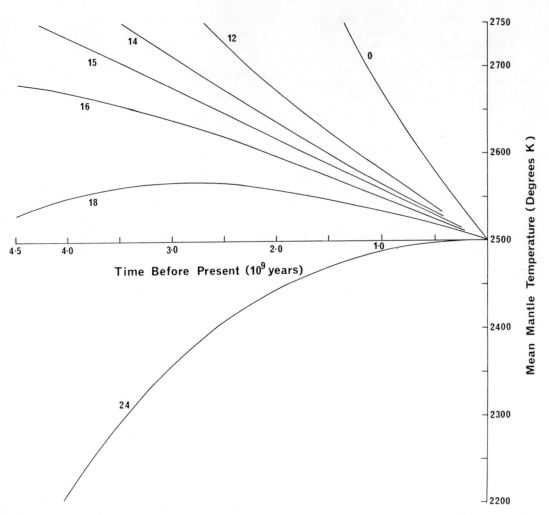

Figure 1. Mean mantle temperature as a function of time for different assumed present day values of mantle radioactive heat generation, $\dot{Q}_{R_O}$, marked on the curves in units of 10^{12} W. All of these curves assume $\theta = 11000$ K, $\dot{Q}_O = 24 \times 10^{12}$ W.

heat flux at any other time, $\dot{Q}$, to the difference between the mantle temperature then and now, with only one composite constant

$$\theta = gT_M/(mn + m - n) \qquad (12)$$

which has the dimension of temperature. The preferred value and plausible range of θ are obtained by noting that $g \approx 20$, and $T_M \approx 2750$ K (mantle average) with probably rather little permitted range compared with m and n. The favoured value of m is 2, and although a substantial departure from this appears implausible, the range $1 \leqslant m \leqslant 3$ is here admitted, to examine the consequences of these extreme assumptions. The creep law (6) allows the possibility of any value of n between 1 and 6. n = 1 corresponds to a Newtonian viscous mantle and higher values to non-Newtonian rheology, although a value as high as 6 seems unlikely. In the absence of a clear observational

lead, n = 3 is selected as the preferred value, with the range 1 to 6 admitted as possible. With these numbers the preferred value of θ is 11000 K with the extreme admissible range 3667 K to 55000 K.

Integration of the Heat Budget Equation

Substituting for $\dot{Q}_R$ by (1) and $\dot{Q}$ by (11) in the heat budget equation (2) we obtain it in the form of a differential equation for T(t)

$$\frac{dT}{dt} = \frac{\dot{Q}_{R_O}}{mC} \exp(-\lambda t) - \frac{\dot{Q}_O}{mC} \exp\left\{\theta\left(\frac{1}{T_O} - \frac{1}{T}\right)\right\} \qquad (13)$$

It makes rather little difference whether we consider $\dot{Q}_R$, $\dot{Q}_O$ and (mC) to refer to the whole Earth or to the mantle. I consider that it is preferable to relate them to the mantle alone, that is to consider that it is the mantle radiogenic heat and the flux of heat from the mantle alone that must be

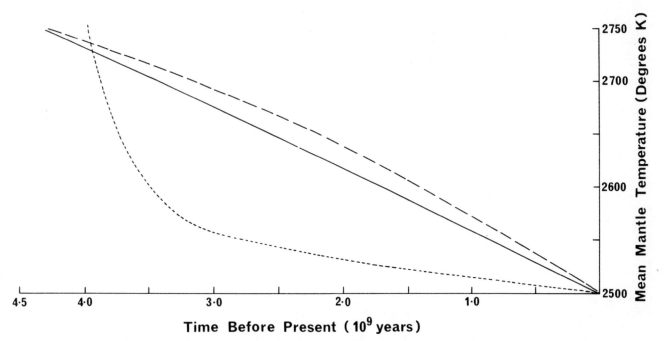

Figure 2. Mantle cooling curves for different assumed values of θ. The solid line is the "preferred" curve ($\theta = 11000$ K, $\dot{Q}_{R_o} = 15 \times 10^{12}$ W). The upper (dashed) line assumes $\theta = 3667$ K, $\dot{Q}_{R_o} = 12.25 \times 10^{12}$ W and the lower dotted line assumes $\theta = 55000$ K, $\dot{Q}_{R_o} = 21.66 \times 10^{12}$ W. In all cases the values of $\dot{Q}_{R_o}$ are selected so that the curves meet the assumed mean mantle melting point (2750 K) shortly after the origin of the Earth.

related, because the core heat can escape to the surface by relatively narrow, rapidly flowing plumes operating semi-independently of the convective process that is needed to remove the mantle heat. This being so, appropriate values are $mC = 5.0 \times 10^{27}$ J K^{-1} (including the gravitational correction for shrinking), $\dot{Q}_o \approx 24 \times 10^{12}$ W and $\dot{Q}_{R_o}$ is an unknown constant, the influence of which is to be examined. The assumed value of $\dot{Q}_o$ is compatible with a total geothermal flux of 38×10^{12} W and disallows 9×10^{12} W of heat diffusing from the continental lithosphere and 5×10^{12} W of core heat. T is taken to be the average mantle temperature and, for the purpose of the present calculation, $T_o = 2500$ K.
The exponential decay of $\dot{Q}_R$ is, of course only an approximation because radiogenic heat is produced by four isotopes (^{238}U, ^{235}U, ^{232}Th, ^{40}K) that have different decay constants. Stacey [1980] favoured a present weighted average value $\lambda = 1.4 \times 10^{-10}$ year$^{-1} = 4.4 \times 10^{-18}$ s^{-1} on the basis of a lower potassium content for the Earth than is generally postulated. This figure will be used here. However, it is emphasised that the result of integrating equation (13) is remarkably insensitive to the choice of all of these constants, within any reasonable bounds.
Figure 1 shows the result of integrating equation (13) backward in time, for $\theta = 11000$ K (the preferred value) and $\dot{Q}_o = 24 \times 10^{12}$ W, with alternative values of $\dot{Q}_{R_o}$ marked on the various curves (in units of 10^{12} W). The lowest of these curves

refers to the assumption that the mantle is now in equilibrium with its internal heat sources, i.e. $\dot{Q}_o = \dot{Q}_{R_o}$. It demonstrates quantitatively that the balance could not have been a permanent feature of the Earth's thermal history, but that, with this assumption, the Earth was cooler than at present through all past time. This is therefore an implausible assumption. Instead we must seek a solution that has the mantle at its melting point shortly after its origin at $t = 4.5 \times 10^9$ years and, as in Figure 1, this solution requires $\dot{Q}_{R_o} = 15 \times 10^{12}$ W, only 63% of the total heat flux, leaving 37% to be accounted for by residual heat.
The generality of the cooling Earth conclusion is examined in Figure 2, in which T(t) curves are plotted for the preferred and extreme values of θ, all constrained to the melting point shortly after the origin of the Earth. The solid line is the $\dot{Q}_{R_o} = 15 \times 10^{12}$ W result of Figure 1 ($\theta = 11000$ K) and the other two are equivalent plots for $\theta = 3667$ K (upper curve) and $\theta = 55000$ K (lower curve). The parameters of these curves and the present cooling rates are summarised in Table 1.

Paleotectonics - a Discussion

The "preferred" solution of Figures 1 and 2 shows an almost linear decrease in mean mantle temperature with time. This is a fortuitous consequence of the parameters in the two exponential terms of equation (13), but it makes a semi-quantitative discussion of tectonic activity in the remote past particularly

Table 1: Parameters of Mantle Cooling Curves

θ (K)	3367	11000	55000
$\dot{Q}_{R_O}$ (10^{12} W)	12.25	15	21.66
$(\dot{Q}_O - \dot{Q}_{R_O})/\dot{Q}_O$ (%)	49	38	10
$\left[-\dfrac{dT}{dt}\right]_{t=0}$ $\left(10^{-8}\,\mathrm{K\ y^{-1}}\right)$	7.4	5.7	1.5

straightforward, because the rate of loss of residual heat remains virtually constant as the radiogenic heat and the total heat flux decrease with time. With these parameters we can put

$$\dot{Q} = -mC\frac{dT}{dt} + \dot{Q}_{R_O}\,e^{-\lambda t} \approx (9 + 15e^{-\lambda t}) \times 10^{12}\ \mathrm{W} \quad (14)$$

where $\lambda = 1.4 \times 10^{-10}$ year^{-1} has been assumed.

Since $\dot{\varepsilon} \propto \dot{Q}^2$ and $\sigma \propto \dot{Q}^{-1}$, with viscosity η^* (starred to avoid confusion with convective efficiency) given by $(\sigma/\dot{\varepsilon}) \propto \dot{Q}^{-3}$, all these quantities are specified functions of time. With an effective weighted half-life of 5×10^9 years for $\dot{Q}_R$, the past rates of convection can be caluclated in terms of $\dot{Q}$ by equation (14). Thus 500 million years ago, $\dot{Q}$ was greater by a factor 1.05, the convective speed or $\dot{\varepsilon}$ was greater by 1.09, σ was smaller by 1.05 and η^* smaller by 1.14. These are quite modest changes. The corresponding factors for 2×10^9 years ago are 1.20, 1.44, 1.20, 1.74 and 4×10^9 years ago 1.47, 2.16, 1.47, 3.17. Thus we see that although convection was faster and the mantle softer in the remote past, the differences are far from startling. The convective speed has decreased only by a factor 2 in 4×10^9 years. Again, these conclusions are not drastically modified by choosing different convection parameters in the present theory, although of course an intense burst of short-lived radioactivity early in the history of the Earth would invalidate an extrapolation that far back.

References

Stacey, F.D. A Thermal Model of the Earth. Phys. Earth Planet. Inter. 15, 341-348, 1977a.

Stacey, F.D. Applications of Thermodynamics to Fundamental Earth Physics. Geophys. Surveys 3, 175-204, 1977b.

Stacey, F.D. The Cooling Earth: A Reappraisal. Phys. Earth Planet. Inter. 22, 89-96, 1980.

Tozer, D.C. The Present Thermal State of the Terrestrial Planets. Phys. Earth Planet. Inter. 6, 182-197, 1972.

Verhoogen, J. Heat Balance of the Earth's Core. Geophys. J. R. Astron. Soc. 4, 276-281, 1961.

Weertman, J. Creep Laws for the Mantle of the Earth. Phil. Trans. Roy. Soc. A, 288, 9-26, 1978.

A REMARK ON VISCOSITY AND CONVECTION IN THE MANTLE

V. P. Trubitsyn, P. P. Vasiljev and A. A. Karasev

Institute of Physics of the Earth, Academy of Sciences USSR, Moscow

Abstract. A simple relationship between the mantle's parameters based on two-dimensional convection is considered. Averaged viscosity of the lower mantle is evaluated as 10^{23} poise. Numerical experiments for convection in the upper mantle show the possibility for existence of areas where the temperature decreases with the depth. Thermal convection in the lower mantle is only now developing and will reach its stationary state after about 10 billion years.

Introduction

Some authors [McKenzie, et al., 1974] follow the point of view that convection in the lower mantle is impossible because the value of its viscosity is greater than $3 \cdot 10^{27}$ poise. The solid lower mantle prevents the descending of the lithosphere. Other authors argued for a smaller value of the viscosity of the whole mantle [Peltier, 1974; Davies, 1977; O'Connell, 1977]. The rheological models [Peltier and Andrews, 1976] for postglacial rebound prescribed the value of $\eta \sim 10^{22}$ poise for the mean viscosity of the whole mantle. Schubert and Young [1976] carried numerical calculation of the Earth's cooling for adopted values of the same parameters. They concluded that $\eta \geq 4 \cdot 10^{24}$ poise.

Our paper is devoted to the investigation of a possible interval for the viscosity of the lower mantle with various realistic values of other parameters. A simplified model for two-dimensional convection was analyzed. It allowed us to evaluate the viscosity without solving the differential equations of convection.

The use of this simplified model may be supported in great part by the uncertainty of the mantle parameters in the differential equations, by the clearness of the results and by the possibility of varying these parameters. Some results of these numerical experiments for convection in the upper and the whole mantle are presented in this paper.

The matter of the Earth is viscous-elastic. Fast processes display these elastic properties. When forces act for a long period of time the matter begins to flow like a viscous liquid. The time defined by this process is $t \sim \frac{\eta}{\mu}$, where η is the viscosity, and μ is shear modules. For the Earth's mantle $\mu \sim 10^{12} \text{dyn cm}^{-2}$. Even for $\mu \sim 10^{27}$ poise the processes extending for periods longer than 30 My are described by the equations for viscous liquid in the first approximation.

In the ideal nonviscous liquid layer heated from below the thermal convection starts when the adiabatic temperature gradient is exceeded. Then the lifting and adiabatically expanding liquid element is always lighter than the environment. In the viscous liquid layer one must overcome other viscous forces for convection to begin. Additional critical superadiabatic temperature gradients where the convection begins in viscous liquid layers as defined by the critical value of the Rayleigh number is equal to

$$R = \frac{\alpha g D^3 \Delta T_{SA}}{\kappa \nu} \qquad (1)$$

where ΔT_{SA} is the superadiabatic temperature difference; α, the coefficient of thermal expansion; g, the acceleration of gravity; D, the thickness of layer; κ, the thermometric conductivity; $\nu = \frac{\eta}{\rho}$, the kinematical viscosity; and ρ, the mean density.

The efficiency of a convective heat transport is measured by the Nusselt number N, or

$$N = \frac{q_{SA}}{\rho C p \kappa \Delta T_{SA}} \qquad (2)$$

where q_{SA} is superadiabatic heat flow on the upper surface of the layer and Cp, the specific heat at constant pressure.

The Boussinesq approximation for the equations of convection is

$$\frac{\partial T_{SA}}{\partial t} + V \nabla T_{SA} = \kappa \nabla^2 T_{SA} + \varepsilon \qquad (3)$$

$$\nu \nabla^2 V = \frac{1}{\rho} \nabla P + \alpha T_{SA} g \qquad (4)$$

$$\nabla V = 0 \qquad (5)$$

where V is velocity, T_{SA} is superadiabatic temperature, ε is the thermometric internal heating rate

$\varepsilon = \dfrac{H}{\rho C}$, and H is the rate of internal heat generation.

If the integral of differential equations is known then one can calculate some quantities without direct solution of these equations. For Equations (3)-(5), one of such integrals is the relationship between the Rayleigh and Nusselt numbers. Because of the great complexity of the nonlinear Equations (3)-(5), this relationship is known from numerical experiments only. For convection with a finite amplitude for fixed temperatures on both horizontal boundaries, Moore and Weiss [1968] expressed this relation in the form

$$N = 1.96 \; (\frac{R}{R_C})^{1/3} \pm 1\%, \; \frac{R}{Rc} > 5 \qquad (6)$$
$$N > 3.3$$

For a small amplitude we may express the numerical results of Moore and Weiss [1968] in the form

$$N = 1 + (\frac{R}{R_C} - 1)^{2/3} \pm 10\%, \quad \begin{array}{c} 1.1 < \frac{R}{R_C} < 5 \\[4pt] 1.15 < N < 3.3 \end{array} \qquad (7)$$

Equations (1), (2), (6) and (7) give the partly known relationship between the parameters of a viscous liquid and the heat flux

$$\nu = \frac{7.5}{R_C} \; \frac{\alpha g k^2 (\rho Cp)^3 \Delta T_{SA} D^3}{q_{SA}}, \; \frac{R}{R_C} > 5 \qquad (8)$$
$$N > 3.3$$

finite amplitude convection,

$$1.1 < \frac{R}{R_C} < 5$$

$$\nu = \frac{\alpha g \Delta T_{SA} D}{R_C \kappa (1 + \dfrac{Dq_{SA}}{\rho Cp \kappa \Delta T_{SA}} \cdot -1)^{3/2}} \quad 1.15 < N < 3.3 \qquad (9)$$

small amplitude convection,

$$\nu \geq \nu_c = \frac{\alpha g \Delta T_{SA} D^3}{\kappa R_C} \quad \frac{R}{R_C} \leq 1, \; N = 1 \qquad (10)$$

no convection.

If there is a heat generation within the layer with heating rate H and the heat flux is pre-scribed on the bottom boundary instead of fixed temperature $q_1 = q - HD$, then Equations (1), (2) and (6) are modified [McKenzie, et al., 1974].

$$N_H = a(\frac{R_H}{R_{HC}})^{\frac{1}{4}} \quad ,$$

$$R_H = \frac{\alpha g D^4 q}{\rho Cp \kappa^2 \nu} \quad ,$$

$$N_H = \frac{q - \dfrac{HD}{2}}{\rho Cp \kappa} \cdot \frac{D}{\Delta T} \qquad (11)$$

where R_H and N_H are modified Rayleigh and Nusselt numbers, and q and q_1 are the heat flux on the upper and bottom boundaries. From Equation (11) we obtain a new relationship which differs from Equation (8) by a factor with

$$\nu = (\frac{a}{1 - \dfrac{HD}{2q_{SA}}})^4 \; \frac{\alpha g \kappa^2 (\rho Cp)^3 \Delta T_{SA}^4 D^3}{R_{HC} q_{SA}^3} \qquad (12)$$

Critical Rayleigh numbers R_{HC} were calculated by McKenzie, et al. [1974]. For free boundaries $R_C = 657.5$ or $7.5/R_C = 0.011$ for Equation (8). For Equation (12), $R_{HC} = 385$, a = 1.58 or

$$\frac{a^4}{R_{HC}} = 0.016,$$

if H = 0.

TABLE I. Mantle Parameters

		Upper mantle	Lower mantle
D	(m)	7×10^5	2×10^6
g	(m s^{-2})	10	10
ρ	(Kg m^{-3})	3.7×10^3 (1)	4.5×10^3
Cp	(JKg$^{-1 \circ}$C^{-1})	1.2×10^3 (2)	1.1×10^3
α	($^\circ$C^{-1})	2.5×10^{-5}	1.5×10^{-5} (1
κ	(m^{2}s^{-1})	10^{-6} (3)	$(2-3) \times 10^{-6}$
∇T	($^\circ$C κm^{-1})	~0.4	~0.35
ΔT_A	($^\circ$C)	300	700
ΔT_A	($^\circ$C)	500-600 (4)	1200 (4)
ΔT_{SA}	($^\circ$C)	250	500
q_{SA}	(w m^{-2})	0.06	0.03
H	(w m^{-3})	5×10^{-8}	10^{-8}
$\nu = \frac{\eta}{\rho}$	(m^{2}s^{-1})	1.5×10^{16}	2×10^{18}
η	(g s^{-1}cm^{-1}=poise)	5×10^{20}	10^{23}
R		1.5×10^6	10^5
N		30	10
R_H		5×10^7	10^6

Sources

1 Dulong-Petit value;
2 Trubitsyn, 1979;
3 Schatz and Simmons, 1972; McKenzie and Weiss, 1975;
4 Stacey, 1977; Anderson, 1979.
   All the other parameters are adopted and cal-culated in this paper.

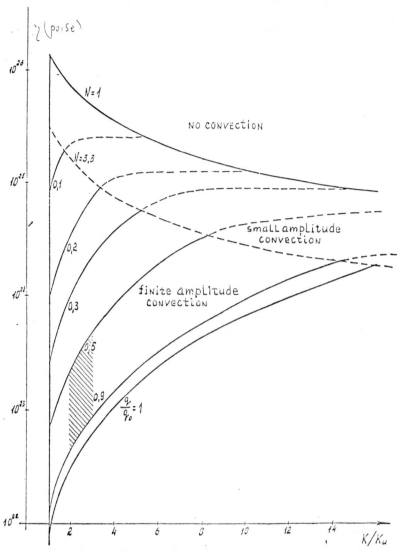

Fig. 1. Mean viscosity of the lower mantle as a function of K and q for thermal convection. K and K_u are mean thermal diffusivity in the lower and upper mantle. q and q_o are heat flux from the lower and upper mantle. T = 3300°K is the adopted temperature on the core-mantle boundary. ΔT_{SA} = 500°K is superadiabatic temperature difference through the lower mantle.

Parameters of the Lower Mantle

Mean viscosity of the mantle may be calculated by Equations (8), (9), (10), or (12) if the values of the other parameters are known. Table I contains both commonly used parameters and parameters which are adopted and evaluated for use in this paper. The thickness of the upper mantle is usually taken by Du = 700 km. Let its upper and lower boundaries be the depths 100 km and 800 km, respectively. Then by Stacey [1977] and Anderson [1979], the temperature difference of the layer is ΔT = T(800) - T(100) = 2100° - 1550° = 550°K. For the lower mantle, where D = 2000 km, the temperature at the core-mantle boundary is probably somewhat higher than the melting temperature of core

matter [Stacey, 1977]. If T(2900) = 3300°K, then for the lower mantle ΔT = 3300-2100 = 1200°K. Acceleration of gravity g, mean density ρ, and specific heat C_p are well known values for the Earth's mantle. The thermometric conductivity κ in the upper mantle is about $10^{-6} m^2 s^{-1}$ [Schatz and Simmons, 1972; McKenzie and Weiss, 1975]. Viscosity η is known with precision of order of the magnitude only. Mean adiabatic temperature gradient is determined from the well-known relationship

$$\Delta T_a = \alpha g \frac{\overline{T}}{C_p} ,$$

where $\overline{T}$ is averaged temperature in the layer. Superadiabatic temperature difference is equal

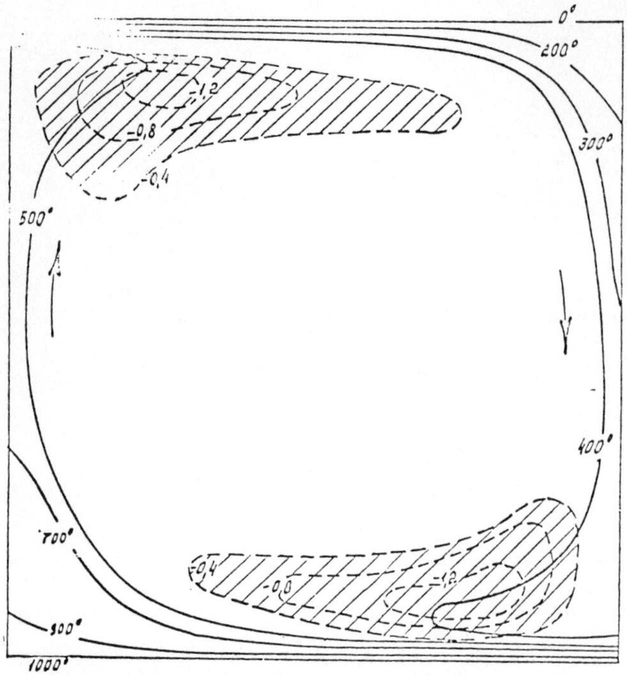

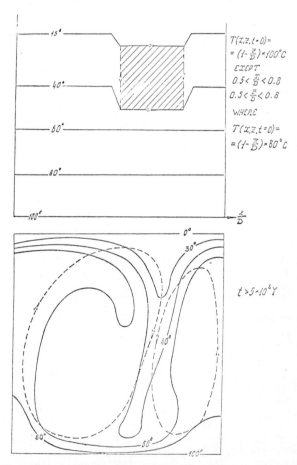

$$q_o = q_A + q_{SA} + q_\lambda \qquad (13)$$

where $q_A = \rho C_p \kappa \nabla T_a$ is defined as the conductive heat flux due to the adiabatic temperature gradient ∇T_a. $q_{SA} = \rho C_p \kappa \nabla T_{SA} + \rho C_p VT$ is the superadiabatic heat flux, which is transported both conductively and convectively; $q_\lambda = \lambda \rho \overline{V}$ is latent heat flux, which doesn't generate convection. This heat flux exists if endothermic phase transition takes place below the layer and inverse exothermal phase transition occurs above the layer.

The adiabatic heat flux through the upper mantle is equal to $q_A \sim 1.8$ mWm$^{-2} \sim 0.03q_o$. Through the lower mantle $q_A \sim 5$mWm$^{-2} \sim 0.1q_o$. The descending lithosphere melts in the lower mantle and absorbs the heat. Solidification of thickening lithosphere releases the latent heat. Because the asthenosphere is only partly melted, q_λ is small. We adopt $q_\lambda \sim (0.05-0.1)q_o$ so superadiabatic heat flux from the upper mantle is equal to that of the lower mantle; e.g., $q_{SA} = q_o - q_A - q_\lambda \sim 0.9q$.

Fig. 3. Contours of T_{SA} in a convective cell for $\nu = 10^{21}$cm^{2}s^{-1}, $\Delta T_{SA} \cong 100°$K for the initial state $t = 0$ and for the stationary state $t > 5 \times 10^8$y. Dashed line presents the contour of stream function. The flow develops in two rolls immediately for $t > 0$.

Fig. 2. Contours of the temperatures from numerical simulations of convection in the upper mantle: $\nu = 10^{21}$sm^{2}s^{-1}. The temperature on the lower boundary is constant equal to $250°$C. Dashed lines present the contours of constant values of the superadiabatic gradient of temperature

$$\frac{\partial T_{SA}}{\partial h} \ (°Ckm^{-1})$$

The temperature does not increase with the depth h when

$$\left| \frac{\partial T_{SA}}{\partial h} \right| \geq 0.4°Ckm^{-1}$$

to $\Delta T_{SA} = \Delta T - \nabla T_a D$. Some parameters in the lower mantle are known poorly, especially internal heating rate H, heat flux q_{SA} and thermometric conductivity κ. The radiation part of κ appears to be small in the lower mantle, but the lattice part of κ increases slowly with the depth. Mean thermometric conductivity in the lower mantle probably has the value $\kappa_L \sim (2-3)\kappa$.

The heat flux in the upper mantle usually was taken to be the same as Earth's mean surface heat flux q_E, which was supposed to be $q_E = 5.85 \ 10^{-2}$ Wm^{-2} [McKenzie, et al., 1974]. The latest data for Earth's heat flux including the ridges are $q_E = (7-8) \times 10^{-2}$ [Sclater et al., 1979; Turcotte and Burke, 1978; Lubimova, 1979]. Subtracting 15 percent continental crustal radioactivity contribution, we find that heat flux from the upper mantle is equal to $q_o = 6.5 \times 10^{-2}$wm^{-2}.

We may divide the heat flux through layer q_o into three parts

The heat flux from the lower mantle, is probably less than $0.9q_o$. If radioactive heat sources are concentrated mainly in the upper mantle, we may adopt $q_{SA} \sim 0.5q$ at the upper boundary of the lower mantle [McKenzie, et al., 1974].

The Earth's core releases the heat Q_c, which is equal to about 13-25 percent of the total Earth's heat, Q_E [Stacey, 1977; Jeanloz and Richter, 1979]. Hence, the mantle, excluding the core and lithosphere, releases heat to about 70 percent. If the heat sources are distributed uniformly through the whole mantle, then its density is $H_w \sim 2.5 \times 10^{-8}$ wm^{-3}. If one half of the mantle heat flux is released in the upper mantle, then its density is equal to about $H_u \sim 5 \times 10^{-8} wm^{-3}$ for the upper mantle and $H_L \sim 10^{-8} wm^{-3}$ for the lower mantle.

After substituting the values of the parameters from Table I into Equation (12), we find the mean viscosity of the upper mantle, $\eta_u = \rho\nu = 5 \times 10^{20}$ poise. Note that this evaluation was made for the isolated upper mantle without consideration of the lithosphere. If the lithospheric plates intensify the convective flow, then the real viscosity of the upper mantle will be several times greater.

Consider convection in the lower mantle. Instead of matching the velocities and temperatures at the upper-lower mantle boundary, we take into account the perturbation of the upper mantle by replacing R_c with the effective critical Rayleigh number R_c^* in Equations (8), (9), and (10).

Intensity of convection depends on the ratio R/R_c. For constant viscosity, the ratio of the Rayleigh numbers for the whole and lower mantles is

$$\frac{R_w}{R_1} \sim \left(\frac{D_1 + D_u}{D_1}\right)^3 \sim 2.5.$$

But viscosity in the upper mantle is greater than that of the lower mantle and it intensifies convection. Thus, as a first approximation, we may adopt

$$R_c^* \sim \frac{R_c}{3} \sim 2 \cdot 10^2.$$

Figure 1 shows the diagram of viscosity in the lower mantle $\eta = \rho\nu(\kappa,q;\Delta T_{SA} = 500°K)$, obtained from Equations (8), (9), and (10). The units for κ and q are chosen to be $\kappa_u = 10^{-6} m^2 s^{-1}$ and $q_o = 6.5 \times 10^{-2} wm^{-2}$, which correspond to the values in the upper mantle. Likely values of lower mantle viscosity lie within a shaded area, $\eta \sim 10^{23}$ poise. Convection in the lower mantle is intensive and has a finite amplitude.

It is easy to find the minimum value of the mean viscosity of the lower mantle when there is no convection. In this case total heat flux is conductive with $q = \rho Cp(\Delta T/D)$.
Adopting $q \sim 0.6q_o$ and $T = 1000°$ K, we obtain $\kappa_L = 6.5 \times 10^{-6} m^2 s^{-1} \sim 6.5 \kappa_u$. By using formula (10), Table I and Figure I, we find $\eta > 2 \times 10^{25}$ poise.

Diagram I is constructed for $\overline{T}(2900) = 3300°$ K. If we adopt $T(2900) = 4000°$ K. which is closer to

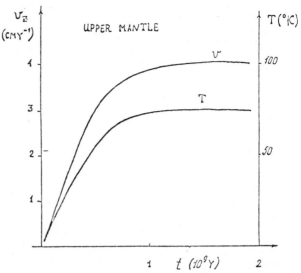

Fig. 4. Smooth velocity V_z and temperature T in the point X = 0, Z = 0.5D in the convective cell as a function of time for the upper mantle: $\nu = 10^{21} cm^2 s^{-1}$, $\Delta T_{SA} = 150°K$.

the melting temperature of matter of the mantle than the core, by Equation (8), the viscosity in the lower mantle will be $(1200/500)^4 \sim 30$ times greater. If we take into account that whole mantle convection has not reached its stationary state, the mean viscosity will be about 10 times less. The spherical effect of the mantle layer gives a correction which is less than uncertainties of the parameters.

Results of Numerical Experiments

Two-dimensional models of the Rayleigh-Benard convection in the Boussinesq approximation was calculated to analyze some of its properties. Anderson [1979] calculated the temperature distribution in the Earth, which may be considered as external for the lithosphere. He obtained a negative value for the temperature gradient $\frac{\partial T}{\partial h}$ at the depths 200-300km. Such temperature profile is obviously nonequilibrium or may exist only in local areas in the mantle.

It was known previously that the temperature in the convective cell may decrease with the depth, but it is only for superadiabatic temperature [McKenzie, et. al., 1974]. We calculated various models of upper mantle convection and found the areas where not only superadiabatic but the total temperature decreases with the depth, i.e. where

$$\frac{\partial T_{SA}}{\partial h} < 0 \text{ and } \left|\frac{\partial T_{SA}}{\partial h}\right| > \frac{\partial T_a}{\partial h} = 0.4°C \ km^{-1}$$

Figure 2 shows the superadiabatic temperature in a convective cell for $\nu = 10^{21} sm^2 s^{-1}$, $\Delta T_{SA} = 250°K$, $D = 7 \times 10^5$ m.

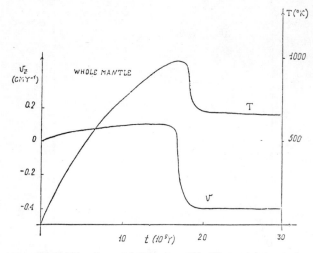

Fig. 5. Smooth velocity V_z and temperature T in the point x = 0, z = 0.5D. For the whole mantle: $\nu = 10^{23} cm^2 s^{-1}$, $q = 6 \times 10^{-2} wm^{-2}$. At $t \sim 2 \times 10^{10} y$, one roll develops into two rolls.

The sizes of anomalous areas where $\frac{\partial T}{\partial h} < 0$ are about 400 km. X 100 km. For the other models it reveals that those sizes increase if the viscosity decreases.

In Figure 2, we see the horizontal temperature differences, which are about $\sim 330°C$. So the density variations in the upper mantle by convection may be equal to $\nabla \rho \sim \alpha \rho \nabla T \sim 2.5 \times 10^{-3} \times 3.7 \times 300 \sim 0.03$ gcm^{-3}.

Figure 3 illustrates the influence, which initial conditions exert on the structure of the convective cells in the upper mantle. The areas with increasing or decreasing initial temperature may cause splitting of convective cells. Instead one will immediately develop two rolls even if $T_{SA} = 100°K$.

Figure 4 and 5 show the amplitudes of temperature and velocities as functions of time in the upper and whole mantle. The time taken to reach a steady (or quasi-steady) state is about 10^9 yr for the upper mantle and about 2×10^{10} yr for the whole mantle. Hence lower mantle convection is only developing in the current geological period.

Conclusions

The available information about heat flows and temperatures at the depths 800 km and 2900 km, respectively, gives us estimates of the mean viscosity $\eta \sim 10^{23}$ poise in the lower mantle. Thermal convection in the whole mantle has a finite amplitude. Intensity of upper mantle convection is high enough to create the areas where the total temperature decreases with the depth. The time taken to reach a steady state for convection in the lower mantle exceeds the age of the Earth.

References

Anderson, O. L. Temperature profiles in the Earth, Proceed. Symposium Physics and Chemistry of the Origin of the Earth, London, Ontario, Canada, Sept. 2-5, 1979.

Davies, G. F. Whole-mantle convection and plate tectonics, Geophys. J. R. Astr. Soc., 49, 459-486, 1977.

Jeanloz, R. and F. M. Richter. Convection, composition and the thermal state of the lower mantle, J. Geophys. Res., 84, 5497-5504, 1979.

Lubimova, G. A. Heat flow history, Proceed. Symposium Physics and Chemistry of the Origin of the Earth, London, Ontario, Canada, Sept. 2-5, 1980.

McKenzie, D. P. and M. Weiss. Speculation on the thermal and tectonic history of the Earth, Geophys. J. R. Astr. Cos., 42, 131-174, 1975.

McKenzie, D. P., J. M. Roberts and M. O. Weiss. Convection in the Earth's mantle: Towards a numerical simulation, J. Fluid Mech., 62, 465-538, 1974.

Moore, D. R. and M. O. Weiss. Two-dimensional Rayleigh-Banard convection, J. Fluid Mech., 58, 289-312, 1968.

O'Connell, R. J. On the scale of mantle convection, Tectonophys., 38, 119-136, 1977.

Peltier, W. R. and J. T. Andrews. Glacial-isostatic adjustment-I, Geophys. J. R. Astr. Soc., 46, 605-645, 1974.

Schatz, J. R. and G. Simons. Thermal conductivity of the earth materials at high temperatures, J. Geophys. Res., 77, 6966-6983, 1972.

Schubert, G. and R. E. Young. Cooling of the Earth by whole mantle subsolidus convection: a constraint on the viscosity of the lower mantle, Tectonophys., 35, 201-214, 1976.

Sclater, J. G., C. Jaupart, and D. Galson. The heat flow through oceanic and continental crust and heat loss of the Earth, Proceed. Symposium Physics and Chemistry of the Origin of the Earth, London, Ontario, Canada, Sept. 2-5, 1979.

Stacey, F. D. A thermal model of the Earth, Phys. Earth and Planet. Int., 15, 341-348, 1977.

Trubitsyn, V. P. Phase transitions, isothermal compressibility and thermal expansion of the Earth, Izv. AN USSR, Fizika Zemli, No. 1, 21-27, 1979.

Turcotte, D. L. and K. Burke. Global sea-level changes and thermal structure of the Earth, Earth and Planet. Sci. Lett., 41, 341-346, 1976.